Nanotechnology-Enabled Sensors

Nanotechnology-Enabled Sensors

Kourosh Kalantar-zadeh
RMIT University
School of Electrical Engineering
Melbourne, Victoria
Australia

Benjamin Fry
RMIT University
Biotechnology and Environmental Biology
Melbourne, Victoria
Australia

Springer

Kourosh Kalantar-zadeh
RMIT University
School of Electrical Engineering
Melbourne, Victoria, 3001
Australia

Benjamin Fry
RMIT University
Biotechnology and Environmental Biology
Melbourne, Victoria, 3083
Australia

Library of Congress Control Number: 2007934285

ISBN 978-0-387-32473-9 e-ISBN 978-0-387-68023-1

Printed on acid-free paper.

9 8 7 6 5 4 3 2 1

springer.com

Preface

Nanotechnology enabled sensors is an exciting field to enter into. It is our intention to provide the readers with a deep understanding of the concepts of nanotechnology enabled sensors, handing them the information necessary to develop such sensors, covering all aspects including fundamental theories, fabrication, functionalization, characterization and the real world applications, enabling them to pursue their research and development requirements.

This book can be utilized as a text for researchers as well as graduate students who are either entering these fields for the first time, or those already conducting research in these areas but are willing to extend their knowledge in the field of nanotechnology enabled sensors. This book is written in a manner that final year and graduate university students in the fields of chemistry, physics, electronics, biology, biotechnology, mechanics and bioengineering, can easily comprehend.

Nanotechnology enabled sensors is multidisciplinary by nature. It is important that the readers are armed with the necessary knowledge of physics, chemistry and biology related to these sensors and associated nanosciences. This book does not assume that its readers are experts in the multidisciplinary world; however, a basic understanding of university level chemistry and physics is helpful.

In this book, the authors present sensors that utilize nanotechnology enabled materials and phenomena. The terminology and concepts associated with sensors are presented which include some of the relevant physical and chemical phenomena applied in the sensor signal transduction system. The role of nanomaterials in such phenomena is also detailed. Throughtout this book, numerous strategies for the fabrication and characterization of nanomaterials and nanostructures, which are employed in sensing applications, are provided and the current approaches for nanotechnology enabled sensing are described. Sensors based on organic and inorganic materials are presented and some detailed examples of nanotechnology enabled sensors are explained.

Acknowledgments

We have been fortunate that many people supported and assisted us during the writing of this book.

First and foremost, our deepest and sincere appreciations go to Dr. Adrian Trinchi and Professor Wojtek Wlodarski. Adrian helped us tremendously at the initial stages of planning and the preparation of the proposal, structural formation of the chapters and the writing of chapters one to five. We would like to express our gratitude to Wojtek for his invaluable feedbacks on chapters. It was also through co-lecturing the course entitled "Nanosensors" at RMIT University and discussions with Wojtek that the seeds of the creation of this book were sown.

Secondly our sincere thanks are extended to Mr. Steve Elliot and all other members of Springer US publications for their assistance and helping us in the one and half years that we spent on writing this book. Steve's faith in us and his tremendous support from the beginning was one of the major reasons that this book has become a reality.

Several other people helped us directly or indirectly in this work. Dr. Kosmas Galatsis for his feedback and moral support, Dr. Anthony Holland for editing several of the chapters, Dr. Tim White for his invaluable feedbacks on chapter five and Prof. Arnan Mitchell for providing us with a number of figures appeared in different chapters. We also have to thank Dr. Wayne Row, Dr. Michelle Spencer, Dr. Kamran Ghorbani, Prof. Yongxiang Li, Prof. Alireza Baghai-Wadji, Prof. Paul Mulvaney, Dr. David Powell, Dr. Samuel Ippollito and Prof. Ali Mansoori for their help and support throughout the course of the preparation of this book.

We would like to thank all students of the "Nanosenors" course at RMIT University in years 2006 and 2007 for their highly valued feedbacks on each and every section of this book. Specially, we would like to express our gratitude to Mr. Blake Plowman and Mr. Chris Feigl for careful reading of the chapters and giving us the students' perspective of the presented material.

We are also grateful to Mr. Michael Breedon for the final editing of the book assisting in the final review and the further crystallization of the structure of the book.

Finally, we would like to thank our friends and family for their understanding and patience in the course of writing this book.

Contents

Preface i

Acknowledgments ii

Chapter 1: Introduction 1
1.1 Nanotechnology 1
1.2 Sensors 6
1.3 Nanotechnology Enabled Sensors 8

Chapter 2: Sensor Characteristics and Physical Effects 13
2.1 Introduction 13
2.2 Sensor Characteristics and Terminology 13
2.2.1 Static Characteristics 14
2.2.2 Dynamic Characteristics 17
2.3 Physical Effects Employed for Signal Transduction 20
2.3.1 Photoelectric Effect 21
2.3.2 Photodielectric Effect 27
2.3.3 Photoluminescence Effect 27
2.3.4 Electroluminescence Effect 31
2.3.5 Chemiluminescence Effect 34
2.3.6 Doppler Effect 34
2.3.7 Barkhausen Effect 36
2.3.8 Hall Effect 36
2.3.9 Nernst/Ettingshausen Effect 38
2.3.10 Thermoelectric (Seebeck/Peltier and Thomson) Effect 38
2.3.11 Thermoresistive Effect 42
2.3.12 Piezoresistive Effect 43
2.3.13 Piezoelectric Effect 46
2.3.14 Pyroelectric effect 47
2.3.15 Magneto-Mechanical Effect (Magnetostriction) 48
2.3.16 Mangnetoresistive Effect 49
2.3.17 Faraday-Henry Law 51
2.3.18 Faraday Rotation Effect 54
2.3.19 Magneto-Optic Kerr Effect (MOKE) 55
2.3.20 Kerrand Pockels Effects 56
2.4 Summary 57

Chapter 3: Transduction Platforms **63**
3.1 Introduction 63
3.2 Conductometric and Capacitive Transducers 63
3.3 Optical Waveguide based Transducers 66
3.3.1 Propagation in Optical Waveguides 67
3.3.2 Sensitivity of Optical Waveguides 69
3.3.3 Optical Fiber based Transducers 71
3.3.4 Interferometric Optical Transducers 72
3.3.5 Surface Plasmon Resonance (SPR) Transducers 74
3.4 Electrochemical Transducers 79
3.4.1 Chemical Reactions 80
3.4.2 Thermodynamics of Chemical Interactions 80
3.4.3 Nernst Equation 84
3.4.4 Reference Electrodes 97
3.4.5 Ion Selective Electrodes 90
3.4.6 An Example: Electrochemical pH Sensors 93
3.4.7 Voltammetry 94
3.4.8 An Example: Stripping Analysis 105
3.5 Solid State Transducers 106
3.5.1 *p-n* Diodes or Bipolar Junction based Transducers 106
3.5.2 Schottky Diode based Transducers 108
3.5.3 MOS Capacitor based Transducers 111
3.5.4 Field Effect Transistor based Transducers 113
3.6 Acoustic Wave Transducers 118
3.6.1 Quartz Crystal Microbalance 119
3.6.2 Film Bulk Acoustic Wave Resonator (FBAR) 121
3.6.3 Cantilever based Transducers 123
3.6.4 Interdigitally Launched Surface Acoustic Wave (SAW) Devices 125
3.7 Summary 129

Chapter 4: Nano Fabrication and Patterning Techniques **135**
4.1 Introduction 135
4.2 Synthesis of Inorganic Nanoparticles 136
4.2.1 Synthesis of Semi-conductor Nano-particles 136
4.2.2 Synthesis of Magnetic Nanoparticles 137
4.2.3 Synthesis of Metallic Nanoparticles 138
4.3 Formation of Thin Films 141
4.3.1 Fundamentals of Thin Film Deposition 141
4.3.2 Growth of One-Dimensional Nano-structured Thin Films 143
4.3.3 Segmented One-Dimensional Structured Thin Films 150
4.4 Physical Vapor Deposition (PVD) 151

4.4.1 Evaporation.......... 151
4.4.2 Sputtering 158
4.4.3 Ion Plating 163
4.4.4 Pulsed Laser Deposition (PLD).......... 164
4.5 Chemical Vapor Deposition (CVD) 164
4.5.1 Low Pressure CVD (LPCVD).......... 168
4.5.2 Plasma-Enhanced CVD (PECVD) 168
4.5.3 Atomic Layer CVD (ALCVD).......... 170
4.5.4 Atmospheric Pressure Plasma CVD (AP-PCVD) 172
4.5.5 Other CVD Methods.......... 173
4.6 Liquid Phase Techniques.......... 173
4.6.1 Aqueous Solution Techniques (AST).......... 173
4.6.2 Langmuir-Blodgett (LB) method.......... 176
4.6.3 Electro-deposition.......... 179
4.7 Casting.......... 182
4.7.1 Spin Coating.......... 182
4.7.2 Drop Casting, Dip Coating and Spraying.......... 184
4.8 Sol-gel.......... 184
4.9 Nanolithography and Nano-Patterning.......... 186
4.9.1 Photolithography 187
4.9.2 Scanning Probe Nanolithography Techniques 190
4.9.3 Nanoimprinting.......... 191
4.9.4 Patterning with Energetic Particles.......... 193
4.9.5 X-Ray Lithography (XRL) and LIGA.......... 197
4.9.6 Interference Lithography 200
4.9.7 Ion Implantation 202
4.9.8 Etching: Wet and Dry 202
4.10 Summary.......... 204

Chapter 5: Characterization Techniques for Nanomaterials 211
5.1 Introduction 211
5.2 Electromagnetic Spectroscopy.......... 211
5.2.1 UV-Visible Spectroscopy.......... 215
5.2.2 Photoluminescence (PL) Spectroscopy 219
5.2.3 Infrared Spectroscopy.......... 223
5.3 Nuclear Magnetic Resonance (NMR) Spectroscopy 228
5.4 X-Ray Photoelectron Spectroscopy (XPS) 232
5.5 X-Ray Diffraction (XRD).......... 237
5.6 Light Scattering Techniques.......... 240
5.6.1 Dynamic Light Scattering (DLS) 241
5.6.2 Raman Spectroscopy 245
5.7 Electron Microscopy.......... 248

5.7.1 Scanning Electron Microscope (SEM) ... 250
5.7.2 Transmission Electron Microscope (TEM) ... 255
5.8 Rutherford Backscattering Spectrometry (RBS) ... 259
5.9 Scanning Probe Microscopy (SPM) ... 263
5.9.1 Scanning Tunneling Microscope (STM) ... 264
5.9.2 Atomic Force Microscope (AFM) ... 267
5.10 Mass Spectrometry ... 270
5.10.1 Matrix-Assisted Laser Desorption/Ionisation (MALDI) Mass Spectrometer ... 272
5.10.2 Time of Flight (TOF) Mass Spectrometer ... 273
5.11 Summary ... 274

Chapter 6: Inorganic Nanotechnology Enabled Sensors ... 283
6.1 Introduction ... 283
6.2 Density and Number of States ... 283
6.2.1 Confinement in Quantum Dimensions ... 284
6.2.2 Momentum and Energy of Particles ... 285
6.2.3 Reciprocal Space ... 286
6.2.4 Definition of Density of States ... 287
6.2.5 DOS in Three-dimensional Materials ... 287
6.2.6 DOS in Two-Dimensional Materials ... 289
6.2.7 DOS in One-Dimensional Materials ... 291
6.2.8 DOS in Zero-Dimensional Materials ... 291
6.2.9 Discussions on the DOS ... 292
6.2.10 Theoretical and Computational Methods ... 296
6.2.11 One-Dimensional Transducers ... 297
6.2.12 Example: One-Dimensional Gas Sensors ... 302
6.3 Gas Sensing with Nanostructured Thin Films ... 304
6.3.1 Adsorption on Surfaces ... 305
6.3.2 Conductometric transducers Suitable for Gas Sensing ... 307
6.3.3 Gas Reaction on the Surface - Concentration of Free Charge Carriers ... 313
6.3.4 Effect of Gas Sensitive Structures and Thin Films ... 319
6.3.5 Effects of Deposition Parameters and Substrates ... 322
6.3.6 Metal Oxides Modification by Additives ... 323
6.3.7 Surface Modification ... 325
6.3.8 Filtering ... 328
6.3.9 Post Deposition Treatments ... 328
6.4 Phonons in Low Dimensional Structures ... 329
6.4.1 Phonons in One-Dimensional Structures ... 330
6.4.2 Electron-Phonon Interactions in Low Dimensional Materials ... 334
6.4.3 Phonons in Sensing Applications ... 337
6.4.3 One-Dimensional Piezoelectric Sensors ... 338

6.5 Nanotechnology Enabled Mechanical Sensors 340
6.5.1 Oscillators based on Nanoparticles 341
6.5.2 One-Dimensional Mechanical Sensors 343
6.5.3 Bulk Materials and Thin Films Made of Nano-Grains 345
6.5.4 Piezoresistors 347
6.6 Nanotechnology Enabled Optical Sensors 348
6.6.1 The Optical Properties of Nanostructures 348
6.6.2 The Optical Properties of Nanoparticles 352
6.6.3 Sensors based on Plasmon Resonance in Nanoparticles 353
6.7 Magnetically Engineered Spintronic Sensors 356
6.7.1 AMR, Giant and Colossal Magneto-Resistors 357
6.7.2 Spin Valves 360
6.7.3 Magnetic Tunnel Junctions 361
6.7.4 Other Nanotechnology Enabled Magnetic Sensors 362
6.8 Summary 363

Chapter 7: Organic Nanotechnology Enabled Sensors 371
7.1 Introduction 371
7.2 Surface Interactions 372
7.2.1 Covalent Coupling 372
7.2.2 Adsorption 379
7.2.3 Physical Entrapment 380
7.2.4 Chemical Entrapment 381
7.2.5 Self-Assembly 381
7.2.6 Layer-by-Layer Assembly 384
7.3 Surface Materials and Surface Modification 386
7.3.1 Gold Surfaces 386
7.3.2 Silicon, Silicon Dioxide and Metal Oxides Surfaces 387
7.3.3 Carbon Surfaces 389
7.3.4 Conductive and Non-Conductive Polymeric Surfaces 390
7.3.5 Examples of Surface Modifications in Biosensors 401
7.4 Proteins in Nanotechnology Enabled Sensors 404
7.4.1 The Structure of Proteins 404
7.4.2 The Analysis of Proteins 409
7.4.3 The Role of Proteins in Nanotechnology 409
7.4.4 Using Proteins as Nanodevices 411
7.4.5 Antibodies in Sensing Applications 412
7.4.6 Antibody Nanoparticle Conjugates 418
7.4.7 Enzymes in Sensing Applications 420
7.4.8 Enzyme Nanopraticle Hybrid based Sensors 425
7.4.9 Motor Proteins in Sensing Applications 427
7.4.10 Transmembrane Sensors 428

7.5 Nano-sensors based on Nucleotides and DNA 436
7.5.1 The Structure of DNA 438
7.5.2 The Structure of RNA 441
7.5.3 DNA Decoders and Microarrays 442
7.5.4 DNA-based Sensors 449
7.5.5 DNA-Protein Conjugate-based Sensors 452
7.5.6 DNA Conjugates with Inorganic Materials 455
7.5.7 Bioelectronic Sensors based on DNA 459
7.5.8 DNA Sequencing with Nanopores 463
7.6 Sensors Based on Molecules with Dendritic Arcitectures 465
7.7 Force Spectroscopy and Microscopy of Organic Materials 467
7.8 Biomagnetic Sensors 469
7.9 Summary 470

Index 482

About the Authors 491

Chapter 1: Introduction

1.1 Nanotechnology

The term *nano* in the SI units means 10^{-9}, or in other words, one billionth. It is derived from the Greek word for dwarf. Materials, structures and devices that have dimensions lying in the nano scale range are encompassed within *nanosciences*. Materials that have at least one dimension less than 100 nm may be considered to be nanodimensional (**Fig. 1.1**).

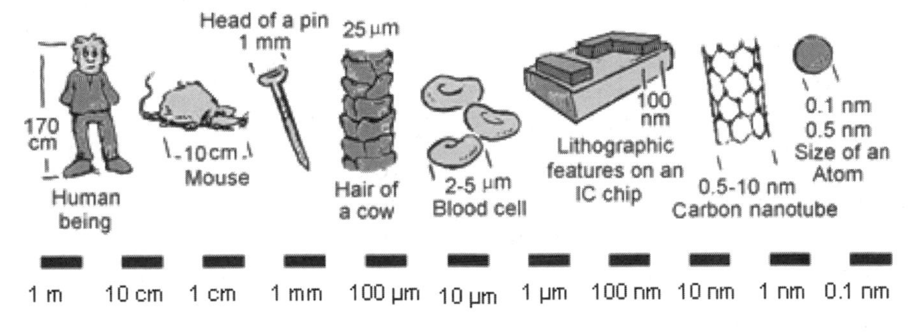

Fig. 1.1 Examples of objects with different dimensions (by Kourosh Kalantar-zadeh).

Nanotechnology comprises technological developments on the nanometer scale. The *United States' National Nanotechnology Initiative* website (http://www.nano.gov) defines nanotechnology as: "The understanding and control of matter at dimensions of roughly 1 to 100 nm, where unique phenomena enable novel applications."

In the nano range, the physical, chemical, and biological properties of materials are unique. Therefore, nanotechnology provides us with tools to create functional and intelligent materials, devices, and systems by controlling materials in the nano scale, making use of their novel phenomena and associated properties.

Nobel laureate Richard Feynman provided one of the defining moments in nanotechnology when in December 1959 he conducted his visionary lecture entitled "There is Plenty of Room at the Bottom" not just "There is Room at the Bottom".[1] In his lecture, Feynman said: "What I want to talk about is the problem of manipulating and controlling things in a small scale … what I have demonstrated is that there is room - that you can decrease the size of things in a practical way. I now want to show that there is plenty of room. I will not now discuss how we arc going to do it, but only what is possible in principle … we are not doing it simply because we haven't yet gotten around to it … arrange the atoms one by one the way we want". What Feynman realized was that "at the atomic level, we have new kinds of forces, new kinds of possibilities, and new kinds of effects. The problems of manufacture and reproduction of materials will be quite different".

Nanotechnology is multidisciplinary in its nature. It not only concerns physics and engineering, it encompasses many other disciplines, in particular chemistry and biology. Consequently, it is essential that people taking an active role in nanotechnology must embrace the disciplines of science, engineering, and even philosophy.

Varied approaches have emerged for the development of nanomaterials, nanostructures and nanodevices. They are generally categorized as *top-down* and *bottom-up* approaches. Top-down approaches are those by which the bulk dimensions of a material are reduced until nanometer size features are produced. A well-known example of the top-down approach is the reduction in dimensions of the transistors on silicon chips which are fast approaching the nanoscale. By contrast, bottom-up approaches involve assembling structures molecule by molecule, or atom by atom, to fabricate structures with nano dimensions such as formation of self-assembled monolayers. Nanotechnology has become more tangible since bottom-up and top-down approaches started to coincide. Clearly, the successful realization of nanotechnology-enabled devices rests on the perfect amalgamation of these two approaches.

Moore's law describes that computing power (in effect the number of transistors on a silicon chip) is doubling every 18 to 24 months. In fact silicon chips have followed this rule quite nicely for four decades. However, due to inherent material properties, no one expects that silicon based electronics can follow Moore's law forever. Nowadays transistor technology features have reached dimension of 50 nm, yet transistors are still larger than the average size of most molecules. Continuing this trend, the silicon-based industry will become stagnant in or around 2015 when there will no longer be a possibility to shrink dimensions.[1]

The predictions are that organic and molecular based transistors will emerge and nanotechnology will play a pivotal role to ensure that Moore's law remains valid.[2,3] This will perhaps be one of the most clear cut examples of how bottom-up and top-down fabrication strategies are meeting in our time, and must be able to coexist and help each other in order to provide solutions for our needs.

We are fortunate to have at our disposal a myriad of scientific and technical tools and processes that are now well established. These include: high resolution characterization techniques, ion and molecular beam fabrication, nano-imprint lithography, atom by atom manipulation, a growing knowledge of cell biology, etc. The proliferation of these tools enable the measurement, fabrication, characterization and manipulation of nanostructures. New instruments with nanoscale resolution are accelerating scientific discovery, providing quality control in the fabrication of nanostructures, and stimulating novel approaches in miniaturization. These tools and processes, among with many others have helped us to delve into this area with greater confidence.

Despite having many tools at our disposal, we are only at the very beginning of our exploration into the nanotechnological realm. There are still many untouched areas in nanotechnology. Nanotechnology researchers with open mind and meticulous ability are required to make observations in all the disciplines available, to allow amalgamating ideas into new theories and developments. Nanotechnology researchers with strong knowledge in different disciplines must be willing to think beyond the realm of their initial training, as being merely an engineer or a scientist is no longer sufficient. An example comes from Albert Einstein (**Fig. 1.2**), who as part of his PhD program was able to calculate the size of a sugar molecule from the experimental data on the diffusion of sugar in water.[1]

Currently nanotechnology is in the forefront of technological discussions, debates and developments as scientists, policy makers and entrepreneurs endeavor to fully harness its capabilities and unleash a broad range of novel products.

It has been proven historically that the emergence and demise of economically powerful and industrial nations depend on their technological prowess. It is likely that countries that are playing a pioneering role in nanotechnology will reap the financial benefits and prosper altering our economical and social balances.

Fig. 1.2 Albert Einstein (Reprinted with permission from Javad Alizadeh – by Javad Alizadeh).

We are already beginning to experience some of the benefits that it has to offer. Nanotechnology enabled sunscreens are already enjoying commercial success and nano magnetic materials are available for the fabrication of highly dense data storage. Carbon nanotubes can be purchased cheaply. Nanoporous and nanostructured thin films have found numerous applications in the building industry and home appliance. Antimicrobial wound dressings, which use nanocrystalline silver to provide a steady dose of ionic silver to protect against secondary infections, are already in the market, as are cosmetics and skin protection products that fully utilize the capabilities of functional nanomaterials. Superior and cheaper products have been realized, and with their initial success, our expectations from nanotechnology are growing.

We are eagerly waiting to see changes for the better in our lives coming from nanotechnology. We have already witnessed the dramatic changes that our day-to-day lives have undergone in the last decades, owing to the emergence of home computers, internet, and mobile phones. As our palettes broaden and continue to grow, so too does our thirst for new products. In such cases, conventional technologies may fail to provide us with the advances that we so desperately crave. It is not beyond the realm of possibility that in the coming decades our lives may once again be revolutionized by products realized though the advances that nanotechnology can provide.

Nanostructures exist naturally in biological systems. Understanding these systems will allow us to improve the way we manage our health care and medical diagnostics. Clearly these days, bio-nanotechnology is among the first areas that is finding real world applications. Labeling and disease markers, drug discovery research, and diagnostic tests are among the pioneering developments.

Nanotechnology has the potential to have enormous impacts on manufacturing and construction industries. Smart nano materials may be employed to resolve the energy problems and provide advanced structures with desired capabilities.

Nanotechnology is in its infancy, and we have just taken the first step into it and consequently our knowledge in this area is still rudimentary. There are still major hurdles that must be surmounted. For example, interfacing between the nano-world and macro-world has not been established properly. Other than extremely expensive tools in the labs, reading tiny signals from the nanomaterials and sending the orders to them remain challenging tasks. There are still many ambiguities as we delve into the nanotechnological realm, as definitions and standards are still vague. What makes it more difficult is that nanotechnology has not been standardized yet. It consists of diverse materials, disciplines and techniques. It is becoming overtly difficult to come up with processes that can be adopted worldwide.

It is needless to say that among the multitude of possibilities that nanotechnology presents, there may be accompanying dangers. The possible negative effects of nanomaterials on our health and on our environmental are still relatively unclear. Our minds may wander on the verge of science fiction when we think about nanotechnology. In Drexler's "Engines of Creation",[4] the author depicted a visionary view of godlike control over materials by creating self-replicating assemblers which produce new creations. Bill Joy, a scientist at Sun Microsystem, drew inspiration from Drexler's book, predicting the possibility of self replicating nanomaterials called "grey goo" which could pose serious danger to the environment **(Fig. 1.3)**. It is with this type of thinking that scientists must act responsibly and tread cautiously when embarking on nanotechnology research. In a similar manner to chemicals such as dichloro-diphenyl-trichloroethane (DDT), which were the origins of terrible chemical pollution, scientist embarking on nanotechnology research should be vigilant to ensure that such disasters are not perpetuated once again. After all, Bill Joy's outlook of "grey goo" may not be so far fetched!

Fig. 1.3 Gray goo! (by Kourosh Kalantar-zadeh).

Despite these potential drawbacks and fears, there is much to look forward to in nanotechnology. With the nanotechnology market predicted to create revenues of over 1 trillion dollars per year by 2015,[5] there is great optimism. There is no doubt that nanotechnology has solid commercial prospects, however, it must be kept in mind that the task of converting basic discoveries into marketable products will be long and hard.[6]

1.2 Sensors

The word *sensor* is derived from the Latin word "sentire" which means to perceive.[7] A sensor is a device that responds to some *stimulus* by generating a functionally related output.[8] Exposure to a certain analyte or change in ambient conditions alters one or more of its properties (e.g. mass, electrical conductivity, capacitance, etc.) in a measurable manner, either directly or indirectly.

Quite simply our motivation for having sensors is so that we will be able monitor the environment around us, and use that information at a latter stage for another purpose. It is through sensors that we make our contact with the world.

A sensor should be sensitive to the *measurand* and insensitive to any other input quantity. It is essential that environmental effects such as temperature, humidity, shock and vibrations be accounted for. All these factors can have a negative impact on the sensors' performance. As a general rule, sensors should be inexpensive yet reliable and durable. They should provide accurate, stable, high resolution, low cost sensing. Each application places different requirements on the sensor and sensing system. However, regardless of the associated application all sensors have the same object: to achieve accurate and stable monitoring of the measurand.

In recent years, the development of sensors has become increasingly important. Sensor technology has flourished as the need for physical, chemical, and biological recognition systems and transducing platforms grow. Nowadays sensors are used in applications ranging from environmental monitoring, medical diagnostics and health care, in automotive and industrial manufacturing, as well as defense and security.[9] Sensors are finding a more prominent role in today's world, as we place strong emphasis on devices aimed at making our lives better, easier, and safer.

We may not even realize it, but sensors are found commonly around the household. They are in electrical devices from surge protectors to automatic light switches, refrigerators and climate control appliances, toasters, and of course in smoke and fire detectors. They are found most toys that have interactive capability. We also encounter sensors in everyday life: entering a department store with automatically opening doors, or in our automobiles, monitoring parameters such as the oil pressure, temperature, altitude and fuel levels. Sensors are installed in gas cook tops, where they determine whether or not the pilot is on, and if not, halt the gas flow preventing the room from being filled with gas. The function of voltage sensitive transistors is not so obvious to us, yet millions of them are contained within central processing units of computers, which are used to convert analogue signals into digital ones.

Many complex machines incorporate sensors. Aircraft are riddled with them as they monitor position, wind speed, air pressure, altitude etc. Another important application is for industrial process control where the sensors continually monitor to ensure that efficiency is maximized, production costs are minimized and that waste is reduced.

Sensors are also an integral part of health care and diagnostics. Sensors can determine whether or not biological systems are functioning correctly and most importantly, direct us to act without delay when something is wrong. For instance, glucose meters are playing a crucial role in determining blood sugar levels in people diagnosed with diabetes.

The area of sensor technology is quite broad, and there is considerable diversity in sensor research. In the last four decades sensor research has grown exponentially, largely due to increasing automation, medical applications and escalating use of microelectronics. Parallel to these developments, the capabilities of sensors are increasingly improving as their prices tumble. Sensors have become a ubiquitous part of life, and now more than ever they are playing an important role in our day-to-day lives.

1.3 Nanotechnology Enabled Sensors

Sensor technology is quite possibly the area in which nanotechnology has had one of the greatest impacts. To meet the increasing demands of industry, new approaches to sensor technology have been taken, and this is where nanotechnology shines. Nanotechnology is enabling the development of small, inexpensive and highly efficient sensors, with braod applications.

It is envisaged that by enhancing the interactions that occur at the nanoscale, nanotechnology enabled sensors may offer significant advantages over conventional sensors. This may be in terms of greater sensitivity and selectivity, lower production costs, reduced power consumption as well as improved stability. The unique properties of nanoscale materials make them ideal for sensing. Such materials could be integrated into existing sensing technologies or could be used to form new devices. Not only does nanotechnology enable us to enhance existing materials, it also enables us to fabricate novel materials, whose properties can be tailored specifically for sensing applications.

There exist possibilities for developing nano-bio-organic elements that are suitable for intracellular measurements (Chap. 7). In particular for sensing applications, nanotechnology allows development of nanostructures and the possibility of forming features, the likes of which cannot be imagined with conventional microtechnologies. The characteristics of nanotechnologically enabled sensors are more favorable for sensing than the classically fabricated systems. For example, sensitivity may increase due to tailored conduction properties, the limits of detection may be lowered, infinitely small quantities of samples can be analyzed, direct analyte detection may be possible without using labels, and specificity may be improved (Chaps. 2,6 and 7). Physical sensors, electro-sensors, chemical sensors and biosensors may all benefit from nanotechnology. Using nanotechnological processes, the density of states in materials can be tuned to develop highly sensitive magnetic sensors (Chaps. 2 and 6) or to create quantum resistance which have enormous applications in electronic industry (Chap. 6). Using nanomaterials, highly efficient Peltier transducers can be fabricated which will change the face of the energy industry in a not distant future (Chap. 2).

Improved sensitivity is a major attraction for developing nanotechnology enabled sensors. At the extreme nanoscale limit, there exist the potential to detect a single molecule or atom. The small size, lightweight, and high *surface-to-volume ratio* of nanostructures are the best candidates for improving our capability to detect chemical and biological species with sensitivity that was previously thought to be unattainable. Additionally, in nanostructures the entire structure can be affected by the analyte and not only the surface as conventional sensors **(Fig. 1.4)**.

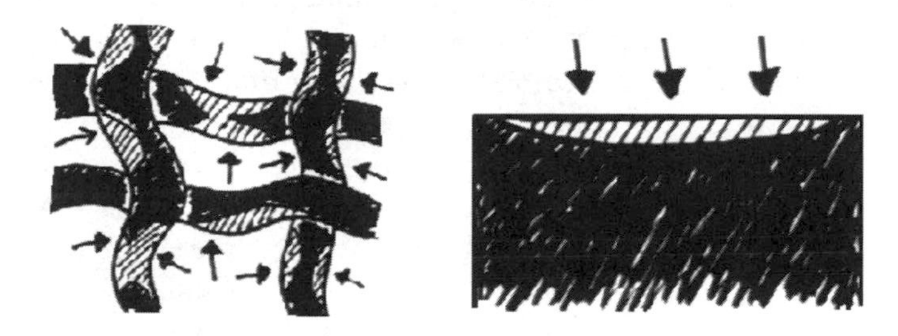

Fig. 1.4 Effect of analytes on nanostructures (left) vs. smooth surfaces (right).

Selectivity is tantamount to sensitivity, yet significantly more difficult to attain. The uses of nanoscale sensors and materials may not implicitly result in greater selectivity; however nanostructuring materials and applying surface modifications and functionalization may greatly assist. Other opportunity which may arise from the employment of nanomaterials includes the deployment of sensor arrays, where multiplicity in the tens to thousands may compensate for the loss of performance of any single measurement.

The speed with which species can be detected is most definitely affected by the sensor's dimensions. Hence, nanoscale modifications present the opportunity for improving the sensor's dynamic performance. Nanostructures minimize the time taken for a measurand to diffuse into and out of that volume **(Fig. 1.4)**. Therefore, this is a key objective of nanotechnology-enabled sensors. For instance, a few seconds may be all the time required to respond to an undesirable and potentially harmful situation. The time taken for the sensor to raise an alarm could be the limiting factor, averting a potential disaster.

Nanotechnology enabled sensors may find applications in numerous fields, however, one of the most significant areas that they will be employed is nano-biotechnology and human health monitoring (Chap. 7).[10] Minimally invasive technologies capable of scanning our bodies for the earliest signs of oncoming of disease are being developed. Their ultimate aim is to create new biomedical technologies that can detect, diagnose and treat diseases inside the human body. Human beings want to live longer, healthier, and happier, be in better control of their bodies, more connected to others and to objects around us. Nanotechnology is one of the tools that may assist humans to reach these goals. *DNA* and *proteins* have been extensively utilized and manipulated by researchers and scientists for biosensing applications (Chap. 7). These natural bio-elements, with embedded intelligence, are ready made nanosized building blocks and tools that can perform pre-programmed functions on demand. They can selectively bind to target molecules and carry out the required alterations. *Redox-ezymatic proteins* are the base of glucose sensors which improve the quality of life for millions afflicted with diabetes (Chaps. 6 and 7).

Other good examples highlighting advantages of nanotechnological applications in sensing resides in the fabrication of chemical sensors. Such sensors have traditionally suffered from limited measurement accuracy, sensitivity and plagued with problems of long-term stability. However, recent advances in nanotechnology have resulted in novel classes of nanostructured thin films, similar to those of polyaniline and TiO_2 thin films shown in **Fig. 1.5**. As will be seen in Chaps. 6 and 7, the nanostructured polyaniline, which is a conductive polymer, thin film can be utilized for the fabrication of optical biosensors as well as gas and liquid phase conductometric sensors with ultra fast responses. TiO_2 nanostructured thin films can be employed as gas sensitive film in conductometric sensors, as an efficient photocatalyst in optical sensors and cells and as metal oxide which provide superhydrophobicity for the immobilization of proteins.

Such nanostructured thin films enhance chemical sensing properties via an increased surface area to volume ratio, improving the active sensing area available for the interaction with the target molecules (Chaps. 2-7). Additionally, strong photon and phonon quenching and amplification are also observed for such surfaces that cannot be seen in conventional bulk materials (Chaps. 6 and 7). With such alterations the optical and electronic properties can be tailored to suit the applications.

Using the nanoparticles, it is possible to tune and amplify the response of optical sensors for narrow frequency bands which makes them more accurate and selective. They can resonate with the same stimuli at different frequencies, a property which can be highly useful in medical imaging for differentiating discerning between different targets. Surface of nanoparticles can be functionalized for specific biosensing applications.

The market for nanotechnology enabled sensors is constantly growing. Advances in technology will further facilitate the nanotechnologically enabled sensors' incorporation with sophisticated electronics signal processing with innovative transducers and actuators, electronic components, communication circuits and in medical sciences.

There are already many nanotechnology enabled sensors in the market. However, in the following decades the smarter, cheaper and more selective and sensitive sensors will influence our lives much more and their applications will become more pronounced in our daily lives.

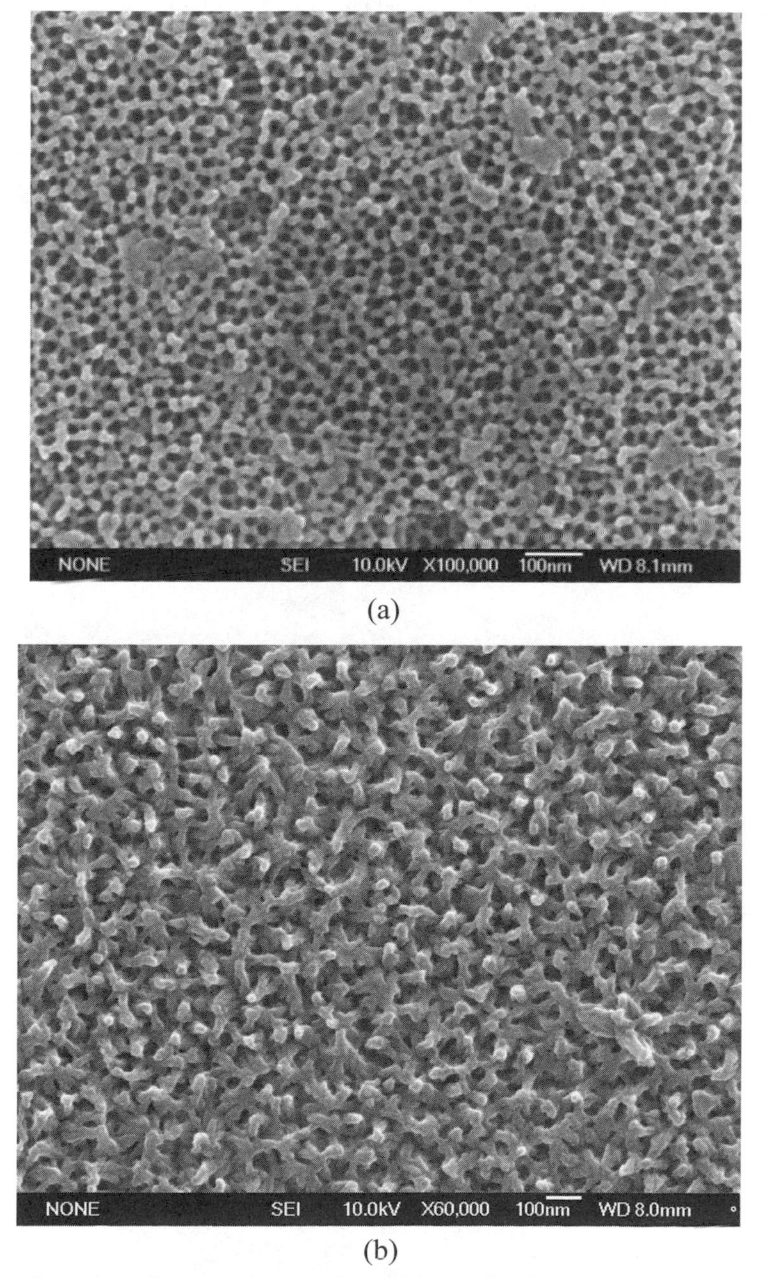

Fig. 1.5 Scanning electron micrographs of (a) anodized nanoporous TiO_2 (b) polyaniline nanofibers electrodeposited on gold.

References

[1] S. Fritz and M. L. Roukes, *Understanding Nanotechnology* (Warner Books, New York, USA, 2002).

[2] G. Horowitz, Advanced Materials **10,** 365-377 (1998).

[3] A. Dodabalapur, L. Torsi, and H. E. Katz, Science **268,** 270-271 (1995).

[4] E. Drexler, *Engines of Creation* (Anchor Books, Garden City, USA, 1988).

[5] L. DeFrancesco, Nature Biotechnology **21,** 1127-1129 (2003).

[6] L. Mazzola, Nature Biotechnology **21,** 1137-1143 (2003).

[7] M. J. Usher and D. A. Keating, *Sensors and transducers: characteristics, applications, instrumentation, interfacing* (Macmillan, London, UK, 1996).

[8] W. Göpel, J. Hesse, and J. N. Zemel, *Sensors: A Comprehensive Survey* (VCH, Weinheim, Germany, 1991).

[9] I. R. Sinclair, *Sensors and transducers* (Newnes, Oxford, UK, 2001).

[10] R. Paull, J. Wolfe, P. Hebert, and M. Sinkula, Nature Biotechnology **21,** 1144-1147 (2003).

Chapter 2: Sensor Characteristics and Physical Effects

2.1 Introduction

The potential of nanotechnology enabled sensors was highlighted in the previous chapter. In this chapter, the fundamental characteristics and terminologies associated with transducers and sensors are introduced. Furthermore, some of the major effects that are utilized in sensing for the conversion of energy from a measurand (the physical parameter being quantified by a measurement) to a measurable signal are described. These effects illustrate the relationship between different physical and chemical phenomena that can be measured using sensors. This will be a prelude to Chap. 3, which focuses on major transduction platforms.

The essence of Chap. 2 is on physical transduction phenomena. The majority of chemical phenomena which are related to nanotechnology enabled sensing can be found in Chaps. 6 and 7.

2.2 Sensor Characteristics and Terminology

A *sensor* is a device that produces a measurable signal in response to a stimulus. A *transducer* is a device that converts one form on energy into another. Generally, a sensing or sensitive layer/medium directly responds to the external stimulus, while the transducer converts the response into an external measurable quantity. As distinct from *detectors,* sensors are employed to monitor and quantify changes in the measurand, whereas detectors simply indicate the presence of the measurand.[1,2]

The characteristics of a sensor may be classified as being either *static*, or *dynamic*. These parameters are essential in high fidelity mapping of output versus input. Static characteristics are those that can be measured after all transient effects have stablized to their final or steady state. They address questions such as; by how much did the sensor's output change in response to the input? what is the smallest change in the input that will give an output reading? and how long did it take for the output value to change to the present value? Dynamic characteristics describe the sensor's transient properties. These typically address questions such as; at what rate is the output changing in response to the input? and what impact would a slight change in the input conditions have on the transient response?

2.2.1 Static Characteristics

Accuracy:

This defines how correctly the sensor output represents the true value. In order to assess the accuracy of a sensor, either the measurement should be benchmarked against a standard measurand or the output should be compared with a measurement system with a known accuracy. For instance, an oxygen gas sensor, which operates at a room with 21% oxygen concentration, the gas measurement system is more accurate if it shows 21.1% rather than 20.1% or 22%.

Error:

It is the difference between the true value of the quantity being measured and the actual value obtained from the sensor. For instance, in the gas sensing example, if we are measuring the oxygen content in the room having exactly 21% oxygen, and our sensor gives us a value of 21.05%, then the error would be 0.05%.

Precision:

Precision is the estimate which signifies the number of decimal places to which a measurand can be reliably measured. It relates to how carefully the final measurement can be read, not how accurate the measurement is.

Resolution:

Resolution signifies the smallest incremental change in the measurand that will result in a detectable increment in the output signal. Resolution is strongly limited by any noise in the signal.

Sensitivity:

Sensitivity is the ratio of incremental change in the output of the sensor to its incremental change of the measurand in input. For example, if we have a gas sensor whose output voltage increases by 1 V when the oxygen concentration increases by 1000 ppm, then the sensitivity would be 1/1000 V/ppm, or more simply 1 mV/ppm.

Selectivity:

A sensor's ability to measure a single component in the presence of others is known as its selectivity. For example, an oxygen sensor that does not show a response to other gases such as CO, CO_2 and NO_2, may be considered as selective.

Noise:

Noise refers to random fluctuations in the output signal when the measurand is not changing. Its cause may be either internal or external to the sensor. Mechanical vibrations, electromagnetic signals such as radio waves and electromagnetic noise from power supplies, and ambient temperatures, are all examples of external noise. Internal noises are quite different and may include:

1. *Electronic Noise*, which results from random variations in current or voltage. These variations originate from thermal energy, which causes charge carriers to move about in random motions. It is unavoidable and present in all electronic circuits.
2. *Shot Noise*, which manifests as the random fluctuations in a measured signal, caused by the signal carriers' random arrival time. These signal carriers can be electrons, holes, photons, etc.
3. *Generation-Recombination Noise*, or *g-r noise*, that arises from the generation and recombination of electrons and holes in semiconductors.
4. *Pink Noise*, also known as *1/f noise*, is associated with a frequency spectrum of a signal, and has equal power per octave. The noise components of the frequency spectrum are inversely proportional to the frequency. Pink noise is associated with self-organizing, bottom-up systems that occur in many physical (e.g. meteorological: thunderstorms, earthquakes), biological (statistical distributions of DNA sequences, heart beat rhythms) and economical systems (stock markets).

Drift:

It is the gradual change in the sensor's response while the measurand concentration remains constant. Drift is the undesired and unexpected change that is unrelated to the input. It may be attributed to aging, temperature instability, contamination, material degradation, etc. For instance, in a gas sensor, gradual change of temperature may change the baseline stability, or gradual diffusion of the electrode's metal into substrate may change the conductivity of a semiconductor gas sensor which deteriorating its baseline value.

Minimum Detectable Signal (MDS):

This is the minimum detectable signal that can be extracted in a sensing system, when noise is taken into account. If the noise is large relative to the input, it is difficult to extract a clear signal from the noise.

Detection Limit:

It is the smallest magnitude of the measurand that can be measured by a sensor.

Repeatability:

Repeatability is the sensor's ability to produce the same response for successive measurements of the same input, when all operating and environmental conditions remain constant.

Reproducibility:

The sensor's ability to reproduce responses after some measurement condition has been changed. For example, after shutting down a sensing system and subsequently restarting it, a reproducible sensor will show the same response to the same measurand concentration as it did prior to being shut down.

Hysteresis:

It is the difference between output readings for the same measurand, when approached while increasing from the minimum value and the other while decreasing from the peak value.

Stability:

The sensor's ability to produce the same output value when measuring a fixed input over a period of time.

Response Time:

The time taken by a sensor to arrive at a stable value is the response time. It is generally expressed as the time at which the output reaches a certain percentage (for instance 95%) of its final value, in response to a stepped change of thc input. The *recovery time* is defined in a similar way but conversely.

Dynamic Range or Span:

The range of input signals that will result in a meaningful output for the sensor is the dynamic range or span. All sensors are designed to perform over a specified range. Signals outside of this range may be unintelligible, cause unacceptably large inaccuracies, and may even result in irreversible damage to the sensor.

2.2.2 Dynamic Characteristics

It is advantageous to use *linear* and *time invariant* mathematical representations for sensing systems. Such representations have been widely studied, they are easy to extract information from and give an overall vision about the sensing systems to the users. The relationship between the input and output of any linear time invariant measuring system can be written as:

$$\begin{aligned} a_n \frac{d^n y(t)}{dt^n} + a_{n-1} \frac{d^{n-1} y(t)}{dt^{n-1}} + \ldots + a_1 \frac{dy(t)}{dt} + a_0 y(t) \\ = b_m \frac{d^{m-1} x(t)}{dt^m} + b_{m-1} \frac{d^{m-1} x(t)}{dt^{m-1}} + \ldots + b_1 \frac{dx(t)}{dt} + b_0 x(t) \end{aligned}, \tag{2.1}$$

where $x(t)$ is the measured quantity (input signal) and $y(t)$ is the output reading and $a_0, \ldots, a_n, b_o, \ldots, b_m$ are constants.

$x(t)$ can have different forms and values. As a simple and commonly encountered example in sensing systems, $x(t)$ may be considered to be a *step change* (*step function*) similar to that depicted in **Fig. 2.1**. However, on many occasions this is an over simplification, as there is generally a rise and fall time for the step input to occur.

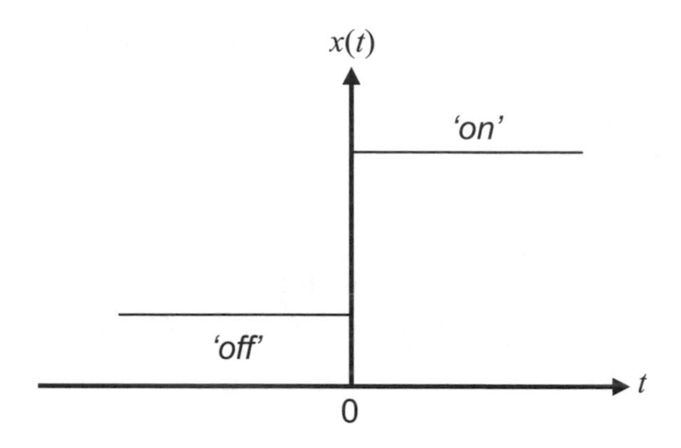

Fig. 2.1 A step change.

When the input signal is a step change, Eq. (2.1) reduces to:

$$a_n \frac{d^n y(t)}{dt^n} + a_{n-1} \frac{d^{n-1} y(t)}{dt^{n-1}} + \ldots + a_1 \frac{dy(t)}{dt} a_{n-1} + a_0 y(t) = b_0 x(t), \tag{2.2}$$

as all derivatives of $x(t)$ with respect to t are zero. The input does not change with time except at $t = 0$. Further simplifications can be made. For instance, if output shows an instantaneous response to the input signal then all $a_1, \ldots, a_n$ coefficients except a_0 are zero, as a result:

$$a_0 y(t) = b_0 x(t) \text{ or simply: } y(t) = Kx(t). \tag{2.3}$$

where $K = b_0 / a_0$ is defined as the static sensitivity. Such a response represents a perfect zero order system. If the system is not perfect and the output does show a gradual approach to its final value, then it is called a first order system. A simple example of a first order system is the charging of a capacitor with a voltage supply, whose rate of charging is exponential in nature. Such a first order system is described by the following:

$$a_1 \frac{dy(t)}{dt} + a_0 y(t) = b_0 x(t), \tag{2.4}$$

or after rearranging:

$$\frac{a_1}{a_0} \frac{dy(t)}{dt} + y(t) = \frac{b_0}{a_0} x(t). \tag{2.5}$$

By defining $\tau = a_1/a_0$ as the time constant, the equation will take the form of a *first order ordinary differential equation* (*ODE*):

$$\tau \frac{dy(t)}{dt} + y(t) = Kx(t)\,. \tag{2.6}$$

This ODE can be solved by obtaining the homogenous and particular solutions. Solving this equation reveals that the output $y(t)$ in response to $x(t)$ changes exponentially. Furthermore, τ is the time taken for the output value to reach 63% of its final value, i.e. $(1\text{-}1/e^{-1}) = 0.6321$, as seen in a typical output of a first order system in **Fig. 2.2**.

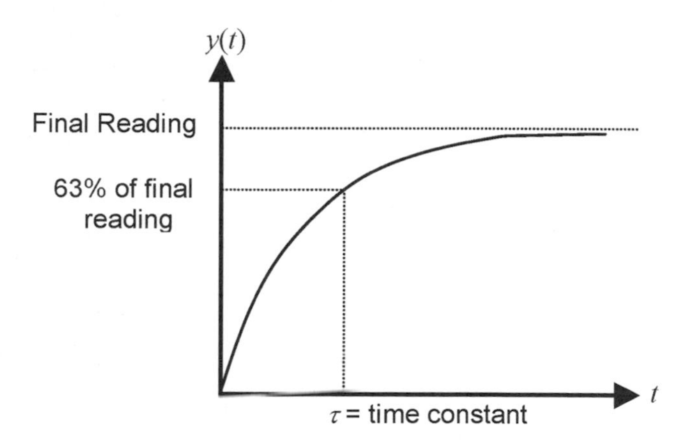

Fig. 2.2 Graphical depiction of a first order system's response with a time constant of τ.

On the other hand, the response of a *second order system* to a step change can be defined as:

$$a_2 \frac{d^2 y(t)}{dt^2} + a_1 \frac{dy(t)}{dt} + a_0 y(t) = b_0 x(t)\,\cdot \tag{2.7}$$

By defining the undamped natural frequency $\omega = a_0/a_2$, and the damping ratio $\varepsilon = a_1/(2a_0a_2)$, Eq. (2.7) reduces to:

$$\frac{1}{\omega^2} \frac{d^2 y(t)}{dt^2} + \frac{2\varepsilon}{\omega} \frac{dy(t)}{dt} + y(t) = Kx(t)\,\cdot \tag{2.8}$$

This is a standard second order system. The damping ratio plays a pivotal role in the shape of the response as seen in **Fig. 2.3**. If $\varepsilon = 0$ there is no damping and the output shows a constant oscillation, with the solution being a sinusoid. If ε is relatively small then the damping is light, and the oscillation gradually diminishes. When $\varepsilon = 0.707$ the system is *critically* *damped.* A critically damped system converges to zero faster than any other without oscillating. When ε is large the response is heavily damped or *over damped.* Many sensing systems follow the second order equations. For such systems responses that are not near the critically damped condition ($0.6 < \varepsilon < 0.8$) are highly undesirable as they are either slow or oscillatory.[3]

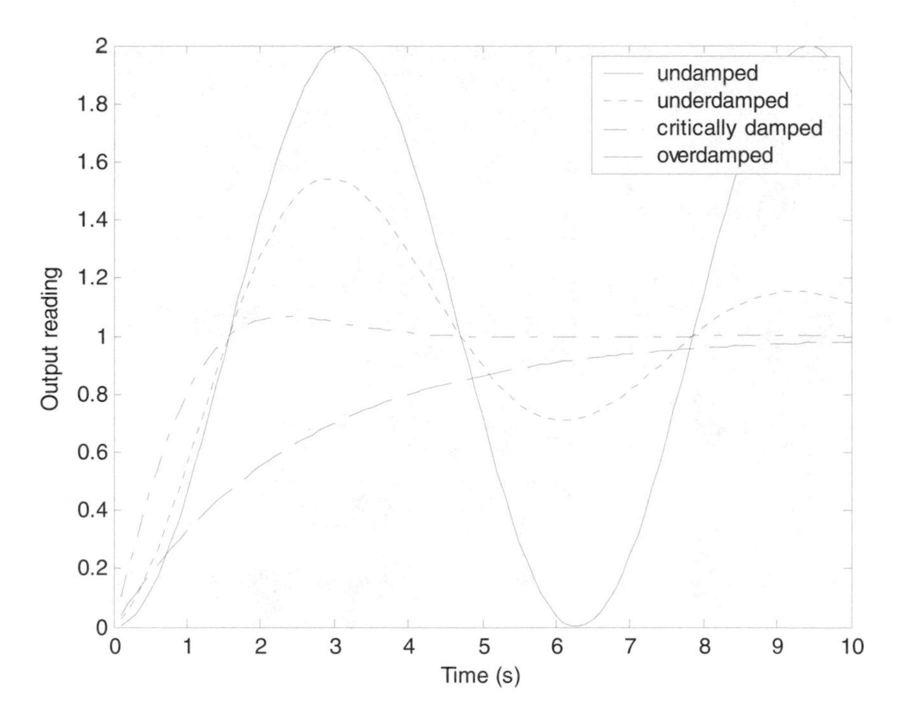

Fig. 2.3 Responses of second order systems to a step input.

2.3 Physical Effects Employed for Signal Transduction

Physical effects employed for signal transduction generally involve the coupling of a material's *thermal*, *mechanical* and *electromagnetic* (including *optical*) properties. **Table 2.1** shows examples of effects that are obtained when thermal, mechanical and electromagnetic properties are coupled with each other, or with themselves.

Within the above mentioned physical effects, *chemical interactions* may also be involved. In chemical reactions thermal, mechanical and electromagnetic energies are released or absorbed.

In the subsequent sections, some of the most widely utilized physical effects in sensor technology, along with several examples relevant to nanotechnology enabled sensing, are provided.

Table 2.1 Some well known physical effects.

Physical effect	Thermal	Mechanical	Electromagnetic (including optical)
Thermal	e.g. heat transfer	e.g. thermal expansion	e.g. thermoresistance
Mechanical	e.g. friction	e.g. acoustic effects	e.g. magnetostriction
Electromagnetic (including optical)	e.g. Peltier effect	e.g. piezoelectricity	e.g. Hall and Faraday effects

2.3.1 Photoelectric Effect

When a material is irradiated by photons, electrons may be emitted from the material. The ejected electrons are called *photoelectrons*, and their kinetic energy, E_K, is equal to the incident photon's energy, $h\nu$, minus a threshold energy, known as the material's *work function* ϕ, which needs to be exceeded for the material to release electrons. The effect is illustrated in **Fig. 2.4** and is governed by Eq. (2.9):

$$E_K = h\nu - \phi\,, \tag{2.9}$$

where h is Planck's constant ($h = 6.625\times10^{-34}$ Js) and ν is the photon's frequency.

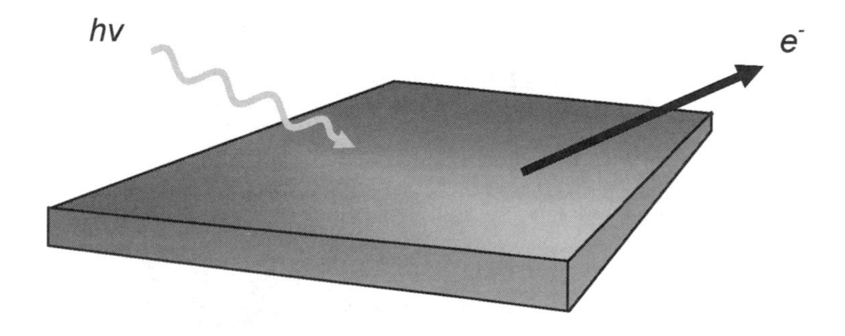

Fig. 2.4 Photoelectric effect.

The photoelectric effect is ideal for use in light sensitive devices. Because the work function depends on the material, sensors may be designed that are tuned to specific wavelengths. Electrodes with nanostructured surfaces have emerged as excellent candidates for use in photoelectric devices and sensors. The work function can be tuned by changing of the material's dimensions. The large surface to volume ratio nanostructures may enhance the photoelectric device's light-to-energy conversion efficiency (Chap. 6). In addition, the release of produced charges is faster in nanomaterials, which translates into a faster device response.[4] Related to this effect are the *photoconductive* and the *photovoltaic* effects.

Photoconductive Effect

Photoconductivity occurs when a beam of photons impinges on a semiconducting material, causing its conductivity to change. The conductivity results from the excitation of free charge carriers caused by the incident photons, which occurs if the light striking the semiconductor has sufficient energy. This effect is widely utilized in electromagnetic radiation sensors, and such devices are termed *photoconductors*, *light-dependent resistors* (*LDR*) or *photoresistors*.

Cadmium sulfide (CdS) and cadmium selenide (CdSe) are the two most common materials for the fabrication of photoconductive devices and sensors (see **Fig. 2.5**).[5] Devices based on CdS can have a wide range of resistance values, from approximately a few ohms when the light has high intensity, to several mega ohms in darkness. They are capable of responding to a broad range of photon frequencies, including infrared, visible, and ultraviolet.

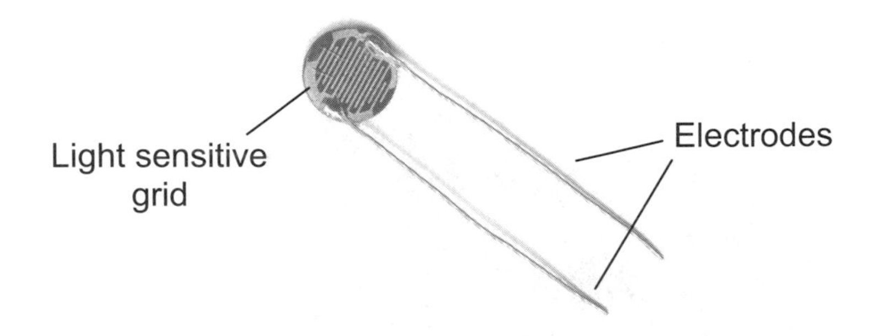

Fig. 2.5 Photo of a commercial LDR based on CdS.

Nanomaterials are currently being employed in photoconductive devices to improve their sensitivity, efficiency and response times. Semiconducting nanomaterials exhibit a charge depletion layer, which extends a few nanometers. This extension of the depletion region changes when exposed to irradiation. Depending on their dimensions and the amount of doping, photosensitive devices may become completely depleted of charge when irradiated. For instance, the photocurrent resulting from the interaction of UV light and semiconducting GaN-nanowires is seen in Fig. 2.6, where a distinct dependence on the nanowires diameter is observed.[6]

The response of photoconductive devices may also be tuned by varying the composition and dimensions of the utilized nanomaterials. This is seen in devices based on CdS and CdSe nanoparticles and nanostructured thin films for applications such as *Tera Hertz* (*THz*) signal monitoring.[7,8] The size-dependent transient photoconductivity of CdSe nanoparticles using time-resolved THz spectroscopy (TRTS) is shown in **Fig. 2.7**, which reveals the response time is reduced to less than 5 ps when the nanoparticle sizes are reduced to approximately 3.5 nm.

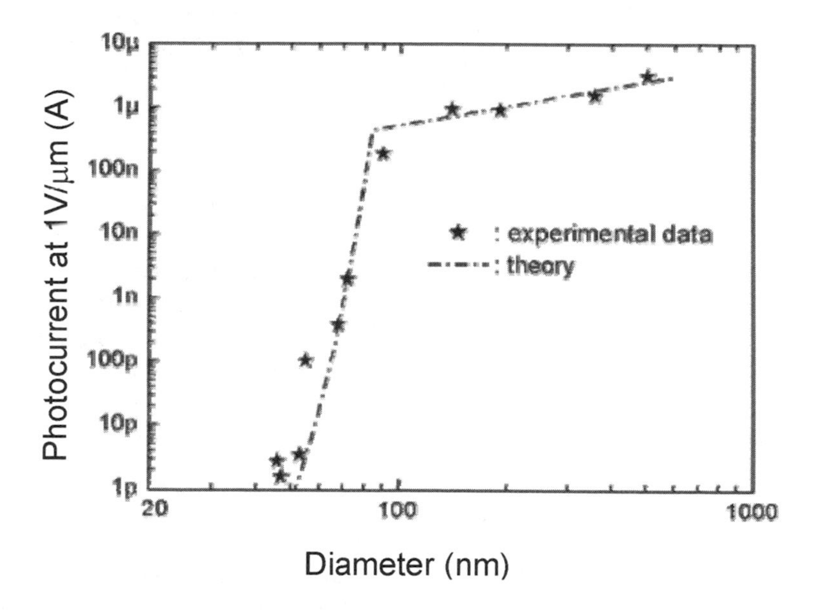

Fig. 2.6 Photocurrent with UV illumination of approximately 15 W/cm^2 versus diameter. The kink in the fitting curve at 85 nm indicates the critical diameter where the surface depletion layer just completely depletes the nanowire. For smaller diameters the photocurrent shows an exponential decrease, for larger diameters the photocurrent is proportional to the diameter. Reprinted with permission from the American Chemical Society publications.[6]

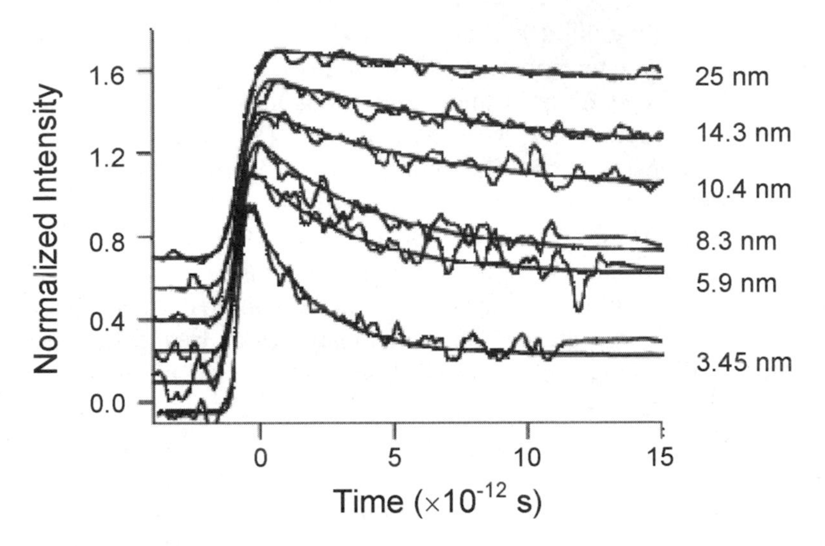

Fig. 2.7 TRTS scans for the eight sizes of CdSe nanoparticles, normalized and offset from the smallest to the largest size nanoparticles. Reprinted with permission from the American Chemical Society publications.[7]

Photovoltaic Effect

In the *photovoltaic effect*, a voltage is induced by the absorbed photons at a junction of two dissimilar materials (*heterojunction*). The absorbed photons produce free charge carriers. The induced voltage in the heterojunction causes the charge carriers to move, resulting in current flow in an external circuit. Materials used for fabricating such heterojunctions are typically semiconductors.

A typical photovoltaic device is seen in **Fig. 2.8**. They generally consist of a large area semiconductor *p-n junction* or diode. A photon impinging on the junction is absorbed if its energy is greater than or equal to the semiconductor's bandgap energy. This can cause a valance band electron to be excited into the conduction band, leaving behind a hole, and thus creating a mobile electron-hole pair. If the electron-hole pair is located within the depletion region of the *p-n* junction, then the existing electric field will either sweep the electron to the *n*-type side, or the hole to the *p*-type side. As a result, a current is created that is defined by:

$$I = I_S[e^{qV/kT} - 1], \tag{2.10}$$

where q is the electron charge (1.602×10^{-19} C), k is Boltzmann's constant (1.38×10^{-23} J/K), and T is the temperature of the *p-n* junction in Kelvin.

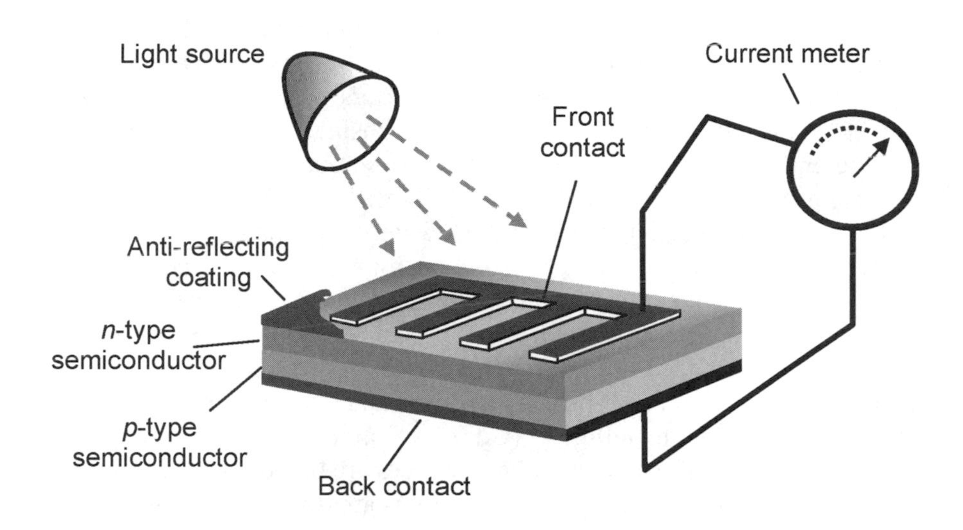

Fig. 2.8 Diagram of a photovoltaic device.

Photovoltaic cells and sensors are commonly made from materials that absorb photons in the visible and UV ranges, such as GaAs (gallium arsenide - band gap 1.43 eV) and compounds thereof. For other wavelength ranges, materials such as: silicon (wavelengths between 190-1100 nm), germanium (800-1700 nm), indium galium arsenide (800-2600 nm), and lead sulfide (1000-3500 nm) are generally used.

Photovoltaic devices can be employed in a wide range of sensing applications. These include use in analytical apparatus such as spectrophotometers, radiation monitors, automatic light adjustment systems in buildings, as light sensors in optical communication systems, etc. Photovoltaic devices are also the basis of photovoltaic cells for the generation of power from solar energy.

Important factors to be considered when designing photovoltaic devices and sensors are efficiency and cost. Such devices are generally inefficient, with efficiency varying from 5% for amorphous silicon-based devices to 35% or higher with *multiple-junction cells* used in research labs.[9] To overcome this, much research is being devoted to multi-junction cells based on thin films with thicknesses measuring a few nanometers. For sensing applications, GaAs is the material of choice as it is relatively insensitive to heat and devices have been made with efficiencies as high as 35.2%.[9]

Research is currently being focused on increasing their sensitivity and performance through the incorporation of nanodimensional structures and materials. The unique advantage of using nanomaterials in photovoltaic devices and sensors include: small effective cross-sections which leads to small capacitance and large mobility of carriers.[10] The combination of very short transit time of the photo-generated carriers with a small capacitance can be implemented for ultra high-speed sensing.[11] A large breakdown voltage and wider depletion region can also be obtained for such sensors because of the spread electric field streamlines.[12]

The range of wavelengths absorbed depends on material's electronic properties. These are not only determined by the material's composition, but also on in its dimensions. For example, devices with tunable absorption wavelengths comprised of carbon nanotubes[13] or semiconducting nanocrystals embedded into polymer matrices[14] have already been developed. Some of the most promising nanomaterials currently being researched are cadmium telluride (CdTe)[15,16] and copper indium gallium selenide (CIGS).[17] These photovoltaic devices are based on a thin film hetero-junctions structures and their efficiency is approximately 19.5%.

Nanostructured organic semiconductors and conductive polymers are also being developed for use in photoconducive devices.[18-20] However, devices and sensors made from them generally suffer from degradation upon exposure to UV light, resulting in short lifetimes, a serious concern for commercial applications. Inorganic semiconductor-based nanomaterials have superior performance due to their intrinsically higher carrier mobilities. Charges may be transported to the electrodes more quickly, reducing current losses through recombination and improving their dynamic performance.[21]

Combining polymeric materials with inorganic semiconductors nanoparticles has been shown to overcome charge transport limitations.[21] Charge transfer is favored between high electron affinity inorganic semiconductors and the relatively low ionization potential inherent in organic molecules. Charge transfer rates can be remarkably increased in the case of organics that are chemically bound to nanocrystalline and bulk inorganic semiconductors, which have a high density of electronic states. The combination of such materials is promising for the development of future generation photovoltaic cells and sensors.

Other developments in photovoltaic cells and sensor include the *dye-sensitized* based devices (also called *photoelectrochemical cells* or *Graetzel cells*), which mimic the photosynthesis process.[22,23] Their structure depends on a layer of nano porous material such as titanium dioxide and dye molecules that absorb light. The photo-generated electrons flow into the TiO_2 layer while the holes flow into an electrolyte on the other side of the dye. Unfortunately, the longevity of such devices is limited, because organic dyes currently utilized suffer form photo-degradation.

2.3.2 Photodielectric Effect

Materials whose dielectric properties change when illuminated by radiant energy are called photodielectric. Photodielectric measurements have been widely employed in photochemistry as in the study of kinetics in photographic materials and semiconductors.[24] It serves as a non-contact approach to measure a material's photoconductivity in an alternating electric field, and can be applied to complex semiconductors for which growth of monocrystals is difficult to monitor.[25] More on photodielectric effect and the integration of nanomaterials will be presented in Chap. 6.

2.3.3 Photoluminescence Effect

In photoluminescence effect, light is emitted from atoms or molecules after they have absorbed photons.[26] The absorbed photons give their energy to the molecule, causing it to change to a higher energy state. Then after some time, the molecule radiates the excess energy back out in the form of a photon, and it consequently returns to a lower energy state. The energy of the emitted light relates to the difference in energy levels of the excited state and the equilibrium state. *Fluorescence* and *phosphorescence* are examples of photoluminescence.

Photoluminescence can be explained with quantum mechanics. It depends on the *electronic structure* of atoms and molecules. Molecules have electronic states, and within each there are different *vibrational levels*, and within each vibrational level there exist *rotational levels*. After accepting energy in the form of a photon, an electron is raised to a permitted electronic state higher up. For most molecules, the *electronic states* can be divided into *singlet* (S) and *triplet* (T) states, depending on the electron spin. After a molecule is excited to a higher electronic energy state, it loses its energy quite rapidly via a number of pathways (**Fig. 3.31**).

In fluorescence, vibrational relaxation brings the molecule to its lowest vibrational energy level, $V' = 1$, in the first excited singlet state, S_1. Consequently, the electron relaxes from the lowest vibrational energy level in S_1 to any vibrational level of S_0.[27,28] For phosphorescence, the electron in S_1 undergoes *intersystem crossing* to T_1 and then relaxes to S_0.[29]

Due to the multiple rearrangements during the process, the phosphorescence has much longer lifetime than the fluorescence. For fluorescence, the period between absorption and emission is typically between 10^{-8} and 10^{-4} s. However for phosphorescence, this time is generally longer (10^{-4} to 10^{2} s).[30]

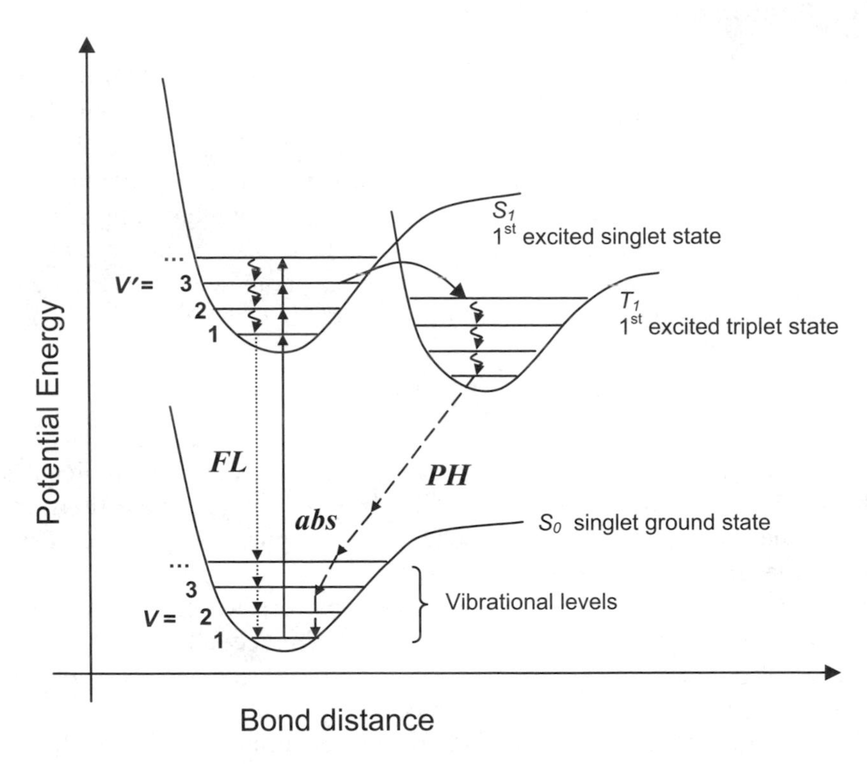

Fig. 2.9 Fluorescence and phosphorescence processes.

Organic molecules that exhibit fluorescence find many applications in nanotechnology enabled sensing. For example, fluorescence probes are used in biotechnology as a tool for monitoring biological events in individual cells. Other molecules are used as *ion probes*, in which after interacting with ions such that are important in neurological processes (e.g. Ca^{2+}, Na^{+}, etc.) their photoluminescence properties such as absorption wavelength, emission wavelength or emission intensity may change. In fact, these changes have been utilized to quantify events that take place in different parts of individual neurons. An example of this can be seen **Fig. 2.10**. Here fluorescence has been employed to image tumor-associated lysosomal protease using the *near-infrared fluorescence* (*NIRF*) probes[31]. Such probes generally have sub-micron dimensions and commercially available in different types and emission spectra.

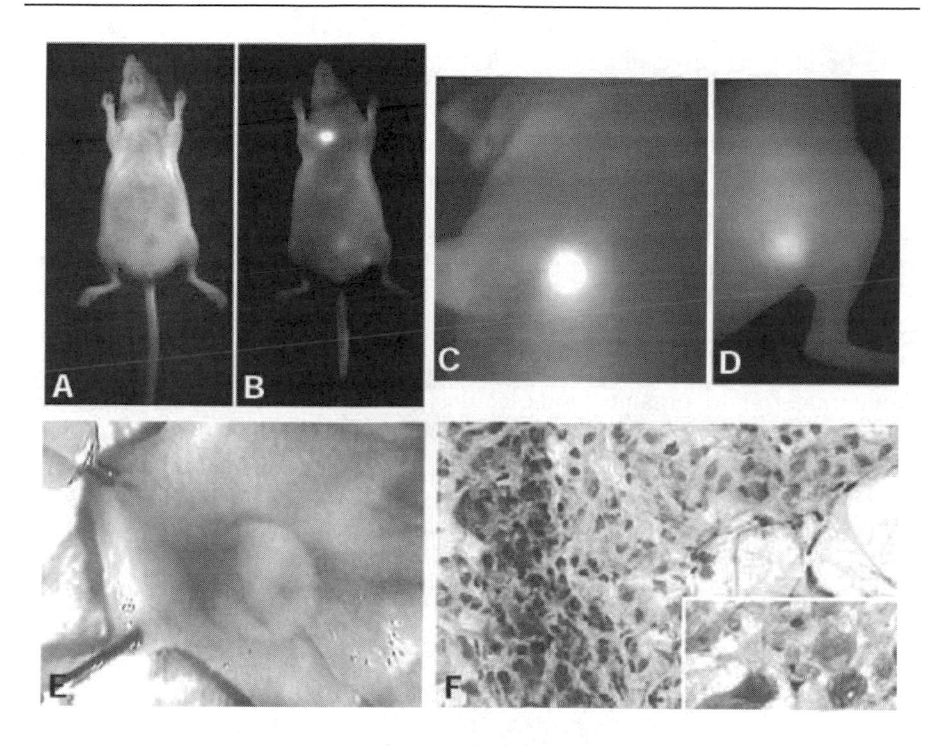

Fig. 2.10 LX-1 tumor implanted into the mammary fat pad of a nude mouse: (A) Light image. (B) Raw NIRF image. Note the bright tumor in the chest. (C) High resolution NIRF image of the chest wall tumor (2 mm). (D) High resolution NIRF image of the additional thigh tumor (<0.3 mm). (E) Dissected tumor in the mammary pad. (F) Hematoxylin-eosin section of the NIRF positive tumor showing malignant and actively proliferating cells (magnification 200; insert 400). Reprinted with permission from the Nature publications.[31]

Many inorganic molecules also exhibit photoluminescence properties that are extremely useful for probing and in nanotechnology enabled sensing. Some of the most popular of such particles are small zero dimensional semiconducting nanocrystals, also called *quantum dots* (*Q-dots*). They exhibit quantum confinement effects when their dimensions are smaller than *Bohr's radius*, which occurs at around 10 nm or less.

One of advantages of utilizing Q-dots in sensing applications is their size and material dependent photoluminescence properties. For example, ZnS nanocrystals have large band gaps and hence lower emission wavelengths whilst CdS nanocrystals have narrower band gaps and higher emission wavelengths. Furthermore, the emission wavelength can be tuned by simply changing the size of the nanocrystals, whilst using the same wavelength for their excitation.[32] Q-dots also have very narrow emission bands, considerably narrower than those of organic photoluminescent molecules, and their emission lifetime can be much longer (> 10^{-4} s). However, they are much larger than conventional fluorescent organic molecules, whose dimensions are generally in the angstrom, not nanometer, range.

Q-dots can be employed for bioimaging,[33,34] and biomolecules such as antibodies (**Fig. 2.11**) can also be attached to them for sensing applications.[35] In such cases the biomolecules carry the Q-dots to specific sites either on the cell surface or inside it, after which they can be probed. In biological applications, they have the added benefit of not being subject to microbial attack. Additionally, by attaching different sized Q-dots of the same material to organic biomolecules, multiple emission wavelengths, colors, can be employed to probe different biological events simultaneously whilst using the a single excitation wavelength. This quality is difficult to achieve with organic photoluminescent molecules, as each of them generally requires a different laser wavelength to activate its fluorescence.

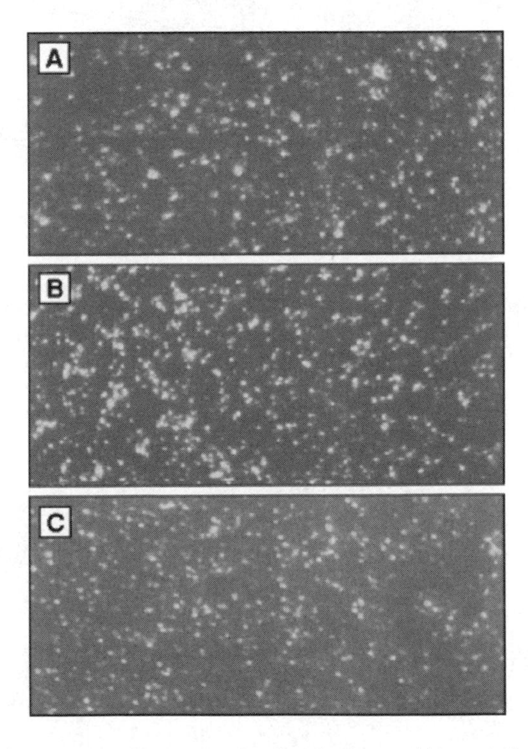

Fig. 2.11 Luminescence images obtained from (A) original Q-dotss, (B) mercapto-solubilized Q-dots, and (C) Q-dotes IgG conjugates (Q-dots conjugates and their interactions with biomaterials will be described in Chap. 7). Reprinted with permission from the Science Magazine publications.[35]

2.3.4 Electroluminescence Effect

Electroluminescence occurs when a material emits light as a result of an electrical current flowing through it, or when subjected to an electrical potential. It is used in the conversion of electrical energy into radiant energy. There are two methods of producing electroluminescence. Firstly, it can occur when a current passes through boundary of highly doped junctions (such as *p-n* junctions of semiconductor materials). Electrons can recombine with holes, causing them to fall into a lower energy level and release energy in the form of photons. Such a device is called *light-emitting diode* (*LED*) and its layout is shown in **Fig. 2.12**.

Electroluminescent devices can be implemented in spectroscopy and integrated sensors. Many new disposable sensors with the light intensity as the measure of a target analyte concentration or a physical change make use of them. This effect is an integrated part of many electrochemical sensing system (electrochemical sensing templates will be presented in Chap. 3). When an electron is generated in an electrochemical interaction it can transformed into a photon via the usage of an electroluminescent device. Consequently, this irradiation can be detected with a photodiode or photo transistor. The use of optical reading reduces the electronic noise and also it is compatible with many standard optical sensing systems.

The wavelength of the emitted light is determined by the bandgap energy of the materials forming the junction. A flow of a current does not guarantee electroluminescence. For example, in diodes based on indirect bandgap materials such as silicon, the recombination of electrons and holes is non-radiative and there is no light emission. Materials used in LEDs must have a direct bandgap. Those comprised of group III and V elements of the periodic table are most common used in the fabrication of LEDs. These include GaAs and GaP. The bandgap of these materials, and hence emission wavelength, can be tailored through the addition of impurities. For instance, LEDs made solely from GaP emit green light at 555 nm. However, nitrogen-doped GaP emits at yellow-green light (565 nm), and ZnO-doped GaP emits red light (700 nm).

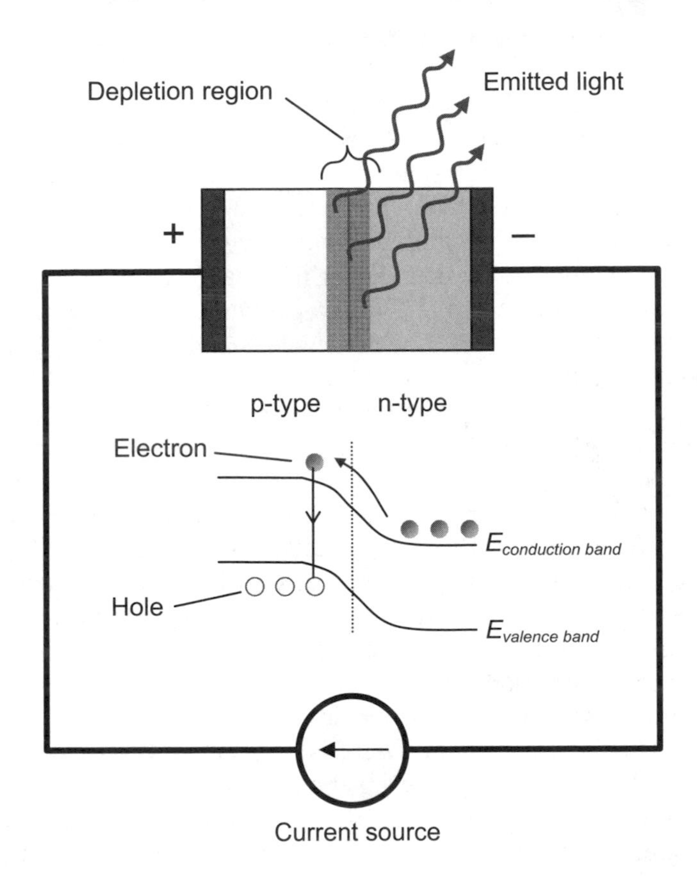

Fig. 2.12 A schematic of an LED. This band diagram illustrates the electron-hole recombination process.

The other way in which electroluminescence occurs is via the excitation of electrons using an electric field that is applied across phosphorescent materials. This type electroluminescence stems from the work of Georges Destriau[36], who in 1936 showed that by applying a large alternating potential across zinc sulfide (ZnS) phosphor powder suspended in an insulating material, electroluminescence was observed. This method for obtaining electroluminescence is the basis of current research into nanotechnology based electro luminescent displays. A typical example of such a device is seen in **Fig. 2.13**.

New *surface-conduction electron-emitter displays* (*SED*) are based on this type of electroluminescence. *Field emission display* (*FED*) is another technology which also uses phosphor coatings as the emissive medium. A FED uses a large array of fine metal tips or carbon nanotubes, with many positioned behind each phosphor dot, to emit electrons through field emission process.

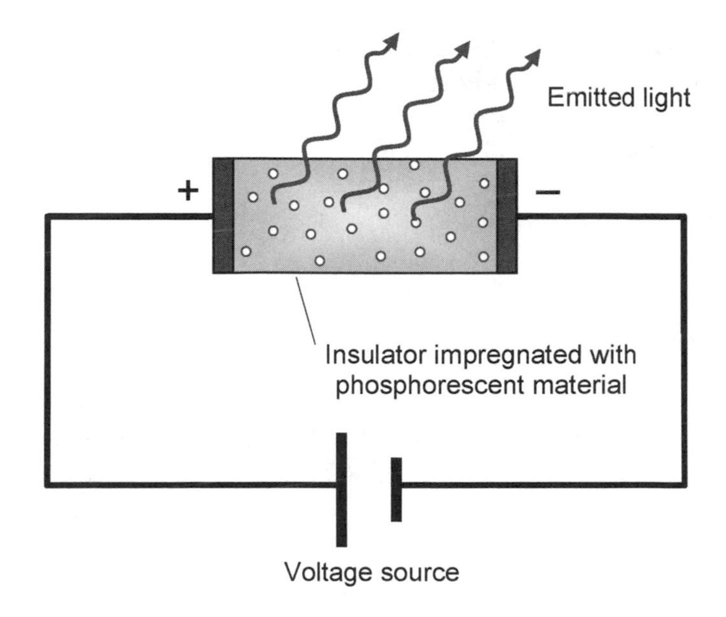

Fig. 2.13 Schematic of an electroluminescence device based on a phosphorous material.

Electroluminescent devices have many applications in chemical sensing. Example of such sensors are given by Poznyak et al.[37] They demonstrated how the electroluminescence of TiO_2 film electrodes could be utilized for measuring of hydrogen peroxide and peroxydisulphate ion concentrations in aqueous solutions. In this system, the analyte molecules are reduced on the electrode's surface, resulting in an electroluminescence whose intensity is proportional to the concentration of the measurand.

The current trend in electroluminescent device and sensor research is to utilize low dimensional nanomaterials, such as Q-dots. These nanomaterials may considerably reduce the scattering of electrons caused by defects in the bulk and reduce the non-radiative recombination rate.[38,39] They also increase in overlap of the wavefunctions for electrons and holes, and increase the electronic density of states (DOS) near the band gap of the low dimensional structures when compared with bulk materials. This leads to higher recombination rates and a narrowing of the gain spectrum.[40] As a result, LEDs incorporating low dimensional nanomaterials exhibit higher sensitivity to the applied charges. Other electroluminescence technologies include *liquid crystal display* (*LCD*), and *organic light-emitting diode*

(*OLED*), which are composed of organic thin films, such as conductive polymers are emerging.

Electroluminescent devices and sensors based on nanomaterials offer distinct advantages over those based on planar bulk materials.[41] For example, planar silicon is poorly suited to many photonic applications since it has a poor efficiency for light emission.[42] Such a deficiency can be selectively eliminated by making nanopores on the surface which provides phonon quenching or amplification capabilities. Such structures can be produced with nano-fabrication strategies, which will be discussed in the Chaps. 6 and 7.

2.3.5 Chemiluminescence Effect

Luminescence that occurs as a result of a chemical reaction is known as chemiluminescence. It is commonly observed at wavelengths from the near ultraviolet to the near infrared. Chemiluminescence can be described by the following reaction:

$$[\mathrm{A}]+[\mathrm{B}] \longrightarrow [\Diamond] \longrightarrow [\mathrm{Products}]+\mathit{light} \tag{2.11}$$

where A and B are reactants yielding an excited intermediate ◊, which is comprised of reaction products and light. Chemiluminescence has been observed for metal and semiconductor nanoparticles in chemical or electrochemical reactions.[43,44,45] When chemiluminescence takes place in living organisms, it is called *bioluminescence*.[28] Bioluminescence has emerged as an important and powerful tool in biological and medical investigations. Examples of molecules and nanomaterials that exhibit chemiluminescence employed in nanotechnology enabled sensing applications will be presented in Chaps. 6 and 7.

2.3.6 Doppler Effect

The *Doppler effect* is the apparent change in a wave's frequency as a result of the observer and the wave source moving relative to each other. If the observer and wave source are moving toward each other, the wave appears to increase in frequency and is said to be *hypsochromically* (or blue) shifted (**Fig. 2.14**). Conversely, if the wave source and observer are moving away from each other, then the wave appears to decrease in frequency and becomes *bathochromically* (or red) shifted.

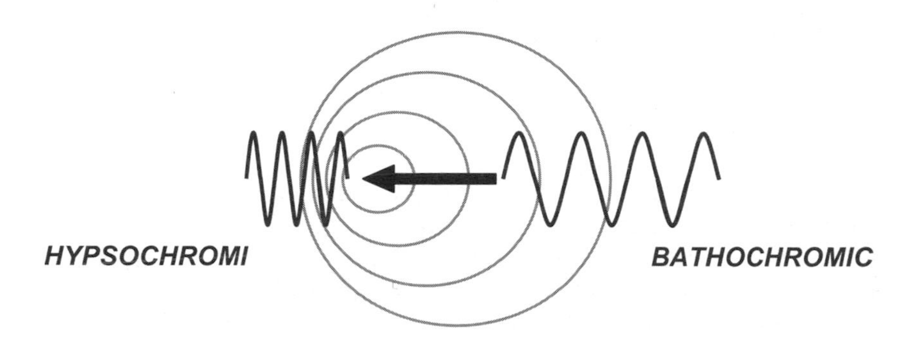

Fig. 2.14 Hypsochromic and bathochromic frequency shifts occurring as a result of the Doppler effect.

The observed Doppler shift in frequency is given by:

$$f_{observed} = \left(\frac{v}{v + v_{source}} \right) f_{source} \tag{2.12}$$

where v is the speed of the wave in the medium, v_{source} is the speed of the source with respect to the medium, and f_{source} frequency of the source wave. If the wave source approaches the observer, then v_{source} is negative, and conversely, if the wave is receding, then it takes on a positive value. A familiar examples of the Doppler effect include the changing pitch of an ambulance siren as it approaches and then drives past the observer.

Common examples of the Doppler effect in sensing include speed monitoring devices and ultrasounds. Hypsochromic and bathochromic shifts are used in measurement of large and distant bodies such as stars, galaxies and gas clouds as their motion and spectrum can be studied with respect to the observer. The Doppler effect also plays an important role in radar and sonar detection systems.

The Doppler effect can play a significant role in the sensing and characterization of nanomaterials. It is known that Doppler broadening (broadening of spectral lines in UV-vis spectroscopy) is caused by the thermal movement of small particles.[46] Doppler broadening generally places severe constraints on precise spectroscopic measurements. However, the signatures of the broadened spectrum (such as its bandwidth and shape) can be utilized to extract information about the presence of atoms/molecules in nanostructures, as well as providing information on their morphologies: by decreasing the temperature, or by employing measurement methods such as Doppler-free saturation spectroscopy, a reference for the characterization of materials can be obtained.[47]

2.3.7 Barkhausen Effect

In 1919 Heinrich Barkhausen found that applying a slowly increasing, continuous magnetic field to a ferromagnetic material causes it to become magnetized, not continuously, but in small steps. These sudden and discontinuous changes in magnetization are a result of discrete changes in both the size and orientation of ferromagnetic domains (or of microscopic clusters of aligned atomic magnets) that occur during a continuous process of magnetization or demagnetization.

This effect generally should be reduced in the operation of magnetic sensors as it appears as a step noise in measurements. This effect is also observed in nanosized ferromagnetic materials.[48]

2.3.8 Hall Effect

Discovered in the 1880s by Edwin Hall, when a magnetic field is applied perpendicularly to the direction of an electrical current flowing in a conductor or semiconductor, an electric field arises that is perpendicular to both the direction of the current and the magnetic field. It is one of the most widely used effects in sensor technology, particularly for monitoring magnetic fields.

In **Fig. 2.15**, a magnetic field is applied perpendicularly to a thin sheet of material that is carrying a current. The magnetic field exerts a transverse force, F_B, on the moving charges and pushes them to one side. Whilst these charges build up on one side, charges of the opposite polarity build up on the other side. This charge separation creates an electric field that generates an electric force, F_E. This electric force balances the magnetic force, preventing further charge separation. As a result, there is a measurable voltage between the two sides of the material, called the Hall voltage, V_{Hall}, and is calculated using:

$$V_{Hall} = \frac{IB}{ned}, \tag{2.13}$$

where I is the current flowing through the material, B is the magnetic field, n is the charge carrier density of the material, e is the elementary electronic charge 1.602×10^{-19} C and d is the thickness of the material.

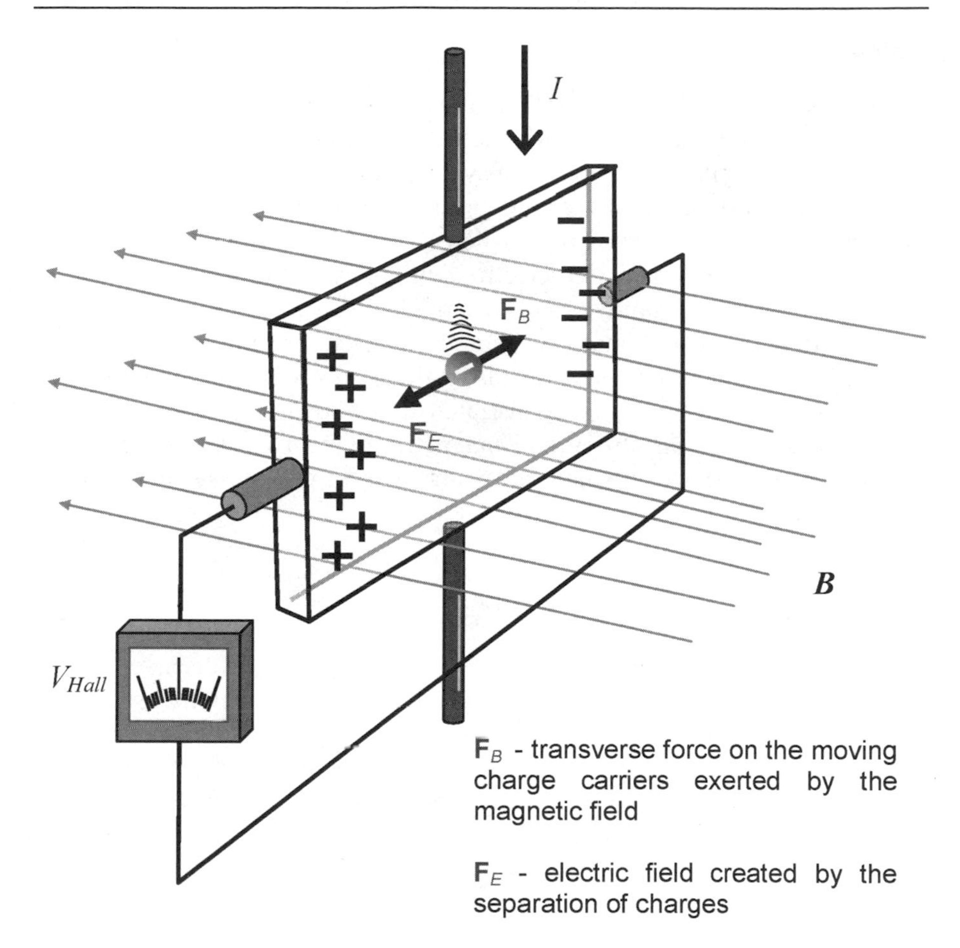

Fig. 2.15 The Hall effect.

Commercial Hall effect sensors are utilized in sensing fluid flow, power, and pressure sensing, yet they are most often employed for the measurement of magnetic fields. In fact, planar Hall sensors have been used to monitor magnetic fields in the nano-tesla range.[49] Being able to detect such low magnetic fields, the Hall effect can be implemented in the development of sensing systems which utilize nano-magnetic beads which generate very small magnetic fields. A good example of this is provide by Ejsing et al who have developed nano-magnetic bead sensors with sensitivities in the order of 3 μV/Oe mA. Their sensor response to an applied magnetic field of 250 nm magnetic beads which are commonly used for biological applications [50,51]

Spin Hall Effect

The *Spin Hall Effect* (*SHE*) refers to the generation of a spin current that is transverse to an applied electric field in such materials, resulting in an accompanying spin imbalance in the system. It occurs in paramagnetic materials as a consequence of the spin–orbit interactions. It was theoretically predicted in 1971 by Yakonov and Perel.[52] The generation, manipulation and detection of spin-polarized electrons in nanostructures are some of the current challenges of spin-based electronics.

This effect has an enormous potential to be used for sensing applications when applied to nano-magnetic beads or thin films with nano thicknesses. For instance Gerber et al[53] demonstrated that SHE can be used to sense magnetocrystalline anisotropy and magnetic moment of far-separated Co nanoparticles arranged in single-layer arrays with thicknesses as small as 0.01 nm.

2.3.9 Nernst/Ettingshausen Effect

Walther Hermann Nernst and Albert von Ettingshausen discovered that an electromotive force (e.m.f.) is produced across a conductor or semiconductor when it is subjected to both a temperature gradient and a magnetic field. The direction of the e.m.f. is mutually perpendicular to both the magnetic field and the temperature gradient. The effect can be quantified by the Nernst coefficient |N| as:

$$|\mathrm{N}| = \frac{E_Y / B_Z}{dT / dx}. \tag{2.14}$$

If the magnetic field component is in the *z*-direction, B_Z, then the resulting electric field component will be in the *y*-direction, E_Y, when subjected to a temperature gradient of dT/dx.

In spite of its potential, this effect is yet to be fully investigated for its application in nanotechnology enabled sensing. However, the authors believe this effect offers the exciting possibility of performing temperature and magnetic field strength measurements on a single nanoparticle, for example, by determining the electrical potentials across carbon nanotubes.

2.3.10 Thermoelectric (Seebeck/Peltier and Thomson) Effect

An e.m.f. is generated at the junction of two dissimilar conducting or semiconducting as a result of a temperature gradient. It was first observed in metals in 1821 by Thomas Johann Seebeck and the effect named after him. As seen in **Fig. 2.16**, for two dissimilar materials, A and B, a voltage difference V is generated when two junctions are held at different temperatures. The voltage difference is proportional to the temperature difference, $\Delta T = T_2 - T_1$, and the relationship is given by:

$$V = (S_A - S_B)\Delta T\,, \tag{2.15}$$

where S_A and S_B are the Seebeck coefficients of material A and B, respectively. This phenomenon provides the physical basis for *thermoelements* or *thermocouples*, the standard devices for measuring temperature.

In 1834 Jean Charles Athanase Peltier found the exact opposite, observing that a temperature difference will arise at a junction of two dissimilar metals when a current is passed through them (**Fig. 2.16**). The heat per unit time, Q, absorbed by the lower temperature junction is equal to:

$$Q = (\Pi_A - \Pi_B)I\,, \tag{2.16}$$

where Π_A and Π_B are the Peltier coefficients of each material and I is the current.

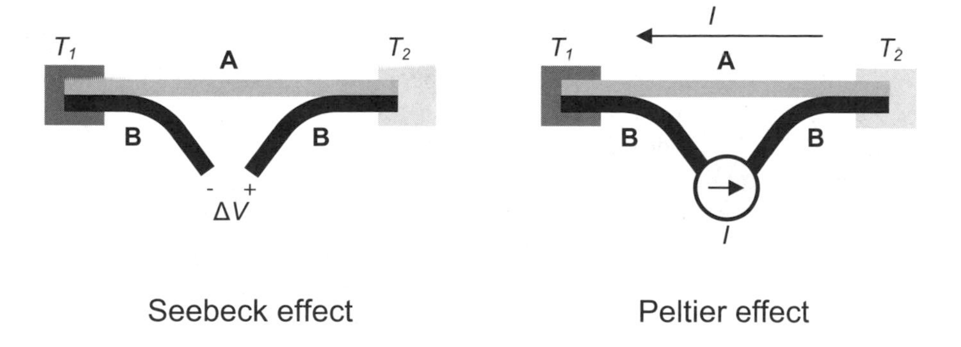

Fig. 2.16 Two dissimilar materials A and B in intimate contact, with either ends are held at different temperatures (T_1 and T_2).

William Thomson (Lord Kelvin), in 1854, discovered that an electric current flowing along a material that has a temperature gradient along its length will cause it to either absorb or release heat. The forced absorption or emission of heat is a result of energy conservation, because if a temperature gradient exists across the length of a material, an e.m.f. may be generated across this length.[5,54]

Many sensing systems incorporate temperature sensors based on the thermoelectric effect, and there are a variety of them available that find application in medical and scientific research, as well as in industrial process control and food storage, etc. There are several types of such devices, called *thermocouples*, and among the most popular are listed in **Table 2.2**.

Table 2.2 Some common types of thermocouples.

Type	Materials	Temp. Range (°C)
K	Chromel/Alumel (Ni-Al alloy)	−200°C to +1200
E	Chromel/Constantan	−110 to 140
J	Iron/Constantan	−40 to +750
N	Nicrosil (Ni-Cr-Si alloy)/Nisil (Ni-Si alloy)	

The different metals and alloys utilized in thermocouples result in different properties and performance. Some commonly utilized alloys are chromel (approx. 90% nickel and 10% chromium) and constantan (approx. 40% nickel and 60% copper). Type K is perhaps the most widely used thermocouple as it operates over a wide temperature range from −200 to +1200°C. This type of thermocouple has a sensitivity of approximately 41 μV/°C. Some type E thermocouples can have a narrower operating range than type K, however, their sensitivity is much higher (68 μV/°C). Type N (Nicrosil (Ni-Cr-Si alloy)/Nisil (Ni-Si alloy)) thermocouples have high stability and resistance to high temperature oxidation, making them ideal for many high temperature measurements. Other thermocouple types: B, R, and S are all made of expensive noble metals, and are the most stable for high temperature, but have low sensitivity (approximately 10 μV/°C).

Thermoelectric materials, particularly those based on semiconducting materials with large Peltier coefficients, can be used to fabricate on chip temperature sensors. [55] They are also used to make heat pumps meeting a growing demand in many products including charge coupled device (CCD) cameras, laser diodes, microprocessors, blood analyzers, etc. may employ thermoelectric coolers. The performance of thermoelectric devices in terms of the ability to convert thermal energy into electrical energy, and vice versa, depends on the *figure of merit* (*ZT*) of the material's utilized, and is given by:

$$ZT=(S^2T)/(\rho K_T), \tag{2.17}$$

where S, T, ρ and K_T are the *Seebeck coefficient*, absolute temperature, electrical resistivity and total thermal conductivity, respectively. Generally, the larger the thermoelectric material's Seebeck coefficient (to generate the maximum voltage difference) and thermal conductivity (so it does not allow the exchange of heat at two junctions), whilst lowering its electrical resistivity (so the internal resistance does not generate heat), the more efficient the thermoelectric device can be.

Bismuth telluride (Bi_2Te_3) and antimony telluride (Sb_2Te_3) are semiconductor materials with high Seebeck coefficients, having *ZT* of approximately equal to unity at room temperature. In the 50s, an Australian researcher, Julian Goldsmid confirmed that bismuth telluride displays a very strong Peltier effect.[56,57] However, until only recently have crystals with higher *ZT*s have been found.

In the early 90s, it was theoretically proven that nanosized materials could dramatically enhance the performance of *Peltier modules* and devices.[58] Nanodimensional materials such as superlattices, segmented one dimensional nano-wires and zero-dimensional quantum dots are excellent candidates for development of high performance thermoelectric structures. The nano-dimensions result in confinement of the charge carriers and scattering of phonons which increase the electrical conductivity and decrease the thermal conductivity. As a result, they will have an increased value for the figure of merit. It has been calculated that for quantum wires of Bi_2Te_3 a *ZT* as large as 14 can be obtained[58,59] when the radius of wire is reduced to 0.5 nm. For a superlattice structure of the same materials with the quantum well of width of 1 nm the best calculated figure of merit is 2.5 and for a 0.5 nm quantum well it is 5.[58,59]

In a major breakthrough for the fabrication of thermoelectric superlattices, Venkatasubramanian et al.[60] reported significant enhancement in *ZT* which is almost equal to the theoretical values. They achieved a *ZT* of 2.4 for p-type Bi_2Te_3/Sb_2Te_3 superlattice devices. *ZT* of several recently reported materials has been shown in **Fig. 2.17**.

With the emergence of more efficient thermoelectric materials, in the near future, such devices may be sought for power generation and the transformation of waste thermal energy into electrical energy.

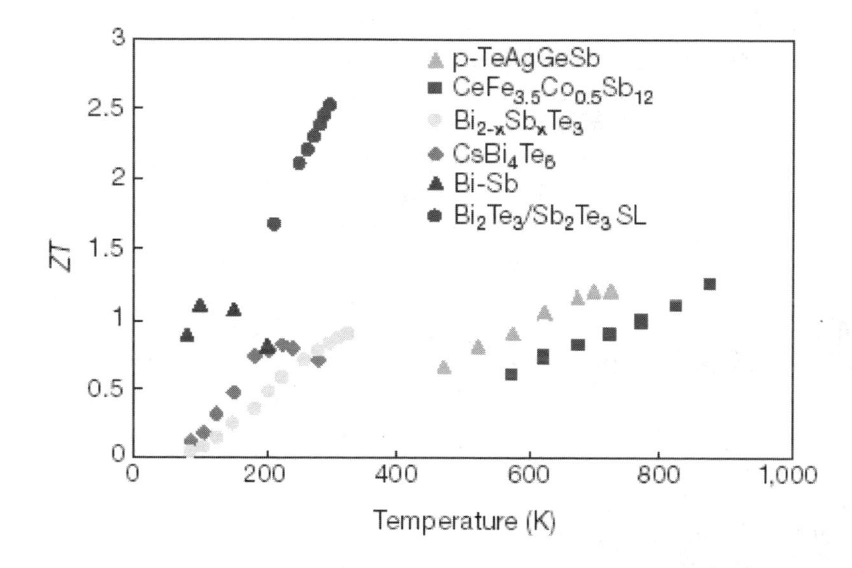

Fig. 2.17 Temperature dependence of *ZT* of 1 nm/5 nm p-type Bi_2Te_3/Sb_2Te_3 superlattice compared to those of several recently reported materials. Reprinted with permission from the Nature publications.[60]

2.3.11 Thermoresistive Effect

Thermoresistivity is concerned with the change in a material's electrical resistance with temperature and is widely used in temperature sensing applications. This effect is the basis of temperature sensing devices such as *resistance thermometers* and *thermistors*. The electrical resistance R, is determined by the formula:

$$R = R_{ref}\left(1 + \alpha_1 \Delta T + \alpha_2 \Delta T^2 + \ldots + \alpha_n \Delta T^n\right) \tag{2.18}$$

where R_{ref} is the resistance at the reference temperature, $\alpha_1 \ldots \alpha_n$ are the material's temperature coefficient of resistance, $\Delta T = (T\text{-}T_{ref})$ is the difference between the current temperature T and the reference temperature T_{ref}. The equation suggests that resistance increases with temperature. This is not the case for all materials, for if the material has a *positive temperature coefficient* (*PTC*) then its resistance increases with temperature, conversly if it has a *negative temperature coefficient* (*NTC*),) then it decreases.

In many cases, materials exhibit a linear relationship between the temperature and resistance, and hence the higher order terms in Eq. (2.18) can be disregarded. However, this linearity is valid only for a limited range of temperatures (**Table 2.3**).

Table 2.3 The temperature range of some metals used as thermoresistive based temperature sensors.

Material	Linear temperature range (°C)
Copper	–200–260
Platinum	–260–1000

Nanomaterials can be implemented for the fabrication of thermistors with desired positive and negative temperature coefficients. Additionally, in conventional materials only the surface of the bulk or thin films is exposed to the environmental changes such as ambient temperature alterations. However, nanostructured materials with a larger surface to volume ratio are more efficiently exposed to the environmental changes. This both increases the sensor response magnitude and reduces its response time.

For instance, Saha et al[61] successfully prepared nano-size powders of $(Mn_xFe_{1-x})_2O_3$ by citrate–nitrate gel method with sintering at high temperature of 1500°C. The developed materials were found to have NTC sensitivity index (β- the measure of sensitivity for a thermoresistive material) as high as 6000 K (in the temperature range of 50–150°C) which is appreciably higher than that of conventional NTC materials. The electrical characteristics of conductive/nonconductive polymer composites mixed with nano-sized particles can also be utilized for the fabrication of thermistors.[62]

2.3.12 Piezoresistive Effect

The piezoresistive effect describes the change in a material's electrical resistivity when acted upon by a mechanical force. This effect takes its name from the Greek word *piezein*, which means to squeeze. It was first discovered in 1856 by Lord Kelvin who found that the resistivity of certain metals changed when a mechanical load was applied. Piezoresitance can be described by the following equation for semiconductors:

$$\frac{\Delta R}{R} = \pi\sigma , \tag{2.19}$$

where π is the *tensor element of the piezoresistive coefficient*, σ is the mechanical stress tensor, and R and ΔR are the resistance and the change in resistance, respectively.

In 1954, C. S. Smith[63] discovered that semiconductors such as silicon and germanium displayed a much greater piezoresistive effect than metals.

There are two phenomena attributed to silicon's resistivity changes: the stress dependent distortion of its geometry; and the stress dependence of its resistivity.

The piezoresistive effect in semiconductors and metal alloys is exploited in sensors. Most materials exhibit some piezoresistive effect. However, as silicon is the material of choice for integrated circuits, the use of piezoresistive silicon devices, for mechanical stress measurements, has been of great interest.[64] The most widely used form of piezoresistive silicon based sensors are diffused resistors.[65] The effect also can be used in cantilever based sensors.[66]

The effects of stress are far more significant on crystalline materials' with nanometer thick planes than they are on bulk materials. For example, for one and two dimensional nanomaterials, the effect of an external force is limited in one and two dimensions which confine the movement of phonons. Additionally for nanomaterials the area for which the external force can act on is greatly reduced. As a result, the effective force per area (i.e. stress), is amplified. Therefore, stress acting on a nanocrystal of a piezoresistive material can be translated into a large change in the crystal's conductivity.

Nanosized piezoresitive elements are developed for both chemical and physical sensing applications. The major advantage of such systems is the ease of measurement, which is basically a voltage or current. A fine example of such sensors is demonstrated in the work by Stampfer et al.[67] They reported the fabrication and characterization of pressure sensors based on individual single-walled carbon nanotubes (SNWT) as electromechanical transducing elements. Their sensor consists of an individual electrically connected SWNT adsorbed on top of a 100-nm-thick atomic layer deposited circular alumina (Al_2O_3) membrane with a radius in the range of 50-100 μm. They performed electromechanical measurements on strained metallic SWNTs adhering to this membrane and found a piezoresistive gauge factor of approximately 210 for SWNTs. The fabrication process of such a device is shown in **Fig. 2.18**. The schematic of such a device is shown in **Fig. 2.19**.

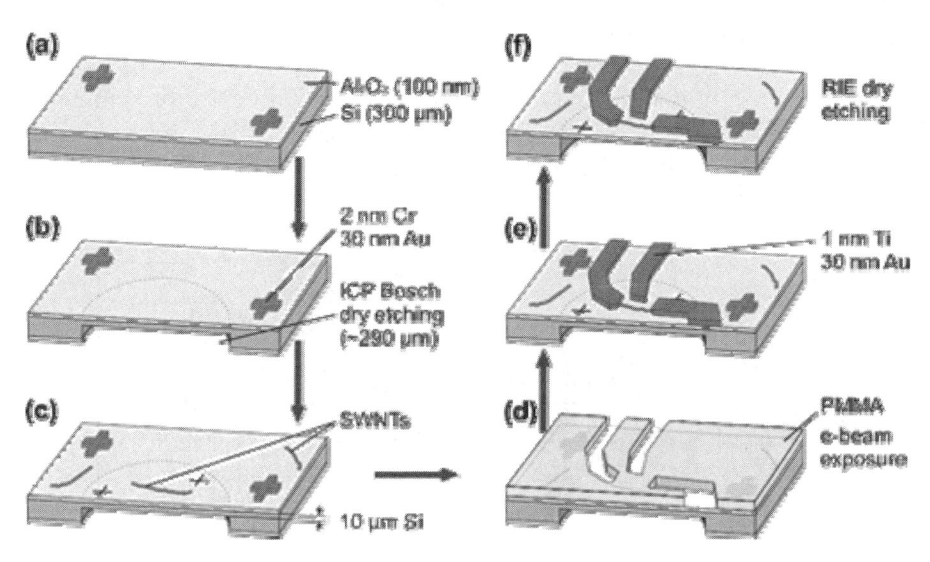

Fig. 2.18 Process flow to fabricate single-walled-carbon-nanotube-based pressure sensors: (a) 100 nm of Al_2O_3 is deposited by atomic layer deposition on a 300-μm-thick Si substrate. Photolithography and lift-off processes are used for patterning. (b) The membrane openings are patterned using infrared backside alignment and anisotropically dry etched from the backside. (c) SWNTs are dispersed from a solution onto the Al_2O_3. (d) PMMA (a polymer) spin coating and e-beam exposure. (e) Metalization and lift-off to electrically connect the SWNTs, and the final dry etch membrane release (f). Reprinted with permission from the American Chemical Society publications.[67]

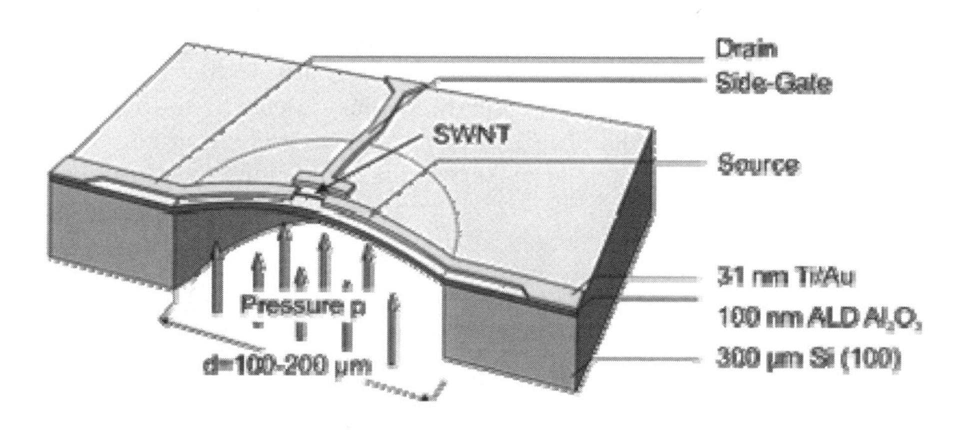

Fig. 2.19 Schematic of the SWNT based piezoresistive device. Reprinted with permission from the American Chemical Society publications.[67]

2.3.13 Piezoelectric Effect

Piezoelectricity is the ability of crystals that lack a centre of symmetry to produce a voltage in response to an applied mechanical force, and vice versa (**Fig. 2.20**). It was discovered by the Curie brothers in 1880.

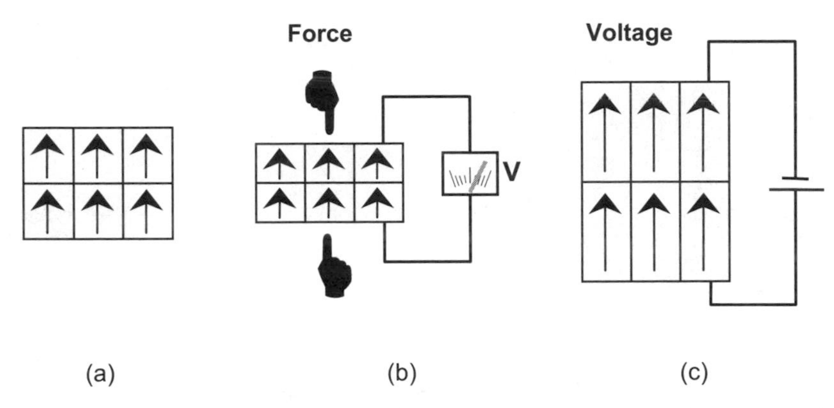

Fig. 2.20 (a) A piezoelectric material. (b) A voltage response can be measured as a result of a compression or expansion. (c) An applied voltage expands or compresses a piezoelectric material.

Out of thirty-two crystal classes, twenty-one do not have a centre of symmetry (non-centro-symmetric), and of these, twenty directly exhibit piezoelectricity (except the cubic class 432). The most popular piezoelectric materials are quartz, lithium niobate, lithium tantalite, PZT and langasite. Many piezoelectric materials are ferroelectric ceramics, which become piezoelectric when poled with an external electric field (**Fig. 2.21**). Piezoelectric crystallites are centro-symmetric cubic (isotropic) before poling and after poling exhibit tetragonal symmetry (anisotropic structure) below the Curie temperature. Above this temperature they lose their piezoelectric properties.

Polymers such as rubber, wool, hair, wood fiber and silk also exhibit piezoelectricity to some extent. Polyvinylidene fluoride (PVDF) is a thermoplastic material that when poled exhibits piezoelectricity several times greater than quartz.

Piezoelectric materials are an extremely popular choice for a broad variety of sensing applications. In Chap. 3, several transducers exploiting the piezoelectric effect will be presented and examples of piezoelectric materials employed in nanotechnology enabled sensing applications will be given in Chaps. 6 and 7.

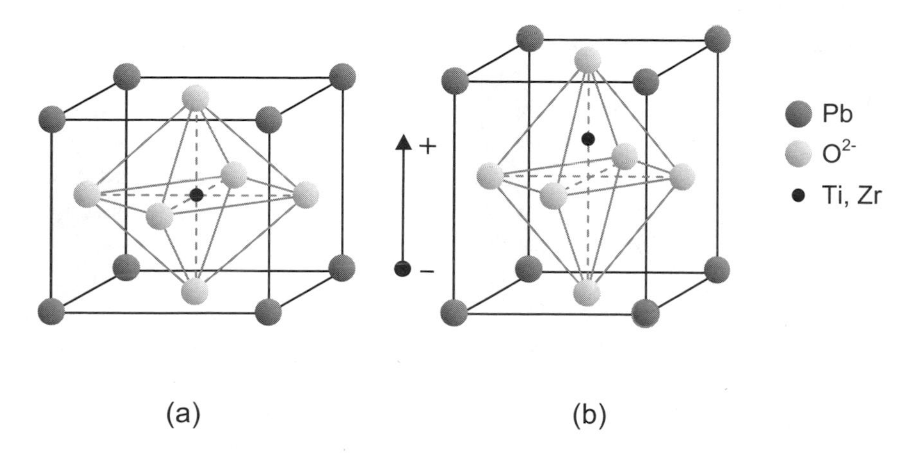

Fig. 2.21 A schematic of piezoelectricity in a PZT crystal: (a) before poling (b) after poling.

2.3.14 Pyroelectric effect

When heated or cooled, certain crystals establish an electric polarization, and hence generate an electric potential. The temperature change causes positive and negative charges to migrate to the opposite ends of a crystal's polar axis. Such polar crystals are said to exhibit *pyroelectricity*, which takes its name from the Greek word *pyro* which means fire.

The pyroelectric materials are employed in radiation sensors, in which radiation incident on their surface is converted to heat. The increase in temperature associated with this incident radiation causes a change in the magnitude of the crystal's electrical polarization. This results in a measurable voltage, or if placed in a circuit, a measurable current given by:

$$I = pA\frac{dT}{dt}, \tag{2.20}$$

where p is the pyroelectric coefficient, A is the area of the electrode and dT/dt is the temperature rate of change.

Pyroelectric effect can be used to generate strong electric fields (giga-V/m) in some materials, by heating them from −30°C to +45°C in a few minutes. Researchers at UCLA headed by Brian Naranjo, recently observed the nuclear fusion of deuterium nuclei in a tabletop device.[68]

Irradiation sensors based on the pyroelectric effect are commercially available, such sensors respond to a wide range of wavelengths. Pyroelectric sensors, which are fabricated from pyroelectric materials such as lithium tantalate and PZT, generate electric charges with small temperature changes caused by irradiation of the surface of the crystal.[69]

Most of the measurement standards for radiometry are based on thermal detectors. These devices employ some form of thermal-absorber coating such as carbon-based paint or diffuse metals such as gold.[70] Potentially, coatings with nanomaterials present an alternative to these technologies which provides higher sensitivity for measurements.[70]

It is possible to increase the pyroelectric coefficient of polymers by adding nano-particles. Zhang et al[71] showed that nanocrystalline calcium and lanthanum powder incorporated into a polyvinylidene fluoride-trifluoroethylene [P(VDF-TrFE)] copolymer matrix have a higher pyroelectric coefficient (by ~35%) than those of the P(VDF-TrFE) film of a similar thickness. It is also possible to indirectly use nanomaterials to enhance the performance of pyroelectric sensors. For instance, Liang et al[72] uatlized porous SiO_2 film as a thermal-insulation layer to block the diffusion of heat flow from the pyroelectric layer to the silicon substrate in multilayer pyroelectric thin film IR detector. This improves the energy confinement within the pyroelectric sensing layer, resulting in an enhanced performance of the sensor.

2.3.15 Magneto-Mechanical Effect (Magnetostriction)

Magnetostriction, also called the *magneto-mechanical effect*, is the change in a material's dimensions when subjected to an applied magnetic field, or alternatively it is a change in a material's magnetic properties under the influence of stress and strain. It was first identified in 1842 by James Joule while examining a sample of nickel. Its name originates from the Greek word, *magnet* and the Latin word *strictus* (meaning compressed, pressured, tense).

The mechanism that occurs in magnetostriction is illustrated in **Fig. 2.22**. As can be seen, the domains arrange themselves randomly in a non-magnetized material. Upon magnetizing, the material's domains orient with their axes which changes the length of the structure.

Magnetostrictive materials convert magnetic energy into kinetic energy, and vice versa. Therefore, they are regularly used for sensing and actuation. Interestingly, this effect causes the familiar humming sound that is heard in electrical transformers.

Magnetostriction is defined by the magnetostrictive coefficient, Λ. It is defined as the fractional change in length as the magnetization of the material increases from zero to the saturation value. The coefficient, which is typically in the order of 10^{-5}, can be either positive or negative. The reciprocal of this effect is called the *Villari effect*, where a material's susceptibility changes when subjected to a mechanical stress.

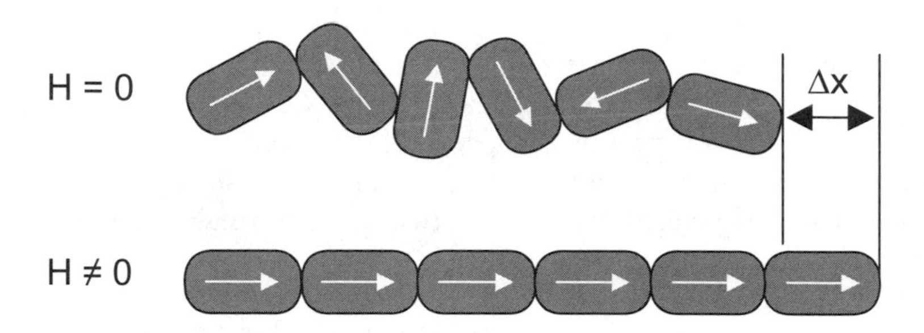

Fig. 2.22 A schematic of the Magnetostriction effect: un-aligned magnetic domains (*top*) will align causing the structure to expand under the influence of an applied magnetic field (*bottom*).

Magnetostriction is defined by the magnetostrictive coefficient, Λ. It is defined as the fractional change in length as the magnetization of the material increases from zero to the saturation value. The coefficient, which is typically in the order of 10^{-5}, can be either positive or negative. The reciprocal of this effect is called the *Villari effect*, where a material's susceptibility changes when subjected to a mechanical stress.

An element of the periodic table that exhibits the largest room temperature magnetostriction is cobalt. However, the most advanced magnetostrictive materials, called *Giant Magnetostrictive* (*GM*) materials, are alloys composed of iron (Fe), dysprosium (Dy) and terbium (Tb). Many of which were discovered at Naval Ordnance Lab and Ames Laboratory in mid 1960s.[73] Their Λ values can be three orders of magnitude larger than those of pure elements. The GM effect can be used in the development of magnetic field, current, proximity and stress sensors.[74]

2.3.16 Mangnetoresistive Effect

Magnetoresistance is the dependence of a material's electrical resistance on an externally applied magnetic field. The applied magnetic field causes a Lorentz force to act on the moving charge carriers, and depending on the field's orientation, it may result in a resistance to their flow. It was first observed by Lord Kelvin in 1856. The effect has become much more prominent owing to the discoveries of *anisotropic magnetoresistance* (*AMR*) [75] and *giant magnetoresistance* (*GMR*) .[76]

AMR is an effect that is only found in ferromagnetic materials, where the electrical resistance increases when the direction of current is parallel to the applied magnetic field. The change in the material's electrical resistivity depends on the angle between the directions of the current and the ferromagnetic material's magnetization. It is possible to develop sensors for monitoring the orientation of magnetic fields based on the AMR effect. In these sensors, the resistance changes when the current parallel to the magnetic moment alters or passes near the magnetic field. However, the resistance change associated with AMR effect is quite small and typically only of the order of one or two percent.

GMR is playing a significant role in nanotechnology enabled sensing. It was first discovered independently in 1988 by research teams led by Peter Grünberg[77] of the Jülich Research Centre and Albert Fert[76] of the University of Paris-Sud. It relies on the quantum nature of materials, and is mostly observed in layered structures that are composed of alternating ferromagnetic and nonmagnetic metal layers, with the thickness of the nonmagnetic layer being a nanometer or so. The effect is illustrated in **Fig. 2.23**. Electron scattering at the ferromagnetic/nonmagnetic interfaces depends on the whether the electron spin is parallel or antiparallel to the magnetic moment of the layer. An electron has two spin values; up and down. The first magnetic layer allows electrons in only one spin state to pass through easily. If the magnetic moments of the adjacent layers are aligned, then only electrons with that matching spin value can easily pass through the structure, and the resistance is low. If the magnetic moments of the adjacent layers are not aligned, then electrons with mismatching spin cannot pass through the structure easily and the electrical resistance is much greater than when the alignments are parallel.

The GMR effect can be utilized to monitor magnetic fields, for in the presence of an applied magnetic field, the relative orientations of the magnetic moments in the alternating ferromagnetic layers change and hence a change in resistance is observed. Currently, research is focused on employing multilayered nanowires (which offer greater sensitivity than the thin films now used in hard drives' reader/writers), which also exhibit GMR. It is largely used in read heads of the magnetic discs for computers information storage for sensing magnetic fields, among other sensing applications.[78] Further sensing examples utilizing this effect will be presented in Chap. 7.

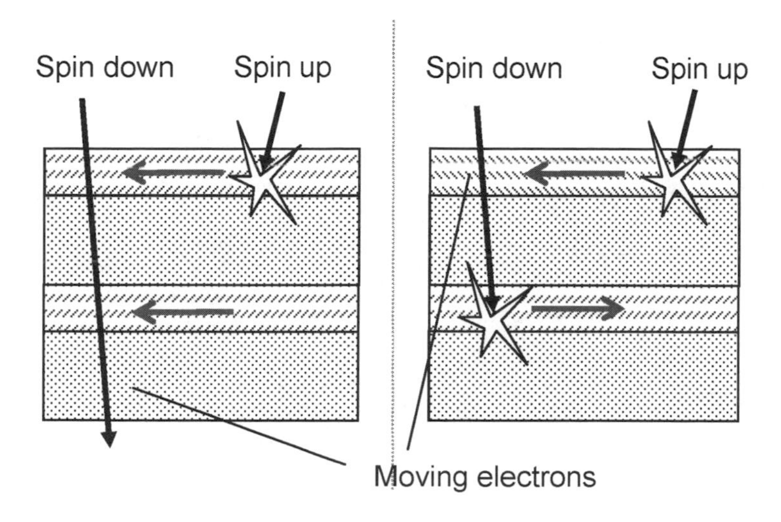

Fig. 2.23 A schematic of the GMR effect after and before applying the magnetic field.

2.3.17 Faraday-Henry Law

The Faraday-Henry law is a fundamental law of electromagnetism, and expresses that an electric field is induced by changing the magnetic field (**Fig. 2.24**). Michael Faraday and Joseph Henry both independently discovered the electromagnetic phenomenon of self and mutual inductance. Their work on the magnetically induced currents was the basis of the electrical telegraph, which was jointly invented by Samuel Morse and Charles Wheatstone later on. Early acoustic sensors and devices (such as microphones), analogue current/voltage meters, and reed-relay switches make use of this effect.

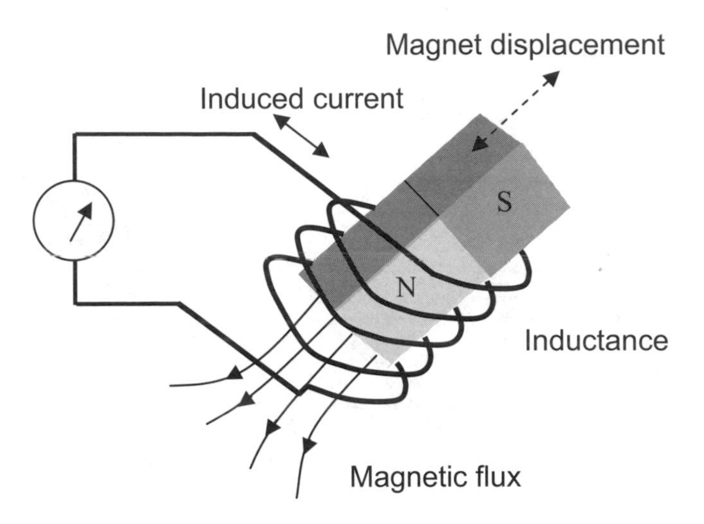

Fig. 2.24 A schematic of the Faraday-Henry effect.

The relation between the electric field, **E**, and the magnetic flux density, **B**, is defined by:

$$\oint_C \mathbf{E} \cdot d\mathbf{s} = -\frac{d}{dt} \oint_S \mathbf{B} \cdot dA, \tag{2.21}$$

or in differential form:

$$\nabla \times \mathbf{E} = -\frac{\partial \mathbf{B}}{\partial t}, \tag{2.22}$$

This law governs antennas, electrical motors and a large number of electrical devices that include relays and inductors in telecommunication circuits. Almost all *radio frequency identification* (*RFID*) tags and sensing systems, which are currently used in warehouses, are based on the Faraday-Henry effect. Such tags are commonly used in supermarkets, shops and offices for identifying the products, improving inventory and logistical efficiency.

Researchers have developed relay switches and actuators based on carbon nanotubes.[79] Such actuators can be used as sensing templates by functionalizing regions with sensitive elements. A conceptual drawing and SEM image of the nanoactuator has been shown in **Fig. 2.25** . As can be seen, a metal plate rotor (R) is attached to a multi-walled carbon nanotube (*MWCNT*) which acts as a support shaft and is the source of rotational freedom. Electrical contact to the rotor plate is made via the MWCNT and its anchor pads (A1, A2). Three stator electrodes, two on the SiO_2 surface (S1, S2) and one buried beneath the surface (S3), provide additional voltage control elements. The SiO_2 surface has been etched down to provide full rotational freedom for the rotor plate. The entire actuator assembly is integrated on a Si chip. A scanning electron microscope (SEM) image of nanoactuator is also shown.

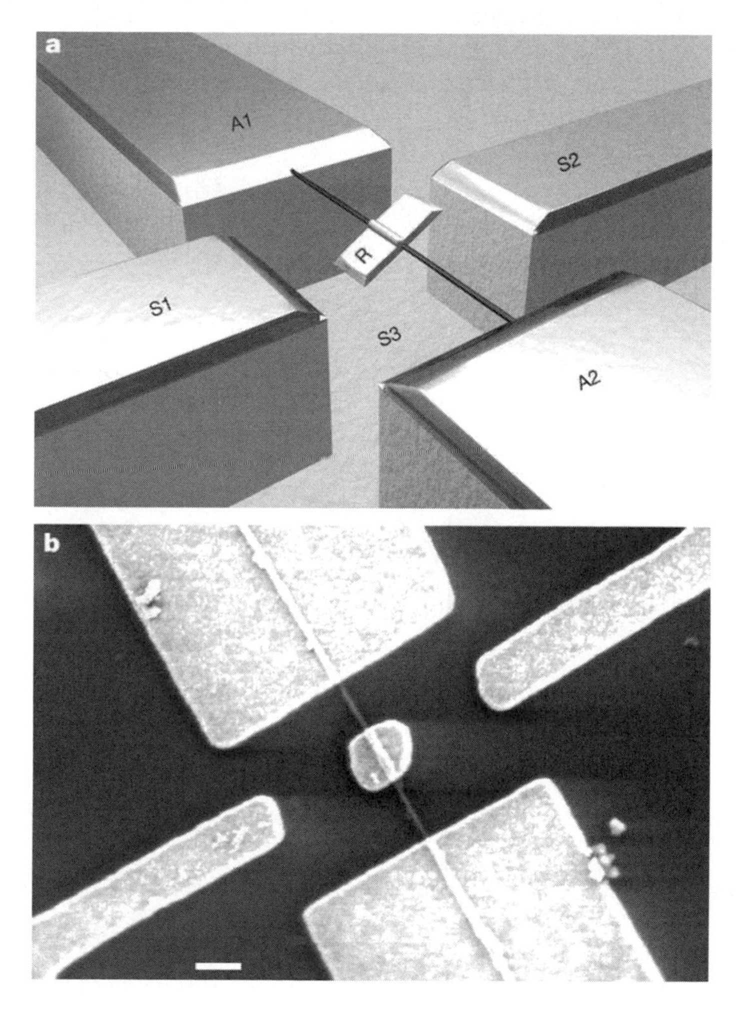

Fig. 2.25 (top) Conceptual drawing and SEM image of the nanoactuator. (bottom) Scanning electron microscopy. The scale bar is 300 nm. Reprinted with permission from the Nature publications.[79]

2.3.18 Faraday Rotation Effect

Discovered by Michael Faraday in 1845, it is a magneto-optic effect in which the polarization plane of an electromagnetic wave propagating through a material becomes rotated when subjected to a magnetic field that is parallel to the propagation direction. This rotation of the polarization plane is proportional to the intensity of the applied magnetic field. This effect was the first experimental evidence that light is an electromagnetic wave and was one of the foundations on which James Clerk Maxwell developed his theories on electromagnetism. The angle of rotation is defined by the equation:

$$\theta = \mathrm{V}\mathbf{B}l \, , \tag{2.23}$$

where **B** is the magnetic flux density, V is the *Verdet constant* and l is the length of the material through which the light is passing.

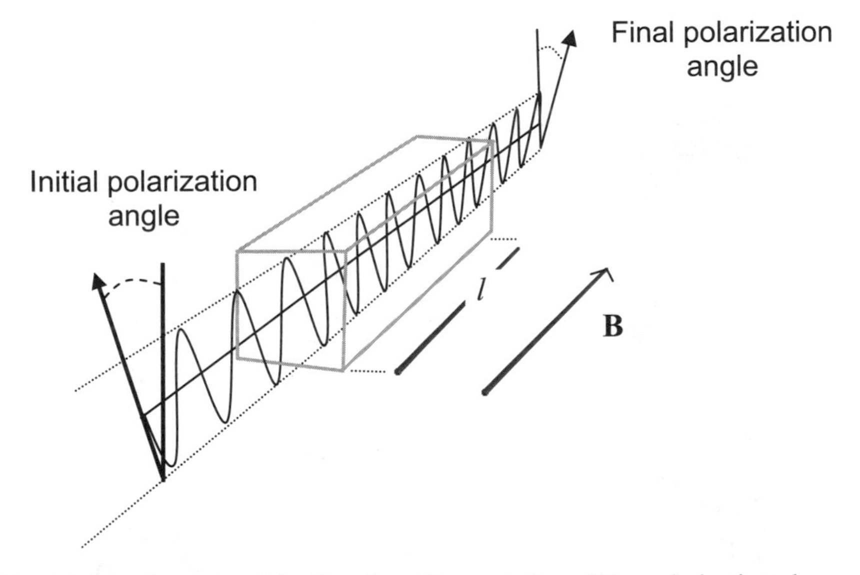

Fig. 2.26 A schematic of the Faraday effect: rotation of the polarization plane as a result of an external magnetic field.

As the polarized beam enters the material, *birefringence* occurs, in which the wavefront is split into two circularly polarized rays. This is caused when the light travels parallel to the magnetic field lines, the absorption line splits into two components which are circularly polarized in opposite directions. This is the *Zeeman effect*, where the splitting of spectral lines into multiple components in a magnetic field is produced. These circularly polarized waves will propagate at different velocities due to the difference in the refractive indices of the two rays. As the rays emerge from the material, they will recombine. However, owing to the changes in the difference in propagation speed, and hence refractive indices, the recombined wave will possess a net phase offset which will result in a rotation of the angle of linear polarization.

Most substances do not show such a difference without an external magnetic field, except optically active substances such as crystalline quartz or a sugar solution. Also, the refractive index in the vicinity of an absorption line does changes with frequency.

There are several applications of Faraday rotation in measuring instruments. For instance, it has been used to measure optical rotatory power,[80] for amplitude modulation of light, and for remote sensing of magnetic fields.[81]

The Verdet constant is a figure of merit used to compare this effect between materials, and has units of angular rotation per unit of applied field per unit of material length.[82] A common magneto-optical material for field sensing is terbium gallium garnet, which has a Verdet constant of 0.5 min/(G cm). Along with a relatively high Verdet constant, this material also can take on a permanent magnetization. It is possible to construct magneto-optical magnetometers with a sensitivity of 30 pT for the detection of magnetic nano-beads for sensing applications. The unique advantage that the magneto-optical sensor has over other magnetic sensors is its quick response time. Sensors with gigahertz responses have been fabricated.[82]

2.3.19 Magneto-Optic Kerr Effect (MOKE)

In 1877, John Kerr discovered that the polarization plane of a light beam incident on a magnetized surface is rotated by a small amount after it is reflected from that surface. This is because the incoming electric field, E, exerts a force, F, on the electrons in the material, and consequently they vibrate in the plane of polarization of the incoming wave. If the material has some magnetization, M, then the reflected wave gain a small electric field component (called the Kerr component, K),[83] as seen in **Fig. 2.27**. As a result, the reflected wave is rotated with respect to the incident wave. The amount of rotation depends on the magnitude of M.[84,85] Both the magneto-optic Kerr and Faraday rotation effects occur because the magnetization in the material produces a change its dielectric tensor.

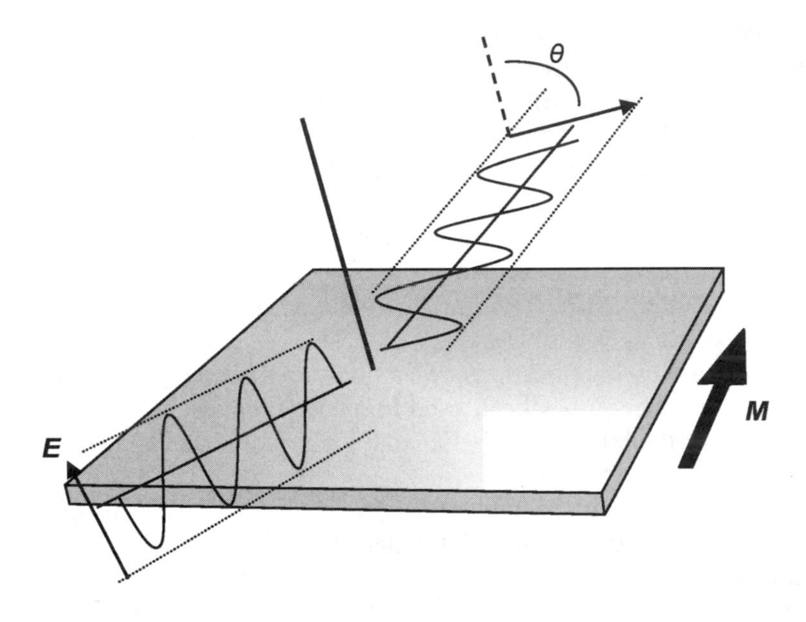

Fig. 2.27 Rotation of the polarization plane on a magnetized surface as a result of the magneto-optic Kerr effect.

The Kerr effect can be used to fabricate sensors for various applications. For instance, Karl et al [86] developed a pressure sensor based on a micromembrane coated with a magnetostrictive thin-film. The pressure difference across the diaphragm causes deflection and thus stress in the magnetostrictive layer. This leads to a change in the magnetic properties of the thin-film, which can be measured as a change in the MOKE properties. It is widely utilized for determining the magnetization of materials. MOKE can also be used to study the magnetic anisotropy of deposited ferromagnetic thin films. Magnetic properties of such films are closely related to their morphology and micro/nano structures. [87]

2.3.20 Kerrand Pockels Effects

Discovered by John Kerr in 1875, it is an electro-optic effect in which a material changes its refractive index in response to an electric field. Here, birefringence is induced electrically in isotropic materials.[88] When an electric field is applied to a liquid or a gas, its molecules (which have electric dipoles) may become partly oriented with the field.[5] This renders the substance anisotropic and causes birefringence in the light traveling through it. However, only light passing through the medium normal to the electric field lines experience such birefringence, and it is proportional to the square of the electric field. The amount of birefringence due to the Kerr effect can be given by:

$$\Delta n = n_o - n_e = \lambda_o K E^2 , \tag{2.24}$$

where E is the electric field strength, K is the Kerr-Pockels constant and λ_o is the wavelength of free space. The two principal indices of refraction, n_o and n_e are the ordinary and extraordinary indices, respectively. The effect is utilized in many optical devices such as switches and monochromators, modulators. This effect is analogous to the Faraday effect but for electric fields.

Pockels effect is a similar effect to the Kerr effect, with differences of the birefringence being directly proportional to the electric field, not its square (as in the Kerr effect). Only crystals that lack a center of symmetry (20 out of the 32 classes) may show this effect.

Optical sensors based on the Pockels effect are being implemented in industrial applications. Pockels voltage sensors have been incorporated into electric power networks and power apparatus such as gas-insulated-switchgear. Pockels field sensors are applied to the measurement of not only the electrostatic field but also the space charge field in electrical discharges subjected to dc, ac, lightning impulse, or switching impulse voltages.[89]

2.4 Summary

The terminology and parameters frequently encountered in sensing were presented in this chapter. In addition, some of the most widely utilized physical effects for signal transduction related to nanotechnology enabled sensors were introduced. Several examples of nanomaterials exhibiting these effects were also provided. It was shown that these effects became dramatically enhanced by the use of nanomaterials. It should be noted that many other effects are also known and widely employed in sensing, yet are not commonly utilized in nanotechnology enabled sensing applications. However, novel nanotechnology enabled sensors are emerging rapidly, and effects previously believed to be irrelevant are finding acceptance and novel applications.

In next chapter, several transduction platforms that are used in conjunction with nanostructured materials for nanotechnology enabled sensing will be introduced.

References

[1] W. Göpel, J. Hesse, and J. N. Zemel, *Sensors: A Comprehensive Survey* (VCH, Weinheim, Germany, 1991).

[2] M. J. Usher and D. A. Keating, *Sensors and transducers: characteristics, applications, instrumentation, interfacing* (Macmillan, London, UK, 1996).

[3] R. Pallas-Areny and J. G. Webster, *Sensors and Signal Conditioning* (Wiley, New York, USA, 1991).

[4] H. Lin, T. Jin, A. Dmytruk, M. Saito, and T. Yazawa, Journal of Photochemistry and Photobiology a-Chemistry **164,** 173-177 (2004).

[5] J. Daintith, *A Dictionary of Physics.* (Oxford University Press, London, UK, 2000).

[6] R. Calarco, M. Marso, T. Richter, A. I. Aykanat, R. Meijers, A. V. Hart, T. Stoica, and H. Luth, Nano Letters **5,** 981-984 (2005).

[7] M. C. Beard, G. M. Turner, and C. A. Schmuttenmaer, Nano Letters **2,** 983-987 (2002).

[8] F. A. Hegmann, R. R. Tykwinski, K. P. H. Lui, J. E. Bullock, and J. E. Anthony, Physical Review Letters **89** (2002).

[9] R. W. Miles, K. M. Hynes, and I. Forbes, Progress in Crystal Growth and Characterization of Materials **51,** 1-42 (2005).

[10] A. S. Achoyan, A. E. Yesayan, E. M. Kazaryan, and S. G. Petrosyan, Semiconductors **36,** 903-907 (2002).

[11] N. Tsutsui, V. Ryzhii, I. Khmyrova, P. O. Vaccaro, H. Taniyama, and T. Aida, Ieee Journal of Quantum Electronics **37,** 830-836 (2001).

[12] V. M. Aroutionian, S. G. Petrosyan, and A. E. Yesayan, Thin Solid Films **451-52,** 389-392 (2004).

[13] R. P. Raffaelle, B. J. Landi, J. D. Harris, S. G. Bailey, and A. F. Hepp, Materials Science and Engineering B-Solid State Materials for Advanced Technology **116,** 233-243 (2005).

[14] T. J. Bukowski and J. H. Simmons, Critical Reviews in Solid State and Materials Sciences **27,** 119-142 (2002).

[15] G. Khrypunov, A. Romeo, F. Kurdesau, D. L. Batzner, H. Zogg, and A. N. Tiwari, Solar Energy Materials and Solar Cells **90,** 664-677 (2006).

[16] J. R. Sites and X. X. Liu, Solar Energy Materials and Solar Cells **41-2,** 373-379 (1996).

[17] F. Kessler, D. Herrmann, and M. Powalla, Thin Solid Films **480,** 491-498 (2005).

[18] A. G. MacDiarmid, Synthetic Metals **125,** 11-22 (2001).

[19] A. G. MacDiarmid, Reviews of Modern Physics **73,** 701-712 (2001).

[20] A. G. MacDiarmid and A. J. Epstein, Makromolekulare Chemie-Macromolecular Symposia **51,** 11-28 (1991).
[21] W. U. Huynh, J. J. Dittmer, and A. P. Alivisatos, Science **295,** 2425-2427 (2002).
[22] B. Oregan and M. Gratzel, Nature **353,** 737-740 (1991).
[23] C. B. Cohen and S. G. Weber, Analytical Chemistry **65,** 169-175 (1993).
[24] J. P. Spoonhower, Photographic Science and Engineering **24,** 130 (1980).
[25] R. Janes, M. Edge, J. Robinson, J. Rigby, and N. Allen, Journal of Photochemistry and Photobiology a-Chemistry **127,** 111-115 (1999).
[26] E. N. Harvey, *A History of Luminescence* (American Philosophical Society, Philadelphia, USA, 1957).
[27] B. J. Clark, T. Frost, and M. A. Russell, *UV spectroscopy : techniques, instrumentation, data handling* (Chapman & Hall, London, UK, 1993).
[28] N. W. Barnett and P. S. Francis, in *Encyclopedia of Analytical Science*, edited by C. F. Poole, A. Townshend, and P. J. Worsfold (Academic Press, New York, USA, 2004), p. 305-315.
[29] G. Blasse and B. C. Grabmaier, *Luminescent Materials* (Springer-Verlag, New York, USA, 1995).
[30] T. H. Gfroerer, in *Encyclopedia of Analytical Chemistry*, edited by R. A. Meyers (John Wiley & Sons Ltd., Chichester, UK, 2000), p. 9209-9231.
[31] R. Weissleder, C. H. Tung, U. Mahmood, and A. Bogdanov, Nature Biotechnology **17,** 375-378 (1999).
[32] A. P. Alivisatos, Science **271,** 933-937 (1996).
[33] P. N. Prasad, *Introduction to Biophotonics,* (Wiley Interscience, Hoboken, USA, 2003).
[34] C. Seydel, Science **300,** 80-81 (2003).
[35] W. C. W. Chan and S. M. Nie, Science **281,** 2016-2018 (1998).
[36] G. Destriau, Journal de Chemie Physique **33,** 587-625 (1936).
[37] S. K. Poznyak and A. I. Kulak, Talanta **43,** 1607-1613 (1996).
[38] D. Huang, M. A. Reshchikov, and H. Morkoc, in *Quantum Dots*, edited by E. Borovitskaya and M. S. Shur (World Scientific, Singapore, 2002), p. 79.
[39] E. Borovitskaya and M. S. Shur, in *Quantum Dots*, edited by E. Borovitskaya and M. S. Shur (World Scientific, Singapore, 2002), p. 1.
[40] G. B. Stringfellow, in *High brightness light emitting diodes* (Academic Press, San Diego, USA, 1997).

[41] Y. Huang, X. F. Duan, and C. M. Lieber, Small **1,** 142-147 (2005).
[42] P. H. Zhang, V. H. Crespi, E. Chang, S. G. Louie, and M. L. Cohen, Nature **409,** 69-71 (2001).
[43] S. K. Poznyak, D. V. Talapin, E. V. Shevchenko, and H. Weller, Nano Letters **4,** 693-698 (2004).
[44] H. Cui, Z. F. Zhang, and M. J. Shi, Journal of Physical Chemistry B **109,** 3099-3103 (2005).
[45] G.-F. Jie, B. Liu, J.-J. Miao, and J.-J. Zhu, Talanta (2006).
[46] P. J. Chantry, Journal of Chemical Physics **55,** 2746& (1971).
[47] E. S. Polzik, J. Carri, and H. J. Kimble, Physical Review Letters **68,** 3020-3023 (1992).
[48] A. Zhukov, J. Gonzalez, J. M. Blanco, M. Vazquez, and V. Larin, Journal of Materials Research **15,** 2107-2113 (2000).
[49] F. N. Van Dau, A. Schuhl, J. R. Childress, and M. Sussiau, Sensors and Actuators A **53,** 256-260 (1996).
[50] L. Ejsing, M. F. Hansen, A. K. Menon, H. A. Ferreira, D. L. Graham, and P. P. Freitas, Journal of Magnetism and Magnetic Materials **293,** 677-684 (2005).
[51] L. Ejsing, M. F. Hansen, A. K. Menon, H. A. Ferreira, D. L. Graham, and P. P. Freitas, Applied Physics Letters **84,** 4729-4731 (2004).
[52] M. I. Dyakonov and V. I. Perel, Physical Letters A **35,** 459-460 (1971).
[53] A. Gerber, A. Milner, J. Tuaillon-Combes, M. Negrier, O. Boisron, P. Melinon, and A. Perez, Journal of Magnetism and Magnetic Materials **241,** 340-344 (2002).
[54] M. W. Zemansky and R. H. Dittman, *Heat and thermodynamics: an intermediate textbook*, 6th ed. (McGraw-Hill, New York, USA, 1981).
[55] H. Baltes, O. Paul, and O. Brand, Proceedings of the IEEE **86,** 1660 - 1678 (1998).
[56] H. J. Goldsmid, *Thermoelectric Refrigeration* (Plenum, New York, USA, 1964).
[57] H. J. Goldsmid and G. S. Nolas, in *A review of the New Thermoelectric Materials*, 2001, p. 1-6.
[58] L. D. Hicks and M. S. Dresselhaus, Physical Review B **47,** 12727-12731 (1993).
[59] A. R. Abramson, W. C. Kim, S. T. Huxtable, H. Q. Yan, Y. Y. Wu, A. Majumdar, C. L. Tien, and P. D. Yang, Journal of Microelectromechanical Systems **13,** 505-513 (2004).
[60] R. Venkatasubramanian, E. Siivola, T. Colpitts, and B. O'Quinn, Nature **413,** 597-602 (2001).
[61] D. Saha, A. D. Sharma, A. Sen, and H. S. Maiti, MATERIALS LETTERS **55,** 403-406 (2002).

[62] J. C. Kim, G. H. Park, S. J. Suh, Y. K. Lee, S. J. Lee, S. J. Lee, and J. D. Nam, Polymer Korea **26,** 367-374 (2002).
[63] C. S. Smith, Physical Review **94,** 42-49 (1954).
[64] G. Gerlach and R. Werthschutzky, Tm-Technisches Messen **72,** 53-76 (2005).
[65] S. M. Sze, *Physics of semiconductor devices*, 2nd ed. (Wiley, New York, USA, 1981).
[66] H. Jensenius, J. Thaysen, A. A. Rasmussen, L. H. Veje, O. Hansen, and A. Boisen, Applied Physics Letters **76,** 2615-2617 (2000).
[67] C. Stampfer, T. Helbling, D. Obergfell, B. Schoberle, M. K. Tripp, A. Jungen, S. Roth, V. M. Bright, and C. Hierold, Nano Letters **6,** 233-237 (2006).
[68] B. Naranjo, J. K. Gimzewski, and S. Putterman, Nature **434,** 1115-1117 (2005).
[69] R. Kohler, N. Neumann, N. Hess, R. Bruchhaus, W. Wersing, and M. Simon, Ferroelectrics **201,** 83-92 (1997).
[70] J. Lehman, E. Theocharous, G. Eppeldauer, and C. Pannell, Measurement Science & Technology **14,** 916-922 (2003).
[71] Q. Q. Zhang, H. L. W. Chan, and C. L. Choy, Computers part A-Applied Science and Manufacturing **30,** 163-167 (1999).
[72] L. Liang, Z. Liangying, and Y. Xi, Ceramics International **30,** 1843-1846 (2004).
[73] A. E. Clark and H. S. Belson, Physical Review B **5,** 3642-3644 (1972).
[74] M. Vazquez, M. Knobel, M. L. Sanchez, R. Valenzuela, and A. P. Zhukov, Sensors and Actuators A **59,** 20-29 (1997).
[75] T. R. McGuire and R. I. Potter, Ieee Transactions on Magnetics **11,** 1018-1038 (1975).
[76] M. N. Baibich, J. M. Broto, A. Fert, F. N. Vandau, F. Petroff, P. Eitenne, G. Creuzet, A. Friederich, and J. Chazelas, Physical Review Letters **61,** 2472-2475 (1988).
[77] G. Binasch, P. Grunberg, F. Saurenbach, and W. Zinn, Physical Review B **39,** 4828-4830 (1989).
[78] G. A. Prinz, Journal of Magnetism and Magnetic Materials **200,** 57-68 (1999).
[79] A. M. Fennimore, T. D. Yuzvinsky, W. Q. Han, M. S. Fuhrer, J. Cumings, and A. Zettl, Nature **424,** 408-410 (2003).
[80] K. Kurosawa, S. Yoshida, and K. Sakamoto, Journal of Lightwave Technology **13,** 1378-1384 (1995).
[81] D. M. Le Vine and S. Abraham, Ieee Transactions on Geoscience and Remote Sensing **40,** 771-782 (2002).
[82] J. Lenz and A. S. Edelstein, IEEE Sensors Journal **6,** 631-649 (2006).
[83] C. Hunt and S. Sahu, The IRM Quarterly **2,** 1-8 (1992).

[84] Z. Q. Qiu and S. D. Bader, Review of Scientific Instruments **71,** 1243-1255 (2000).

[85] Z. Q. Qiu and S. D. Bader, Journal of Magnetism and Magnetic Materials **200,** 664-678 (1999).

[86] W. J. Karl, A. L. Powell, R. Watts, M. R. J. Gibbs, and C. R. Whitehouse, Sensors and Actuators a-Physical **81,** 137-141 (2000).

[87] F. Tang, D. L. Liu, D. X. Ye, T. M. Lu, and G. C. Wang, Journal of Magnetism and Magnetic Materials **283,** 65-70 (2004).

[88] A. Yariv, *Optical electronics* (Oxford University Press, New York, USA, 1991).

[89] K. Hidaka, IEEE Electrical Insulation Magazine **12,** 17-23.
28 (1996).

Chapter 3: Transduction Platforms

3.1 Introduction

In this chapter, some of the major transduction platforms utilized in sensing are presented. The focus is on platforms which are fabricated utilizing micro/nano-fabrication processes. Integrating them with nanomaterials for sensing applications can enhance their performance and consequently lead to increased sensitivity towards measurands. The transduction platforms, presented in this chapter, include: conductometric and capacitive, optical, electrochemical, solid state, and acoustic wave based. In Chaps. 6 and 7, examples of integrating nanomaterials based sensitive layers to such platforms, along with various sensing examples will be presented.

3.2 Conductometric and Capacitive Transducers

Conductormetric (or *resistive*) and *capacitive* transducers are among the most commonly utilized devices in sensing applications. This is largely due to their simple and inexpensive fabrication and set-up. They involve placing a material between conducting electrodes, to which a voltage is applied and the electrical conductivity or capacitance is subsequently measured. A typical set-up is shown in **Fig. 3.**1.

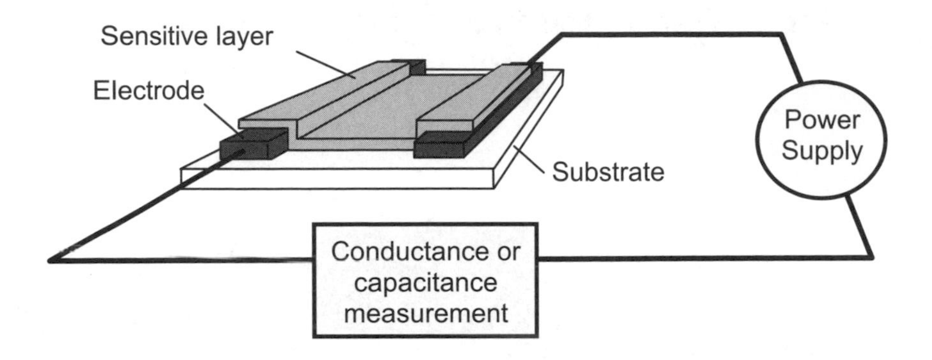

Fig. 3.1 Typical setup for capacitive or conductometric transducers.

For conductometric measurements, a voltage is applied across the electrodes, which generates an electric field. The electrical conductivity of the material between the electrodes can then be determined from the measured current that flows. The current density, J, electric field, E, and electrical conductivity, σ, are related through *Ohm's law*:

$$J = \sigma E, \tag{3.1}$$

or in its more common form, Ohm's law is written as:

$$V = IR, \tag{3.2}$$

where V is the voltage, I is this current, and R is the electrical resistance. Many materials exhibit nonlinear electrical conductivity, and therefore careful consideration of the biasing conditions must be made during measurements. Generally the applied voltages and currents are selected such that the conductivity remains in a relatively linear region.

As an example, **Fig. 3.2** shows the change in electrical resistance of a nanostructured SnO_2 thin film when exposed to different concentration of O_2 gas at an operating temperature of 400°C.[1] The response, S, which is defined as the ratio of electrical resistances in the presence and absence of O_2 gas, exhibits a log-linear relationship to the O_2 gas concentration.

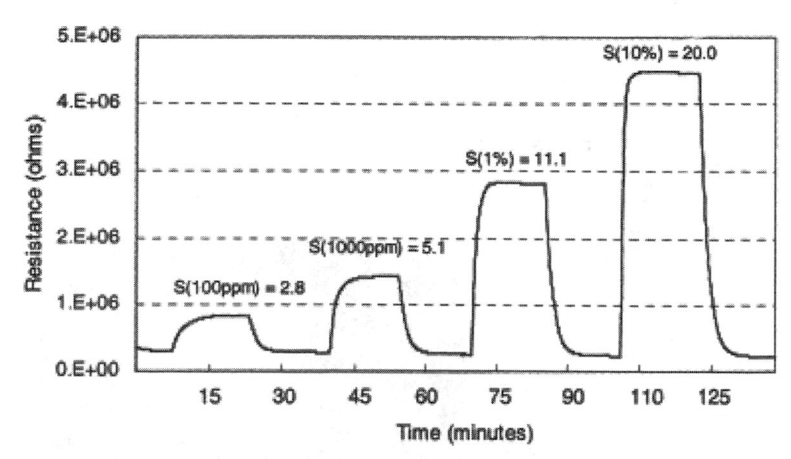

Fig. 3.2 Dynamic response of the SnO_2 thin film to O_2 operating at 400°C. Reprinted with permission from Elsevier publications.[1]

For capacitance measurements, a build up of charge, Q, across the electrodes is related to the capacitance, C, and voltage, which is described by the following empirical relationship:

$$Q = CV . \tag{3.3}$$

Between the electrodes is a dielectric material. The electrical field that arises between the electrodes strongly depends on the materials dielectric properties. For example, in a parallel plate capacitor, where the electric field is simply the voltage divided by the distance between the electrodes, the capacitance is given by:

$$C = \varepsilon \frac{A}{d} . \tag{3.4}$$

where A is the electrode area, d is the distance between the electrodes, and ε is the dielectric constant.

Both conductometric and capacitive measurements can be carried out in either DC or AC conditions. A material's dielectric properties exhibit a strong frequency dependence and therefore operating frequency can have a major impact on measurements.

Conductance and capacitance measurements are generally obtained with an *interdigital transducers* (*IDTs*), similar to the one shown in **Fig. 3.3**, in which a sensitive layer is deposited over the IDT electrodes. The sensitive layer can be a nanostructured thin film. The electrodes are comprised of noble metals, such as Pt and Au, and are deposited onto insulating and inert substrates such as alumina, sapphire or different polymers. The substrate and electrode materials are chosen such that they do not interact with any measurand during the measurements.

The effect of nanostructured sensitive layers in enhancing the performance of capacitive and conductometric transducers will be presented in Chaps. 6 and 7 in details.

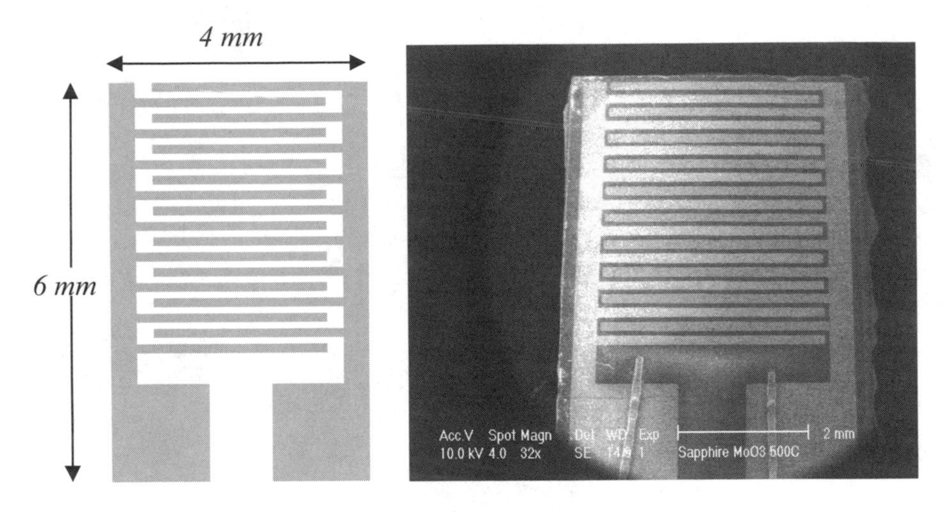

Fig. 3.3 Inter-Digital Transducer (IDT). *Left*: Schematic; *Right*: SEM image.

3.3 Optical Waveguide based Transducers

Optical waveguide based transducers are among the most utilized transducers in nanotechnology enabled sensing. Such sensors utilize interactions of optical waves, generally in the visible, infrared, and ultraviolet regions, with the measurand. These interactions cause the properties of the waves to change, e.g. intensity, phase, frequency, polarization etc., and these changes are then measured.

There are many types of optical sensors which are not based on waveguide structures; rather they make use of *optical spectroscopy* for characterization of target analytes. Such devices are concerned with the measurement of spectrums, which may include UV-visible and infrared wavelengths, parameters. These measurements were discussed in Chap. 2 and will be further explained in the Chaps. 5 and 6. In this section, several optical waveguide based transducers commonly used in sensing will be presented. These encompass different waveguides with various geometries, and are categorized into two major types: transducers based on the propagation of optical waves and transducers based on *surface plasmon* (*SP*) waves. Prior to presenting them, information regarding light propagation and the sensitivity of such waveguides is first provided.

3.3.1 Propagation in Optical Waveguides

The propagation of optical waves is generally *evanescent*, meaning that they decay exponentially with distance from the point at which they are sourced. However, we usually refer to waves as *propagating waves*, if they can propagate for relatively long distances before losing their intensity. The optical waves are *transverse*, as they oscillate perpendicularly to the direction of propagation, as in **Fig. 3.4**. An *optical waveguide* is a path which confines optical waves within one or two dimensions. Depending on the waveguide, only certain propagating waves, or *guided modes*, are possible. All optical modes may consist solely of the electrical or magnetic wave components, or they may be a combination of both.

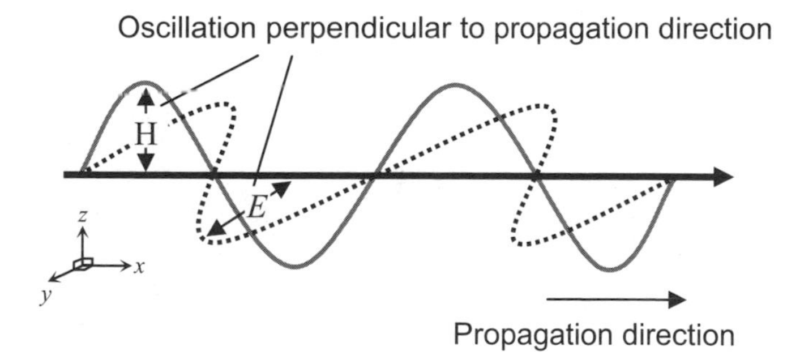

Fig. 3.4 Propagation of a transverse optical waves.

The guided waves in a *planar optical waveguide*, confined in two dimensions, are either TM_m (*transverse magnetic* or *p-polarized*) or TE_m (*transverse electric* or *s-polarized*), where m is an integer called the mode number.[2] At any time, t, a single frequency (*monochromatic*) propagating mode in the x direction may be represented by the function[3]:

$$f(x,t) = e^{i(k_x x - \omega t)}, \tag{3.5}$$

where $\omega = 2\pi f$ is the angular frequency and k_x is the propagation constant in x direction.

One of the most important parameters in sensing applications is the *effective refractive index, N*, as the device's sensitivity directly depends on it. It is defined as $N = k_x/k$ where $k = \omega/c = 2\pi/\lambda$ and λ is the wavelength of the wave when propagating in vacuum. The parameter itself is a function of the polarization mode, mode number, the waveguide thickness d_f, and the refractive indices of layered media.

A cross section of a planar optical waveguide is shown in **Fig. 3.5**. The waveguide is fabricated on a substrate and the sample media, which contains the measurand, is placed on the top of the waveguide. The wave propagates in the x direction and the waveguide can be considered to have either infinite or finite dimensions in the y direction. The thickness of the waveguide, d_f, is finite. Also seen in this figure is the field distribution of TE_m and TM_m modes component in z direction, propagating in the waveguide in the x direction. Understanding this distribution allows the device to be optimized for sensing applications. Firstly, the TE_m mode is characterized by the y component of its electric field, and is given by:

$$E_y(t) = u_m(z)e^{i(k_x x - \omega t)}, \tag{3.6}$$

where $u_m(z)$ is the transverse electric field distribution of the mth mode. Similarly, a TM_m mode is characterized by the y component of its magnetic field, and is given by:

$$H_y(t) = v_m(z)e^{i(k_x x - \omega t)}, \tag{3.7}$$

where $v_m(z)$ is the transverse magnetic field distribution of the mth mode. By choosing materials with the appropriate properties, it is possible to design transducers such that the field distribution is largest at the surface of the waveguide, as in **Fig. 3.5**. Ultimately, the distribution of these waves determines the extent of the interaction with the measurand and hence the device sensitivity.

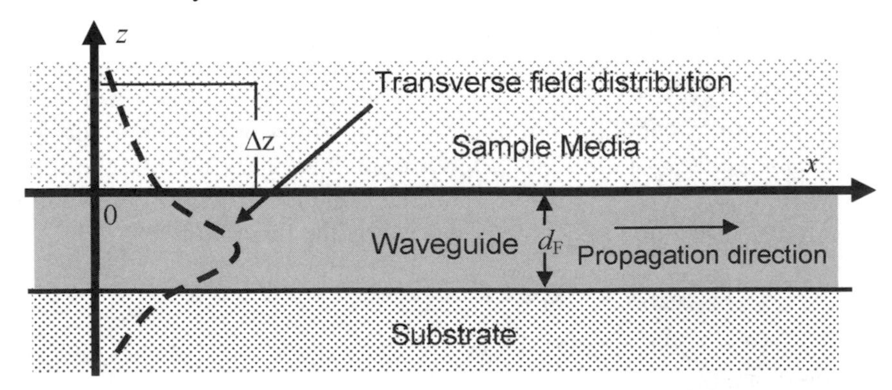

Fig. 3.5 Distribution of an evanescent wave in z direction.

As the propagating waves are evanescent in z direction, their amplitude decays exponentially when entering the sample media. The decay is described by the *penetration depth,* Δz, into the sample media which can be defined as[3]:

$$\Delta z = \frac{\lambda}{2\pi}\left(N^2 - n_C{}^2\right)^{-1/2}, \tag{3.8}$$

where n_C is the refractive index of sample media (**Fig. 3.5**). For a guiding mode to exist, the refractive index of the waveguide must be larger than those of the substrate and the sample medium. In such a case, the effective refractive index, N, is larger than the refractive indices of the substrate and sample media, yet smaller than that of the waveguide. From the penetration depth, the field distribution in the z direction may be defined with:

$$u_m(z) = u_m(0)e^{-z/\Delta z}, \tag{3.9}$$

$$v_m(z) = v_m(0)e^{-z/\Delta z}. \tag{3.10}$$

If the refractive index of the sample media changes, then the penetration depth will also change. This in turn results in a measurable change in the field distribution, and is the basis of affinity sensing with optical waveguides.

When fabricating nanotechnology enabled sensors based on optical waveguide transducers, materials are chosen such that the penetration depth generally lies between a few to several hundred nanometers. These penetration depths are utilized for detecting analyte molecules whose dimensions are in the order of nanometers. Such analyte materials include proteins and DNA strands.

3.3.2 Sensitivity of Optical Waveguides

The sensitivity of an optical waveguide based sensor strongly depends on the interaction between the measurand and the surface confined guided mode in the sample media. Analyte molecules may diffuse into or out of the evanescent region, they may become immobilize onto the boundary, or they may move along the surface by convection. Each of these interactions can change the effective refractive index, and as a result, produce a response. The change in effective refractive index can be calculated using perturbation theory. For TM modes, the result is expressed as[3]:

$$\Delta(N^2) = \frac{\left(\frac{\int_{-\infty}^{+\infty} \Delta\varepsilon(z) \left(\frac{dv(z)/dz}{\varepsilon(z)} \right)^2}{k^2} - N^2 \int_{-\infty}^{+\infty} \Delta \left(\frac{1}{\varepsilon(z)} \right) |v(z)|^2 \, dz \right)}{\int_{-\infty}^{+\infty} \frac{|v(z)|^2}{\varepsilon(z)} dz}, \tag{3.11}$$

and for TE modes it is:

$$\Delta(N^2) = \frac{\int_{-\infty}^{+\infty} \Delta\varepsilon(z) |u(z)|^2 \, dz}{\int_{-\infty}^{+\infty} |u(z)|^2 \, dz}, \tag{3.12}$$

which can be calculated using the computational methods. $v(z)$ and $u(z)$ are field distributions for the TM and TE modes, respectively.

From Eqs. (3.11), (3.12) the sensitivities for optical waveguides and surface plasmon sensors can be obtained by measuring the change in effective refractive index with respect to the change of the waveguide thickness ($\partial N/\partial d_F$) upon the interaction o the sensitive layer with target molecules and change of refractive index in the sample media ($\partial N/\partial n_C$).

Generally a layer that is chemically sensitive to the measurand is deposited on top of the waveguide. Typically, the sensitivity can be increased by reducing the waveguide's thickness. Producing a single propagation mode with large difference between the refractive index of the substrate and waveguide is another option. These conditions improve the confinement of the energy at near the surface of the waveguide.

Many examples of waveguide based optical sensors are found in *immunosensing* (this will be explained in details in Chap. 7). For instance, if a layer of protein that has a refractive index of approximately 1.45 is added on top of the guiding layer, the sensitivity for TE_0 and TM_0 modes propagating along the waveguide are shown in **Fig. 3.6.**[3] In this example the refractive indices of substrate and the guiding layer are 1.8 and 1.47 for the solid lines and 2.0 and 1.46 for the dashed lines, respectively. As can be seen, the larger the difference between the refractive indices of the substrate and guiding layer, the larger the sensitivities will be. In this example the refractive index of the sample media is 1.33.

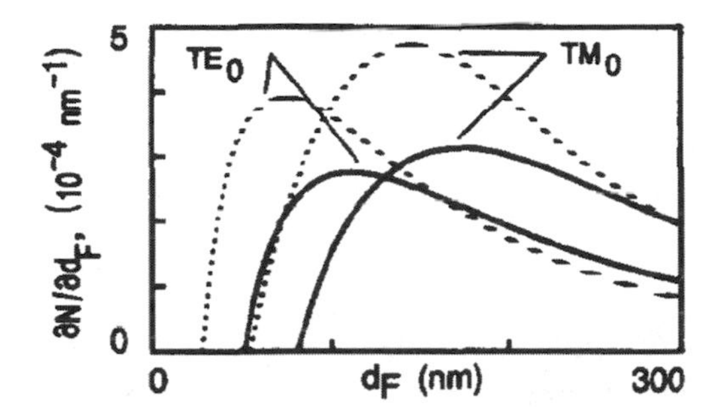

Fig. 3.6 Sensitivity (the change of refractive index with reference to the thickness of the protein layer) of the TE_0 and TM_0 modes. Reprinted with permission from Elsevier publications.[3]

When developing an affinity sensor using optical transducers, it is generally desired to build a system in which maximizes $\partial N/\partial d'_F$, where $\partial d'_F$ is the thickness of the immobilized layer, and minimizes $\partial N/\partial n_C$, where ∂n_C is the change of the refractive index of the sample media. This ensures that the sensor is sensitive to thc measurand, rather than the surrounding environment.

3.3.3 Optical Fiber based Transducers

Optical fibers consist of a solid core which is encased in a cladding. Light waves propagate along the optical fiber provided that the refractive index of the core is greater than that of the cladding.

Optical fiber based transducers may have several configurations, yet two of the most popular are shown in **Fig. 3.7**. In the first, part of the fiber's cladding may be removed, exposing the core directly to the sample medium. Interaction with the sample medium alters the optical wave's properties such as phase and amplitude when compared to the light originally propagating. For chemical and biochemical sensing, the exposed core may also be covered with a sensitive layer to further provide selectivity to a target analyte. In the second configuration, the end of the fiber is exposed to sample medium, and the optical waves reflecting is analyzed. Once again, selectivity to a target can be obtained by depositing a sensitive layer at the fiber's end.

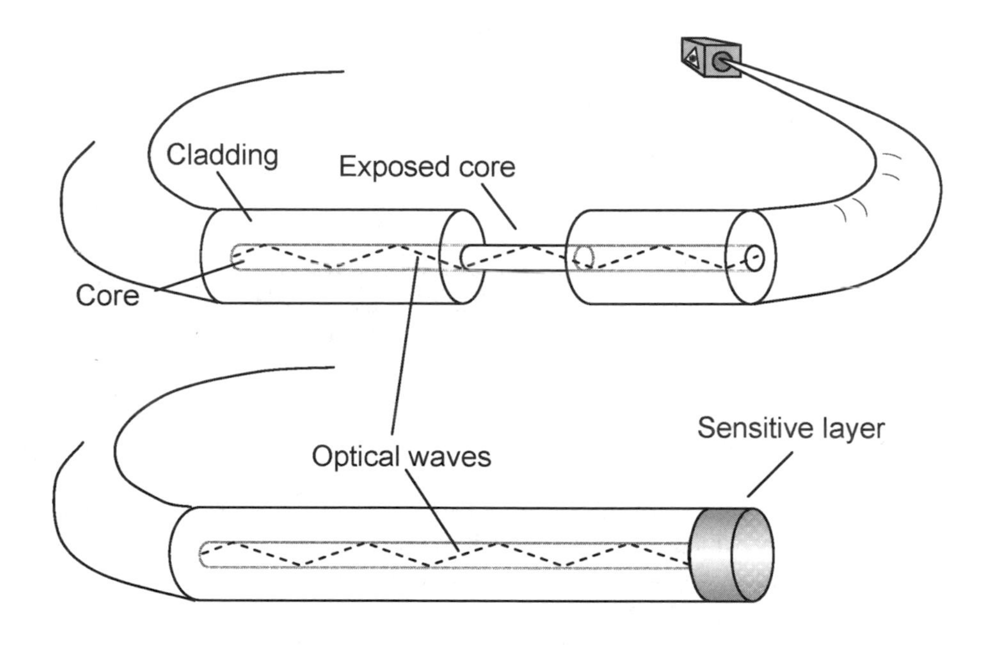

Fig. 3.7 Two types of optical fiber as sensing platforms: (top) when cladding is removed and the core is exposed to the target analyte or the sensitive layer is deposited on this core (bottom) when a sensitive layer is deposited at the end of an optical fiber.

3.3.4 Interferometric Optical Transducers

Interferometric optical transducers are measure the constructive and destructive interference of optical waves. In general, waves traveling through a waveguide is split into two or more beams of equal intensity. After splitting, one of the beams travels unperturbed through the waveguide, whilst the others travel through a waveguide that may be exposed to the sample media. The light beams are then recombined either destructively or constructively, and the optical properties of this recombined beam (e.g. intensity, wavelength, phase, etc) are analyzed.

Interferometric optical sensors are ideal for realizing on-chip optical sensors as they can be fabricated using standard microfabrication techniques. They have a relatively simple input and output coupling, capability for differential on-chip referencing with excellent stability. However, their major shortcomings are their rather large dimensions and fabrication costs with respect to other sensing platforms.[4]

The most common interferometric sensors are based on monolithic and hybrid *Mach-Zehnder* structures. A schematic of a *Mach-Zehnder interferometer* used for sensing is shown in **Fig. 3.8**.

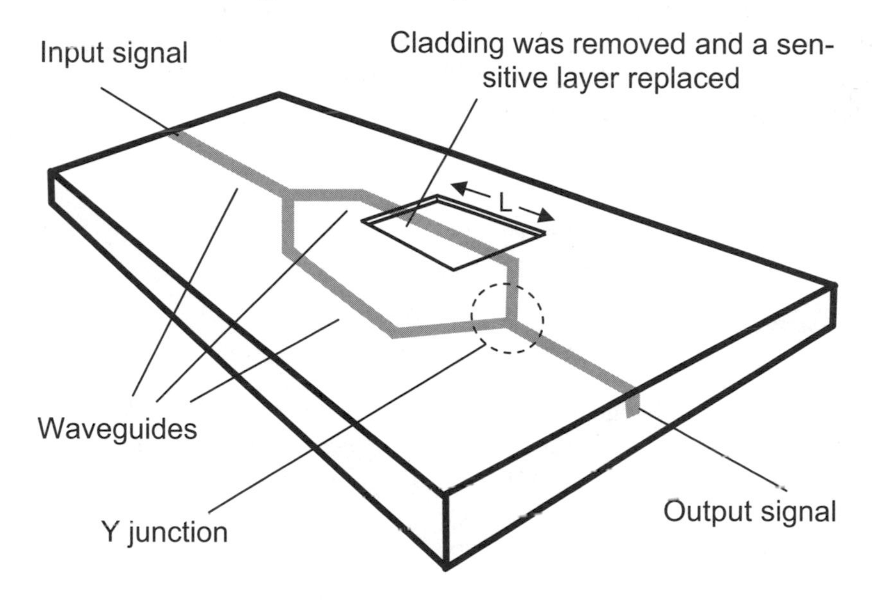

Fig. 3.8 Schematic of a Mach-Zehnder interferometer sensor.

In the Mach-Zehnder interferometer, the field distribution is coupled at the input. The field at the Y junction splits into two parts. If it is a symmetric interferometer, and if S is the distance between the two arms, then:[4]

$$F_{total}(x,y) = F_1(x,y) + F_2(x,y) = \frac{1}{2}F(x+\frac{1}{2}S,y) + \frac{1}{2}F(x-\frac{1}{2}S,y)\,. \tag{3.13}$$

A section of cladding in one of the waveguide arms can be removed, and the wave may be directly exposed to the sample media. Also, a sensitive layer may be placed where the section of cladding was removed, to improve the sensitivity. After the beam interacts with the analyte, a phase shift between the light waves traveling through both arms occurs. If each of the arms are single-mode, the output field is given by:[4]

$$F_{total}(x,y) = \frac{1}{2}F(x+\frac{1}{2}S,y)e^{i\varphi} + \frac{1}{2}F(x-\frac{1}{2}S,y), \tag{3.14}$$

in which:

$$\varphi = \left(\frac{2\pi}{\lambda}\right)\Delta NL\,, \tag{3.15}$$

where ΔN is the effective refractive index and L is the length of interaction (or the length of the sensitive layer). Eventually both light beams are recombined in the second Y junction. If the two beams are out of phase, the intensity of the resulting beam will be related to the phase difference:

$$I = 1 + \cos(\varphi)\,. \tag{3.16}$$

The beam intensity change is what accounts for the sensitivity, and the sensitivity can be defined as:

$$\frac{\partial I}{\partial d_F} = \frac{\partial I}{\partial \varphi}\frac{\partial \varphi}{\Delta N}\frac{\Delta N}{\partial d_F}\,, \tag{3.17}$$

where d_F is the change in the thickness of the waveguide. It has been shown that with Mach-Zehnder sensors refractive index changes as small as 4×10^{-6} can be detected.[4] This is almost equivalent to the mass detection limit of 1 pg/mm^2.

Although Mach-Zehnder based sensors are currently quite costly to fabricate, with the emergence of polymeric optical waveguides and devices, the cost of fabrication for such devices is expected to decrease significantly. Additionally, the large refractive index difference of polymers enables such devices to be fabricated with smaller dimensions.

3.3.5 Surface Plasmon Resonance (SPR) Transducers

If waveguide thickness, d_f, decreases to a thin sheet of infinitely small thickness, then the waveguide virtually changes to a boundary between two media, the substrate and the sample media (**Fig. 3.9**). Also, if this waveguide is metallic, then the TM mode waves may become trapped near and around this surface propagating along the boundary. If this occurs, then these waves are known as *Surface Plasmon (SP)* waves, and their subsequent excitation with light defined as *Surface Plasmon Resonance* (*SPR*).

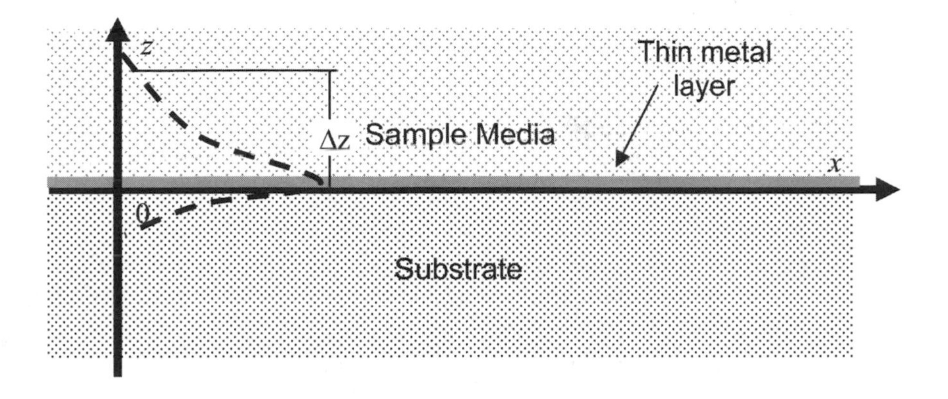

Fig. 3.9 A SP waveguide consisting of a thin metal film.

For SPR to occur, not only is a thin metal surface required, but the real part of the metal's permittivity, defined below, must be negative.

$$\varepsilon_M = \varepsilon'_M + i\varepsilon''_M \tag{3.18}$$

In addition, the magnitude of the real part of the metal's permittivity must be greater than the square of the refractive index of the sample medium:

$$\left|\varepsilon'_M\right| > n_C^2 . \tag{3.19}$$

The SP wave is defined by Eq. (3.7) with the effective refractive index given by:[3]

$$N = \left(n_C^{-2} + \varepsilon'^{-1}_M\right)^{-1/2} . \tag{3.20}$$

The SP wave's field distribution in the sample media is evanescent in the z direction. The penetration depth can be obtained by combining Eqs. (3.8) and (3.20) as follows:

$$\Delta z = \left(\frac{\lambda}{2\pi n_C N}\right)\left(-\varepsilon'_M\right)^{1/2} . \tag{3.21}$$

Due to the adsorption characteristics of light in the metallic layer, SP waves are highly attenuated in the visible spectrum. This attenuation is attributed to the relatively large imaginary part of the metal's permittivity. The intensity decays exponentially as $e^{-\alpha x}$ along the x propagation direction, where the attenuation constant, α, is given by:

$$\alpha = \left(\frac{2\pi}{\lambda}\right) N^3 \left(\frac{\varepsilon''_M}{\varepsilon'^2_M}\right). \tag{3.22}$$

Consequently, the propagation length is defined as:

$$L_\alpha = \frac{1}{\alpha}. \tag{3.23}$$

This propagation length can be in the order of several micrometers. Contrary to SP waves, optical waveguides have much larger propagation lengths.

SP waves propagate with a high attenuation and as a result, they characteristically demonstrate a significant localization of an electromagnetic field around the region they are generated. Therefore, the total surface area available for sensing interactions is limited to the region on the metal surface where the SP waves are excited. In general, SP waves are generated either by irradiating a thin metal film's surface with light, or by coupling a guided wave onto the boundary between the metal thin film and the waveguide layer. The most widely used configurations of SPR sensors are: prism coupler-based SPR systems (*Attenuated Total Reflection – ATR* method); grating coupler-based SPR systems; and optical waveguide-based SPR systems.[5]

Coupling optical waveguides with thin metal films for generating SP waves has several attractive features for sensing. These include low cost, robustness, relatively small device size, and it provides a simple way for controlling the optical wave's path. An example of coupled optical SP sensing set-up is shown in **Fig. 3.10**. The optical wave enters the region with a thin metal overlayer, where consequently it penetrates through the metal layer to form SP waves. The sample media containing the measurand is then placed over this thin metal film. The SP waves and the guided mode have to be well phase-matched to excite SP waves at the outer interface of the metal.

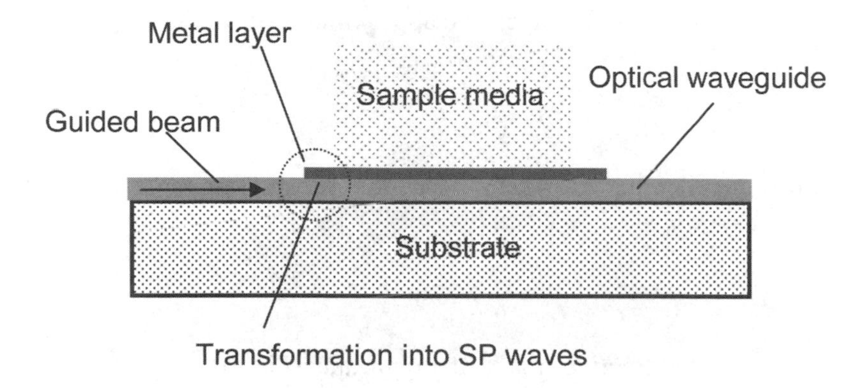

Fig. 3.10 SP waves coupled using an optical waveguide.

The ATR configuration developed by Kretschmann[6] is widely used for the design of the most of the SPR sensing systems (**Fig. 3.11**). ATR occurs when light traveling through an optically dense medium reaches an interface between this medium and a medium of a lower optical density. This produces total reflection. The resonance condition of the light in the prism with the SP at the metal/sample media interface occurs at a critical incidence angle that depends on the parameters of the media utilized. Consequently, this causes the energy from the incident light to produce the evanescent SP waves. This reduces the energy of the reflected light, and the reflected light can be detected by an array of photodiodes or charged coupled detectors (CCDs).

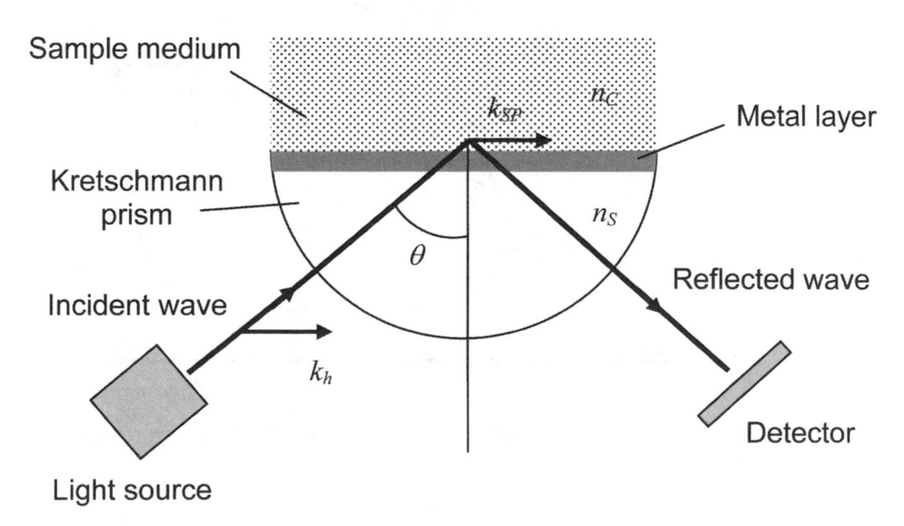

Fig. 3.11 The attenuated total reflectance configuration.

The wavenumber of the SP can be approximated from Eq. (3.20) as:

$$k_{SP} = \frac{2\pi}{\lambda c} \frac{1}{\sqrt{n_C^{-2} + \varepsilon_M'^{-1}}}. \tag{3.24}$$

If this wavenumber is equal to the horizontal component of incident wavenumber (k_H) which is given by:

$$k_H = \frac{2\pi}{\lambda c} n_S \sin\theta, \tag{3.25}$$

where n_S is the refractive index of prism and θ is the angle of incident where resonance occurs. An example of the response of a SP sensor after the immobilization of particles such as proteins is shown in **Fig. 3.12**. The change in the maximum attenuation angle corresponds to the amount of protein attached on the surface.[7]

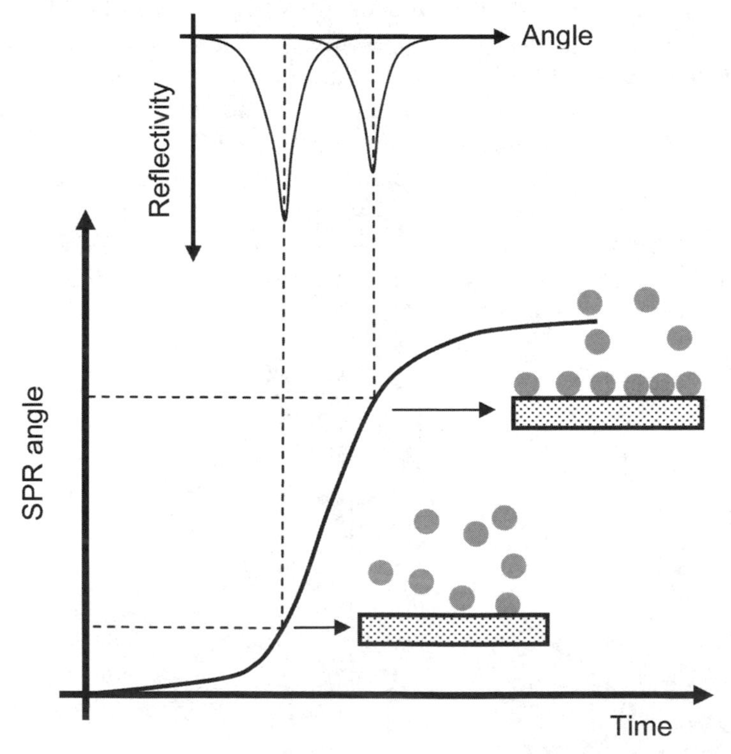

Fig. 3.12 A schematic representation of protein immobilization and the corresponding response in an SPR sensor.

The sensitivities for the SP resonance sensors may also bc calculated similarly in similar manner to that of optical waveguide based sensors. For metals such as gold and silver where $(-\varepsilon'_M >> n'^2_F)$ the sensitivities can be described as:[3]

$$\frac{\partial N}{\partial d_{F'}} \approx \frac{2\pi}{\lambda}\left(\frac{1}{n_C^{\,2}} - \frac{1}{n'^{\,2}_F}\right) N^4 \frac{n_C}{(-\varepsilon'_M)^{1/2}}, \tag{3.26}$$

where λ is the optical wavelength, n_C is the refractive index of the sample media, n'_F is the refractive index of the metal, and ε'_M is the real part of the dielectric constant of the metal. As can be seen, the sensitivity is inversely proportional wavelength and to $(-\varepsilon'_M)^{1/2}$.

Eq. (3.26) describes sensitivity, with reference to a layer added on the surface of the sensor. In addition, the change of the effective refractive index is also a function of the change of refractive index of the sample medium. Using (3.26) it can be shown that the sensitivities of SPR sensors are generally 5-10 times larger than those of optical waveguides. Furthermore, by varying the type of metal thin film, and SPR wavelength, the sensitivity may also be enhanced. For example, an SPR based sensor utilizing a gold surface, and having optical wavelength of 632.8 nm, has a sensitivity that is 1.4 times larger than that of a silver layer device, and it is almost double that of silver at 780 nm.[3]

3.4 Electrochemical Transducers

Electrochemical transducers generate signals that result from the presence and interaction of chemical species. They make use of various chemical effects to monitor concentrations of such species. The two main effects that are utilized in electrochemical sensors are the Volta effect (*voltammetry*); in which two dissimilar metals are brought into intimate contact resulting in the formation of a contact potential, and the Galvanic effect (*amperometry*); in which a potential difference is formed when different conducting materials are placed in an electrolyte solution. Usage of nanomaterials can enhance the performance for both sensor types. Nanostructured thin film can increase the surface area to volume ratio at the sensitive regions of the electrodes, enhance and tailor the electrochemical, optical and mechanical properties of the sensor and sensitive layer, etc. As will be demonstrated, utilizing these effects in various ways also utilized to investigate and quantify the concentration of target analytes and monitoring of associated chemical reactions.

3.4.1 Chemical Reactions

Interaction of the target chemical, X, with the chemical constituents of the sensor, S, can be described by the chemical equation that represents the reaction within the sensor:

$$X + S \underset{k_r}{\overset{k_f}{\rightleftharpoons}} S_X + R \tag{3.27}$$

where S_X represents the chemicals formed within the sensor and R is the chemical byproduct, which leaves the sensor. As indicated by the arrow, this reaction is reversible. The rate of reaction is different in each direction, and is described by the rate constant in the forward reaction k_f, and in the reverse direction k_r. whose unit are $\sec^{-1}$.

Most of the interactions eventually reach a state of *chemical equilibrium* in which a chemical reaction proceeds at the same rate in both directions. When this condition is met, there is no change in the concentrations of the various compounds involved. This process is known as *dynamic equilibrium*. Without energy input chemical reactions always proceed towards equilibrium. For a reaction of the type:

$$aA + bB \rightleftharpoons cC + dD, \tag{3.28}$$

the *reaction quotioent*, Q_P, provides an indications of whether the reaction will shift to the right or left, and is defined as:

$$Q_P = \frac{[\mathrm{C}_i]^c [\mathrm{D}_i]^d}{[\mathrm{A}_i]^a [\mathrm{B}_i]^b}, \tag{3.29}$$

where i denotes the instantaneous concentration at a moment in time. When this reaction is in a state of equilibrium, the concentrations of reactants and products are related by the *equilibrium constant*, K:

$$K = \frac{[\mathrm{C}]^c [\mathrm{D}]^d}{[\mathrm{A}]^a [\mathrm{B}]^b}. \tag{3.30}$$

It is important to note that non-interacting liquids, solvents and solids are not included in these equations, as their concentrations remain constant.

3.4.2 Thermodynamics of Chemical Interactions

A chemical sensor's performance strongly depends on the energies that are released or accepted, during chemical reactions. Many chemical reactions are reversible, however they have a tendency to proceed in a direction in which is energetically favorable, or *spontaneous*. Therefore, knowledge of the spontaneity of a reaction is an indication of the likelihood that reaction will take place. This knowledge is important in electrochemical based sensing, as it provides an indication of the response of a sensing layer towards an analyte species. The operation of an electrochemical sensor can be described using the first and second law of thermodynamics.

First Law of Thermodynamics

The *first law of thermodynamics* expresses the energy conservation, and states that fractional change in the internal energy, E, of a system results from the addition or removal of heat, Q, and amount of work, W, done by the system on the surroundings:

$$\Delta E = \Delta Q + \Delta W\,. \tag{3.31}$$

The energy of the surroundings, $E_{surroundings}$ and the system, E_{system}, represent the total energy of the universe. This energy is constant and given by:

$$E_{\text{universe}} = E_{\text{system}} + E_{\text{surroundings}}. \tag{3.32}$$

A system's *enthalpy*, H, is the sum of the internal energy and the product of pressure, p, and volume, V. The change in enthalpy is given by:

$$\Delta H = \Delta Q + \Delta pV \tag{3.33}$$

For any process at constant pressure, Δp is zero and hence the change of a system's enthalpy, ΔH, is the same as the heat added or removed. This enthalpy change is negative for *exothermic* processes and positive for *endothermic* processes.

Second Law of Thermodynamics

The *second law of thermodynamics* describes that heat moves spontaneously from a cold body to a hot body, and energy spontaneously disperses from being localized. *Entropy*, S, is a measure of the spontaneous dispersion of energy at a specific temperature. The change in entropy at temperature T of a thermodynamic system is given by:

$$\Delta S = \Delta Q/T. \tag{3.34}$$

Ludwing Boltzmann described the entropy relationship quantitatively in terms of probability:

$$S = k \ln N \tag{3.35}$$

where k is the Boltzmann constant (1.38×10^{-23} J/K) and N is the number of states available to a system. The entropy of the universe, which is equal to the entropy of the system and its surroundings, is always positive for a spontaneous process:

$$S_{\text{universe}} = S_{\text{system}} + S_{\text{surroundings}}. \tag{3.36}$$

It is possible to determine whether or not a reaction is spontaneous at a particular temperature by separately measuring the entropy change in the system and it its surroundings. During a chemical reaction some thermal energy ΔQ is transferred between the system and its surroundings. At constant temperature and pressure, ΔQ is the change in system enthalpy, ΔH_{system}. Enthalpy lost from a system is gained by its surroundings, and vice versa as:

$$\Delta H_{surroundings} = -\Delta H_{system}, \tag{3.37}$$

and consequently from Eq. (3.34) can be utilized to obtain that:

$$\Delta S_{surroundings} = -\Delta H_{\text{system}}/T. \tag{3.38}$$

This change in entropy of the surroundings is directly related to the system's heat change. From Eq. (3.36) it is found that:

$$\Delta S_{\text{universe}} = \Delta S_{\text{system}} - \Delta H_{\text{system}}/T. \tag{3.39}$$

Now, if one chemical reaction is occurring in the universe at constant temperature and pressure, then measuring $\Delta S_{\text{universe}}$ would provide information in the system. Therefore during the reaction, the energy changes that has been spread out, as with the entropy. It is seen that a portion of it remains in the system as ΔS, whilst the remainder has been transferred to the surroundings, $\Delta H/T$.

Gibbs Free Energy

From the second law of thermodynamics, the entropy of the universe is always increasing and is greater than zero. Then Eq. (3.39) must be greater than or equal to zero. Therefore:

$$\Delta S_{system} - \frac{\Delta H_{system}}{T} \geq 0 \quad (3.40)$$

or after rearranging:

$$\Delta H_{system} - T\Delta S_{system} \leq 0 \quad (3.41)$$

In a spontaneous chemical reaction, free energy is released from the system, causing it to become more thermodynamically stable. The spontaneity can be deduced by determining the change in free energy available for doing work, which is called the *Gibbs free energy*:

$$\Delta G = \Delta H - T\Delta S\,, \quad (3.42)$$

where ΔH is the change of the system's *enthalpy*, which at constant pressure is the same as the heat added or removed, T is the temperature, and ΔS is the change in the systems *entropy*. The free energy conditions for spontaneity are:

$$\begin{aligned} &\Delta G < 0, \text{ spontaneous (favored) reaction} \\ &\Delta G = 0, \text{ system in equilibrium, no driving force prevails} \\ &\Delta G > 0, \text{ non-spontaneous (disfavored) reaction} \end{aligned} \quad (3.43)$$

From the second law of thermodynamics, ΔS of a chemical reaction that is not in equilibrium will tend to increase. Therefore, to ensure the reaction is spontaneous ($\Delta G < 0$), from Eq. (3.42) it is observed that ΔH must be sufficiently negative. If the free energy of the reactants in a chemical reaction occurring at constant temperature and pressure is higher than that of the products, $G_{reactants} > G_{products}$, then the reaction will occur spontaneously.

The Gibbs free energy at any stage of the reaction can be found through its relationship with the reaction quotient in Eq. (3.29), defined as:

$$\Delta G = \Delta G^0 + RT\ln(Q_P)\,, \quad (3.44)$$

where ΔG^0 is the standard-state free energy of reaction, and R is the gas constant (8.314472 $J \cdot K^{-1} \cdot mol^{-1}$). When the reaction reaches equilibrium, $\Delta G = 0$ and the reaction quotient takes on the valued of the equilibrium constant from Eq. (3.30). The change in free energy at equilibrium becomes:

$$\Delta G^0 = -RT\ln(K)\,. \quad (3.45)$$

As will be demonstrated later, this equation is of fundamental importance to the function of electrochemical sensors. K is a function of analyte concentrations. It will be seen that ΔG^0 is a function of voltage produced by the electrochemical interactions. As a result, the concentrations of target analytes can be obtained using voltage or current measurements.

3.4.3 Nernst Equation

In this section, the fundamental basis about the *electrochemical sensors* will be presented. *Electrochemistry* deals with the transfer of charge from an electrode to its surrounding environment. Electrochemistry uses electrical measurements for analytical applications.[8] During an electrochemical process, chemical changes take place at the electrodes and charges transfer through the media. In fact, the largest and oldest group of chemical sensors are electrochemical sensors. Sensors as diverse as enzyme electrodes, high temperature metal oxide gas sensors in automobiles, fuel cells, etc. are included in this category.

Electrochemistry is primarily concerned with *redox reactions*. A redox reaction involves transfer of electron from one species to another. A species is oxidized when it loses electrons, conversly is reduced when it gains electrons. An *oxidizing agent*, which is also called an *oxidant*, receives electrons from another substance and is reduced in the process. A *reducing agent*, which is also called a *reductant* donates electrons to another substance and *oxidizes*.

For understanding the performance of electrochemical sensors, we first need to become familiar with some basic electrochemical concepts such as: *Galvanic cells*, *reference electrodes*, *salt bridges*, and *standard reduction potentials*.

A Galvanic (or *voltaic*) cell uses spontaneous chemical reactions to generate electricity.[9] In such a cell, one reagent oxidizes and another reduces and a voltage difference is produced as a result of these reactions. If an electrode is placed in an *electrolyte solution* (an electrolyte solution is a substance that dissociates into free ions when dissolved producing an electrically conductive medium), it generates a potential. However, this potential cannot be measured directly. Always a combination of two of such an *electrode-electrolyte* system is needed. Each of the electrode-electrolyte system is called a *half-cell*.[10]

The two half-cells must be connected by means of an electrically conductive membrane or bridge. For many interactions the net reaction is spontaneous but little current flows through the circuit as aqueous ions may react at the other electrode surface. This generates no flow of current through the external circuit. For instance in:

$$Cd(s) + 2Ag^{+}(aq) \rightleftharpoons Cd^{2+}(aq) + 2Ag(s), \tag{3.46}$$

aqueous Ag^{+} ions can interact directly at the Cd(s) surface, which generates no net current. This means that such interactions cannot be monitored in a sensing measurement. In order to avoid this problem, we can separate the reactant into two half-cells by connecting the two half-cells with a *salt bridge*.

A *standard reduction potential* (shown by E°) can be used to predict the generated voltage when different half-cells are connected to each other. The term standard means the activities of all species are unity.[9] A *hydrogen electrode* is generally used for the standard. The electrode consists of a Pt surface in contact with an acidic solution for which A_{H+}=1 mol. A stream of H_2 gas (1 bar pressure and 25°C) is purged through the electrode to saturate the surface of electrode with aqueous H_2. The reaction is:

$$H^{+}(aq) \rightleftharpoons \tfrac{1}{2} H_2. \tag{3.47}$$

A potential of zero is assigned to this *standard hydrogen electrode* (*SHE*).

The generated voltage is the difference between electrode potentials of the two half-cells. The magnitude of potential depends on: (a) the nature of electrodes (b) the nature and concentrations of solutions (c) the liquid junction potential at the membrane (or the salt bridge). An example of a conventional Galvanic cell set-up, with zinc and copper electrodes, is shown in **Fig. 3.13**. If concentrations of electrolytes are 1 mole then the potential is measured to be equal to 1.1 V as:

$$Zn(s) \rightarrow Zn^{2+} + 2e^{-} \qquad +0.763\ V \tag{3.48}$$

$$Cu^{2+} + 2e^{-} \rightarrow Cu(s) \qquad -0.337\ V \tag{3.49}$$

$$Zn(s) + Cu^{2+} \rightarrow Zn^{2+} + Cu(s) \qquad +1.100\ V \tag{3.50}$$

The Gibbs free energy for this reaction is negative (the reaction proceeds spontaneously at room temperature). This cell can be employed as a practical battery.

In **Fig. 3.13** the salt bridge consists of a tube filled with a gel containing high concentration of KNO_3, which does not affect the cells reactions. The ends of the bridge are covered with porous glass disks that allow ions to diffuse but minimize the mixing of solutions inside and outside the bridge. In this case, K^{+} from the bridge migrates into the cathode compartment and a small amount of NO_3^{-} migrates from the cathode into the bridge. Ion immigration offsets the charge build up that would occur. The migration of ions out of electrodes is larger than migration of them into the bridge as the salt concentration is much higher in the bridge than the half-cells.

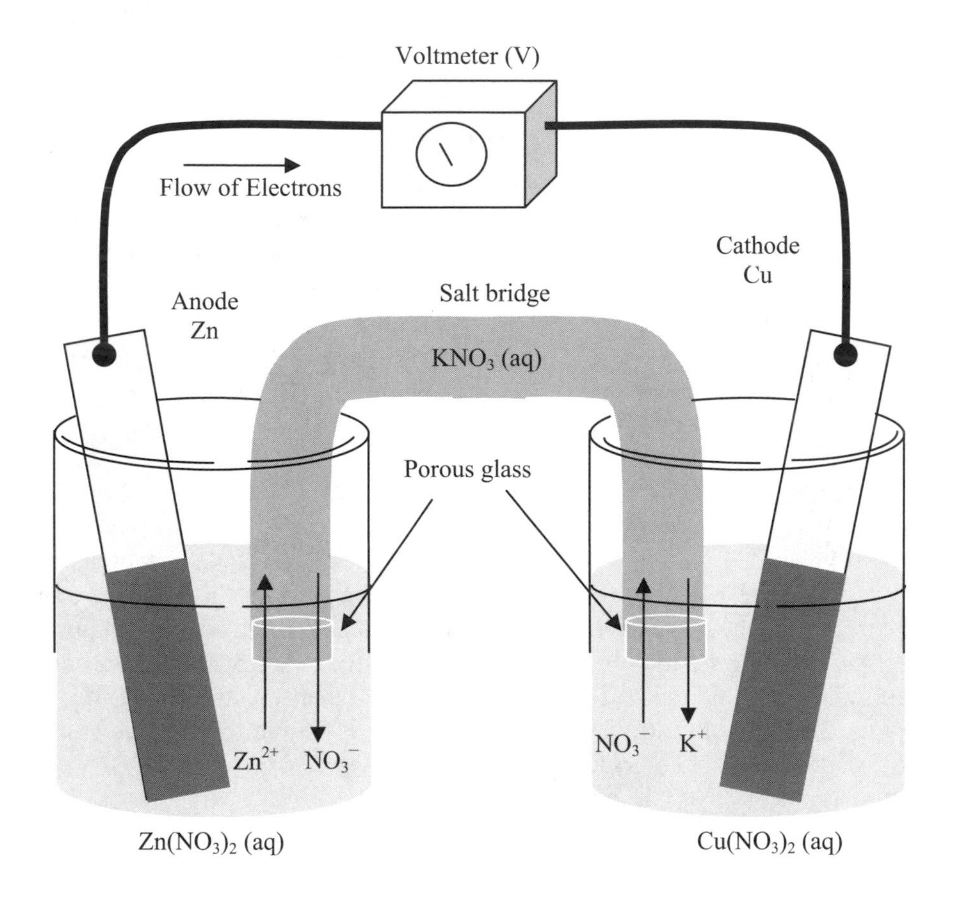

Fig. 3.13 Measurement of the electromotive force of an electrochemical cell.

Electrical work carried out by an electrochemical cell equals the product of the charge flowing and the potential difference across. If we operate the electrochemical cell in liquid at constant pressure and temperature cell, then the work carried out in the cell is:[9]

$$W = -E \times Q, \tag{3.51}$$

where E is the electromotive force (emf) of the cell in Volts and Q is the charge flowing through the cell which is calculated from:

$$Q = n \times N_A \times e, \tag{3.52}$$

where n is the number of moles of electrons transferred per mole of reaction, N_A is Avogadro's Number (6.02×10^{23}), and e is the charge of an electron (-1.6×10^{-19} C).

As $N_A \times e = F$ (Faraday constant which is equal to 96487 $Cmol^{-1}$), thus:

$$W = -nFE. \tag{3.53}$$

The free energy change for a chemical reaction conducted reversibly at constant temperature and pressure equals the electrical work that can be carried out by the reaction on its surroundings:

$$W = \Delta G. \tag{3.54}$$

The Gibbs free energy relates to the voltage of the cell through:

$$\Delta G = -nFE \tag{3.55}$$

From Eq. (3.44) and defining $\Delta G^0 = -nFE^0$ we can obtain:

$$E = E^0 + 2.303\ (RT/\mathrm{nF})\ \log(K) \tag{3.56}$$

or:

$$E = E^0 + (0.0591\ \mathrm{V/n})\ \log(K) \tag{3.57}$$

which is called the *Nernst equation*. It is important to distinguish between two different classes of equilibria: *equilibrium between two half-cells*, and *equilibrium within each half-cell.*[9] If a galvanic cell generates a nonzero voltage, then the net cell reaction is not at equilibrium. As a result, the equilibrium between the two half-cells has not been reached. However, the cell can establish chemical equilibriums within each half-cell.

3.4.4 Reference Electrodes

As an electrochemical cell is always comprised of two electrodes, it is common to utilize a *reference electrode*, which does not participate in interactions and its properties are known, in sensing processes. For sensing applications, in practice, reference electrodes are used which are easy to set-up, are non-polarisable, and give a reproducible electrode potential with low coefficients of variations to temperature. Many varieties of such electrodes are available but two of the most common reference electrodes are: *silver-silver chloride* and the *saturated calomel* electrodes.

Silver chloride is not soluble in water. Consequently, it can be used in many aqueous sensing applications without affecting the target analyte. For the silver-silver chloride reference electrode the half-cell reaction is:

$$\text{AgCl+e}^- \rightarrow \text{Ag + Cl}^- \qquad E^0 = +0.22\text{V} \tag{3.58}$$

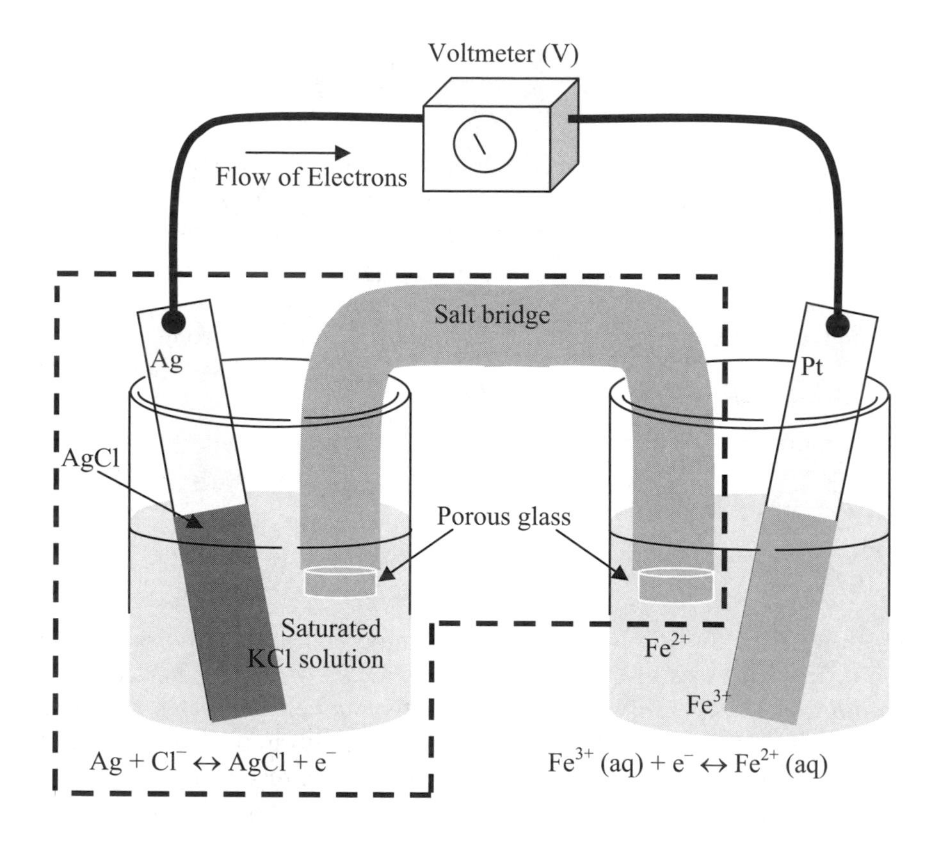

Fig. 3.14 The half cell within the doted space is called an silver-silver chloride reference electrode.

Consider the cell shown in **Fig. 3.14**. In this example, we intend to measure the relative concentrations of Fe^{2+} and Fe^{3+}. Pt is used as it does not interact with the aqueous Fe ions. The two half reactions can be written as follow: [9]

$$Fe^{3+} + e^- \rightleftharpoons Fe^{2+} \qquad E^0 = 0.77\text{ V} \tag{3.59}$$

$$\text{AgCl(s)} + e^- \rightleftharpoons \text{Ag (s)} + \text{Cl}^- \qquad E^0 = 0.22\text{ V} \tag{3.60}$$

From Eq. (3.57) the electrode potentials will be:

$$E_+ = 0.77\text{ V} - 0.0591\text{ V} \log\left([Fe^{2+}]/[Fe^{3+}] \right), \tag{3.61}$$

$$E_{-} = 0.22 \text{ V} - 0.0591 \text{ V} \log ([Cl^{-}]). \tag{3.62}$$

The concentration of Cl^- remains constant, maintained by the saturated KCl solution. Therefore, the measured differential voltage $E = E_{-} - E_{+}$ only changes when ratio of $[Fe^{2+}]/[Fe^{3+}]$ changes. As a result, obviously, this system can be used as a Fe ion sensor.

A commercial silver-silver chloride reference electrode generally consists of a silver wire or silver substrate coated with silver chloride which is dipped in a solution of a salt such as potassium chloride (usually 1M). The deposition of the silver chloride layer is commercially viable practice. For instance, it can be obtained by making the silver plate the anode of an electrochemical cell with a platinum cathode and potassium chloride as electrode. Several minutes of electrolyzation with a positive potential, as small as 0.5 V, oxidizes the silver surface to silver ions. The silver ions will attract chloride ions to form the silver chloride films in the process.

The voltage of a reference electrode with saturated KCl is approximately +0.197 V (slightly smaller than the standard solution voltage). **Fig. 3.15** shows an Ag/AgCl reference electrode set-up.

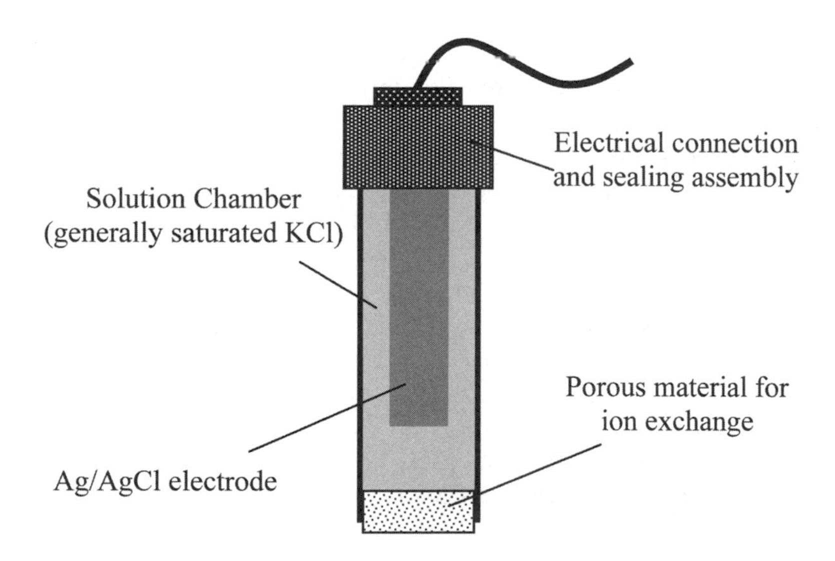

Fig. 3.15 A schematic of a commercial silver-silver chloride reference electrode.

The saturated calomel electrode (mercury chloride) is another common reference electrode in electrochemical sensors. The standard half-cell reaction is:

$$Hg_2Cl_2(s) + 2e^- \rightarrow 2Hg + 2Cl^- \qquad E^0 = 0.268 \text{ V} \qquad (3.63)$$

The electrode voltage with saturated KCl is approximately 0.241 V which is slightly smaller than the standard voltage.

When any two dissimilar materials are in contact, a voltage difference, *junction potential* is developed at their interface due to the diffusion of free charges. In salt bridges, this voltage is generally very small and in the range of millivolts.[9] This junction potential can create a lack of accuracy in standard electrochemical sensors. However, such voltages can be of great interests in nanotechnology enabled sensors as they are related to areas that extended only several nanometers to several micrometers from the junctions. Voltage sweeps and the observation of the current-voltage curves can be utilized to study such potentials relating them to near surface interactions. Such curves and their relation to surface layers will be explained more in the following sections.

3.4.5 Ion Selective Electrodes

Ion selective electrodes (*ISEs*) are membrane electrodes that respond selectively to target ions in the presence of other ions. They are employed to measure specific ions in a solution or in a gas phase. These sensors are generally made of a membrane-based electrochemical set-up. The membrane is somehow acts as a replacement for the salt bridge which also filters ions. Generally, in the development of sensors a membrane is chosen that makes the electrode selective for a particular ion.

These ion-selective membranes are fundamentally different from metal electrodes as they do not involve in a redox process themselves. A voltage difference develops across the membrane (due to the generation of the junction potentials) when it is placed in a solution. To measure this voltage, an ISE is used in combination with an internal or external reference electrode (**Fig. 3.16**).

The relationship between the measured potential, E, and the ion activity is also mathematically described by the Nernst equation.

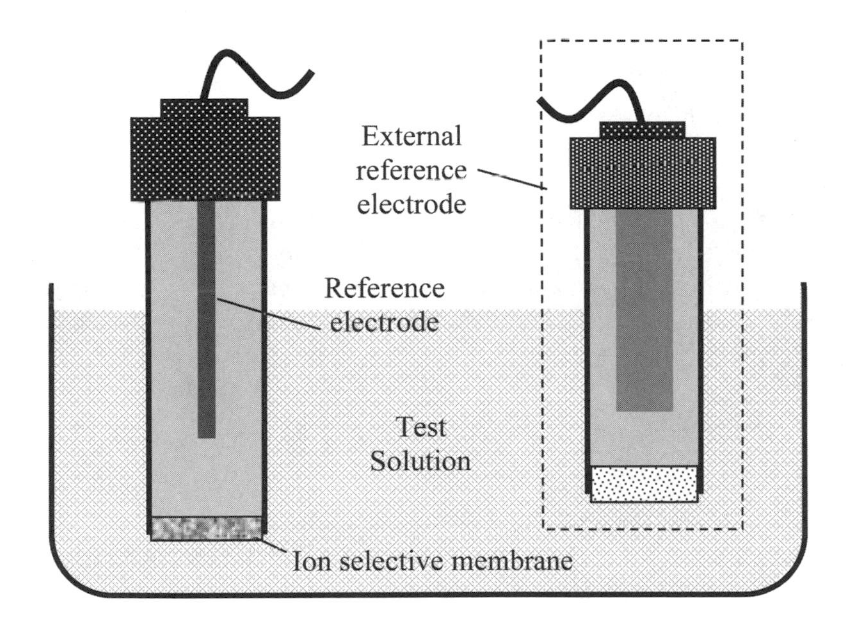

Fig. 3.16 An ion selective electrode with an external reference electrode.

ISEs play an important role in diagnostic medical sciences. Four of the chemical components in *chem-7 tests* are analyzed with ion selective electrodes.[9] In this monitoring process, which represents a large number of common hospital laboratories tests, Na^+, K^+, Cl^-, CO_2, glucose, urea, and creatinine are measured. No electrode responds selectively to only one kind of ion; however, some are fairly selective. For instance, the glass pH electrode is among the most selective membranes to hydrogen ions. In this case, sodium ions are the main interfering species, limiting th practical applications of pH measurements.

ISEs can be categorized to four different types, depending on the material of the membrane:

Glass membrane: It is used for measuring ions such as Na^+ or measuring pH. Glass membrane electrodes are generally formed by chemically doping a silicon dioxide glass matrix. The most common glass membrane electrodes are the pH electrodes.

Solid-state membranes: They are used for measuring ions such as Pb^{2+}. Solid-state electrodes are made of relatively insoluble inorganic materials within a membrane. Other ions, which can be measured, include cupric, cyanide, thiocyanate, chloride and fluoride, etc.

Polymeric membranes: They are used for measuring ions such as potassium, calcium, fluoroborate, nitrate, perchlorate, and even other applications such as water hardness. Polymer membrane electrodes generally consist of various ion-exchange materials incorporated into an inert matrix of polymer such as polyethylene, PVC, polyurethane and silicone.
Gas permeable membrane: Sensing electrodes are available for the measurement of gas species such as ammonia, carbon dioxide, dissolved oxygen, nitrogen oxides, sulfur dioxide and chlorine gas. These electrodes have a gas permeable membrane, selectively filtering target analyte gases.

Currently electrochemical sensors are widely used for determining the oxygen content of gas species for industrial applications especially in the automotive industry. They are particularly utilized to analyze exhaust gases in a combustion process as they have relatively short response time in comparison with other gas sensors.[11]

Zirconium oxide (*zirconia*) *electrochemical oxygen sensor* is one of the most common gas sensors. In such a sensor, the electrolyte is typically made of zirconium oxide (**Fig. 3.17**). The sensor electrodes can be made of platinum which also operates as a catalyst.

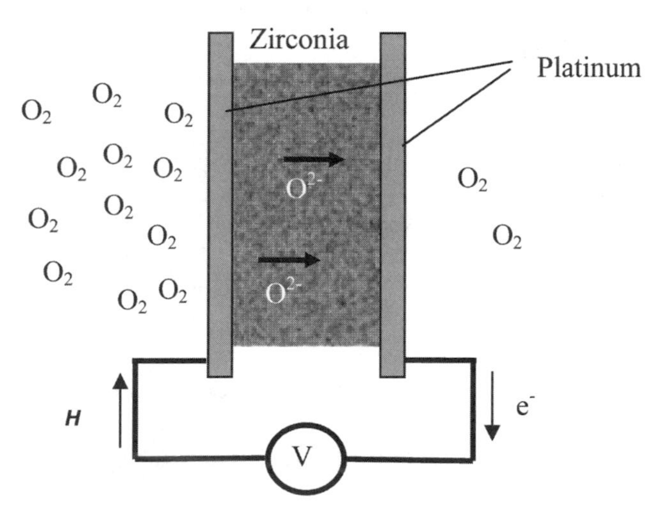

Fig. 3.17 A schematic of zirconia oxygen sensor.

This sensor operates at elevated temperatures of higher than 400°C. Based on the partial pressure of oxygen, the porous zirconium allows the movement of oxygen ions from a higher concentration to a lower concentartion. The diffusion of oxygen ions produces a voltage across the device, which is proportional to the partial pressure. Schematic of a commercial oxygen sensor installed in a gas exhaust is shown in **Fig. 3.18**.

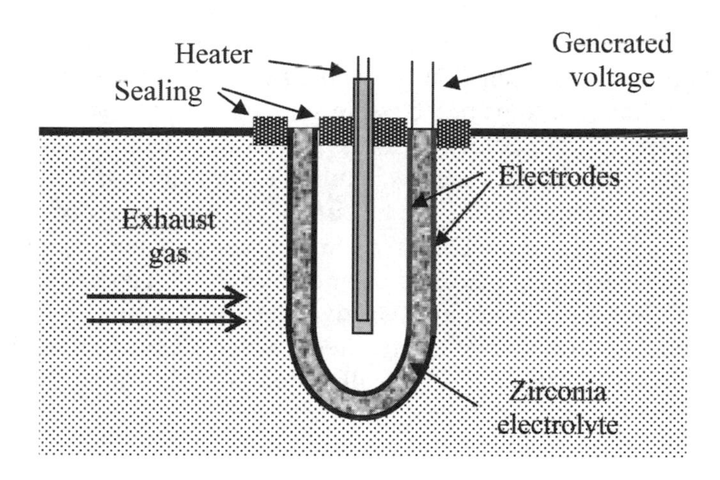

Fig. 3.18 Schematic of an industrial zirconia sensor installed in the exhaust an engine.

3.4.6 An Example: Electrochemical pH Sensors

For industrial and medical applications, pH is an important parameter to be measured and controlled. The pH of a solution indicates how acidic or basic (alkaline) it is. In electrochemical pH measurements, the selective membrane is generally made of glass. When measuring pH, we measure the negative log of hydrogen activity as:

$$\mathrm{pH} = -\log\,[\mathrm{H}^{+}{}_{\mathrm{activity}}], \tag{3.64}$$

$$[\mathrm{H}^{+}{}_{\mathrm{activity}}] = 10^{-\mathrm{pH}}. \tag{3.65}$$

The pH readings range is from 0 to 14. Using an electrochemical pH sensor, the potential develops across the membrane when in contact with a solution is measured. Generally, a reference electrode is used. The Nernst equation can be used for the calculation of pH as:

$$E = E^{0} - 0.05916\,\log\,([\mathrm{H}^{+}]), \tag{3.66}$$

or:

$$E = E^0 + 0.05916\,\text{pH}. \tag{3.67}$$

As can be seen, the output voltage changes linearly with the pH of the environment. One pH unit corresponds to 59.16 mV at 25°C, the standard voltage and temperature to which all calibrations are referenced.[11] The pH of any solution is also a function of temperature. The temperature of the solution determines the slope of the response.

Industrial applications aside, pH sensors are also widely used in biosensing applications. In such sensors, antibodies, enzymes or DNA are incorporated into the structure of the electrodes. Several types of enzymes are able to produce H^+ or OH^- ions when they interact with target molecules. These ions can then be electrochemically sensed, generally with the assistance of a *mediator*, by an ion-selective electrode.

Several other applications of electrochemical sensors and the utilization of nano-materials in their fabrication will be presented in Chaps. 6 and 7.

3.4.7 Voltammetry

In *Voltammetry* techniques the relationship between currents-voltages (*I-V*) are observed during an electrochemical process. Consequently, sensing information can be derived from these *I-V* characteristics. Voltammetric sensors are finding increasing use in medical applications, in the analysis of very low concentrations of pharmaceuticals and their metabolites as well as detecting environmental pollutants. Generally, electroanalytical sensing systems are simple and inexpensive. They can provide information on electrochemical redox processes, and chemical reactions. Since transient responses can be obtained by voltammetry, such responses can be used for studying very fast reaction mechanisms. In addition, the electrodes can be used as tools for producing reactive species in a small layer surrounding their surfaces to monitor chemical reactions involving target species.

Linear sweep voltammetry (*LSV*), *cyclic voltammetry* (*CV*) and *square wave voltammetry* (*SWV*) are the most widely used signaling schemes in voltammetric techniques.

A voltammetric set-up (*voltagram*) is shown in (**Fig. 3.19**). It consists of a reference electrode and a working electrode where redox interactions occur. It also consists of a programmable voltage source and a system for monitoring the *I-V* characteristics. The cell contains an *indifferent electrolyte* (or *supporting electrolyte*) along with an oxidisable or *reducible species* (*electroactive species*). The indifferent electrolyte does not participate in the interaction and only makes the solution conductive.

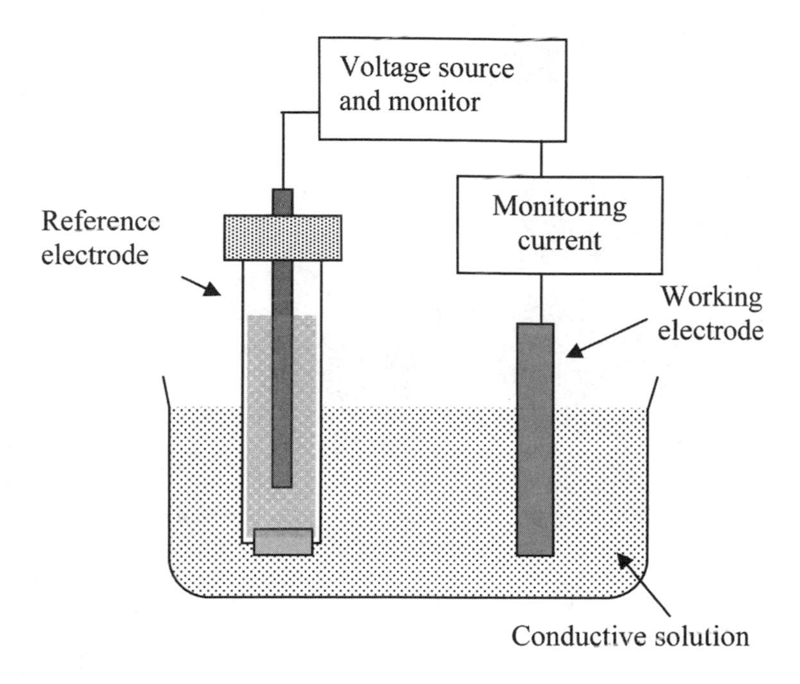

Fig. 3.19 A voltammetric sensing system.

When the power supply forces electrons into and out of an electrode, the charged surface of the electrode attracts ions of opposite charge.[12] The charged electrode and the oppositely charged ions next to it form an *electric bi-layer* (**Fig. 3.20**). The first layer of molecules at the surface of the electrode is adsorbed by van der Waals forces. The next layer is established when ions are attracted by the electrode's charge. This region, in which the composition is different from the bulk solution, is called the *diffuse part* of the double layer and can be from a few nanometers to a few micometers thick. Any given solution has one *potential of zero charge* (*POZC*) at which there is no excess charge on the electrode.[12] By changing the applied potential and observing the measured current, the POZC can be obtained.

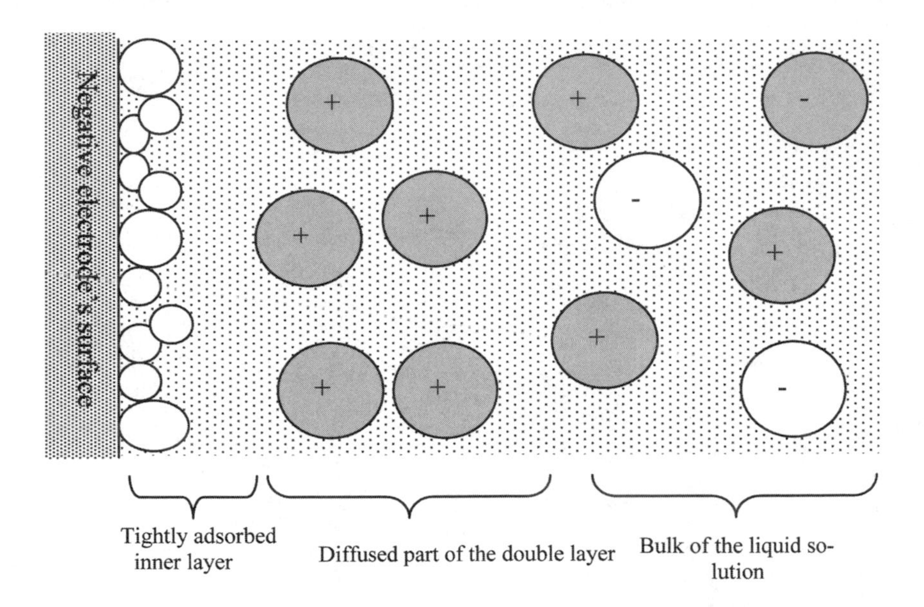

Fig. 3.20 The electric bi-layer.

LSV and CV techniques were proposed at the beginning of the 1950s when their related theories were described.[9] However, the use of these methods only received considerable attention in 80s and 90s when capabilities for interpreting the relevant responses, fast data acquisition systems and availability of computational tools for processing experimental data became widely available.

The electrochemical process and the shape of *I-V* characteristics obtained during voltammetry depend on *diffusion* and *capacitive* currents.

Diffusion current

The following are the most common conditions in the observation of diffusion current:

(a) *The voltammetric system is mixed up constantly.* In this case, the thickness of the diffusion layer remains constant. The diffusion current depends on the difference of target analyte concentrations on the surface of the electrode and in bulk of the solution as well as the thickness of the diffusion layer, δ.

The diffusion current can be calculated from the general Faradiac relationship: [12]

$$i(t) = (dN / dt)nF , \tag{3.68}$$

where N is the number of moles of the target analyte, n is the number of electrons take part in the redox interaction and F is the Faraday number. From *Fick's first law of diffusion*:[13]

$$\frac{(dN/dt)}{A} = -D((c_a - c_s)/\delta), \tag{3.69}$$

where A is the electrode surface area, D is the diffusion coefficient, c_s and c_a are the concentrations of analyte near the surface and in the bulk of analyte, respectively.

By changing the applied voltage, before reaching POZC, the difference between the two concentrations increases. As the solution is constantly mixed, the bulk analyte concentration remains constant. However, the concentration of ions on the surface of electrodes, c_s, is altered by changing the applied voltage. When we reach:

$$i_d = \frac{nFADc_a}{\delta}, \tag{3.70}$$

which is called the *diffusion limiting current* and occurs when the voltage at the surface is so large that all ions exchange electrons and an ion free surface meaning $c_s = 0$ is generated. As can be observed, this current is proportional to the value of c_a. As a result, it is a key parameter in sensing for the determination of the concentration of the target analyte. In the process of increasing the applied voltage, i_d is the maximum value for the measured current. When current is one half of the value of the limiting current $i = i_d/2$, the voltage is called the *half-wave potential,* $E_{1/2}$. The *I-V* characteristic a constantly mixed system, with a small rate constant, is shown in **Fig. 3.21**. As can be seen, the *I-V* characteristic in region 2 is similar to a diode's *I-V* characteristic. However, it tapers and eventually reaching saturation in region 3.

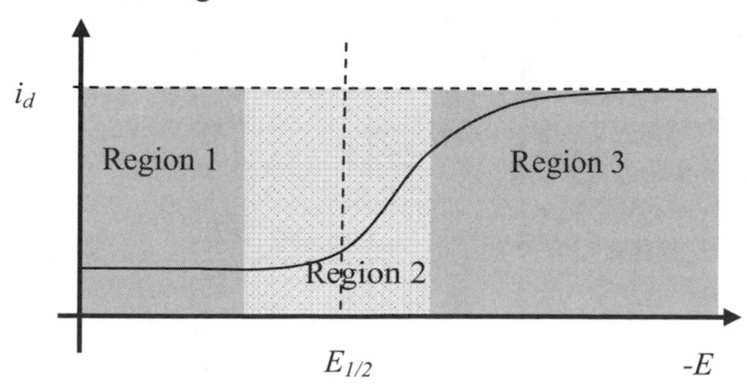

Fig. 3.21 The *I-V* characteristic of an electrochemical system when the system's rate constant is small.

(b) *When the concentration of the analyte is low or the mixing is not aggressive,* the thickness of the diffusion layer alters with voltage change. In this case, the *Fick's second law of diffusion* is used:[13]

$$\frac{\partial c_a}{\partial t} = D\left(\frac{\partial^2 c_a}{\partial x^2}\right). \tag{3.71}$$

If voltage is changed from the initial condition where the concentration of the analyte at the electrode's surface is equal to thc bulk concentration, $c_s = c_a$, to where the analyte concentration at the surface of the electrode is zero, $c_s = 0$, the diffusion layer thickness increases. In this case, the gradient dc_a / dx, analyte concentration gradient, results from the combination of two changes; the diffusion layer thickness change dx and the concentration change dc_a. They can change in different directions with different rates and the effect of one may dominate the other one.

If the diffusion layer increases linearly with time constant $(\pi Dt)^{1/2}$, the current magnitude will follow *Cottrell*'s *equation*[14] reaching a maximum value at E_p with its corresponding i_p:

$$i_p(t) = nFAD^{1/2}c_a(\pi\ u_t)^{-1/2}, \tag{3.72}$$

where u_t is the scan rate.

The change of current as a function of voltage is shown in **Fig. 3.22**. The current initially increases as a result of an increase in the difference of analyte concentration and the surface concentration. However, after the peak, when voltage increases further, current decreases as result of the increase of the diffusion layer thickness.

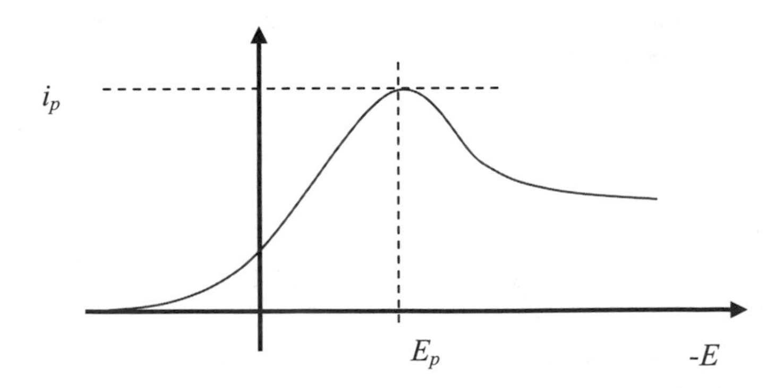

Fig. 3.22 *I-V* characteristic of an electrochemical system when the system's rate constant is large.

Linear Sweep Voltammetry

LSV or *direct current polarography* (*DCP*) was the first electrochemical voltammetric sensing technique used.[12] Although, many sophisticated voltammetric techniques were developed to replace LSV addressing its deficiencies, its relative simplicity makes it an attractive choice for electrochemical sensing systems.[15,16]

In LSV, the applied voltage between the working electrode and a reference electrode is scanned from a low voltage to a higher voltage as current is simultaneously monitored. The rate of voltage change can be in the order of 0.01-100 mV/s. However, it depends on the concentration of the target analyte as well as materials and dimensions of electrodes. The characteristics of the LSV depend on three major factors: (1) the rate of the electron transfer, (2) the chemical reactivity of the species, (3) and the voltage scan rate.

In voltammetric experiments, the current response is generally plotted as a function of voltage rather than time. For example, for the Fe^{3+}/Fe^{2+} redox system:

$$Fe^{3+} + e^{-} \rightleftharpoons Fe^{2+}, \tag{3.73}$$

a single voltage scan voltammogram is observed in **Fig. 3.23**. Such systems are important in the release of Fe ions in Ferritin protein within our body regulating the body's iron storage.[17] Similar electrochemical processes are used for measuring the Fe ion concentrations in our body.

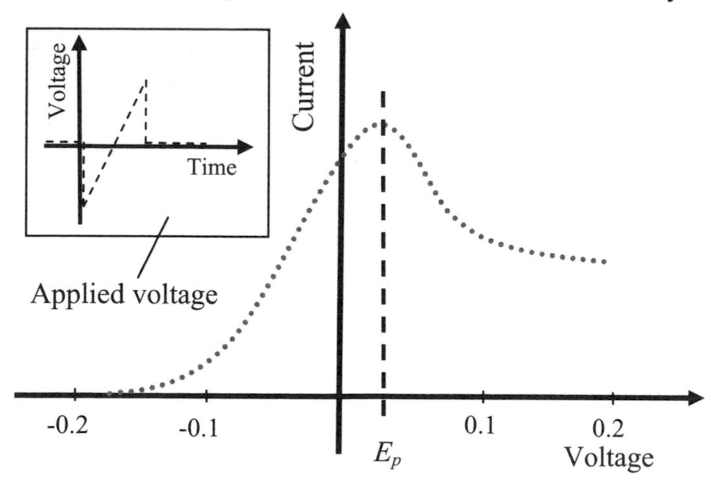

Fig. 3.23 IV characteristics of a Fe^{3+}/Fe^{2+} redox system.

If in such a Fe^{3+}/Fe^{2+} redox system the concentration of the analyte is high, for a voltage lower than −0.2 V no current can be measured. As the voltage increases the current increases and reaches a peak value (E_p). Further increase in the voltage decreases the current. As has been described previously, the decrease occurs when the diffusion layer thickness increases.

The *sweep rate* is an important factor. It determines how fast the electrons can diffuse and interact at the surface of the electrodes. The reversibility of the process is also determined by the sweep rate. As a result, the *I-V* characteristics changes when the scan rate is altered. The rates of the charge transfer at a specific potential are calculated from:[13]

$$k_{red} = k^0 \exp\left(\frac{-\alpha nF(E-E^0)}{RT}\right), \tag{3.74}$$

$$k_{ox} = k^0 \exp\left(\frac{(1-\alpha)nF(E-E^0)}{RT}\right), \tag{3.75}$$

where k_{red} and k_{ox} are the reduction and oxidation rates of the charge transfer, respectively, k^0 is the *standard heterogeneous rate constant*, α is the charge transfer coefficient, n is the number of electrons involved, F is the Faraday constant, T is the temperature in Kelvin and R is the gas constant. In these equations, E^0 is the reference electrode potential and E is the potential of the working electrode.

Fig. 3.24 shows the change in the linear sweep voltammograms as the scan rate is increased. As it can be observed, the curves remain proportionally similar. However, the current increases with the increasing the scan rate. This effect is attributed to the change in the diffusion layer thickness.

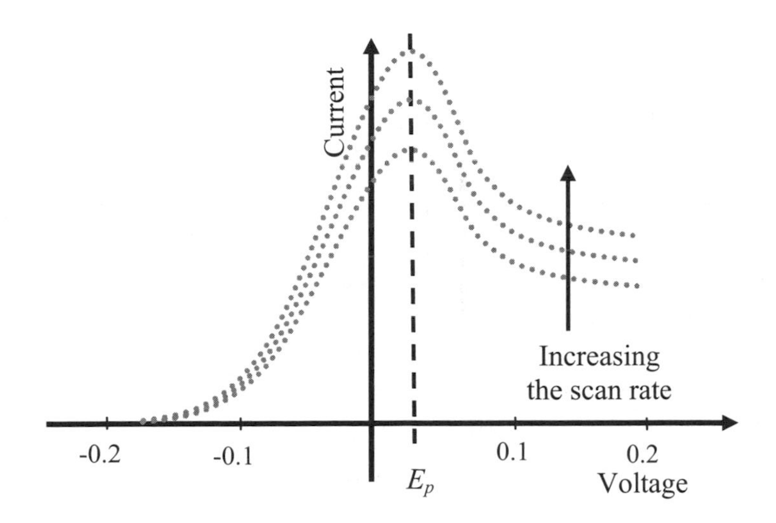

Fig. 3.24 Changing the scan rate in a Fe^{3+}/Fe^{2+} redox system.

The value of the peak currents are proportional to the square root of the scan rate according to Eq. (3.72), $i_p \propto \sqrt{u_t}$. In slow voltage scans, the diffusion layer grows further from the electrode surface than in fast scans. Consequently, the ion flux to the electrode surface becomes smaller decreasing the current magnitude.

For the Fe^{3+}/Fe^{2+} redox high concentration system, the electron transfer kinetics is fast. A rapid system is generally a *reversible electron transfer* system. Conversely, for a slow electron transfer system, the *I-V* characteristics depict *quasi-reversible* or *irreversible electron transfer* systems.

Reversible systems are often encountered in sensing measurements when the concentration of the target analyte is high in the environment in comparison of the concentration of the molecules interacting on the surface of the electrodes. In such cases, a repeating cycle gives the same response and reversing the voltage sweep produces a mirrored *I-V* curve. Sensors based on reversible reactions can be reused as their surface dose not change.

In **Fig. 3.25** the IV curves for the Fe^{3+}/Fe^{2+} redox system voltamogramms are recorded as the reduction rate constant (k^0) is changing at a constant applied voltage change rate.

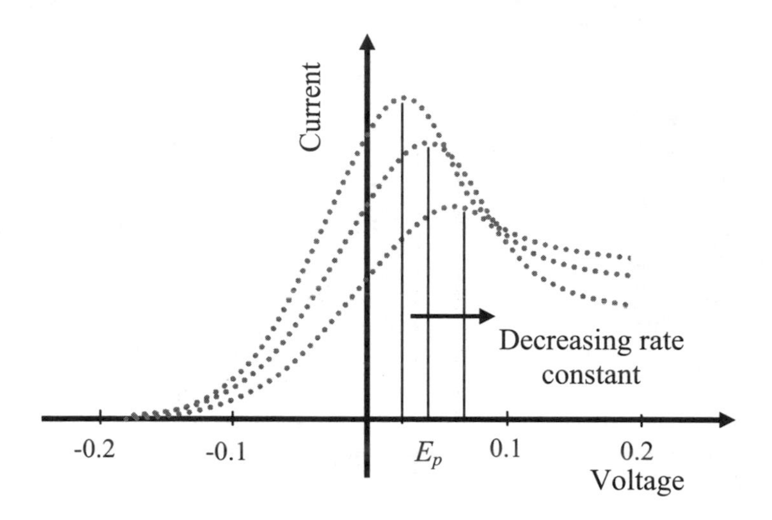

Fig. 3.25 The Fe^{3+}/Fe^{2+} redox system voltammograms when the reduction rate constant is changing at a constant applied voltage change rate.

Decreasing the rate constant decreases the concentrations of ions at the electrode surface and slows down the reaction kinetics. In this case, the equilibrium is not established rapidly. As a result, the position of the current peak shifts to the higher voltages upon the reduction of the rate constant. However, decreasing the rate constant makes the system less reversible.

Generally charge transfer reactions are reversible if $k^0 > 0.1 - 1$ cm/s and irreversible for $k^0 < 10^{-4} - 10^{-5}$ cm/s. They are referred as quasi-reversible for values that fall between reversible and irreversible.

If the target analyte concentration, during the sensing measurements or the surface of electrodes, alters, it results in a non reversible electrochemical interaction. Many disposable electrochemical sensors for medical applications are based on irreversible interactions.

Capacitive current

In addition to the diffusion current, capacitive currents also exist in electrochemical processes. The double layer forms a capacitive dielectric region between the bulk analyte area and the surface of the electrode. This capacitance is known as the *double layer capacitance* and its value is proportional to the surface area of plates. The capacitive current is essentially unwanted in most of the sensing applications as the sensing information is generally extracted from the diffusion current curves.

In addition to the capacitive current, there are also interferences such as adsorption currents caused by adsorption/desorption of molecules on electrodes. The most common interferent molecules are surfactants.

Cyclic voltammetry is similar to LSV, with a difference that the potential applied to a working electrode is frequently altered in time.(a triangular waveform is shown in **Fig. 3.26**). The voltage change rate is in the range of 0.1 to 10000 mVs^{-1} for electrodes with the area of approximately 1 mm^2. [12] This voltage repeatedly oxidizes and reduces the species located within the diffusion layer near the surface of electrode.

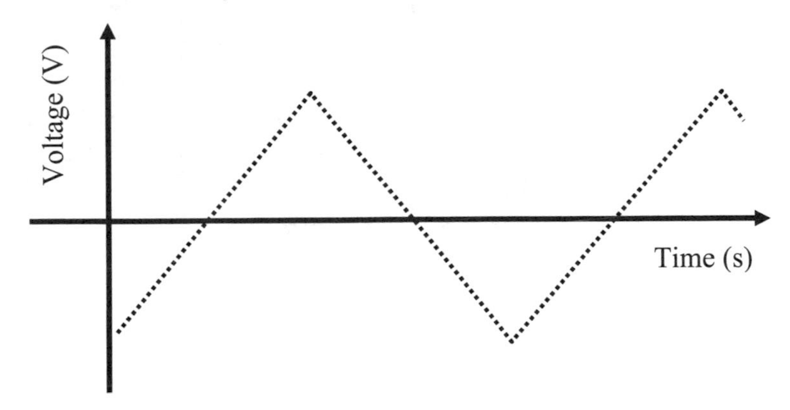

Fig. 3.26 A typical waveform used in cyclic voltammetry.

Recently, the emergence of powerful data acquisition systems, measurement equipment and associated microelectrodes, has made it possible to increase the voltage change rate and measure extremely small currents. As a result it is now possible to identify species that exist for just a few nanoseconds, and even measure individual electron transfer reactions.[12] The *I-V* characteristics also provide a powerful means to study the redox reaction energies to investigate the dynamics and reversibility of the electron transfer, as well as the rates of coupled chemical reactions.[12] These advantages have ensured that *I-V* measurements are widely adopted for studying biochemical/chemical reactions, environmental sensing, and the monitoring of industrial chemical components.

Fig. 3.27 shows a typical cyclic highly reversible voltammogram. When the voltage is altered, before the onset of the E_i, voltage, no measurable current is observed. At this point, the voltage of the working electrode reaches the threshold value causing the reduction of the target species. This generates a current, associated with the reduction process. Following this, the current increases rapidly as the concentration of near surface free ions decreases where the diffusion current reaches a peak at $E_{p,c}$. As the potential decreases further, the thickness of the diffusion layer increases. This results in decay of the current. In this example the final voltage is centered at 0 V.

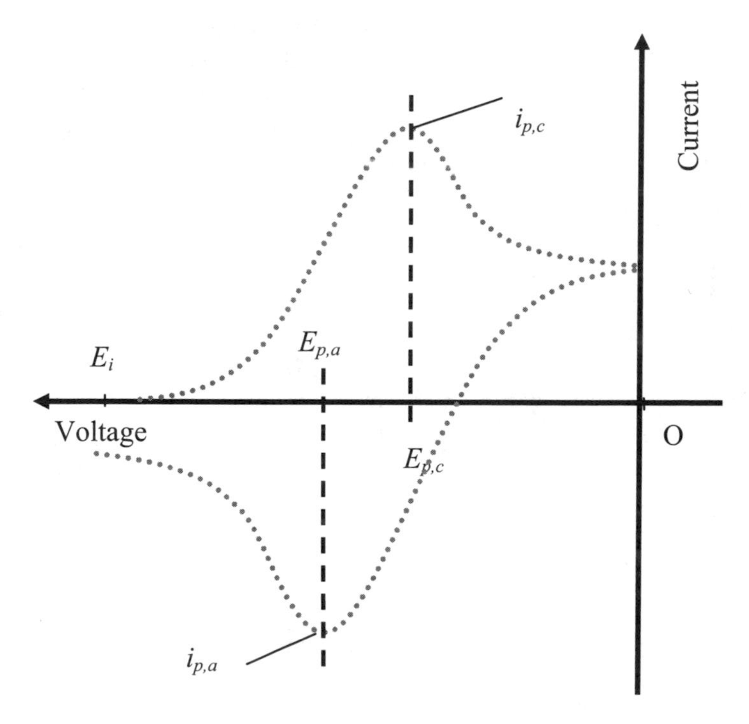

Fig. 3.27 A cyclic voltammogram of a reversible, one-electron redox reaction.

When voltage reaches 0 V the voltage polarity alternates and increases again. However, the value of current is kept constant and a reduction reaction proceeds at the electrode's surface, which is caused by the residual charges within the diffusion layer. This cathodic current continues to decrease. At E_f, the overall number of oxidation interactions becomes equal to the number of the reduced species adjacent to the electrode surface and the current will become zero. Increasing the voltage further depletes of reduced material at the surface of electrode. It reaches a minimum current correspond to a voltage peak of $E_{p,a}$. As the potential returns to E_i, the magnitude of current decreases as the thickness of the diffusion layer increases.

A large number of enzyme-based voltammetric commercial sensors are available. The most famous example is the glucose sensor which is extensively used in medical tests. Cyclic voltammetry can also be utilized to interpret more complex behavior of electrochemical interactions. The nature of the double layer changes if the electrodes are coated with nanomaterials or nanomaterials are deposited during the electrochemical process. The effects of nanostructured thin films on the surface of electrodes and the interactions which occur at their surface are prominent as they are located within the double layers.

Fig. 3.28 shows the cyclic voltammetry of electrodes made of polyaniline nanofibres (which is a conductive polymer) that was obtained on a three-electrode system in 1 M HCl solution containing 1 M NaCl.[18] The scanned potential started from 0.5 to –0.5 V versus saturated Ag/AgCl reference electrode with a scan rate of 10 mV/s. In the potential range of –0.5 to 0.5 V two cathodic peaks (P_1 and P_2) are observed. The value of P_1 diminished in successive scan cycles, while P_2 increased. The cyclic voltammogram of polyaniline nanofibres also exhibits two anodic peaks, P_3 near 0.15 V and P_4 near 0.3 V. Similar to the two cathodic peaks, P_3 diminished in the sequence scans while P_4 increased. Such a tendency can be ascribed to possible changes of the layer structure during continuous potential cycling.

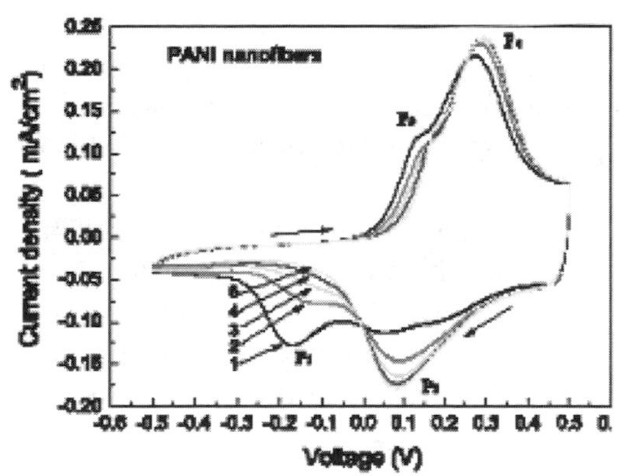

Fig. 3.28 Cyclic voltammograms of PANI nanofibres. Curves 1–5 correspond to different scan cycles. The scan rate is 10 mV/s. Reprinted with permission from the Institute of Physics Journals publications.[18]

3.4.8 An Example: Stripping Analysis

In *stripping analysis*, analyte from a diluted solution is adsorbed into a thin film of Hg or other electrode material, usually by electro-deposition. The electroactive species is then striped from the electrode by reversing the direction of the voltage sweep. Current measured during the oxidative removal is proportional to the analyte's concentration. Stripping is one of the most sensitive methods for the sensing of heavy metal ions.

In one of the early examples Florence[19] developed an anodic stripping technique in which a very thin mercury film is formed on a polished carbon substrate. He added mercuric nitrate to the sample solution and electrodeposited mercury and trace metals simultaneously. The trace metals were then anodically stripped from the mercury.

3.5 Solid State Transducers

Solid state transducers are identified as devices containing semiconducting and insulating materials in a solid form. They are made of *metal-semiconductor* and/or *semiconductor-semiconductor junctions* and work by monitoring changes in the device's *electrical field distribution* in the presence of the measurand. Parameters that can be directly measured include voltage, current, capacitance and impedance, and from them, a myriad of electrical properties (such as electrical conductivity, barrier height, carrier concentrations, etc.) can be derived.

Solid state transducers are built upon well established micro-fabrication strategies, originating from the silicon microelectronics industry's development of complex systems incorporating millions of semiconducting devices.

Semiconductor based devices such as diodes and transistors are naturally sensitive to environmental changes, when the numbers of free electrons and holes change or the electric field distribution is altered in response to external stimuli. In this section, some of the most popular solid state transducers are introduced and their application in sensing will be presented.

3.5.1 *p-n* Diodes or Bipolar Junction based Transducers

The *p-n diodes* or *bipolar junction transducers* (*BJTs*) are based on semiconductor-semiconductor junctions. When implemented in sensing applications, electrical properties, such as barrier height, carrier concentration, etc. can be altered by the presence of a measurand. These changes result in a change in relationships between current, voltage and accumulated charge. Such junction type devices are based on semiconductors which are heavily doped so that the materials have a majority of free electrons (*n*-type) or free holes (*p*-type).

When *p*-type and *n*-type semiconductors are placed next to each other to form a *p-n* junction, the majority carries in the *p*-type material (holes) diffuse a certain distance into the *n*-type material, and similarly at the *n*-type material, majority carries (electrons) diffuse into the *p*-type material. Once this diffusion is balanced, a depletion region forms (**Fig. 3.29**). A potential barrier then exists between the *p*- and *n*-doped materials, which must be exceeded for the current to conduct. Devices having one *p-n* junction are termed diodes, whilst those contained two, in the form of *p-n-p* or *p-n-p* junctions, are referred to as BJTs. BJT's have the added advantage of internal current amplification as well.

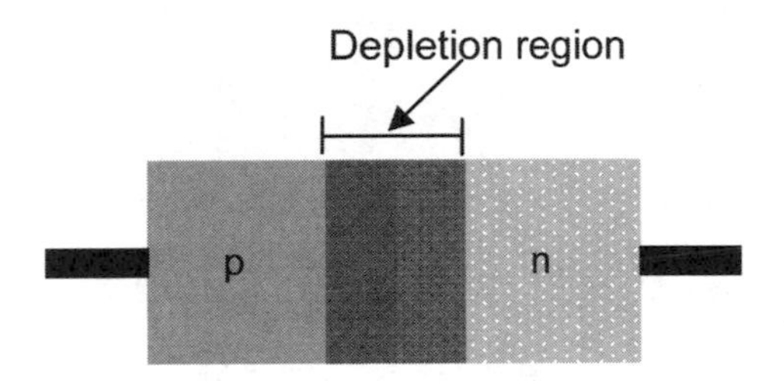

Fig. 3.29 A Schematic reorientation of a *p-n* junction.

From the *Shockley equation*, the current flowing is given by:[20]

$$I(V) = I_{saturation}(e^{qV/nkT} - 1), \tag{3.76}$$

where q is the electron charge, V is the voltage, n is the ideality factor, k is Boltzmann's constant, T is the temperature in Kelvin, and $I_{saturation}$ is the saturation current of the device..

Devices such as diodes and BJTs are commonly employed for monitoring charge[21] and widely used in sensing electromagnetic irradiation. The application of these devices in monitoring irradiation has already been described in Chap. 2 in the section pertaining to photoconductive and photovoltaic devices. They may also be employed, albeit less commonly, for chemical and pressure sensing applications.

Diode and transistor based transducers are extensively used in irradiation spectroscopy,[22] which will be described in further detail in Chap. 5. In many such spectroscopic measurements, the amount of light to be monitored may be very low. Using a photodiode in conjunction with an amplifier may rectify such limitations. However, low irradiation intensity produces small currents which are inherently difficult to amplify externally, if electrical noise is to be avoided. In such situations, phototransistor may be used as an alternative to external signal amplification. A transistor can amplify a current internally through its base (**Fig. 3.30**) owing to its internal current gain, according to:

$$I_{collector} = \beta I_{base} \,. \tag{3.77}$$

where β is the current gain. Irradiation impinges directly at the *p-n* junctions. One of the junctions is reversed biased which generates a large current from the changes of the forward biased *p-n* junction current. Owing to recent developments in fabrication technology, electrical noise in such transducers has been dramatically reduced.

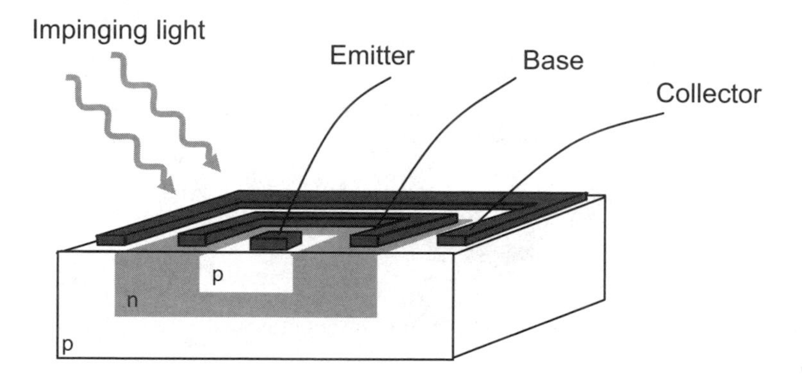

Fig. 3.30 A schematic of a phototransistor.

3.5.2 Schottky Diode based Transducers

A diode constructed from a metal-semiconductor junction is called Schottky diode. The metal-semiconductor junction forms a rectifying barrier that only allows current to flow in one direction. As shown in **Fig. 3.31**, for an Schottky diode of *n*-type semiconductor, the barrier height, ϕ_b, depends on the difference between the work function of the metal, ϕ_m, and electron affinity of the semiconductor, χ_s.[20] In most cases, the barrier height is controlled by the density of surface states located in the metal-semiconductor interface.[23]

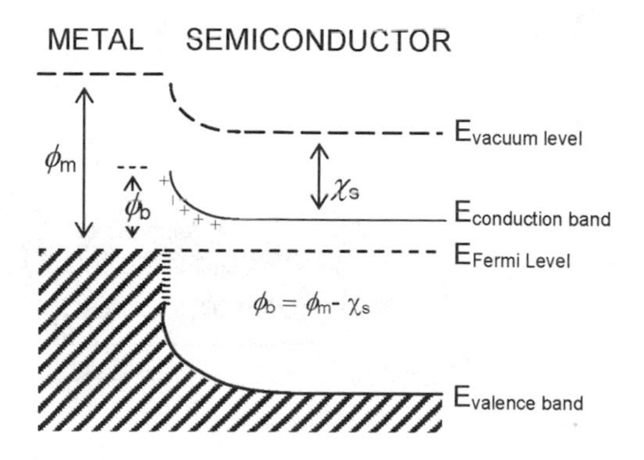

Fig. 3.31 Energy band diagram of an ideal Schottky diode.

A Schottky junction device also follows the Shockley equation with the saturation current defined as:[20]

$$I_{saturation} = SA^{**}T^2 \exp(-q\phi_b / kT), \tag{3.78}$$

where S is the area of the metal contact (cm^2), A^{**} is the effective Richardson's constant ($Acm^{-2}K^{-2}$) and ϕ_b is the barrier height An example of the set-up of a Schottky diode based transducer can be seen in **Fig. 3.32**. In typical operations, the current and voltage of the Schottky diode are measured simultaneously to produce a current-voltage (*I-V*) characteristic.

Depending on the selection of the materials utilized in the fabrication of the diode, a change in pressure, ambient temperature, or presence of different gases (which changes the deletion region distribution) causes the *I-V* characteristic of the Schottky diode change, as seen in **Fig. 3.33**. The response may be obtained when operating the device at a constant current and thereby measuring the voltage shift, or by operating it a constant voltage and measuring the current change. Also, by correlating the measured *I-V* characteristics to Eq. (3.78), the change in barrier height in the presence of the measurand can also be experimentally derived.

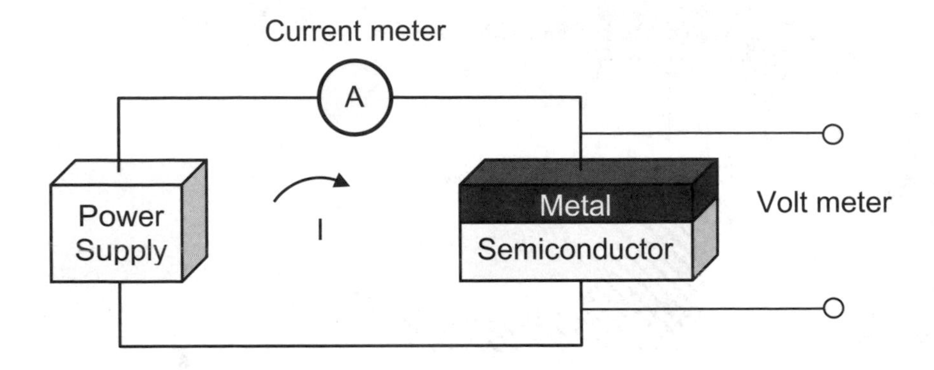

Fig. 3.32 Example of a Schottky diode based sensing system.

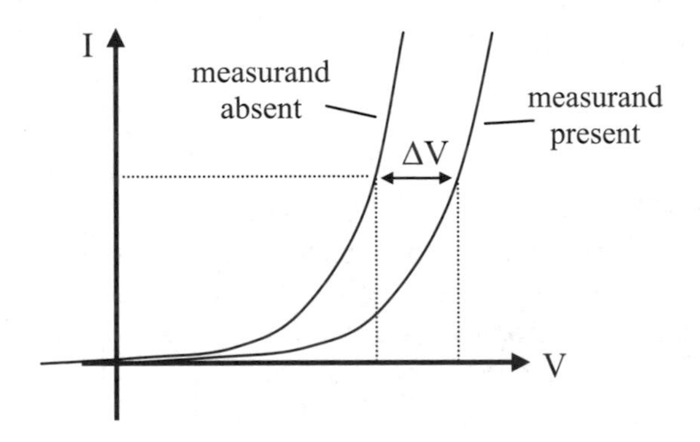

Fig. 3.33 A typical the response of a Schottky diode based device when current is held constant.

The semiconducting materials commonly employed for Schottky diodes include silicon, gallium arsenide and silicon carbide, whilst metals utilized as the Schottky contact include from Pd, Pt and Al. Quite often in chemical sensing applications, a very thin layer, typically a few nm, is added between the metal-semiconductor junction. This serves to increase the sensitive and selectivity towards the target analyte.[24] Semiconducting metal oxide layers, such as SnO_2, Ga_2O_3, WO_3, are popular choices for such a layer, as they show high sensitivity towards gases like CO, CH_4, H_2 and O_2. Oxidation or reduction of this metal oxide layer, which is caused by the exposure to the target gas, changes the properties of Schottky junction which results in the change of device's electrical charateristics.

3.5.3 MOS Capacitor based Transducers

The *metal oxide semiconductor* (*MOS*) *capacitor* consists of a metal deposited over a thin oxide layer, which is in turn deposited over a semiconductor.[20] A typical example of a MOS capacitor based transducer can be seen in **Fig. 3.34**. The oxide layer is typically a native oxide of the semiconductor (e.g. SiO_2 on Si). When a voltage is applied across it, the device appears like a parallel plate capacitor. The dielectric properties of the oxide and semiconductor change when the device is exposed to different environmental conditions. Capacitance is a function of the dielectric properties, and hence capacitance-voltage (*C-V*) characteristics of such devices are generally obtained during a sensing experiment.

A time varying AC signal (**Fig. 3.34**) is applied to the MOS device and the impedance is measured, from which the capacitance is obtained.

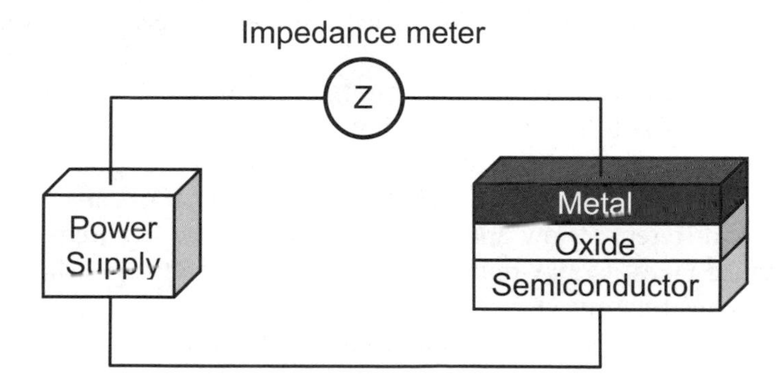

Fig. 3.34 Example of a MOS capacitor based transducer system.

When positive or negative voltages are applied to MOS capacitors, three regimes may exist on the semiconductor surface. These are *inversion, depletion* and *accumulation*. **Fig. 3.35** shows these regimes for an *n*-type semiconductor. Applying a positive voltage to the metal attracts electrons (majority carriers for *n*-type semiconductors) from the substrate to the oxide-semiconductor interface, causing them to accumulate there. The top of the conduction band bends downwards and current flows through the structure. Applying a small negative voltage causes the bands to bend upwards, repelling electrons until they are depleted from the interface. Decreasing the voltage further causes holes (minority carriers for *n*-type semiconductors) to accumulate at the interface, and the device is said to be in the inverted regime.[20]

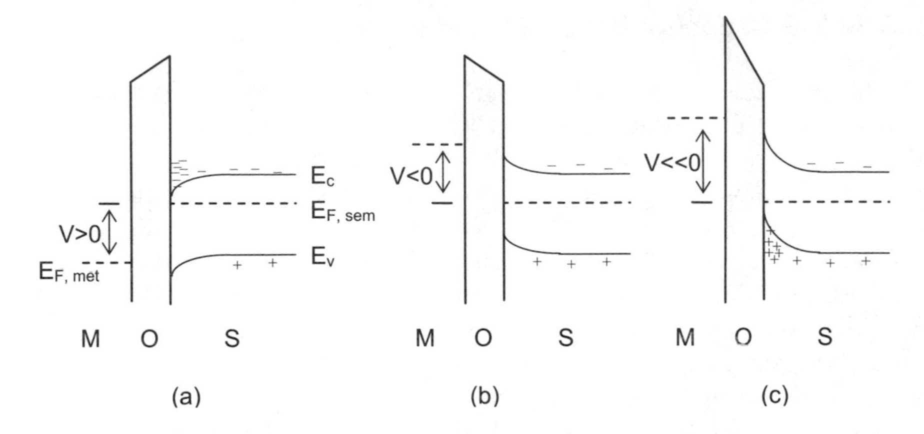

Fig. 3.35 Energy band diagrams for ideal MOS devices under different bias conditions, showing the three regimes (a) accumulation, (b) depletion, and (c) inversion (E_c, $E_{F,sem}$ and E_v are the conduction, Fermi, valance energy levels, respectively).

As a consequence of these three operating regimes, the *C-V* relation of MOS capacitors is non-linear. Furthermore, the *C-V* relation is frequency dependent because of the time required for the generation of minority carriers. As seen in **Fig. 3.36**, the *C-V* characteristic of a MOS capacitor device is vastly different at low and high frequencies. For low frequencies, the capacitance takes its maximum value for most bias voltages, however during inversion it approaches a minimum. For high frequencies, the capacitance transitions from a minimum value during inversion to a maximum value in accumulation.

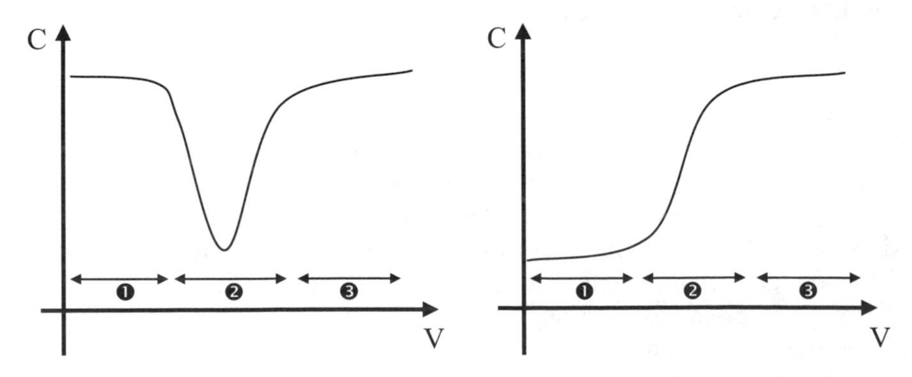

Fig. 3.36 *C-V* relations for MOS devices in the three regimes (1) inversion, (2) depletion, and (3) accumulation. The *C-V* curve was measured at: *Left:* a low frequency; *Right:* a high frequency.

Exposure to measurands, such as temperature, humidity, electromagnetic radiation, chemical species, mechanical stress etc. may result in a voltage shift in the *C-V* characteristics of the device, as illustrated in **Fig. 3.37** for a C-V curve obtained at a high frequency.

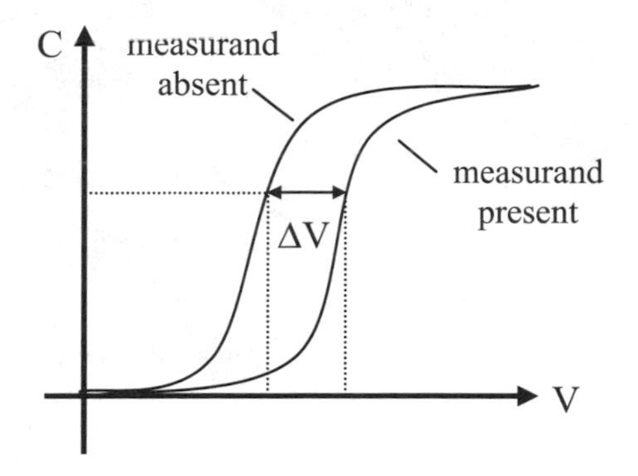

Fig. 3.37 A typical response of a MOS capacitor based device operating at a high frequency.

Similar devices can be fabricated in which the oxide layer is replaced with an insulating material, and such devices are called *metal insulator semiconductor* (*MIS*). Under DC conditions, MOS and MIS devices can be operated as Schottky diodes.

3.5.4 Field Effect Transistor based Transducers

A *field effect transistor* (*FET*) is a transducer in which the flow of current, between two of its terminals, is controlled by an applied voltage at a third terminal. As a transducer, it can be employed for converting chemical, physical, and electromagnetic signals into a measurable current.

The most well known FET is the *metal oxide field effect transistor* (*MOSFET*)) shown in **Fig. 3.38**. The current flowing between the *drain* and *source* semiconducting electrodes is controlled by the electric field generated by a third electrode, the *gate*, which is located between them. The gate electrode is insulated against drain and source by an oxide layer. The device shown is an *n-channel enhancement MOSFET*. If the gate is made sufficiently positive with respect to the source (larger than the transistor's threshold voltage – $V_{threshold}$), electrons are attracted to the region below the gate, and a channel of *n*-type material is created. This may trigger by the application of a voltage, the adsorption of ions on the gate surface, etc.

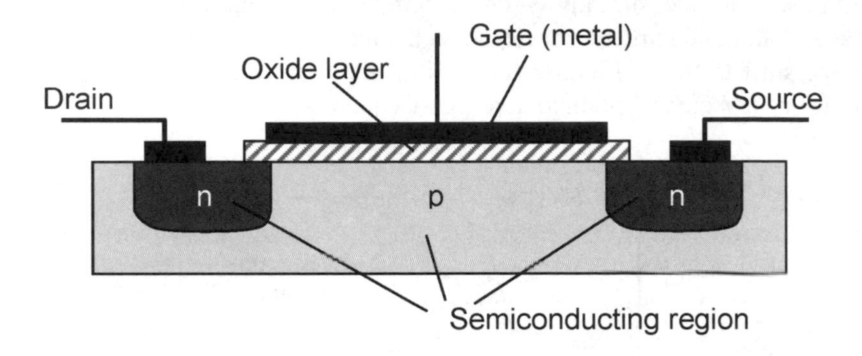

Fig. 3.38 Schematic representation of an n-channel enhancement MOSFET.

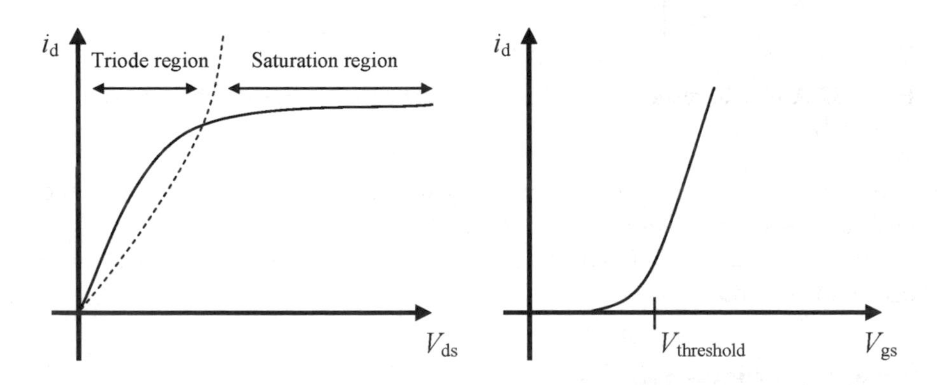

Fig. 3.39 Input and output characteristics of an *n*-channel enhancement MOSFET.

The drain-source current characteristics of an *n*-channel enhancement MOSFET for different gate-source and drain-source voltages (V_{gs} and V_{ds}) are shown in **Fig. 3.39**. This current, I_{ds}, takes on different expressions for different operating regions, and is defined below:

$$I_{ds} = \frac{1}{2} C_{ox} \mu \frac{W}{L} \left(V_{gs} - V_t\right)^2 \quad \text{(saturation region)}, \tag{3.79}$$

$$I_{ds} = C_{ox} \mu \frac{W}{L} \left[\left(V_{gs} - V_t\right) V_{ds} - \frac{1}{2} V_{ds}^2 \right] \quad \text{(triode region)}, \tag{3.80}$$

where with C_{ox} is the capacitance of the oxide per unit area, W and L are the width and the length of the channel, respectively, and μ is the electron mobility in the channel. V_{gs} and V_{ds} are the current applied between gate and source and drain and source, respectively. Furthermore, the threshold voltage, V_t, is defined as: [25,26]

$$V_t = \frac{\phi_m - \phi_s}{q} - \frac{Q_{ox} + Q_{ss} + Q_b}{C_{ox}} + 2\phi_f , \qquad (3.81)$$

where ϕ_m and ϕ_s are the metal and semiconductor workfunctions respectively. Q_{ox} Q_{ss} and Q_b are the accumulated charges in the oxide, oxide/semiconductor interface, and the depletion charge in the semiconductor, respectively. Finally, the last term, ϕ_f, depends on the doping level of the semiconducting material. As a consequence V_t being dependent on charge and capacitance, a FETs *I-V* characteristics are strongly affected by changes in humidity, ions, chemical species and the presence of dielectric materials in the environment.

FETs may be implemented for applications in liquid media, as shown in **Fig. 3.40**, in which the metal gate is replaced with a reference electrode in a liquid touching the gate metal oxide layer. Insulating layers are added to isolate the connections to the drain and source from the liquid. A sensitive layer is deposited over the oxide layer, forming the gate, and as a result of chemical reactions occurring on this surface when free electrons or holes in the semiconductor become attracted or repelled to the gate surface.

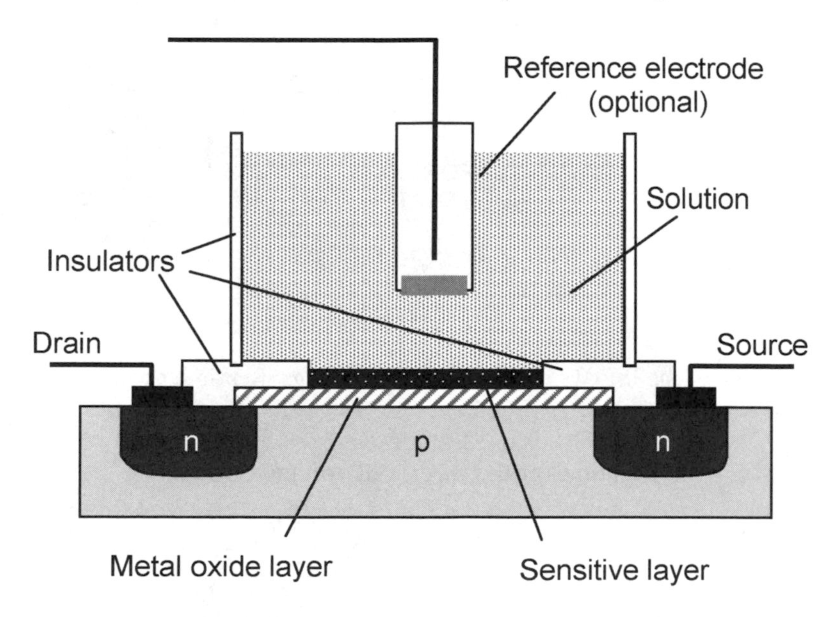

Fig. 3.40 A schematic representation of an ISFET.

The first FET sensors were developed in early 1970 at the Stanford University by Wise et al.[27] **Fig. 3.41** shows one of the original design drawings.[25] They used the device for the measurement of pH. As the device is very sensitive to the presence of ions in the analyte, for measuring the redox interactions on the sensitive layer a reference electrode is also needed. A MOSFET incorporating a reference electrode is called an *ion-sensitive field effect transistor* (*ISFET*).[26] Hence, the incorporation of a reference electrode is deemed necessary to make an electrochemically sensitive device.

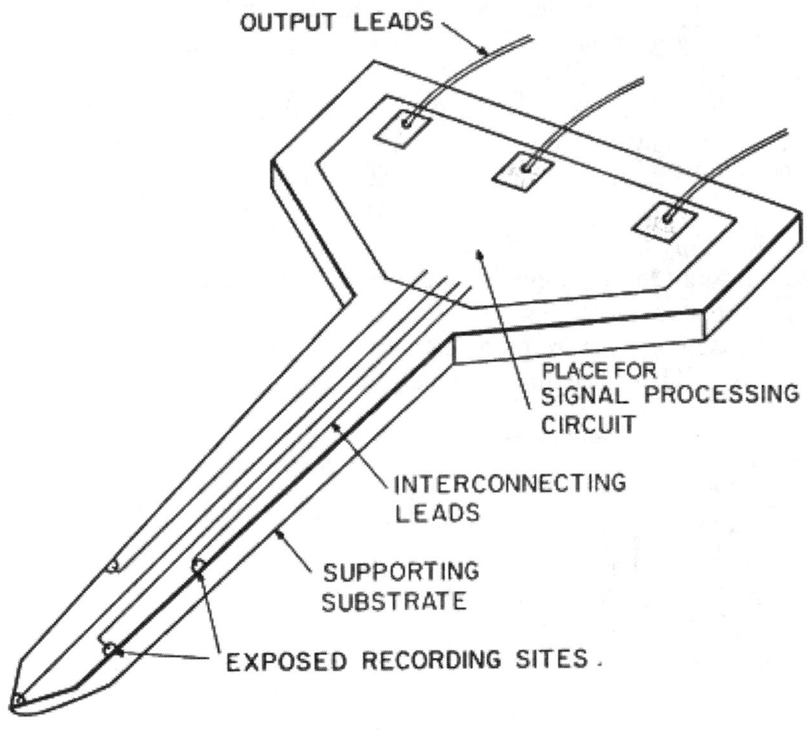

Fig. 3.41 One of the ISFETs original design drawings. Reprinted with permission from Elsevier publications.[25,27]

For ISFET the equation for the threshold voltage changes to:[25,26]

$$V_t = E_{ref} - \psi + \chi^{sol} + \frac{\phi_s}{q} - \frac{Q_{ox} + Q_{ss} + Q_b}{C_{ox}} + 2\phi_f, \tag{3.82}$$

There are two new parameters that can be seen in this equation: the interfacial potential, E_{ref}, and the potential at the solution/oxide interface, and $\psi+\chi^{sol}$. The term $\psi+\chi^{sol}$ consists of ψ, which is a function of the analyte's pH and χ^{sol} which is the surface dipole potential. ϕ_m, the workfunction of the gate metal, does not appear in this equation anymore and its effect is included in E_{ref}.

The I_d-V_{ds} curves of an ISFET as function of the pH of the solution is shown in **Fig. 3.42**.[27] The sensor response is due to the fact that ψ is a function of pH, which is the chemical input parameter as $\psi(pH)$.

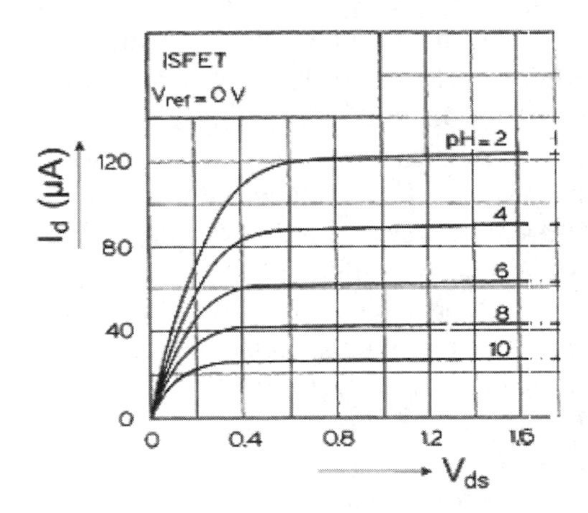

Fig. 3.42 I_d-V_{ds} curves of an ISFET vs the pH change. Reprinted with permission from Elsevier publications.[27]

Bousse et al.[28] developed an empirical equation for $\psi(pH)$ derived from site-dissociation and double-layer models. The equation is function of pH_{pzc}, which is the value of the pH when the metal oxide surface is electrically neutral and β, which can be measured from device's sensitivity.

$$\psi(pH) = 2.3\frac{kT}{q}\frac{\beta}{\beta+1}(pH_{pzc} - pH). \tag{3.83}$$

This equation describes a sub-Nernstian response for the device. It is a function of $[H^+]_s$ which represents the surface concentration of H^+ ions. van Hal and Eijkel[29,30] described the relation for β using the capacitance equation Q=CV. As a result, ψ can be expressed in a sensitivity factor α, resulting in the equation:

$$\psi(pH) = -2.3\frac{RT}{F}\alpha(pH_{bulk} - pH_{bulk-initial}), \tag{3.84}$$

with:

$$\alpha = \frac{1}{\left(2.3\frac{kT}{q^2}\right)\left(\frac{C_s}{\beta_s}\right)+1}, \tag{3.85}$$

where β_s is the surface buffer capacity (the ability of the oxide surface to deliver or take up protons), and C_s is the differential double-layer capacitance. Eq. (3.85) describes that as α approaches 1, the maximum sensitivity of 59 mV for one unit of pH (at 298°K) can be obtained. $\alpha = 1$ is reached for the large values of surface buffer capacity β_s and/or low values of C_S. For $\alpha < 1$ a reduction in sensitivity is expected. It is obvious that if the ion concentration at the surface $[H^+]_s$ changes with the bulk ion concentration the sensitivity will decrease. $[H^+]_s$ should be independent of the bulk pH, which translates as a surface with a large buffer capacity.

Many biosensors based on the chemical reactions on the surface of ISFETs are enzyme based (enzyme based sensors will be described in Chap. 7), which are called *Enzyme field efect transistors* (*ENFET*). By applying an enzyme-entrapping membrane on top of the ISFET gate, the device can perform as an enzyme sensor.

Considerable effort has been made to develop different types of ENFETs since the introduction of an ENFET for penicillin.[31] The performance of ENFET biosensors is greatly affected by the integration mechanism of the enzymes with the ISFET. Immobilizing enzymes on the gate surface of ISFETs by cross-linking with bovine serum albumin and glutaraldehyde is one of the most popular ways to establish a sensitive layer that can be employed to create cross-linked enzyme matrices.[32-36] To extend the dynamic range and/or increase the selectivity, additional membranes can be added.[37-41] Another method for ENFET preparation is the entrapment of enzymes with polymers.[42,43]

3.6 Acoustic Wave Transducers

Acoustic wave transducers employ piezoelectric materials in which mechanical waves are launched. As described in Chap. 2, piezoelectric phenomenon occurs in crystals, which do not have a center of symmetry. Applying stress to such crystals deforms their lattices and electrical fields are produced, and vice versa. It is believed that the first acoustic wave device was developed in 1921 when Cady utilized a quartz resonator for stabilizing electronic oscillators.[44]

Generally acoustic waves in solids can be placed in two categories: *Bulk acoustic waves* (*BAW*s) in which acoustic waves propagate in the bulk of a solid, and *surface acoustic waves* (*SAW*s) in which their propagation is confined to the region near the surface of the solid.

When employed for sensing, perturbations in the acoustic wave's fields are measured. As the acoustic wave propagates through the bulk or on the surface of the material, any changes to the characteristics of the propagation path affect the velocity and amplitude of the wave. These changes can be monitored by measuring the frequency or phase characteristics. They can be correlated to the corresponding physical or chemical quantity being measured.

3.6.1 Quartz Crystal Microbalance

The *quartz crystal microbalance* (*QCM*) is a BAW resonator that is based on the quartz crystal, which was the first piezoelectric material to be discovered. A schematic representation of a QCM is shown in **Fig. 3.43**. Generally, quartz substrate is machined into thin disks with metal pads deposited on both sides, at which electrical signals are applied.

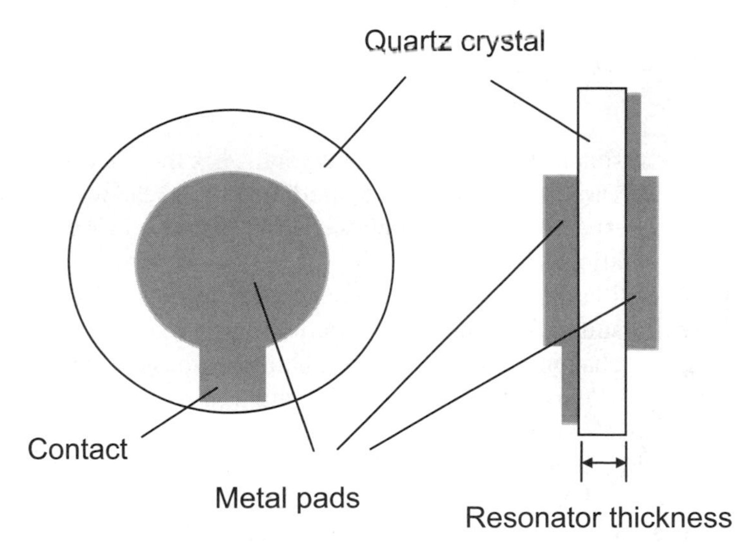

Fig. 3.43 Top and side views of a QCM.

The piezoelectric crystal transforms the applied electric signal on the metal pads to acoustic waves. These acoustic waves are trapped within the top and bottom boundaries of the crystal, reflecting back and forth in the bulk of the quartz, resulting in resonation. The addition of mass on the surface of a quartz crystal increases its thickness. As a result, the resonator will hold longer wavelength standing waves, which accommodates smaller resonant frequencies (**Fig. 3.44**).

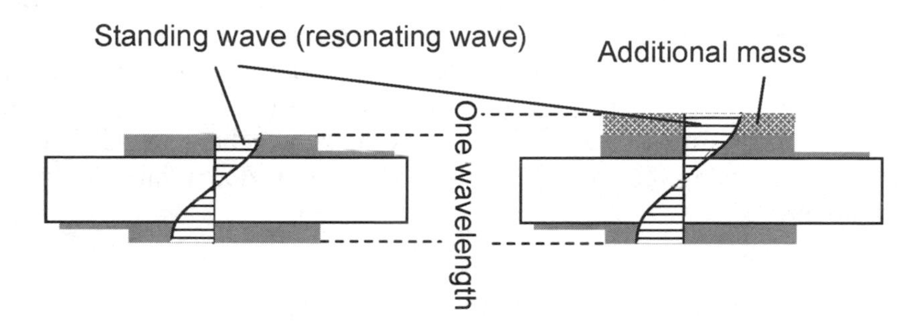

Fig. 3.44 The effect of resonator thickness increase on the oscillation cavity.

The relationship between the change in the oscillation frequency, Δf, of a QCM to the change in mass added to the surface of the crystal, Δm, is given by the *Sauerbrey equation*:[45]

$$\Delta f = \frac{-2\Delta m f_0^{\,2}}{A\sqrt{\rho\mu}} = \frac{-2\Delta m f_0^{\,2}}{A\rho\ v}, \tag{3.86}$$

where f_0 is the resonant frequency of the crystal, A is the area of the crystal, and ρ, μ and v are the density, shear modulus and shear wave velocity of the substrate, respectively. As can be seen, any increase in Δm results in a decrease in operational frequency Δf.

Clearly the oscillating frequency's dependence on mass change makes the QCM ideally suited for sensing applications. The mass sensitivity can be defined as the change in frequency per change in mass on the unit area of the device. QCM mass sensitivity can be enhanced by adding a sensitive layer on its surface. As observed in Eq. (3.86), increasing the operational frequency (or the reduction in the crystal thickness) will increase the QCM's sensitivity. For a 10 MHz device, the mass detection limit of a QCM can be calculated to be approximately less than 1 ng/cm^2.

In addition to sensing mass changes, the electrical boundary condition perturbations also change the piezoelectric properties of the device. This will therefore change its resonant frequency. Consequently, such a device can be used for monitoring mass changes as well as conductivity changes.

QCMs have been employed for measuring mass binding from gas-phase species for moisture and volatile organic compounds sensing[46,47] and environmental pollutants.[48] They are also employed for monitoring gas concentrations, redox reactions occurring on metal oxide or polymeric sensitive layers deposited on top of them.[49-51] QCMs operating in liquid media were developed in 80s to measure viscosity and density.[52] They have also been successfully realized as commercially available biosensing applications. Further examples concerning these types of biosensors will be presented in Chap. 7.

The efficiency with which a piezoelectric material can transform mechanical waves to electromagnetic waves, and vice versa, is found from its piezoelectric coupling coefficient, k^2. In the past few decades, several crystals have emerged that exhibit stronger piezoelectric effects than quartz, These include: lithium niobate ($LiNbO_3$), lithium tantalite ($LiTaO_3$) and recently lanagsite (LNG). $LiNbO_3$ and $LiTaO_3$ have k^2s that are almost an order of magnitude larger than that of quartz. These materials, however, do have drawbacks, such as not being suitable for fabricating bulk type devices due to their fragile structure and low quality factor (Q). However, they are being rigorously investigated for surface acoustic wave applications.

3.6.2 Film Bulk Acoustic Wave Resonator (FBAR)

Film bulk acoustic wave resonators (*FBARs*) represent the next generation of bulk acoustic wave resonators. They are comprised of thin films and have much smaller dimensions than QCMs. They have a relatively high operating frequency, which is the main reason for their higher mass sensitivity. In addition, their fabrication is compatible with current Micro-Electromechanical System (MEMS) standard technologies.[53,54] A typical FBAR is built on a low-stress, inert membrane (such as silicon nitride) with a piezoelectric film sandwiched between two metallic layers (**Fig. 3.45**).

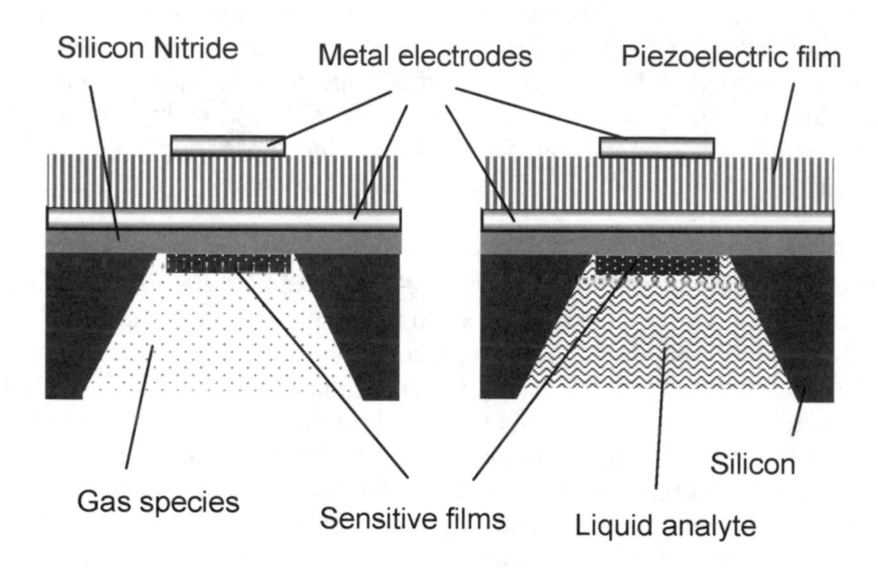

Fig. 3.45 Cross-sectional view of a typical FBAR used for mass sensing in: (a) gas media, (b) liquid media.

FBARs generally have a resonant quality factor Q in a range of 20–1000 at operational frequencies of 1 GHz which is much lower than their QCM counterpart. Unfortunately, this low Q translates into a noisier and less stable oscillation. If the membrane thickness is small with respect to the acoustic wavelength, then the frequency change Δf is resulting from added mass, Δm, of adsorbed target analyte, can be found from the Sauerbrey–Lostis approximation by:[45,55,56]

$$\frac{\Delta f}{f_0} = -\frac{\Delta m}{m_0}, \tag{3.87}$$

where $m_0 = \rho Sd$, is the mass of the resonator for which ρ is its density, S is its surface area, and d is its thickness, and $f_0 = v_p/2d$, with v_p being the acoustic wave phase velocity. The fractional frequency change $\Delta f/f_0$, produced by the mass of analyte adsorbed per unit surface ($\Delta m/S$), is equal to:

$$\frac{\Delta f}{f_0} = -\frac{1}{\rho d}\left(\frac{\Delta m}{S}\right). \tag{3.88}$$

As a result, the frequency shift is:

$$\Delta f = -\frac{v_p}{2\rho_p d^2}\left(\frac{\Delta m}{S}\right). \tag{3.89}$$

From Eq. (3.92), for a given $\Delta m/S$, the frequency shift magnitude increases with increasing the acoustic wave velocity in the piezoelectric resonator medium and decreasing its density. In addition, the frequency shift magnitude also increases as the membrane thickness decreases.

The density, phase velocity and the frequency shift produced by 1ng/cm^2 of analyte for three different piezoelectric materials (AT–cut quartz, c–AlN and ZnO) are presented in **Table 3.1**.[56] The materials comprised of bulk (AT-quartz), thin-film (AlN, ZnO) resonators, with the thickness of d=100 and 1 µm, respectively. The ZnO and AlN thin film resonators show an increase in the frequency, of approximately 16000 and 5000, respectively, in comparison with quartz resonator. It was practically demonstrated that an FBAR with a structure of Al/ZnO/Al/SiN (0.2 µm/2.0 µm/0.2 µm/0.2 µm), has a mass sensitivity of approximately more than 50 times greater that of a QCM operating at 6 MHz.[57]

These examples obviously show the advantage of using FBARs as transduction platforms over conventional QCMs. It is predicted that the FBAR technology will become popular in the near future.

Table 3.1 Acoustic phase velocity and density for three different piezoelectric materials; frequency shift produced by 1 ng/cm^2 of adsorbed analyte ($v_p/2\rho_p d^2$), calculated for a thickness d of the resonator.[56]

Material	v_p (m s^{-1})	ρ_p (kg m^{-3})	(cm^{2}Hz/ng)
AT-Quartz (d=100 µm)	3 750	2648	0.708
c-AlN (d=1 µm)	11 345	3260	34 800
c-ZnO (d=1 µm)	6 370	5665	11 244

3.6.3 Cantilever based Transducers

These transducers consist of micro cantilever that resonate as the result of environmental stimuli or adhesion of particles on the cantilever's surface. The resonant frequency changes as a result of the measurand's presence. These frequencies are typically in the range of 100 MHz to 5 GHz, and depend on the cantilever's dimensions and materials properties. Micro cantilevers can be employed for monitoring parameters such as mass, temperature, heat, electromagnetic radiation, and stress. The shift in resonance as a function of the added mass can be calculated with the following equation:

$$\Delta m = \frac{K}{4\pi^2}\left(\frac{1}{f_1^{\ 2}} - \frac{1}{f_0^{\ 2}}\right), \tag{3.90}$$

where f_0 and f_1 are the operating frequencies before and after adding the mass, respectively, and K is the spring constant for the cantilever.

The cantilever may be composed of layers having different thermal expansion coefficients, in which case a temperature change causes it to bend. Such cantilevers are capable of measuring temperature changes as small as 10^{-5} K.[58] A shift in resonance frequency may also arise from changes in medium viscosity or mass added to the cantilever's surface. Increasing the viscosity, as well as adding mass, will dampen the cantilever's oscillation, lowering its operational frequency.[59] Examples of such changes are shown in **Fig. 3.46.**[60] SEM images of five microfabricated (poly)silicon rectangular cantilevers of different lengths is shown in **Fig. 3.47.**[60]

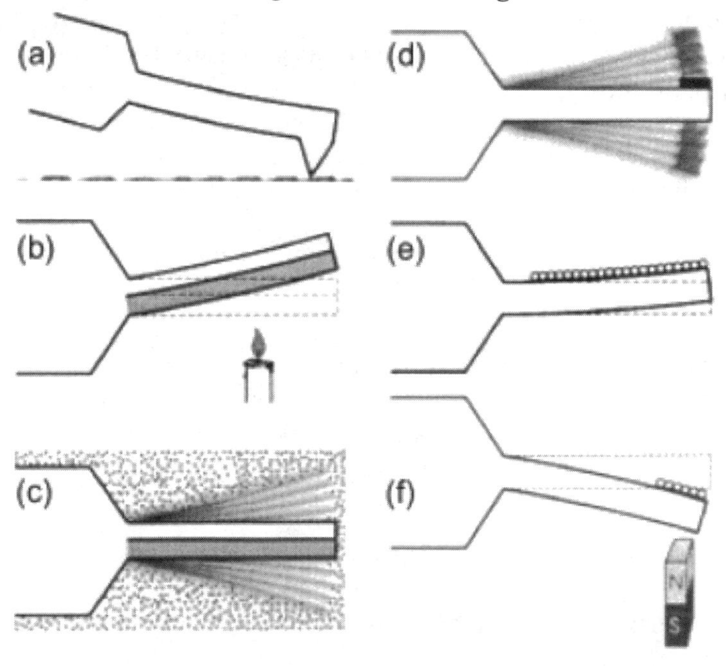

Fig. 3.46 Possible uses of a cantilever transducer (side view) for the measurement of different properties: (a) force; (b) temperature, heat; (c) medium viscoelasticity; (d) mass (end load); (e) applied stress and (f) magnetic measurements of magnetic beads on its surface. Reprinted with permission from Elsevier publications.[60]

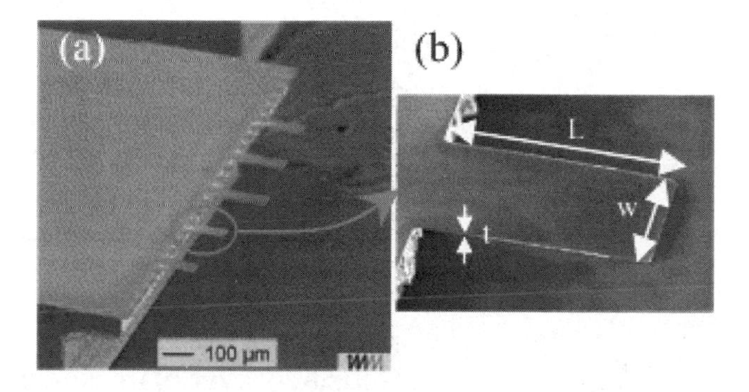

Fig. 3.47 SEM images: (a) five cantilevers (b) zoomed image of a cantilever with length L=100 µm, width w=40 µm and thickness t=0.5 µm. Reprinted with permission from Elsevier publications.[60]

Micro cantilevers can be employed for monitoring biological interactions such as: enzymatic and antigen–antibody interactions, as well as interactions of complementary DNA strands.[60] The minimum detectable mass of a cantilever can be as low as $\approx 10^{-15}$ g which is excellent in comparison to other acoustic wave devices.[61] Unfortunately, when operated in liquid media, the resonance frequency shifts toward lower values and the quality factor decreases dramatically as a result of damping caused by the liquid.[62]

3.6.4 Interdigitally Launched Surface Acoustic Wave (SAW) Devices

Surface acoustic waves (*SAWs*) can be launched using *inerdigital transducers* (*IDTs*) first described introduced by White and Voltmer.[63] Metal thin-film IDTs are patterned on the surface of a piezoelectric crystal that is cut in a manner that allows the propagation of SAWs. As shown in **Fig. 3.48**, an alternating voltage is applied to the input IDT, launching the acoustic wave. It travels along the surface to the output IDT, where the acoustic wave is converted back into an electrical signal. A sensitive layer may be deposited on the active are of the device to provide sensitivity to a target analyte.

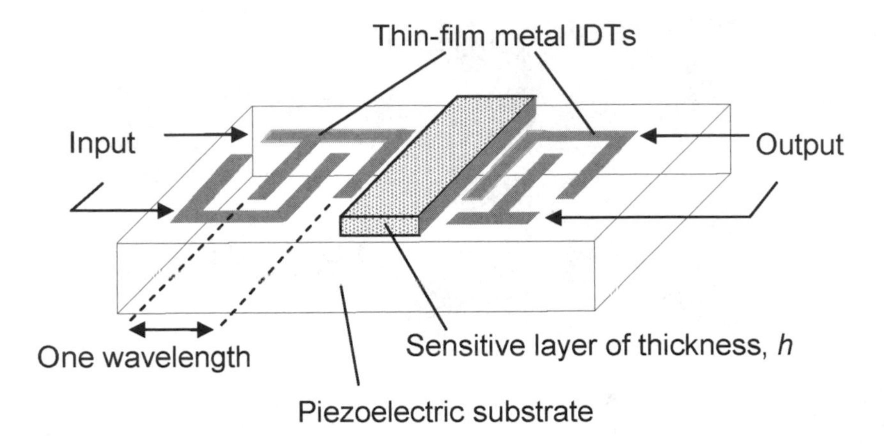

Fig. 3.48 Basic layout of a SAW device.

Depending of the crystal cut and IDT geometry, different propagation modes are possible. The particle displacements can be transverse or longitudinal For *Rayleigh waves*, the particles in the surface move elliptically along the propagation direction, normal to the surface. For *shear horizontal waves*, particles in the surface are displaced parallel to the surface.

In recent years, the discovery and utilization of novel piezoelectric crystals have coincided with the discovery of new SAW operational modes. These include: *leaky SAW* (*LSAW*), *Surface-Skimming Bulk Wave* (*SSBW*) which both are shear horizontal SAWs (SH-SAWs). Other major acoustic propagation modes that can be launched and received using IDTs include: *acoustic plate mode* (*APM*) and *Lamb wave* (or *flexural plate wave- FPW*) which are both bulk acoustic waves.

Through the deposition of layers onto the piezoelectric substrate the SSBW and LSAW modes of operation can be altered. If the propagation speed of the SH waves in the deposited layer is less than that of the substrate, the SH wave can become a near surface confined wave (**Fig. 3.49**). This mode of operation is referred to as Love mode.[44] The Love mode SAW sensors are the most sensitive and most robust devices in the acoustic wave family. Sensitivity of a Love mode SAW sensor, based on 90° rotated ST-cut quartz crystal operating at 100MHz, can be two orders of magnitude greater than a QCM operating at 10 MHz.

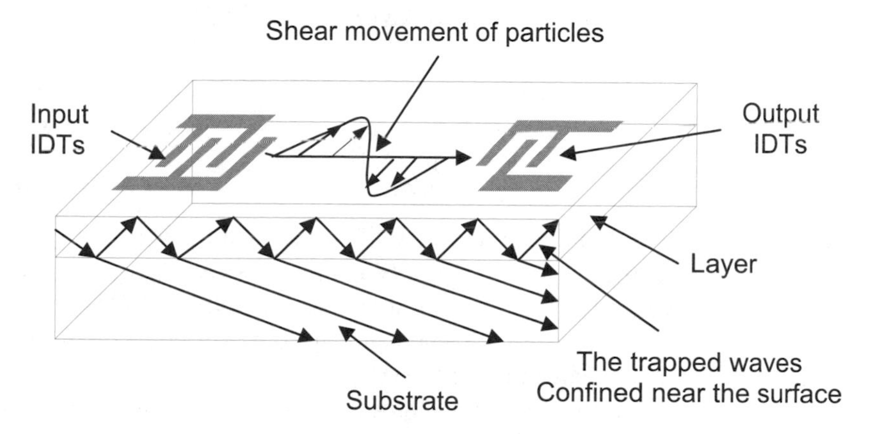

Fig. 3.49 Basic layout of a Love mode SAW device.

SAW devices can be utilized for sensing various physical and chemical parameters including temperature, acceleration, force, pressure, AC/DC high voltages, electric fields, magnetic fields, ionic concentration, gas flow, vapour concentration, viscosity, and in biosensing.[51]

SAW devices can be employed as affinity sensors through the addition of a mass sensitive layer on their active surface. The following equation describes the frequency shift, Δf, as a result of added mass as a coated thin, isotropic, non-conducting film on the active area of the device (**Fig. 3.48**):[64,65]

$$\Delta f = (k_1 + k_2) f_0^2 h \rho - k_2 h f_0^2 \left[\frac{4\mu}{V_r^2} \left(\frac{\lambda + \mu}{\lambda + 2\mu} \right) \right], \tag{3.91}$$

where f_0 is the operational frequency, k_1 and k_2 are material constants of the substrate, V_r is the Rayleigh wave velocity in the piezoelectric substrate, h is the added coating thickness, ρ is the added coating density, λ is the Lamb constant for the added coating, and μ is the shear modulus of the added coating. It can be seen from the equation that there is a linear relationship between the thickness of the added coating and the frequency shift. The frequency shift per mass added describes the sensitivity of such a SAW device.

SAW devices can also be used to monitor changes in conductivity of a sensing layer. Many metal oxides and conductive polymers change their conductivity in response to different oxidizing or reducing gas species. Depositing such materials over the active area of the SAW device can transform it into a gas sensor. In this case, the shift in operational frequency, f, after adding the conductive layer is given by:[65]

$$f = f_0 \frac{k^2}{2} \frac{1}{1+\left(\frac{\sigma_{SH}}{\sigma_{OR}}\right)^2}, \tag{3.92}$$

where f_0 is the operational frequency, k^2 is the electromechanical coupling coefficient, σ_{SH} is the sheet conductivity of the sensitive layer, σ_{OR} is the product of SAW mode velocity and substrate permittivity. Obviously, a SAW device with a higher electromechanical coupling coefficient generates a larger frequency shift in response to gases.

The point at which the maximum frequency shift can be obtained is calculated from the slope of the graph which is obtained by plotting f in Eq. (3.92) against σ_{SH}, as can be seen in **Fig. 3.49**. Maximum sensitivity is obtained when σ_{OR} and σ_{SH} are equal.

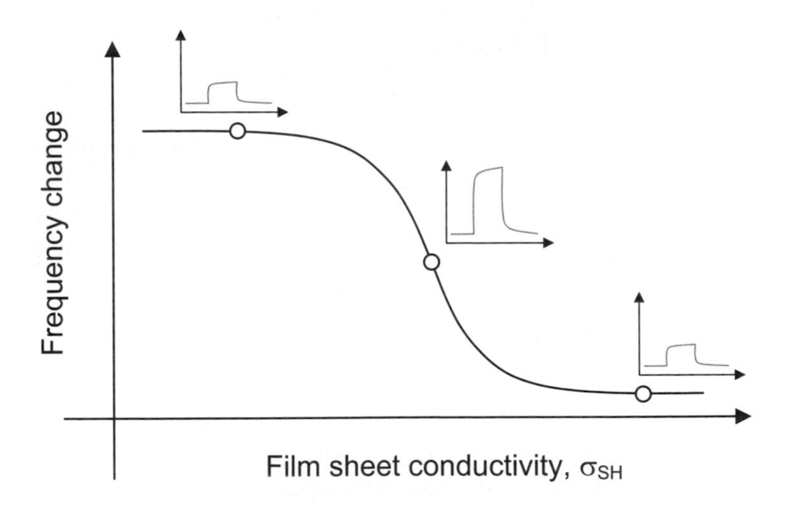

Fig. 3.50 Maximum response occurs when sheet conductivity value is matched to product of SAW mode velocity and substrate permittivity.

For sensing in liquid media, SH-SAW modes are preferred as they suffer much less attenuation when liquids come in contact with the propagating medium.[66] As the movements of particles are shear horizontal, not normal to the sensing surface, the contact liquid cannot damp the movements, except in cases where the liquid has high viscousity. This makes them ideally suited for biosensing applications. In such devices, a bio-selective layer is deposited on the surface of the device.

IDTs can also be designed for launching and receiving Lamb waves. Lamb waves propagate in plates (diaphragm). They are composed of two Rayleigh waves propagating on each side of the plate.[67] Two groups of Lamb waves can propagate through the plate independently, which include symmetrical and asymmetrical waves (**Fig. 3.51**).

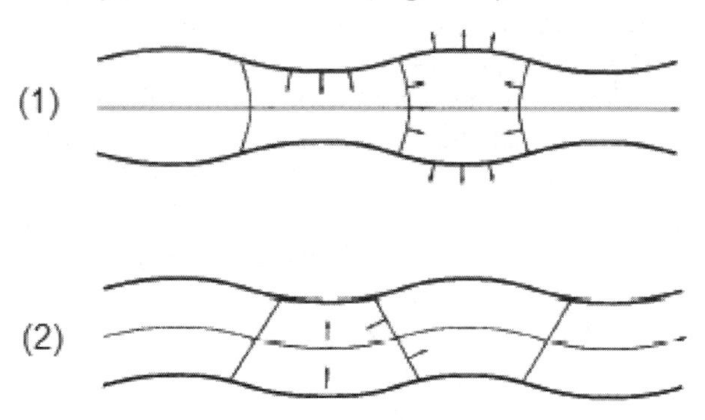

Fig. 3.51 (1) Symmetric and (2) asymmetric Lamb waves.

Due to the low phase velocities required to enable low-loss wave propagation, the operating frequency of these Lamb waves falls with the range of 5-20 MHz. Similar to FBAR, the mass sensitivity of the Lamb mode sensors is given by:

$$S = -1/(2\rho d), \quad (3.93)$$

where d is the plate thickness and ρ is the density of the diaphragm. Mass detection limit of a 10 MHz device with the thickness of 2 μm can be as low as 200 pg/cm^2.

3.7 Summary

Some of the most popular transduction platforms were introduced in this chapter. Many of them consist of interdigitated electrodes, between which a sensitive layer located. The addition of a sensitive layer makes them useful for conductometric and capacitive sensing measurements.

Optical waveguide based transducers were introduced, and it was shown that such transducers have various configurations. The most popular optical transducers are SPR based. In sensing applications, a sensitive layer is generally deposited over the waveguide layer. As was shown, the sensitivity is a function of the effective refractive index of the layer/sample medium that is in contact with the waveguide. Optical transducers are widely used in chemical sensing for medical and biochemical applications. SPR technology has become quite mature nowadays, and there are many commercially available SPR systems.

Electrochemical transducers, which are the most commonly employed in sensing, are suitable over a wide range of applications, including ion concentration measurements in our body; sensing environmental pollutants; and gas concentrations in vehicle exhaust systems. They are quite robust, and the implementation of ion selective membranes can make them favorable for chemical sensing applications. They are utilized in blood glucose sensing, which is the most prevalent in medical and healthcare industries. Such transducers not only offer a myriad of sensing applications, they also provide a means for investigating the many physiochemical interactions that occur during chemical interactions, in particular those occurring at the surfaces of the electrodes.

Solid state transducers are versatile and consist of devices based on junctions between metals, semiconductors and insulators. They can be arranged into numerous configurations, such as BJT, MOS capacitor, MOSFET, ISFET etc. Changes in the ambient, both physical and chemical, result in changes in the junction depletion region or the distribution of field. Semiconductor-semiconductor junction diodes and transistors are extensively used as optical sensors and metal-semiconductor junction devices in gas sensing applications. Devices exploiting field effects are becoming popular for chemical and biochemical applications as practical drawbacks of the technology are being addressed.

Acoustic wave transducers are useful when low concentrations of an analyte is the sensing target, as their interaction with the sensitive layer adds some mass to their surface. Different configurations, including SAW, BAW, are suitable for different ambient conditions. They are particularly useful for sensing application in liquid media. However, they are also useful in conductometric, thermal and mechanical measurements. Although it is many years since their introduction, only quartz crystal based devices have been commercialised, as its operational mechanisms and performance is extremely well understood. Love mode devices and thin film bulk resonators are becoming popular as their fabrication costs decrease, offering much higher sensitivities than QCMs.

In the next chapter, techniques for the fabrication of nanotechnology-enabled sensors will be presented. Some of them can be implemented for the fabrication of transduction platforms and others in the implementation of sensitive layers.

References

[1] L. M. Cukrov, P. G. McCormick, K. Galatsis, and W. Wlodarski, Sensors and Actuators B-Chemical **77,** 491-495 (2001).

[2] M. N. O. Sadiku, *Elements of Electromagnetics* (Saunder College Publishing, New York, USA, 1989).

[3] W. Lukosz, Biosensors & Bioelectronics **6,** 215-225 (1991).

[4] R. G. Heideman, R. P. H. Kooyman, and J. Greve, Sensors and Actuators Bl **10,** 209-217 (1993).

[5] J. Homola, S. S. Yee, and G. Gauglitz, Sensors and Actuators B-Chemical **54,** 3-15 (1999).

[6] Kretschm.E, Zeitschrift Fur Physik **241,** 313-& (1971).

[7] R. J. Green, R. A. Frazier, K. M. Shakesheff, M. C. Davies, C. J. Roberts, and S. J. B. Tendler, Biomaterials **21,** 1823-1835 (2000).

[8] D. C. Harris, *Harris' Quantitative Chemical Analysis*, 7th ed. (W H Freeman & Co, New York, USA, 2006).

[9] D. A. Skoog, D. M. West, and F. J. Holler, *Fundamentals of Analytical Chemistry*, 5th ed. (Saunders College Publishing, New York, USA, 1988).

[10] R. H. Petrucci, F. G. Herring, W. S. Harwood, and J. D. Madura, *General Chemistry: Principles and Modern Applications*, 9th ed. (Prentice Hall, Upper Saddle River, USA, 2006).

[11] R. A. Macur, *Miniature multifunctional electrochemical sensor for simultaneous carbon dioxide-pH measurements* (United States Patent 3957613).

[12] F. G. Thomas and G. Henze, *Introduction to voltammetric analysis: theory and practice* (CSIRO Publishing, Melbourne, Australia, 2001).

[13] R. S. Nicholson, Analytical Chemistry **37,** 1351-1355 (1965).

[14] F. G. Cottrell, Zeitschrift Für Physikalische Chemie **42,** 385-431 (1903).

[15] J. Heyrovsky and J. Kuta, *Principles of Polarography* (Academic. Press, New York, USA, 1966).

[16] R. C. Kapoor and B. S. Aggarwal, *Principles of polarography* (Wiley, New York, USA, 1991).

[17] K. Kalantar-Zadeh and G. H. Lee, Clinical Journal of the American Society of Nephrology **1,** S9-S18 (2006).

[18] X. F. Yu, Y. X. Li, N. F. Zhu, Q. B. Yang, and K. Kalantar-zadeh, Nanotechnology **18** (2007).

[19] T. M. Florence, Journal of Electroanalytical Chemistry **27,** 273-& (1970).

[20] S. M. Sze, *Physics of semiconductor devices*, 2nd ed. (Wiley, New York, USA, 1981).
[21] J. Rius and J. Figueras, Journal of Electronic Testing-Theory and Applications **9,** 295-310 (1996).
[22] R. M. Mihalcea, D. S. Baer, and R. K. Hanson, Applied Optics **36,** 8745-8752 (1997).
[23] A. M. Cowley and S. M. Sze, Journal of Applied Physics **36,** 3212-3220 (1965).
[24] K. D. Schierbaum, U. Weimar, W. Göpel, and R. Kowalkowski, Sensors and Actuators B **3,** 205-214 (1991).
[25] K. D. Wise, J. B. Angell, and A. Starr, IEEE Transactions on Biomedical Engineering **BM17,** 238-& (1970).
[26] P. Bergveld, IEEE Transactions on Biomedical Engineering **BM17,** 70-71 (1970).
[27] P. Bergveld, Sensors and Actuators B **88,** 1-20 (2003).
[28] L. Bousse, N. F. Derooij, and P. Bergveld, IEEE Transactions on Electron Devices **30,** 1263-1270 (1983).
[29] R. E. G. van Hal, J. C. T. Eijkel, and P. Bergveld, Advances in Colloid and Interface Science **69,** 31-62 (1996).
[30] R. E. G. van hal, J. C. T. Eijkel, and P. Bergveld, Sensors and Actuators B-Chemical **24,** 201-205 (1995).
[31] S. Caras and J. Janata, Analytical Chemistry **52,** 1935-1937 (1980).
[32] K. Y. Park, S. B. Choi, M. Lee, B. K. Sohn, and S. Y. Choi, Sensors and Actuators B-Chemical **83,** 90-97 (2002).
[33] L. T. Yin, J. C. Chou, W. Y. Chung, T. P. Sun, K. P. Hsiung, and S. K. Hsiung, Sensors and Actuators B-Chemical **76,** 187-192 (2001).
[34] J. G. Liu, L. Liang, G. X. Li, R. S. Han, and K. M. Chen, Biosensors & Bioelectronics **13,** 1023-1028 (1998).
[35] V. Volotovsky and N. Kim, Analytica Chimica Acta **359,** 143-148 (1998).
[36] A. S. Poghossian, Sensors and Actuators B-Chemical **44,** 361-364 (1997).
[37] S. V. Dzyadevich, Y. I. Korpan, V. N. Arkhipova, M. Y. Alesina, C. Martelet, A. V. El'Skaya, and A. P. Soldatkin, Biosensors & Bioelectronics **14,** 283-287 (1999).
[38] V. Volotovsky and N. Kim, Biosensors & Bioelectronics **13,** 1029-1033 (1998).
[39] C. Jiménez, J. Bartrol, N. F. deRooij, and M. KoudelkaHep, Analytica Chimica Acta **351,** 169-176 (1997).
[40] A. P. Soldatkin, D. V. Gorchkov, C. Martelet, and N. JaffrezicRenault, Materials Science & Engineering C-Biomimetic Materials Sensors and Systems **5,** 35-40 (1997).

[41] V. Volotovsky, A. P. Soldatkin, A. A. Shulga, V. K. Rossokhaty, V. I. Strikha, and A. V. Elskaya, Analytica Chimica Acta **322,** 77-81 (1996).
[42] C. Puig-Lleixa, C. Jimenez, J. Alonso, and J. Bartroli, Analytica Chimica Acta **389,** 179-188 (1999).
[43] J. Munoz, C. Jimenez, A. Bratov, J. Bartroli, S. Alegret, and C. Dominguez, Biosensors & Bioelectronics **12,** 577-585 (1997).
[44] A. Ballato, in *Piezoelectricity: history and new thrusts*, 1996, p. 575-583.
[45] G. Z. Sauerbrey, Zeitschrift für Physik **155,** 206-222 (1959).
[46] W. H. King Jr., Analytical Chemistry **36,** 1735-1739 (1964).
[47] G. G. Guilbault, Analytical Chemistry **55,** 1682-1684 (1983).
[48] G. G. Guilbault and J. M. Jordan, Crc Critical Reviews in Analytical Chemistry **19,** 1-28 (1988).
[49] J. Q. Hu, F. R. Zhu, J. Zhang, and H. Gong, Sensors and Actuators B-Chemical **93,** 175-180 (2003).
[50] J. W. Grate, S. J. Martin, and R. M. White, Analytical Chemistry **65,** A987-A996 (1993).
[51] J. W. Grate, S. J. Martin, and R. M. White, Analytical Chemistry **65,** A940-A948 (1993).
[52] T. Nomura and M. Okuhara, Analytica Chimica Acta **142,** 281-284 (1982).
[53] H. Zhang and E. S. Kim, in *Air-backed Al/ZnO/Al film bulk acoustic resonator without any support layer*, 2002, p. 20-26.
[54] H. Zhang and E. S. Kim, Journal of Microelectromechanical Systems **14,** 699-706 (2005).
[55] P. Lostis, **38,** 1 (1959).
[56] M. Benetti, D. Cannata, F. Di Pietrantonio, V. Foglietti, and E. Verona, Applied Physics Letters **87,** 1735041-1735043 (2005).
[57] C. S. Lu and O. Lewis, Journal of Applied Physics **43,** 4385-& (1972).
[58] J. K. Gimzewski, C. Gerber, E. Meyer, and R. R. Schlittler, Chemical Physics Letters **217,** 589-594 (1994).
[59] P. I. Oden, G. Y. Chen, R. A. Steele, R. J. Warmack, and T. Thundat, Applied Physics Letters **68,** 3814-3816 (1996).
[60] R. Raiteri, M. Grattarola, H. J. Butt, and P. Skladal, Sensors and Actuators B **79,** 115-126 (2001).
[61] P. I. Oden, Sensors and Actuators B-Chemical **53,** 191-196 (1998).
[62] H. J. Butt, P. Siedle, K. Seifert, K. Fendler, T. Seeger, E. Bamberg, A. L. Weisenhorn, K. Goldie, and A. Engel, Journal of Microscopy-Oxford **169,** 75-84 (1993).
[63] R. M. White and F. W. Voltmer, Applied Physics Letters **12,** 314& (1965).

[64] D. S. Ballantine and H. Wohltjen, Analytical Chemistry **61,** A704-& (1989).
[65] A. J. Ricco, S. J. Martin, and T. E. Zipperian, Sensors and Actuators **8,** 319-333 (1985).
[66] F. Josse and Z. Shana, Journal of the Acoustical Society of America **84,** 978-984 (1988).
[67] M. J. Vellekoop, Ultrasonics **36,** 7-14 (1998).

Chapter 4: Nano Fabrication and Patterning Techniques

4.1 Introduction

Nanotechnology enabled sensors are fabricated from nanoparticles, one dimensional nanomaterials, thin films of nanoscale thicknesses and/or thin films comprised of nanostructures. These nanostructures can be used both in the fabrication of the sensing layers and/or the fabrication of the transducer structures.[1] Some of the micro/nano fabrication technologies are very mature and widely used, such as photo-lithography, whilst others are in their infancy falling into niche applications.

Knowledge of nano-material synthesis, nano-structured thin film deposition, and nano-patterning techniques is fundamental for the development of nanotechnology enabled sensors and their sensitive layers. Another the driving force of nanofabrication and nanopatterning is that the dimensions of transducers, which are fabricated using standard microtechnology techniques, are becoming smaller and smaller. They are rapidly approaching nano dimensioned feature sizes. At such scales, unique technologies for the synthesis of nano-materials and nanopatterning techniques are imperative for the fabrication of transducer structures and their connections to the macro world.

In this chapter, we will present current and emerging methods in the synthesis of nanoparticles, one-dimensional nanomaterials and nanostructured thin films, as well as different approaches for fabricating nanosized features and patterns. This chapter will deal with both organic and inorganic materials; however most of the issues involved in the formation of organic and biological nanostructures will appear in Chap. 7.

4.2 Synthesis of Inorganic Nanoparticles

Inorganic nanoparticles are nano-dimensional particles synthesized from inorganic compounds such as metals and metal oxides. For applications in a sensing system, they may be suspended in liquid or gas phase (e.g. *colloidal suspensions*) or they may form ordered arrays on the surface of the transducer.

The organization of such nanoparticles on substrates plays an important role in the development of nanotechnology enabled sensors and micro/nano devices. Their small dimensions make them suitable for colloidal applications and for applications involving ordered arrays. They find significant applications in optical devices due to their high radiation-induced quantum yield. Additionally, the ability to join these one-dimensional materials into complex assemblies, with or without organic particles, creates many opportunities for novel scientific applications in biosensing.[2]

There are numerous methods for synthesizing nanoparticles. The type of method to be used depends strongly on the type of particle to be synthesized, the particle's functional medium, and the surface to which they will be attached. Many synthesis methods are conducted in liquid to form colloidal suspensions. The dimensions of these nanoparticles can be tailored, as can their surface functionalities. It is also possible to influence some level of control of their shape. This is achieved through tight control of the conditions and parameters during synthesis.[3,4] One of the greatest challenges faced in nanoparticles synthesis is the instability of the nanoparticles and their tendency to easily aggregate or precipitate. These effects are avoided through the use of stabilizing agents that adhere to the surface of the nano-particle. Generally, these stabilizing agents can manipulate the solubility, growth and surface charge of the particles. As a result, they keep the particles separated and suspended in the liquid environment during their synthesis.[5]

In this section, our attention will be focused on the synthesis of inorganic nano-particles, which consist of semi-conducting, magnetic and noble metal constituents.

4.2.1 Synthesis of Semi-conductor Nano-particles

Semiconductor nanoparticles and nanocrystals[2] (often called *quantum dots* or *Q-dots,* if their dimensions are less than 10 nm) exhibit optical, thermal, mechanical and electrical properties that are strongly dependant on the dimensions of the particle. Their unique properties arise from their ability to confine the conduction band electrons, valence band holes, vibrations and other particle activities in all three spatial directions. When the dimensions of these nanoparticles are in the order of a few nano-meters they possess discrete, quantized, molecular-like energy spectrums caused by quantum confinement effects. Their energy spectrum can be engineered by manipulating their dimensions, shapes, and other related properties.

There are many methods for the synthesis of semiconductor nanoparticles. Generally, they are based on initiating and stopping the growth of nano-particles in either a gaseous or liquid environment with a high degree of control. Low cost routes capable of mass-producing nanoparticles of desired dimensions are generally performed via colloidal synthesis.[6]

The use of *reverse micelles* as nano-scale reaction vessels is one the most common methods in colloidal synthesis. After coming into contact, reverse micelles exchange their contents. This process has been used to make nanosized materials via either the chemical reduction of metal ions or co-precipitation reactions.[7] An example is the synthesis of the CdS nano-particle.[8] CdS has a direct band-gap of 2.42eV, and is widely studied for applications in both photoresistive and photovoltaic devices. In this process, reverse micelles containing two different precursors are prepared: one with metal salts and the other with a sulphide source (such as Na_2S). The solutions are mixed and nanoparticles are formed as the different micelles exchange their contents. After the formation of these nano-particles, thiol molecules can be employed to stabilize them as they bond to the surface. Other nano-particles such as CdSe, PbS, TiO_2, etc can also be obtained using different colloidal synthesis processes.

Other synthesis methods include electro-deposition[9], formation in gas phase[10] and pyrolysis[11], to name a few.

4.2.2 Synthesis of Magnetic Nanoparticles

Amorphous and nano-crystalline *soft and hard magnetic* (whether they are non-permanent or permanent) materials have many applications including transformers, inductive devices and sensors, etc. Additionally, the development of uniform magnetic nanoparticles is typically an important issue in ultrahigh-density magnetic storage devices and sensors. Magnetic alloys are materials such as cobalt and Iron. Magnetic type oxides are materials such as magnetite (Fe_3O_4), cobalt ferrites, nickel ferrites, zinc ferrites (MFe_2O_4 - with M to be metals such as Co, Ni, Zn) and the mixed nickel and cobalt ferrites.

It is known that the magnetic anisotropy of nanoparticles within an *exchange length* dictates the magnetic softness/hardness of such materials. The magnetic exchange length is typically measured in nano-meters or tens of nano-meters which highlights the underlying importance of the nano-metric dimensions in the properties of magnetic systems.

Several synthetic approaches of magnetic nano-particles include: thermal [12] and sonochemical [13] decomposition of organometallic precursors, high-temperature reduction of metal salts, [14] and reduction inside reverse micelles [15] to name a few.

4.2.3 Synthesis of Metallic Nanoparticles

There has been an enormous amount of research carried out into synthesizing metallic nanoparticles especially noble metals such as gold and platinum. The reason for such widespread efforts is the novel use of noble metal nanoparticles in a wide variety of applications, such as medicine, electronics and optics. In sensors, nanoparticles of this type can be put to use as catalysts, biomaterial tags, optical resonators and many other functional devices. Due to their inert nature, noble metals are generally not hazardous to living tissue and hence they are becoming attractive in biosensing and biotechnology. They exhibit properties that are markedly different from those of their bulk counterparts, are very stable, and their size can be easily controlled via the synthesis approach chosen.

Colloidal gold nanoparticles are among the most stable metal nanoparticles and have many fascinating properties. For example, their size and quantum confinement effects cause them to absorb and scatter light several orders of magnitude greater than other materials.[5] Certainly, the breakthroughs reported by Schmid[16] and other research groups[17] have caused research into *gold nanoparticles* into the spotlight.

One of the conventional methods for the synthesis of gold nano-particles is the reduction of gold(III) derivatives. In 1951, Turkevitch et al[18] suggested the use of citrate for reducing $HAuCl_4$ in aqueous conditions to synthesize gold nanoparticles with dimensions of approximately 20 nm. Later Frens et al[19] showed that gold nano-particles of predetermined dimensions (16 and 147 nm) could be obtained by controlling the ratio between the reducing and stabilizing agents (the trisodium citrate-to-gold ratio). This method still remains very popular and is especially useful for applications where the surface of the gold particles are functionalized or capped with ligands. As an example, **Fig. 4.** shows the preparation of sodium 3-mercaptopropionate-stabilised gold nanoparticles,[20] where citrate salt and amphiphilic surfactant molecules were simultaneously added.

Reductive synthesis of metal nano-particles is by no means limited to gold, as many noble metals can be synthesized similarly.

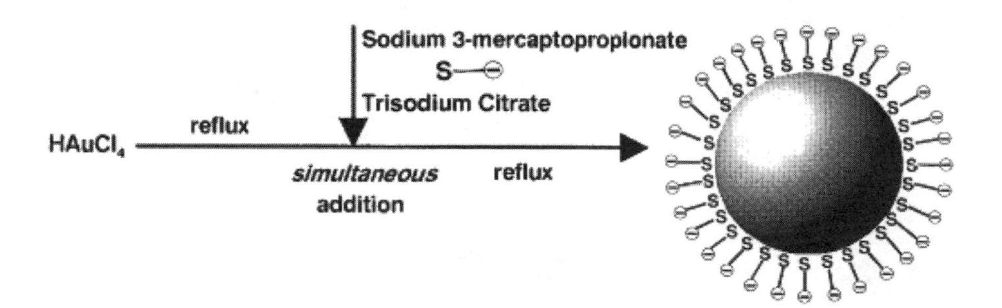

Fig. 4.1 The procedure for preparing anionic mercapto-ligand-stabilized gold nanoparticles in aqueous media. Reprinted with permission from Elsevier publications.[20]

Templates can also be employed during the synthesis of nano-particles. For instance, Zhao and Crooks reported the use of dendrimers (dendrimers will be described in Chap. 7) as templates for the synthesis of platinum and other metals.[21,22] The nanoparticles were prepared by sequestering metal ions within supramolecular organic assemblies, namely polyamidoamine (PAMAM) dendrimers,[23] followed by chemical reduction to yield the corresponding metallic nanoparticles. The dendrimers served as both the template and stabilizer. The dimensions of such particles depend on the number of metal ions initially loaded into the dendrimer. By preloading a dendrimer nanotemplate with suitable metal ions and then chemically reducing this composite in-situ, a dendrimer-encapsulated metal nanocluster may be synthesized as shown in **Fig. 4.2**.[24]

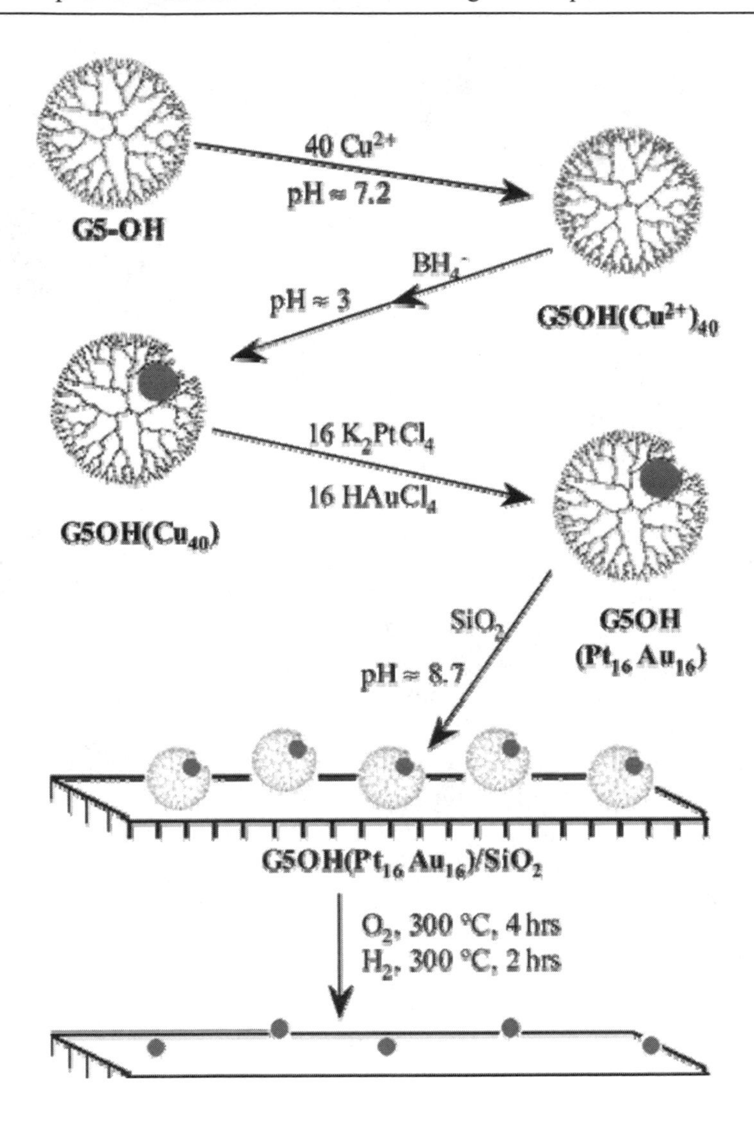

Fig. 4.2 Hydroxy-terminated generation 5 PAMAM dendrimers were used to prepare Cu nanoparticles (NPs). The Cu NPs were subsequently used to reduce K_2PtCl_4 and $HAuCl_4$, resulting in stabilized bimetallic Pt-Au NPs with a 1:1 stoichiometry. The stabilized NPs were adsorbed onto a silica substrate and thermally activated to remove the dendrimers. Reprinted with permission from American Chemical Society publications.[24]

4.3 Formation of Thin Films

Thin film deposition is an important part of the fabrication of many sensors and associated sensing platforms. Metallic electrodes are integrated as part of many sensors to receive and transmit electrical signals. Optical and acoustic waveguides in sensor templates are made from thin layers. Magnetoresistive sensors and many photo-receivers are made of nanodimensioned thin films. Standard thin film deposition and lithographic patterning techniques, which are the basis of the field effect transistors and microchips, are widely utilised in the fabrication of sensors.

4.3.1 Fundamentals of Thin Film Deposition

In most circumstances, in the deposition of thin films, the first few atomic layers of the depositing material grow like islands, centered about *nucleation sites*. These nucleation sites can remain separated during the deposition in a process known as *Volmer-Weber*[25] growth, or they can be connected via thin layers in a *Stranski-Krastanov* growth process.[26] These islands can grow until they touch each other. This point is known as the *percolation threshold.* Depending on the material, this percolation threshold can typically be between a few nanometres to submicron ranges. If the deposition occurs layer-by-layer, then it is called a *Frank-Van der Merwe* growth.[27] This is an ideal *epitaxial growth*. The term *epitaxy* originates from the Greek words, "epi" which means "equal" and "taxis" meaning "in ordered manner". It describes the ordered crystalline growth on a substrate. The three different modes of thin film growth are illustrated in **Fig. 4.3**.

A typical thin film growth may include the following steps:[28]
(a) adsorption, (b) surface diffusion, (c) incorporation of growth species onto the surface structure (and surface interaction) and (d) desorption of the by-products (**Fig. 4.4**).

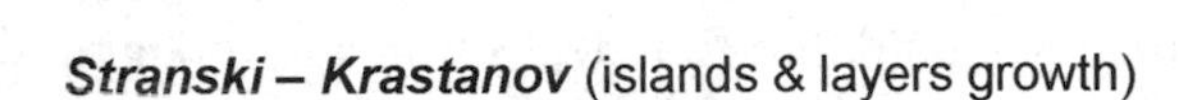

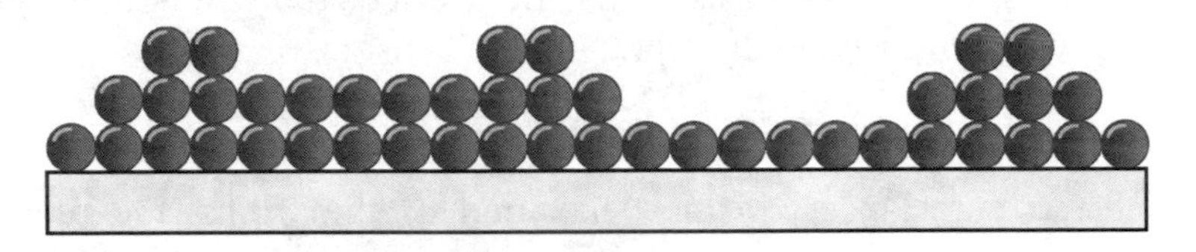

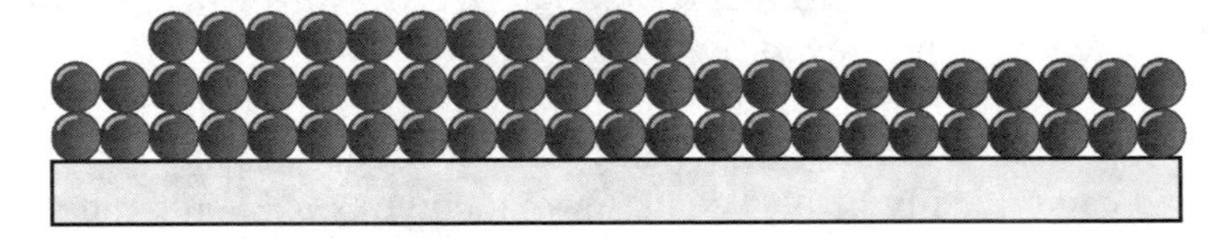

Fig. 4.3 Modes of thin film growth.

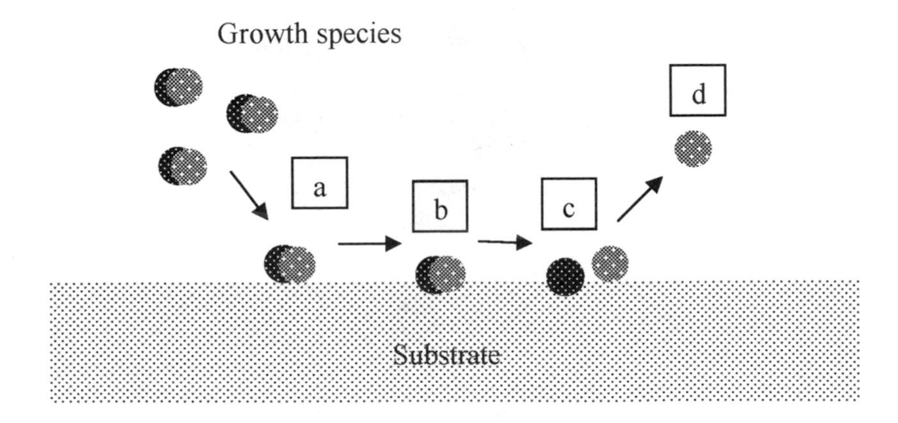

Fig. 4.4 The typical thin film growth mechanism.

There are numerous thin film deposition techniques available which are categorized as either *vapour phase deposition* or *liquid phase deposition.*

The deposition technique determines whether the film is *conformal* or *non-conformal*, with respect to the underlying substrate's surface morphology. A conformal coating covers all surfaces to a uniform depth, following (conforming to) the substrate's shape. A non-conformal coating is non-uniform and does not conform to the substrate's surface (**Fig. 4.5**).

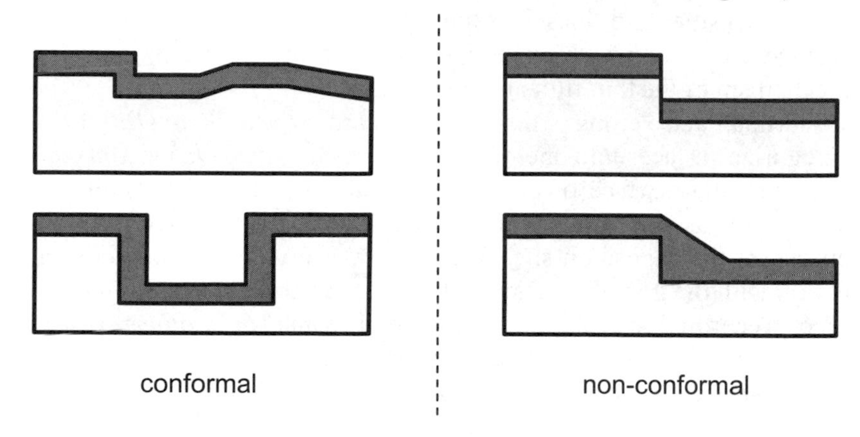

Fig. 4.5 Conformal and non-conformal coatings.

This section is an introduction of how nano-structured thin films can be formed. In the following sections, we will describe different vapor and liquid phase deposition techniques that are commonly utilized in both the nanotechnology and sensor technology fields.

4.3.2 Growth of One-Dimensional Nano-structured Thin Films

Thin films made from *one-dimensional nanostructures* are useful for the fabrication of both gas and liquid phase sensors. Their surface-to-volume ratio is large and they can be deposited using either standard or low cost microfabrication techniques. *Nanobelts, nanorods, nanotubes* and *nanofibers* are all examples of one-dimensional nanostructures. They can be deposited spontaneously without the help of any templates using the following techniques:[28] (a) *evaporation-condensation* and *vapour-liquid-solid* (*VLS*) growth in the gas phase (b) *dissolution-condensation* and *solution-liquid-solid* (*SLS*) growth in the liquid phase. Spontaneous growth on a surface can be realized through reduction of the *Gibbs free energy* associated with the surface. Phase transformation, stress changes, and chemical reactions can all be used to control the Gibbs free energy.

It is possible to change the Gibbs free energy locally through the introduction of surface anomalies such as impurities, kinks and ledges. In such cases, the anomalies will play an important role as the *nucleation sites* where the *growth species* are initially adsorbed. Different facet of the substrate crystal, addition of a catalyst on the surface and changing the local pressure can also alter the Gibbs free energy of the deposition sites.

The existence of different faces on a surface substrate produces different atomic densities and 'loose' or 'unsatisfied' bonds. This results in isolated surface energies, which in turn determines the growth rate and growth mechanism of the thin film at the facet site.

Hartman and Perdok [29] developed *periodic bond chain* (*PBC*) *theory* to categorize surfaces into one of three types: *flat, stepped*, and *kinked.*

Even a flat surface on the atomic scale is not truly flat, due to interatomic effects and localization. Such discontinuities change the Gibbs free energy of the deposition site. Sites of low energy are favourable for interaction with the growth species and are responsible for growth on a flat surface. According to PBC theory adsorbed atoms (or *adatom*s) of a growth species form a single chemical bond with flat surfaces, while stepped and kinked surfaces may form three and four chemical bonds, respectively. [29]

Evaporation-condensation growth

This method (also referred to *Vapor-Solid, VP*) is based on the decrease of Gibbs free energy produced by surface clusters forming or the decrease in *supersaturation* (change of pressure on the surface). The condensation on the surface can be purely physical; however, chemical reactions can also participate in the process. Generally nanostructures which are grown using this process, are fairly single crystalline. However, condensation on substrates of similar materials with different faces can also result in nanostructures with varied dimensions and crystals structures. Impurities and imperfections on the substrate may also produce special growth directions for the nanostructures.

The growth rate is defined with the *condensation rate*, J. This rate depends on the vapour pressure P, the *accommodation coefficient* α, the *supersaturation of growth species* in vapour $\sigma = (P-P_0)/P_0$ (with P_0 the equilibrium vapour pressure), the substrate temperature T in Kelvin, the atomic mass of the growth species m, and Boltzmann's constant k and is defined as:[28]

$$J = \frac{\alpha\sigma P_0}{\sqrt{2\pi mkT}}. \tag{4.1}$$

The following affect the condensation rate:

(a) If the accommodation coefficient, α, is different in various faces, the result can be an asymmetric growth.

(b) J linearly depends on α at low concentrations of the deposition species and low J. However, at high value for J the increase in α does not cause any increase in J anymore as the Eq. (4.1) is no longer valid.

(c) High deposition pressure, P, increases, J, and at the same time may increase the probability of producing defects. This can be beneficial in the formation of nanostructures. Generally, we would like to produce gaps which result in separate nanostructures during the deposition process. High concentrations of the growth species may also generate secondary nucleation. The secondary nucleation can act as a seed for the growth of multiple crystal facets and branching (**Fig. 4.6**).

Fig. 4.6 The formation of nucleation sites and the subsequent branching of nanorods from these sites during the thermal evaporation of molybdenum oxide onto the surface of alumina.

The residence time, τ_s, and diffusion distance, D_s, for the growth of a species on a substrate are two important parameters which determine the formation and the dimensions of nanostructures during the deposition process. τ_s is given by the equation:

$$\tau_s = \frac{\exp(E_{des} / kT)}{\nu}, \tag{4.2}$$

where E_{des} is the desorption energy of the desorbing species and v is the vibrational frequency of the adsorbed species on the surface. D_s is described by:

$$D_s = \frac{a_0 v \exp(-E_s / kT)}{2}, \tag{4.3}$$

where a_0 is the growth species' largest dimension, and E_s is surface diffusion activation energy. Consequently, the mean diffusion distance, X, is defined as:

$$X = \sqrt{2} D_s \tau_s . \tag{4.4}$$

X plays an important role in the formation of nanostructures as it is used for the calculation of accommodation coefficient, α (see Eq. (4.)).

As we learned earlier, kinks, steps and other types of anomalies on a surface tend to reduce the surface energy and thus making them energetically favorable for the adsorption of the growth species. If X is longer than the distance between the anomalies, the growth species has a chance to adsorb on the surface of the substrate. However, if the distance between anomalies is shorter then the growth species, it can escape the surface and materials growth will not occur. In the formation of one-dimensional structures, either anomalies with the right separations from each other or crystals with different surface facets are needed. After the formation of a seed nucleus on the surface, the growth will continue on the facet (or facets) with the lower energy.

Several theories have been proposed for describing the formation of one-dimensional structures in the evaporation-condensation process such as: *axial screw dislocation*, *formation of micro-twins*, *stacking faults* etc. However, most of them fail to present a general idea which can predict the growth in different situations.[30]

Interestingly, the growth rate of nanowires exceeds the theoretical condensation rate as calculated using Eq. (4.) with the accommodation coefficient of one. This translates into the fact that the growth of such nanowires is the fastest type of growth which can occur on a surface. A *dislocation-diffusion theory* as proposed by Dittmar et al has been proposed to describe such a phenomenon.[31]

It is believed that one of the first reports on the evaporation-condensation growth of mercury nanowires was presented by Sears in 1955.[32] He developed mercury nanowires with the cross sectional diameters of 200 nm and the length of several mm using a condensation temperature of -50°C. The growth rate was 1.5 μm/sec for a supersaturation, α, value of 100.

In most of the one-dimensional deposition experiments, a lower supersaturation result in more anisotropic structures. At higher supersaturation secondary or inhomogenous nucleation may occur, resulting in secondary growth effects such as bulk deposition and branching.

The pioneering and encompassing work of Wang et al are the first examples of the deposition of one-dimensional metal oxides and other semiconductors using simple low vacuum deposition in horizontal ovens.[33] They used commercially available nano-powders of metal oxides and evaporated them at temperatures much lower than their actual bulk evaporation. The materials were evaporated at a high temperature and condensed at lower temperature sites along the horizontal tube.[34] The Transmission electron microscopy (TEM) micrographs of ZnO nanobelts deposited by Wang et al are shown in (**Fig. 4.7**).

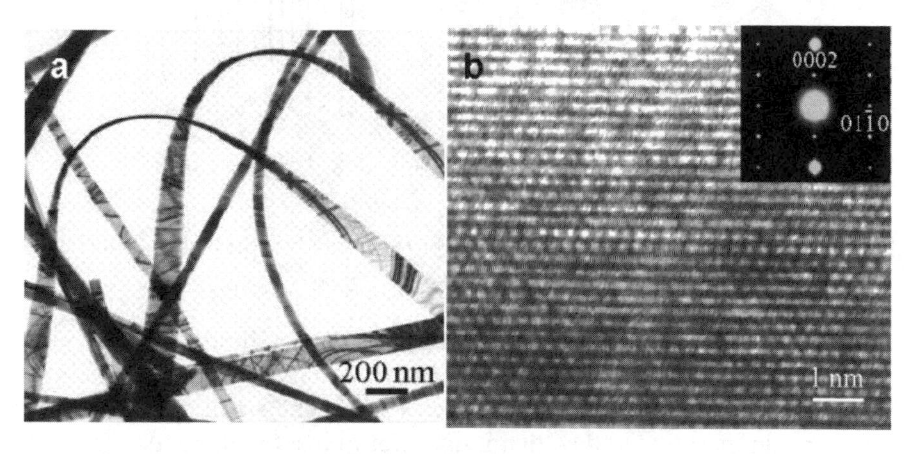

Fig. 4.7 (*a*) Transmission electron microscopy (TEM) image of the as-synthesized ZnO nanobelts. (*b*) High-resolution TEM image recorded with the incident electrons arriving normal to the top surface of the nanobelt. Reprinted with permission from the Annual Review of Physical Chemistry.[34]

Vapour-Liquid Solid (VLS) growth:

In this method the vapour is condensed on the surface with the help of a second phase which acts as a catalitic (or an impurity) nucleation site.

The catalyst directs and confines growth within a predetermined direction (**Fig. 4.8**). The catalyst acts as a trap for the growth species as it can either amalgamate within or make a liquid phase on the substrate surface during the deposition process. The growth species, which adsorbs onto the catalyst, can participate within it and encourage a gradual one directional growth. The theory of VLS was first described by Wagner et al in the 60s.[35] They claimed that formation of one dimensional structures can not be simply defined using the evaporation-condensation equations which were presented earlier. In this method, a liquid-like sphere is always present at either the tip or at the starting point (stem) of the one-dimensional structures.

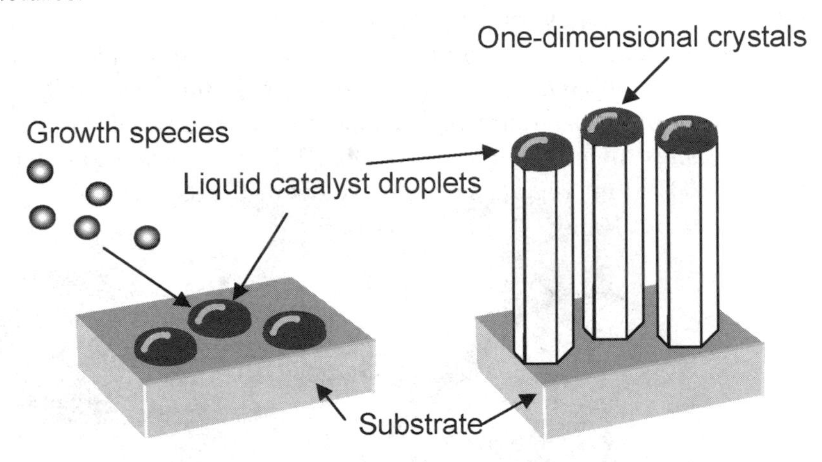

Fig. 4.8 Vapour-liquid-solid (VLS) growth: (left) catalyst droplet and nucleation (right) growth of one-dimensional structures.

In this deposition process, the catalyst forms a liquid droplet when heated, with the distribution coefficient of the catalyst being less than one. The vapour pressure of the growth species should be small and the catalyst should be an inert material so that it does not to react with the crystals. The wettability of the surface by the catalyst is also very important. If the wetting contact angle is large then the catalyst droplet will be small (**Fig. 4.9**). However, if the surface is highly wettable by the catalyst then, due to the spread of the liquid, the base diameter will become large. This will result in larger base size one-dimensional structures.

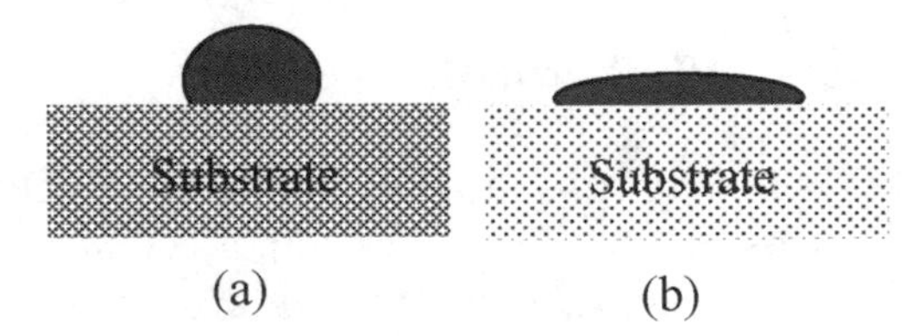

Fig. 4.9 The wettability of the substrate: (a) a large contact angle on a less wettable surface (b) the spread of the same droplet volume on a more wettable surface.

As can be seen in **Fig. 4.10**, Gudiksen et al [36] employed nano-sized catalysts to define both the nanowire diameter (**Fig. 4.10** (a)) and to initiate the growth of InP nanowires using a vapour-liquid-solid mechanism process (they have used laser ablation for evaporating the semiconductor source material). By varying the deposition time, it is possible to control the length of the nanowires (**Fig. 4.10** (b)).

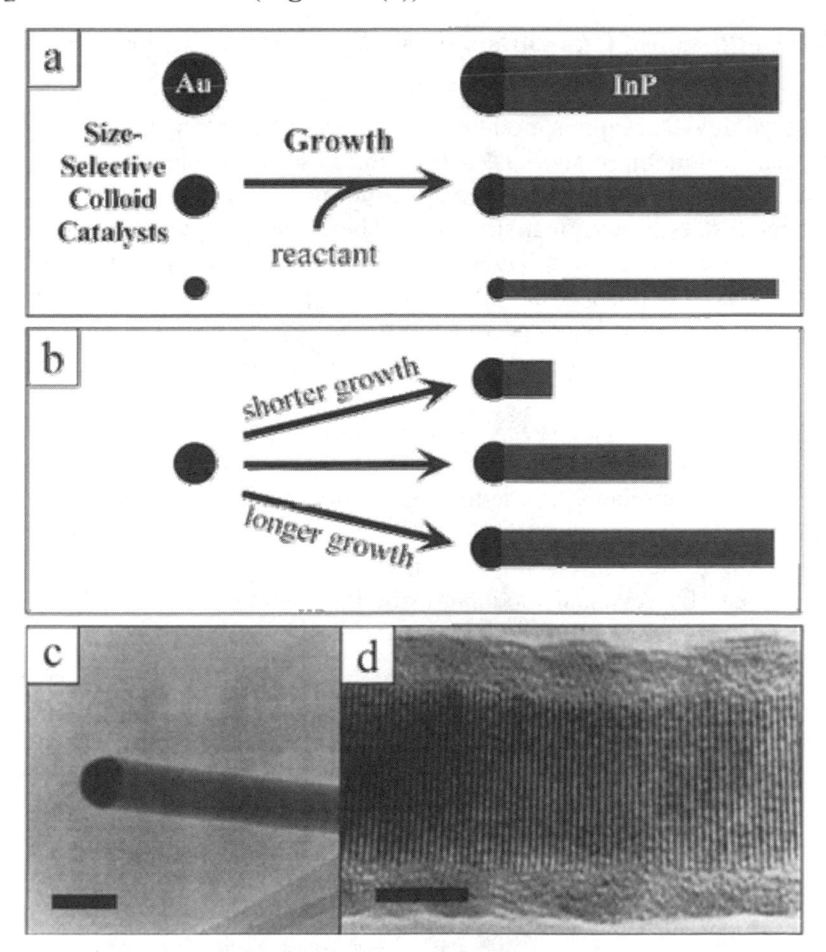

Fig. 4.10 Schematic depicting the use of colloidal catalysts for the: (a) diameter-selective InP nanowires, (b) varying the growth time of nanowires to obtain different lengths. TEM image in (c) shows the colloidal catalyst at the end of an InP nanowire grown from a 20 nm Au nanoparticle (scale bar is 50 nm). High-resolution TEM image (d) shows the crystalline core of the InP nanowire grown from a 10 nm Au nanoparticle (scale bar is 5 nm). Reprinted with the permission from the American Chemical Society publications[36]

Dissolution-condensation and solution-liquid-solid (SLS)

These methods differ from gas phase processes in that the growth media is a liquid. The growth species are dissolved in a liquid and then deposited onto the surface via chemical reactions. These methods will be described in detail in the following sections.

4.3.3 Segmented One-Dimensional Structured Thin Films

One-dimensional light-emitting diodes, lasers, complementary and diode logic devices, Peltier modules, which are made of stacks of *n*- and *p*-type semiconducting materials, will be the base of efficient transducers in the near future. Utilizing the current technologies, it is possible to fabricate segmented one-dimensional structures. There are many methods available for the fabrication of such layered structures. Methods such as molecular beam epitaxial depostion and different types of chemical vapour deposition techniques, which will be described in the following sections, are ideal for this task as they can be used for the deposition of layers on the atomic scale.

It is possible to synthesize segmented one-dimensional semiconductors such as one dimensional superlattices, by either using the one-dimensional structures as templates or modifying the techniques that have been described in the previous sections (e.g. by alternating the deposition materials), As can be seen in **Fig. 4.11**, Gudiksen et al fabricated one-dimensional superlattices by repeated modulation of the vapour-phase semiconductor reactants during the growth.[37]

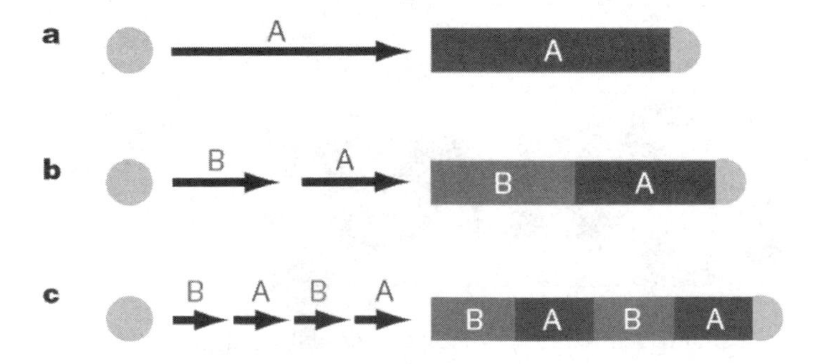

Fig. 4.11 (a) A nanocluster catalyst nucleates and directs the one-dimensional growth of a semiconductor nanowire growth with the catalyst remaining at the terminating end of the nanowire. (b) Upon completion of the first growth step, a different material can be grown from the end of the nanowire. (c) Repetition of steps (a) and (b) leads to a compositional superlattice within a single nanowire. Reprinted with permission from the Nature publications.[37]

4.4 Physical Vapor Deposition (PVD)

PVD encompasses a group of vacuum techniques used for depositing thin films of different materials onto various substrates by physical means. In general, it can be categorized into the following groups:

- *Evaporation*
- *Sputtering*
- *Ion plating*
- *Pulsed laser deposition* (laser ablation)

In a PVD process, the material to be deposited is placed in an energetic environment. This energetic environment evaporates the source material in the forms of particles such as molecules and ions. The escaping particles are directed towards the surface of a substrate. This substrate surface draws energy from these particles as they arrive, allowing them to form nucleation sites or thin layers. The whole system is kept in a vacuum chamber, which allows the particles to travel as freely as possible, arriving on the surface with a high energy promoting adhesion. The vacuum also encourages a thin film deposition free from the contaminants interferences.

4.4.1 Evaporation

Evaporation is one of the most common methods of thin film deposition. Although it is one of the oldest physical techniques, it is still widely used in the laboratory and in industry. In this process, a vapor is generated by evaporating or subliming a source material, which is subsequently condensed as a solid film on a substrate.

A broad range of materials, with different reactivity and vapor pressures, can be used in this technique. In addition, a large diversity of source components such as resistance-heated filaments, electron beams, crucibles heated by conduction, radiation, RF-induction, arcs, exploding wires, and lasers can be employed. The following details of some of the most commonly used evaporation techniques.

Thermal evaporation

In *thermal evaporation*, the target material is melted and evaporated or sublimed using an electric resistance heater. In this process, the vapor pressure in the chamber (**Fig. 4.12**) is raised from the initially very low pressure to one that allows the material to be deposited on the substrate.

This initial high vacuum is necessary because it allows the vapor to reach the substrate without reacting with or scattering by other atoms in the chamber. In addition, the initial low pressure pumps out unwanted residual gas that is in the chamber. By doing this, it reduces the chance of incorporating impurities in the deposited film.

It is important that we avoid melting or subliming the target holder and heater. Naturally, only materials with a much higher vapor pressure than the materials of the heating element and sample holder can be evaporated to obtain a highly pure thin film. Typical metals used as heating element include tantalum (Ta), molybdenum (Mo) and tungsten (W). The evaporation temperature is typically in a range of 1000-2000°C. This technique is straightforward and suitable for depositing metals and compounds that have low fusion temperatures such as Al, Ag, Au, and AgCl.

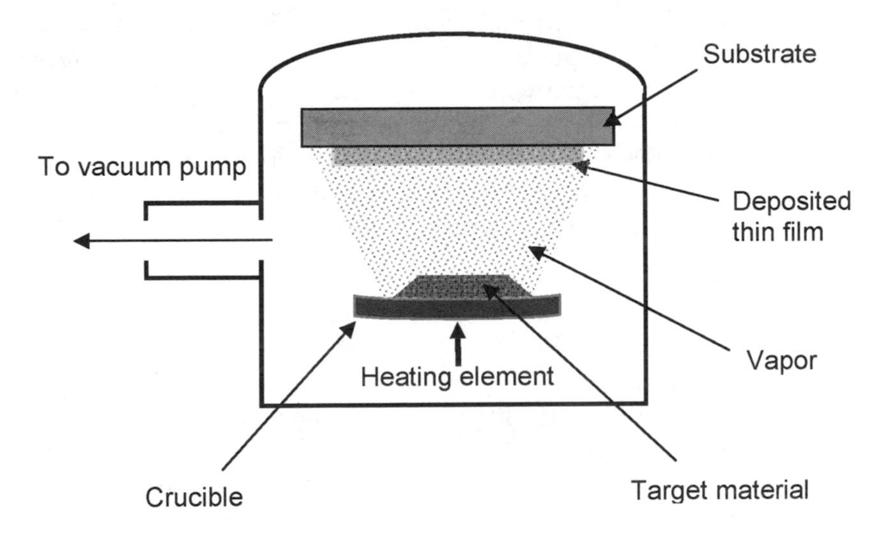

Fig. 4.12 A representation of a typical of thermal evaporation system.

Nanostructured thin films can be synthesized by a solid-vapor process.[34] In this case, a powder source material is vaporized at elevating temperatures and is usually conducted in a horizontal tube furnace, as shown in **Fig. 4.13**. Such systems are comprised of an alumina or fused quartz tube, a rotary pump system, a gas supply and control systems.

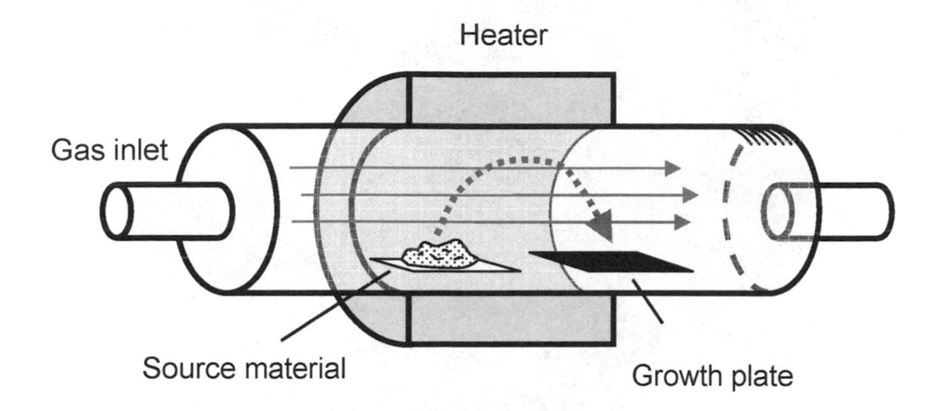

Fig. 4.13 Schematic representation of an experimental apparatus for the growth of oxide nanostructures.

As previously described, the evaporation method can be efficiently utilized for the fabrication of nanostructured thin films. **Fig. 4.14** shows nanostructured MoO_3 thin films consisting of randomly oriented nanorods deposited by thermal evaporation on an alumina substrate.

Electron beam evaporation

In an electron beam evaporator, an electron gun generates a high-energy beam to evaporate a small spot of the source material. The evaporation occurs under a very high vacuum (less than 10^{-4} torr). A voltage of in the range of 1 to 100 kV is normally applied across a tungsten filament (**Fig. 4.15**), which heats it up to a point where the *thermionic emission* of electrons takes place. Thermionic emission is the flow of electrons from a charged metal surface. The charge produces a thermal vibrational energy that overcomes the electrostatic forces that keep electrons at the surface.

Fig. 4.14 MoO_3 thin films consisting of randomly oriented nanorods deposited by thermal evaporation on an alumina substrate, at two different magnifications.

As seen in **Fig. 4.15** the electron beam is generally placed at an angle of 270°, outside the deposition zone. The electron beam is then focused, bent and directed towards the crucible, which contains the target material, by magnets. This eliminates virtually all ions generated from the filament, and prevents them from contaminating the sample. This is one of the main advantages of electron beam evaporators over conventional thermal evaporators. Typical deposition rates for electron beam evaporation range from 1 to 10 nm/s, depending on the material and instrument settings.

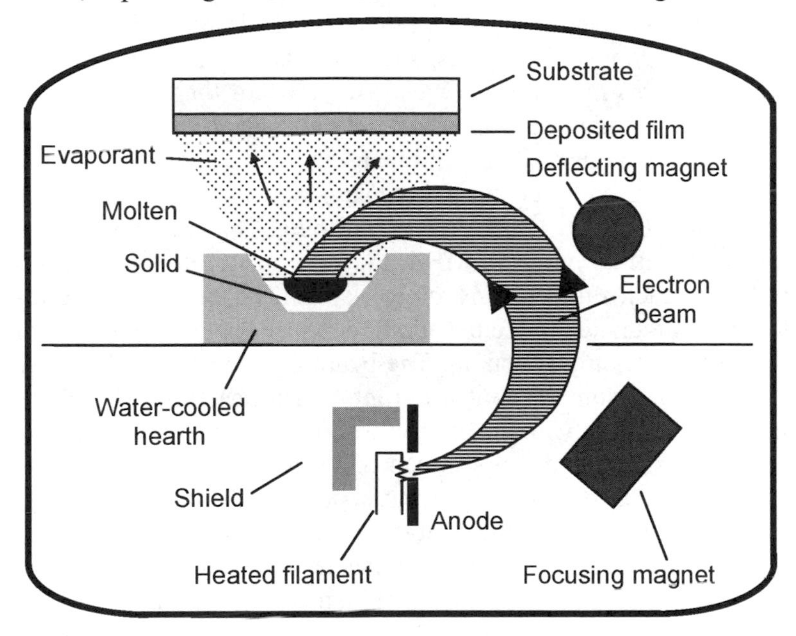

Fig. 4.15 Schematic of an electron beam evaporation system.

The electrons colliding with the target material's surface cause it to heat up, as these electrons transfer their kinetic energy of motion. Although the energy transferred by a single electron is relatively small, there are many of them. Therefore, the overall energy released is quite high - often more than several million W/cm^2. For this reason, the crucible must be cooled to prevent it from melting.[38] The energy from the electrons evaporates the material and the evaporated material eventually lands on the substrates placed in the chamber.

A large number of wafers can be placed in an evaporation chamber simultaneously. As a result, this thin film deposition method is widely used for the deposition of metal electrodes and metal oxide thin films in the microfabrication industry.

Molecular Beam Epitaxy (MBE)

MBE was first developed by J.R. Arthur and A.Y. Cho in 1968 at the Bell laboratory,[39,40] and is one of the most significant thin film deposition techniques in nanotechnology. It is a technique that grows atomically thick layers. Its slow deposition rate allows the films to grow epitaxially (ideal

Frank – van der Merwe type deposition), with growth rates are typically in the order of a few Å/s. This process takes place in an ultra high vacuum (10^{-11} torr or less). The pumping arrangement is designed to keep the partial pressure of contaminant gases, such as H_2O, CO_2, and CO to less than $\sim10^{-11}$ torr. At this pressure, the *mean free path* of the constituents of the beam is several orders of magnitude greater than the nominal source-to-sample distances of about 20 cm in other vacuum techniques.[41]

MBE is conducted by creating a *molecular beam* of the source material that we wish to deposit. This beam then impinges on the substrate where it is deposited atom by atom. In MBE, the evaporated atoms (beam) do not interact with each other or any other vacuum chamber gases until they reach the substrate, due to the beam's large mean free path length, which is a result of the ultra high vacuum. The beam can be shuttered in a fraction of a second, allowing for almost abrupt transitions from one material to another. MBE is widely employed for producing epitaxial layers of metals, insulators and superconductors. It can be used for the formation of *superlattices*, which are structures that have periodically interchanging semiconducting solid layers, in the nanometer scale. These structures are used for the fabrication of quantum well lasers and giant magneto-resistors, among many other technologically sophisticated electronic and optoelectronic devices. In recent years, MBE has become a popular technique for depositing III-V compound semiconductors as well as several other materials. MBE is used in both research institutions and industry to grow crystals of the highest quality for devices such as *semiconductor laser diodes* (*SLD*s or *LD*s), *light-emitting diodes* (*LED*s), *high electron mobility transistors* (*HEMT*s), *heterojunction bipolar transistors* (*HBT*s), *photodetectors* (*PD*s), and quantum-sized structures (quantum wells, wires, and dots).

A typical MBE system is shown in **Fig. 4.16**. The system generally consists of a *growth chamber* and an *auxiliary chamber*. The auxiliary chamber is used for bringing samples into and out of the ultra high vacuum environment so that the ultra high vacuum of the growth chamber, which is very difficult to obtain, is maintained.

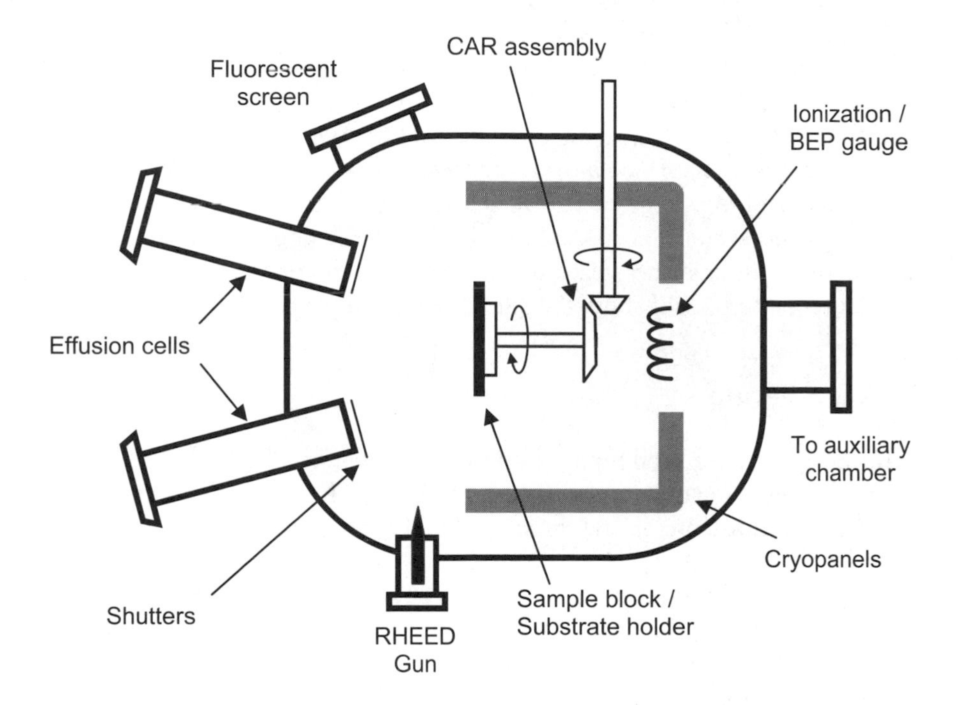

Fig. 4.16 Diagram of a typical MBE system growth chamber.

The substrates are placed onto the growth chamber sample holder/ heater, which rotates on two axes. The layer growth uniformity is improved by continually rotating the sample on the *CAR* (*continual azimuthal rotation)* assembly. The CAR normally has an ion gauge for reading either the chamber pressure or the *beam equivalent pressure* (*BEP*) of the incoming molecular source. Surrounding the CAR assembly is a liquid nitrogen cooled cryo-shroud, which is used for pumping the unreacted beam from the sample, and any contaminants.

Materials such as Ta, Mo, and pyrolytic boron nitride are used to make the substrate holder and assembly parts, because they do not decompose or outgas impurities even when heated to approximately 1500°C.

In solid-source MBE, ultra-pure elements are heated in separate *effusion cells* (also called *K-cells* or *Knudsen Cells*), until they begin to evaporate. The evaporated elements then condense on the substrate, where they form thin films. Temperature control in effusion cells is extremely important as they can change the molecular beam flux dramatically. Computer controlled shutters are placed in front of each of the effusion cells. They can shutter the flux reaching the sample within a fraction of a second.

Reflection high-energy electron diffraction (*RHEED*) is one of the most useful tools for in-situ monitoring of thin film growth, particularly in MBE systems.[42] It can be used to calibrate growth rates, observe removal of oxides from the surface, monitor the substrate temperature, and monitor the arrangement of the surface atoms, among many other things. The RHEED gun emits electrons with energy in the range of 5-20 keV. This beam impinges on the surface at a shallow angle (~0.5-2 degrees). Electrons reflected from the surface strike a phosphor screen forming reflection and diffraction patterns that show the surface crystallography.[43]

4.4.2 Sputtering

In the *sputtering* technique, a solid target material is bombarded with energetic ions (approximately 100 eV or more). This bombardment creates a cascade of collisions in the target material's surface. These multiple collisions eject (or sputter) atoms from the surface into the gas phase. These atoms are then directed towards the target substrate to form a thin film. The number of atoms ejected from the surface per incident ion is called the *sputter yield*. The ions for the sputtering process are produced by plasma which is generated above the target material. The atoms sputtered from the surface of the target enter the plasma where they are excited and emit photons.

A conventional *diode plasma* (or *DC sputterer*) is simply a diode, which consists of an anode and a cathode inside a vacuum system (**Fig. 4.17**). Applying the right voltage across the electrodes and a suitable gas pressure, the gas breakdowns into plasma. Near the cathode a dark space with a very large electric field is formed (**Fig. 4.18**). Ions are accelerated rapidly across this dark region and strike the cathode. These collisions cause sputtering of the target material. Some electrons, known as secondary electrons, are also emitted from the surface and then accelerate back across the dark region, gaining significant energy. This energy is used, through collisions with gas atoms, to form more ions to sustain the plasma.

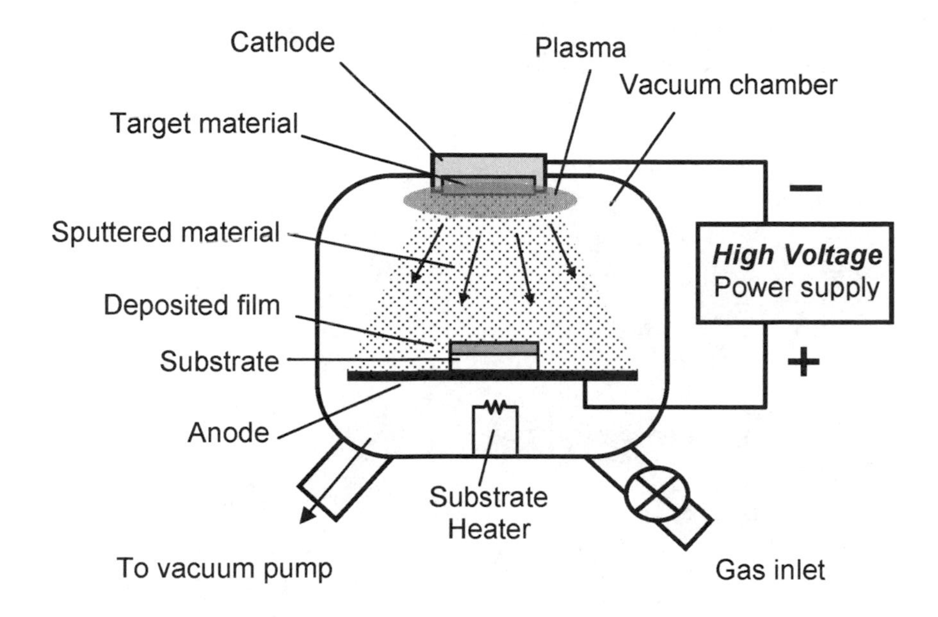

Fig. 4.17 A diagram of a simple diode sputtering system.

The sputtering gas can be inert, such as argon. In this case, the deposited thin film's composition is the same as the target (perhaps with different crystal structure). *Reactive sputtering* takes place when the deposited film is formed by chemical reaction between the target material and a gas, which is introduced into the vacuum chamber. Oxide and nitride films can also be fabricated in this way. However, the deposition rate can be very different. The deposition rate depends on parameters such as the softness of the target, the target to substrate distance, the power applied and the energy applied to the target. The applied energy can be in a range of several Watts to a few thousands of Watts depending on the area of the target and the deposition rate required. The deposition rate can be as small as several nanometers to several micrometers per hour.

In practice, a variety of techniques are used to modify the plasma properties. For instance, the ion density can be tailored to optimize the sputtering conditions with a *radio frequency* (RF - 13.56 MHz) power source and the application of modulating voltage at the target. Furthermore, the target material itself plays a significant role when choosing the sputtering system. For instance, a diode sputterer cannot be used for the deposition of insulating materials due to charge builds up on insulating targets. This problem can be solved by using an RF source, for which the sign of the anode-cathode bias alternates at a high rate. RF sputtering is used to deposit highly insulating oxide or nitride films. However, it requires an impedance matching network which consists of inductors and capacitors.[44]

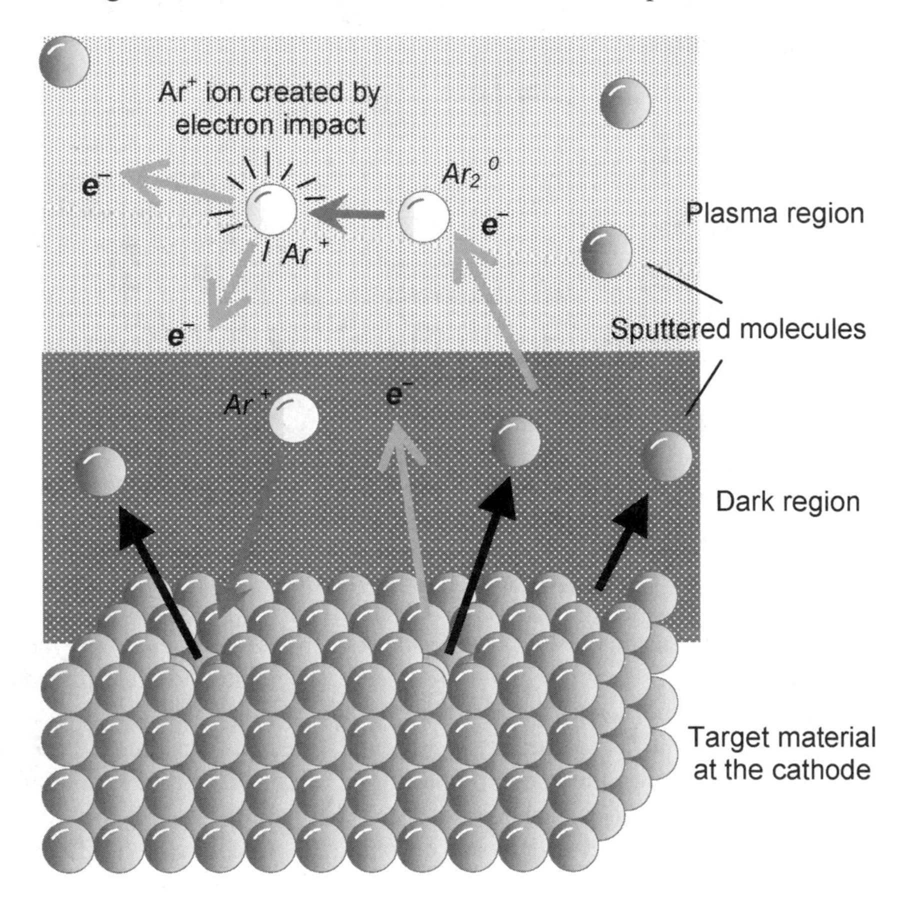

Fig. 4.18 Formation of the dark and plasma regions above the target's surface.

Magnetic fields (*magnetron*) can also be applied to the plasma region in order to improve the sputtering performance. For instance, a static magnetic field at surface of the sputtering target can be produced by a permanent magnet. The magnetic field generated is parallel to the cathode (the target) surface (**Fig. 4.19**). This magnetic field forces the secondary electrons to move perpendicular to both the electric field (normal to the surface) and the magnetic field (parallel to the surface). Causing these electrons to move parallel to the cathode surface, where they form a current loop of drifting secondary electrons (**Fig. 4.19**). As a result, these secondary electrons are trapped in a region close to the target. These electrons eventually lose their kinetic energy through collisions with gas atoms causing ionization, or with other electrons, which generates heat. This results in dense plasma near the target, which increases the deposition rate as it enhances the ionization of sputtering gas. However, this plasma does not affect the ejected atoms, as they have a neutral charge and are unaffected by the magnetic field. The magnetron requires cooling because of the heat generated by the energetic collisions around the target.

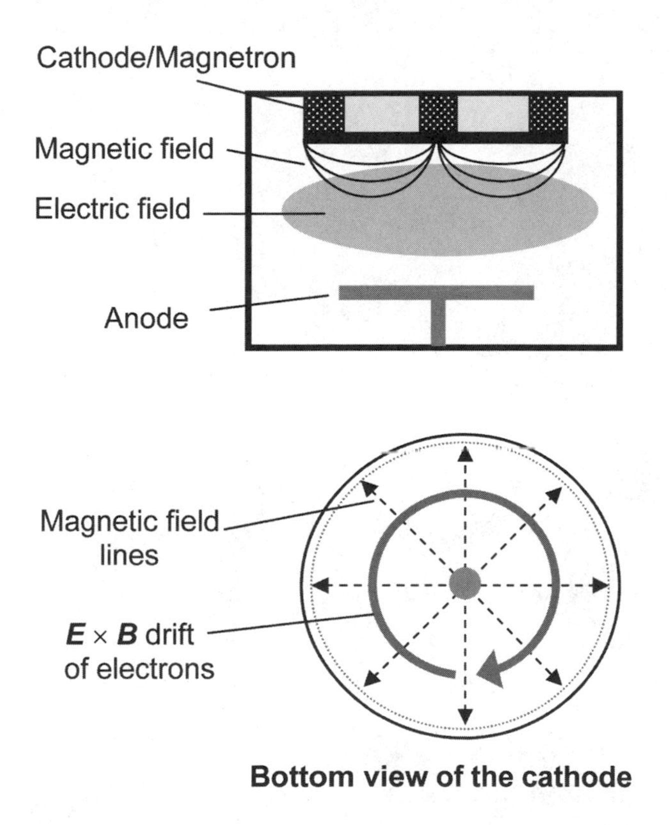

Fig. 4.19 The magnetic field configuration for a circular planar magnetron cathode.

Through sputtering not only is it possible to deposit films with nanosized features, but it is also possible to tailor these features. This can be achieved by adjusting various sputtering parameters such as: the sputtering power, sputtering gas, the magnetron, pressure, distance from the sample, and temperature. Furthermore, crystal structure and surface morphology of the films also depends on the substrates on which they are deposited. For example, in **Fig. 4.20** the deposition nanostructured ZnO by RF magnetron sputtering onto different substrates is shown, where the morphology and crystal structure are markedly different on various substrate surfaces.

Fig. 4.20 (a) ZnO growth on bare $LiTaO_3$ (left) and gold (right); (b) ZnO growth on aluminum (left) and bare $LiNbO_3$ (right).

There are also several modifications that can be made to traditional sputters. Sputters with such modifications include *ion-beam sputtering*

(*IBS*) and *ion-assisted deposition* (*IAD*) , where the substrate is exposed to a secondary ion beam.[44]

4.4.3 Ion Plating

Ion plating is a technique that is classed between thermal evaporation and sputtering. It employs an electron beam evaporator to evaporate the target material (e.g. chromium or titanium). At the same time plasma is also generated near the substrates. After the evaporated particles pass through plasma, some of them become ionized and sputtering effects takes place on the substrate surface. This produces highly energetic ions of the target material near the substrate. As these ions can be implanted deeper in the surface of the substrate, the adhesion of the thin films to the substrate surface is greatly increased. Also at the substrate's surface a different material phase, which is the combination of the deposited film and the substrate, is formed.

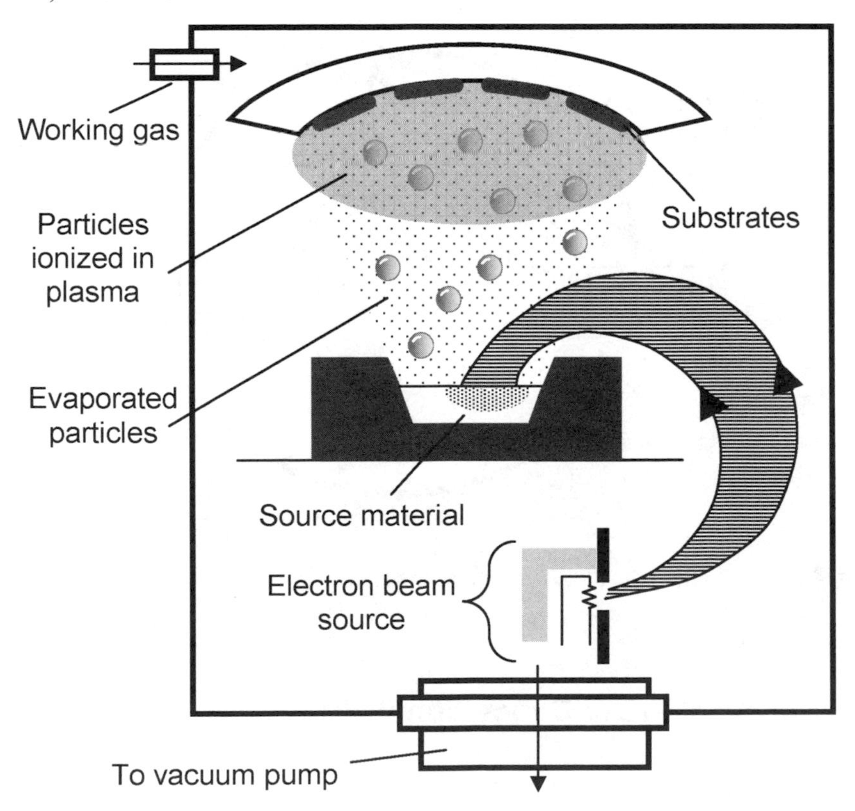

Fig. 4.21 The schematic diagram of an ion plating system.

4.4.4 Pulsed Laser Deposition (PLD)

PLD is a thin film deposition technique in which the target material is ablated by a focused laser beam that is periodically pulsed. This process is generally conducted in a high vacuum (less than 10^{-5} torr). A schematic of a PLD system is shown in **Fig. 4.22**. The source material is transformed from a solid to plasma by consecutive pulses of the focused laser light. Due to thermal expansion, this plasma is directed perpendicularly away from the source's surface towards the substrate. The plasma is cooled due to its expansion and consequently it transforms to a gas. However as the process is carried out at low pressure, the gas has enough momentum to reach the substrate, where it condenses to form a solid thin film.

Thin films produced by PLD can be well controlled as the parameters of the laser can be accurately altered. PLD can be utilized for the deposition of thin films comprising nano-dimensional highly ordered grains.

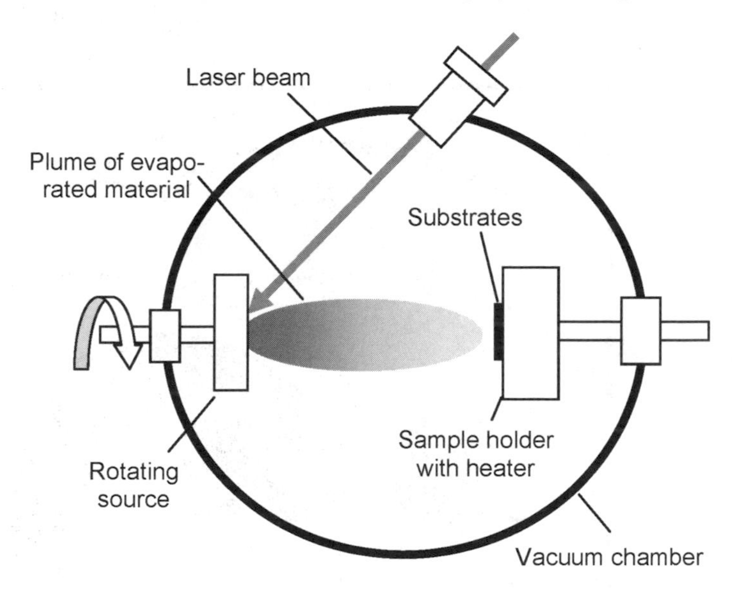

Fig. 4.22 The Schematic diagram of a PLD system.

4.5 Chemical Vapor Deposition (CVD)

*CVD*s are chemical techniques for depositing thin films in vapor phase. They are extensively employed in the semiconductor industry, to deposit films such as; silicon oxide; silicon nitride; polycrystalline, amorphous, and epitaxial forms of silicon; gallium arsenide; synthetic diamond and even carbon nanotubes.

In a typical CVD process (**Fig. 4.23**) the substrate is exposed to one or more volatile *precursors*. A precursor is a substance from which another, usually more stable substance, is formed.[45] The precursors are first vaporized and then transported to the substrate by carrier gases. The precursor then reacts or decomposes on the substrate's surface to produce the desired film upon receiving energies. The desired atoms are released from the precursor onto the surface. There they may interact with other atoms to establish strong binding sites, resulting in nucleation and growth of the thin film. At this point, unwanted chemical groups are released from the surface and carried out of the chamber by the flow of gas. Volatile byproducts are frequently produced as a result of the detachment of unwanted chemicals reacting. CVD can produce uniform coatings that grow conformally on the substrate.

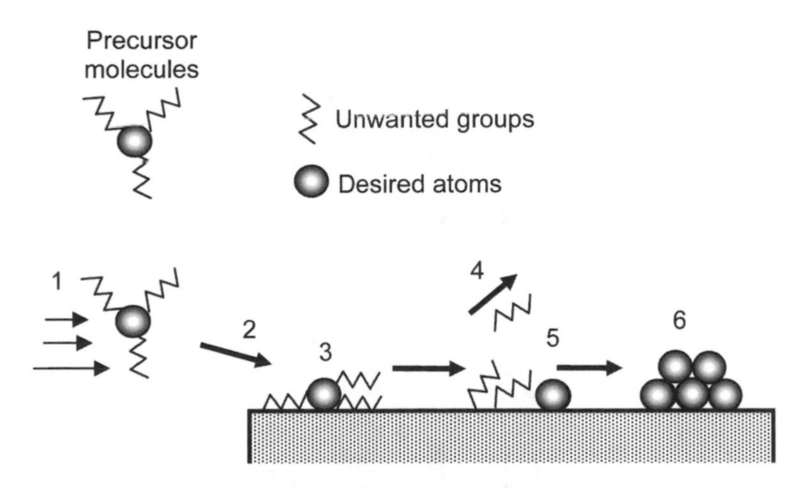

1- Gas phase transportation of the precursor molecules
2- Precursor adsorption
3- Immobilization and reaction of the precursor on the substrate - detachment of weak bonds
4- Desorption of the unwanted groups
5- Diffusion to stronger binding sites
6- Nucleation and growth of thin film

Fig. 4.23 The CVD process.

Thermal CVD relies on high substrate temperatures and catalytic reactions between precursors and substrate surfaces to achieve the dissociation of input gases with binding energies in the 1 to 3 eV range. As a consequence, thermal CVD deposition rates are relatively low (100 to 1000 Å/min).[46]

A hot-wall CVD reactor set-up for the deposition of titanium diboride (TiB_2) is shown in **Fig. 4.24**. In the process, the titanium tetrachloride ($TiCl_4$) precursor is first produced in a pre-reactor, in which chlorine (Cl_2) gas is reacted with the titanium (Ti) metal to give $TiCl_4$ gas:

$$Ti + 2Cl_2 \rightarrow TiCl_4. \tag{5.5}$$

In the CVD process, $TiCl_4$, H_2, and BCl_3 gases react at the surface of the heated substrate to form TiB_2 thin films. The byproduct from this reaction is HCl gas, which is removed through an exhaust. The chemical reaction of this CVD process can be summarized as:

$$TiCl_4 + 2BCl_3 + 5H_2 \rightarrow TiB_2 + 10HCl. \tag{5.6}$$

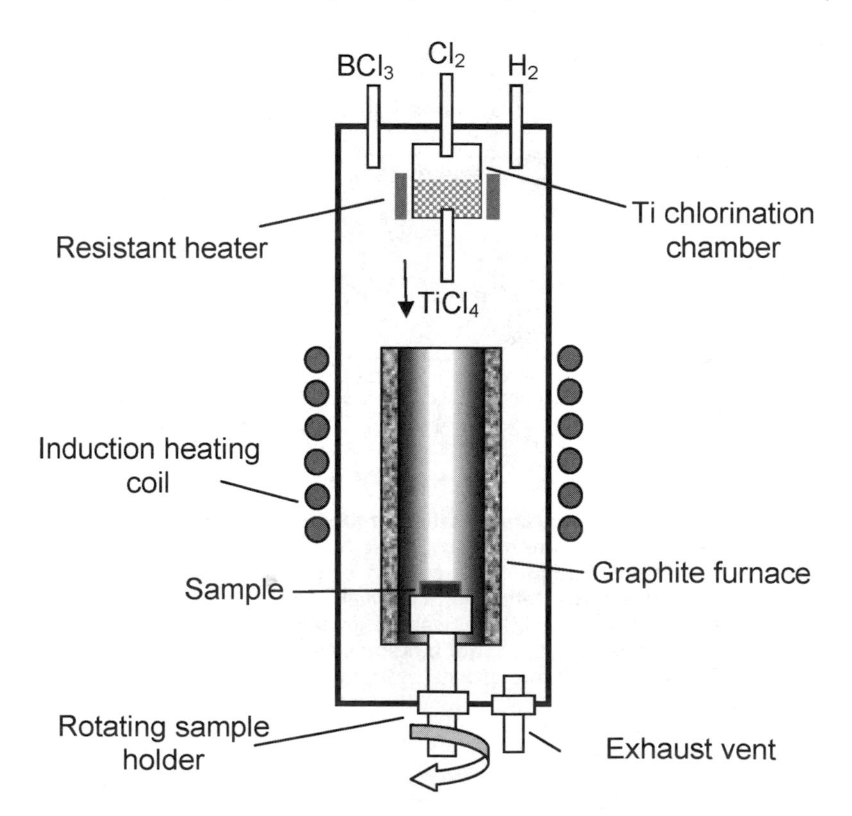

Fig. 4.24 The CVD setup for deposition of titanium diboride.

Some other examples of CVD reactions are:

$$\begin{gathered} WF_6 + 3H_2 \rightarrow W + 6HF. \\ TaCl_5 + CH_4 + \tfrac{1}{2} H_2 \rightarrow TaC + 5HCl. \\ ZrCl_4 + 2H_2O \rightarrow ZrO_2 + 4HCl. \end{gathered} \tag{5.7}$$

Metal-organic CVD (*MOCVD*) or *metalorganic vapour phase epitaxy* (*MOVPE*) is used for epitaxial growth of materials, especially semiconductors, by *co-pyrolysis* (pyrolysis is the chemical decomposition of organic materials in the absence of oxygen or other reagent by heating) of *organometallic compounds* and/or *hydrides* (hydride is a compound formed by the union of hydrogen with other elements). The process is shown in **Fig. 4.25**.

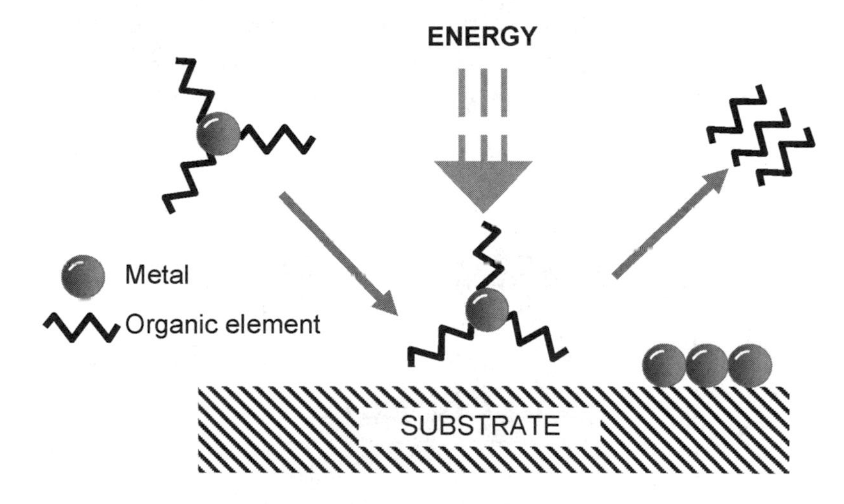

Fig. 4.25 The MOCVD process.

Epitaxially grown thin films of III-V and II-VI compounds find use in many applications such as *lasers*, *PIN photodetectors*, *solar cells*, *phototransistors*, *photocathodes*, *field effect transistors* and *modulation doped field effect transistors*. The efficient operation of these devices requires the deposited films to have a number of excellent materials properties, including purity, high luminescence efficiency, and/or abrupt interfaces.

The general chemical reaction that occurs in a MOCVD process for the deposition of the III-V and II-VI compounds and alloys, may be written as:

$$R^1{}_nM(v) + ER^2{}_n\,(v) \rightarrow ME(s) + nR^1R^2(v), \tag{5.8}$$

where R^1 and R^2 represent organic functional groups such as methyl ($-CH_3$) and ethyl ($-C_2H_5$) (or higher molecular weight organics) radicals or hydrogen (hydride), M is a Group II or Group III metal, E is a Group V or Group VI element, n = 2 or 3 (or higher for some higher molecular weight sources) depending on whether II-VI or III-V growth is taking place, and *v* and *s* indicate whether the species is in the vapor or solid phase.[47] As an example, gallium arsenide can be grown in a reactor on a substrate by introducing trimethyl gallium, $(CH_3)_3Ga$, and triphenyl arsenic, $(C_6H_5)_3As$.[48] A number of forms of CVD are in wide use and are frequently referenced in the literature.

4.5.1 Low Pressure CVD (LPCVD)

LPCVD is one of the most widely used CVD deposition techniques for industrial applications as it produces layers with excellent uniformity of thickness. LPCVD reactors generally operate at medium vacuum, in the range of 10-300 Pa (**Fig. 4.26**). This process requires the substrate to be heated at temperatures in excess of 500°C and the deposition rate is relatively slow.

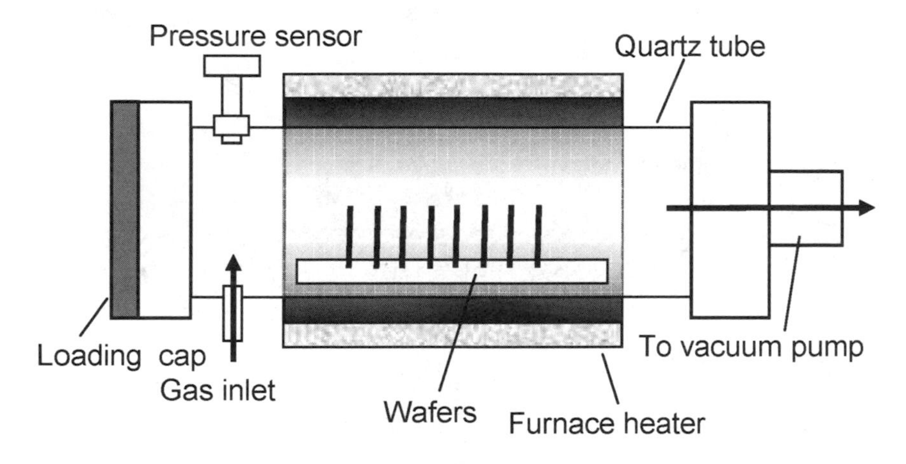

Fig. 4.26 A typical hot-wall LPCVD reactor.

4.5.2 Plasma-Enhanced CVD (PECVD)

In the *PECVD* deposition method plasma is used to enhance the chemical reaction rates of the precursors (**Fig. 4.27**). This process has the advantage of lower deposition temperatures, which is crucial in many semiconductor thin film deposition processes.

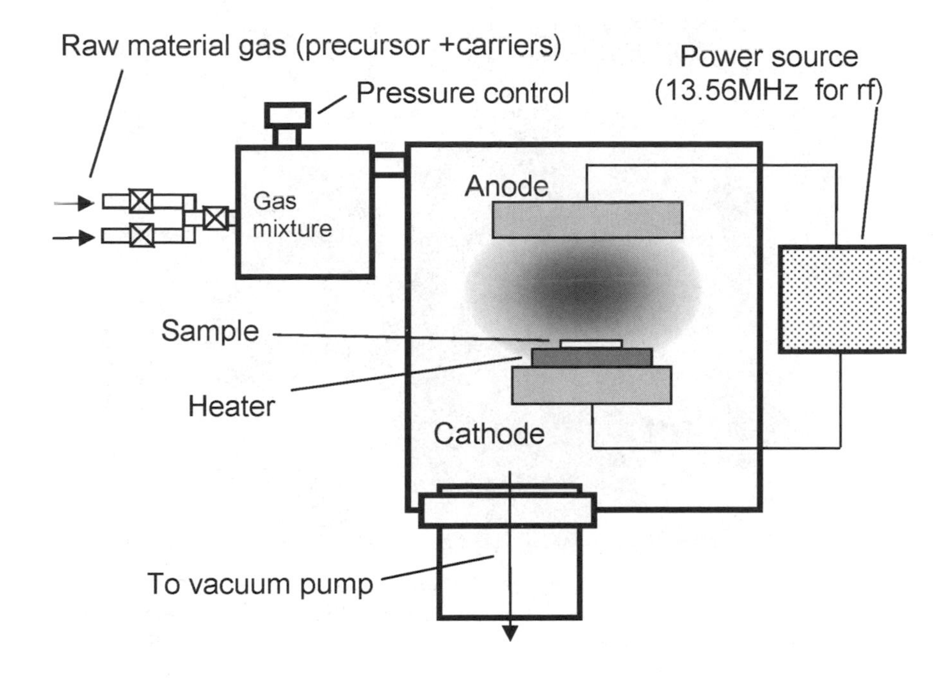

Fig. 4.27 The PECVD process.

Carbon nanotubes production is commonly conducted using the PECVD systems. Such systems are often used to grow free standing vertically aligned MWCNT (multi-walled carbon nanotubes). The plasma field promotes the incident of the precursors on the tip of the nanotubes. Additionally, the field also aligns the nanotubes in vertical positions with reference to the substrates.[49]

Carbon nanotubes can be deposited by PECVD using acetylene (C_2H_2) and ammonia (NH_3) gases in the presence of a nickel catalyst at elevated temperatures, as high as 700°C.[50] Acetylene is the carbon feedstock gas for nanotube growth, while ammonia acts as an etching agent to suppress the deposition of amorphous carbon from the deposited plasma decomposition of acetylene.

A comprehensive discussion on the deposition of carbon nanotubes using PECVD can be found in [51]. Several morphologies of various carbon nano structures that can be deposited by PECVD are shown in **Fig. 4.28.** Milne et al also presented the structural uniformity of CNs produced by plasma enhanced chemical vapour deposition evaluated for field emission applications.

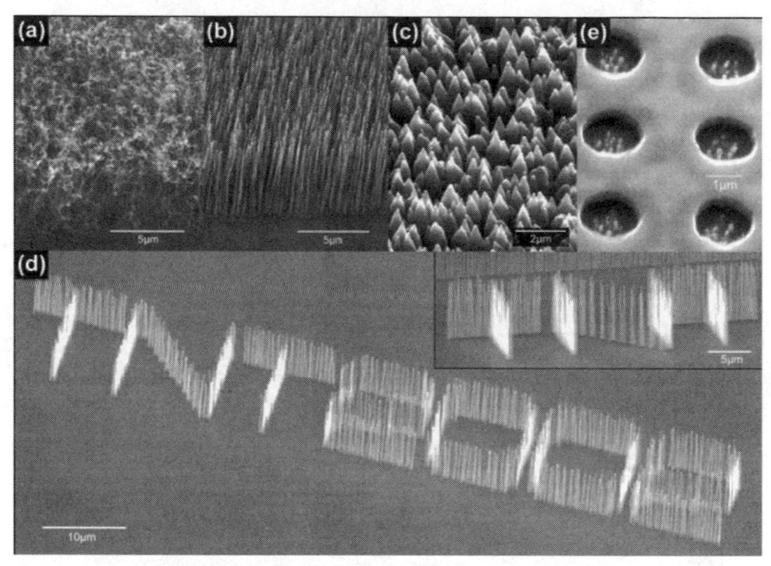

Fig. 4.28 (a) Curly, non-aligned carbon nanotubes deposited by thermal CVD. (b) Vertically aligned carbon nanotubes deposited by PECVD. (c) Conical carbon nanotubes deposited using a high (75%) C_2H_2 ratio in the gas flow. (d) Example of patterned growth of carbon nanotubes. (e) Growth of vertically aligned carbon nanotubes inside a gated field emission microcathode. The tilt in these images were 45° except in the inset of (d) where a 75° tilt was used. Reprinted with permission from Institute of Physics Journal publications.[51]

High frequency Plasma-assisted CVD can also be used for thin film deposition at high rates. At these frequencies, the deposition rate is increased due to the absence of ion bombardment in the growth zone. Very high frequency PECVD (VHF-PECVD) allows for deposition rates higher than 1 nm/s.[52] A major drawback of VHF-PECVD, is that uniform deposition over large areas area is hard to obtain.

Microwave Plasma CVD is currently the method of choice for the deposition of nanocrytalline diamond thin films with smooth surfaces[53] as well as carbon nanotubes.[54]

4.5.3 Atomic Layer CVD (ALCVD)

In *ALCVD*, which is also called *atomic layer epitaxy* and *atomic layer deposition* (*ALD*), two complementary precursors are alternatively introduced into the reaction chamber, forming a deposition cycle (**Fig. 4.29**). Typically, one of the precursors adsorbs onto the substrate surface, until it saturates the surface and further growth cannot occur. The precursor doesn't completely decompose by itself. Rather, the decomposition is activated through the introduction of a second precursor. At the end of each cycle, only a monolayer of the desired film is grown on the substrate.

In ALCVD process, film thickness and uniformity are precisely controlled by the number of deposition cycles rather than the deposition time.

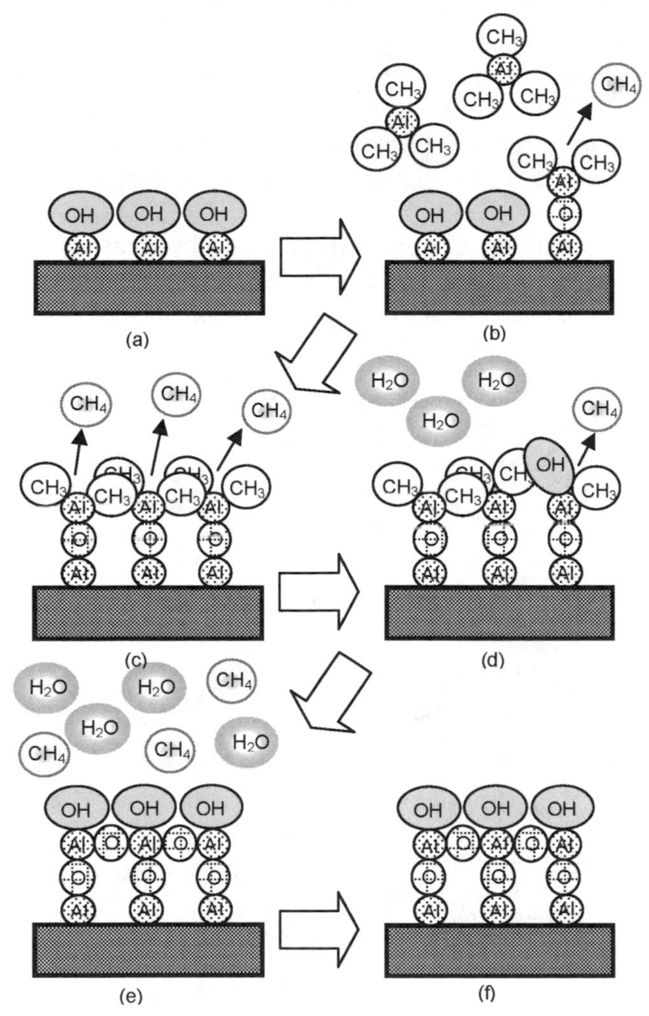

Fig. 4.29 The first half of the cycle in the formation of Al_2O_3 from trimethylaluminium, $Al(CH_3)_3$, and water. (a) starting with a hydroxylated precursor surface, (b) reaction of $Al(CH_3)_3$ and OH to form methane and adsorbed aluminium – precursor fragments, (c) after a complete reaction, and removal of excess CH_4 (e.g. by purging with N_2) (d) interaction with H_2O, (e) complete saturation of the surface, (f) purging gas, the resulting monolayer.

4.5.4 Atmospheric Pressure Plasma CVD (AP-PCVD)

The *AP-PCVD* is conducted at atmospheric pressures. In AP-PCVD, atmospheric pressures and very high frequency (VHF) plasma are employed to obtain ultra-high deposition rates (**Fig. 4.30**). On the contrary, a conventional fixed electrode for generating the plasma at atmospheric pressure will result in poor quality thin films that are non homogenous and non conformal to the substrate. With a high-speed rotary electrode, both the homogeneity and conformation of thin films can be dramatically improved. With such a rotating electrode, and the help of the high speed gas flow, the byproducts in the plasma, which degrade the thin film quality, can be removed.

AP-CVD is becoming attractive for deposition of metallic, insulating and even polymeric coatings in industrial processes. Nanostructured thin films with highly ordered grains can also be obtained using this method.

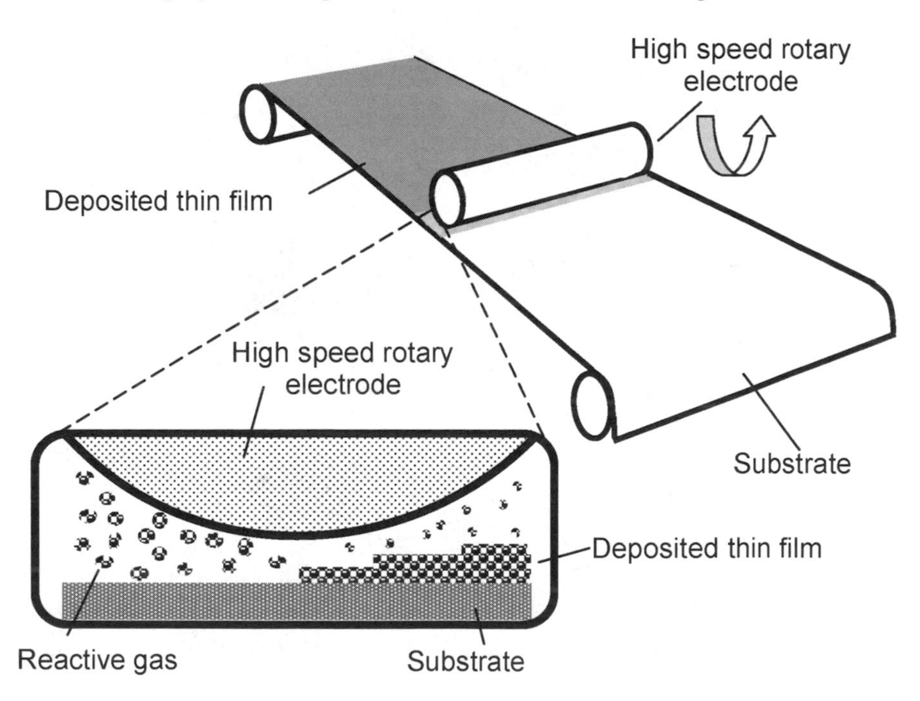

Fig. 4.30 The AP-PCVD set-up and the deposition process.

4.5.5 Other CVD Methods

There are many derivations of the CVD processes, which are basically enhancements of the processes that have already been presented. For instance: *remote plasma enhanced CVD* (*RPECVD*) , which is similar to PECVD except that the substrate is not directly in the plasma discharge region. This means the deposition can be conducted at lower temperatures. Another method is *ultra high vacuum CVD* (*UHV-CVD*) which is similar to LPCVD except that it operates at very low pressures, typically in the range of 1 to 10 Pa.

4.6 Liquid Phase Techniques

There are many techniques for the deposition of thin films in liquid phase such as *Aqueous Solution Techniques* (*AST*), *Langmuir-Blodgett* (*LB*), *electrochemical deposition.*

In these methods: (a) thin films can be deposited on non-planar substrates, (b) the techniques are simple and much less expensive, (c) there is reduced reliance on expensive or sensitive organometallic precursors (such as in MOCVD) and less hazardous byproducts are produced, and (d) thin films of single crystals (for ceramics), or homogenous monolayer thin films (for organic materials) can be obtained.

4.6.1 Aqueous Solution Techniques (AST)

The *AST* are generally used for synthesizing ceramic thin films from aqueous solutions at low temperatures (20–100°C).[55,56] ASTs have not yet reached the same levels of popularity of the universal deposition techniques such as CVD and PVD yet. Nevertheless, their potential to produce ceramic films over large areas at comparatively low cost has already made these techniques attractive and has stimulated a surge of interest in them since the mid 1980's. These techniques are also becoming increasingly popular in nanotechnology as they can be employed for the formation of nanostructured thin films. The main techniques are listed below:

Chemical Bath Deposition (CBD)

Using *CBD*, thin films are produced by immersion of the substrate in a liquid with tight control of the kinetics of formation of the solid. Most CBD reports in the literature concern sulfide and selenide thin film deposition. However, this technique has also been recently used for depositing metal oxide thin films, which have potential applications in semiconductor technology. They can also be efficiently used for the deposition of highly ordered and single crystalline array of nanostructures.

In CBD one or more metals in the form of salts, M^{n+}, and a complexing agent, L^-, in aqueous solution are used. The processes that occur in the CBD solution in general consist of the following steps:

1. Equilibrium between the complexing agent and water;
2. Formation/dissociation of ionic metal-ligand (ligands are the ions or molecules surrounding the metal ion) complexes,
3. Hydrolysis,
4. Formation of the solid.

For instance, during the deposition of ZnO the following reaction takes place:

$$Zn(NO_3)_2 + 2NaOH \rightarrow Zn(OH)_2 + 2NaNO_3 \rightarrow ZnO(s) + 2NaNO_3 + H_2O. \quad (5.9)$$

An example of ZnO nanorods arrays deposited with the CBD method is shown in **Fig. 4.31**. The substrate was 64° YX $LiNbO_3$. Initially a very thin ZnO seed layer was deposited using the laser pulsed deposition technique, and subsequently the substrate was submerged in hot aqueous solutions of $Zn(NO_3)_2$ and NaOH. The growth temperature and deposition time were 70°C and 60 min, respectively.

Fig. 4.31 SEM image of ZnO nanorods deposited on ZnO/64° YX $LiNbO_3$ substrate in 1 μm scale grown by the CBD process.

Successive ion layer adsorption and reaction (SILAR)

The distinguishing characteristic of *SILAR* is the use of alternating aqueous solutions (a metal salt solution, followed by a hydrolyzing or sulfidizing solution). In principle, this allows molecule-by-molecule growth of the compound film via the sequential addition of individual atomic layers.

Liquid phase deposition (LPD)

LPD refers to the formation of oxide thin films from an aqueous solution of a metal–fluoro complex $[MF_n]^{m-n}$ which can be hydrolyzed by adding water to a scavenging agent such as boric acid (H_3BO_3) or aluminum metal. In the LPD process, it is possible to produce thin metal oxide films films directly on a substrate that is immersed in a treatment solution for deposition.[57]

In this process, metal oxide thin films can be formed via a two step reaction. In the first step, which is an equilibrium reaction, a metal-fluoro complex ion ($MF_x^{(x-2n)-}$) reacts with water to form the desired metal oxide plus F^- and H^+ ions. However, the F^- and H^+ ions quickly react with the metal oxide and reform the original metal-fluoro complex:

$$MF_x^{(x-2n)-} + nH_2O \rightarrow MO_n + x\,F^- + 2nH^+. \qquad (5.10)$$

The equilibrium reaction can be shifted towards the right-hand side of the chemical equation through the addition of a scavenging agent such as H_3BO_3. This acid readily reacts with F^- to form HBF_4 and $3H_2O$:

$$H_3BO_3 + 4H^+ + 4F^- \rightarrow HBF_4 + 3H_2O. \qquad (5.11)$$

The advantage of producing HBF_4 is that it does not interact with MO_n, and as a result, the MO_n will be eventually formed on the substrate.[58,59] The block diagram of the process is shown in **Fig. 4.32**.

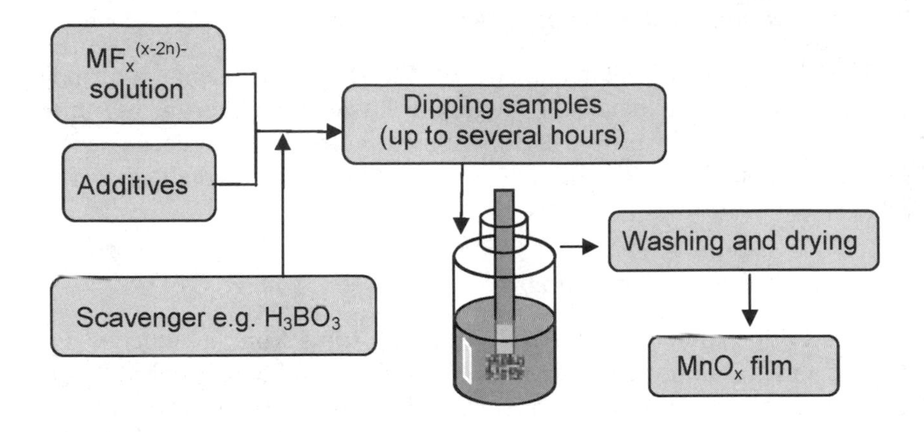

Fig. 4.32 The block diagram of a LPD process.

Electroless deposition (ED)

In the *ED* method the thin films are produced without the use of a counterelectrode or a connection to an external electrical power source[55]. This method involves a change in the oxidation state of the metal cations (dissolved in aqueous solution) to the insoluble neutral metallic state.

The substrate and subsequently the film surface participate in the redox process, usually as a path for transferring the electrons from the site of oxidation to the site of reduction. In this process, a catalyst is generally required to initiate and sustain the process.

4.6.2 Langmuir-Blodgett (LB) method

A *Langmuir-Blodgett* layer is deposited from the surface of a liquid onto a solid (**Fig. 4.33**), at the air-water interface, by immersing (or emersing) the solid substrate into (or from) the liquid. With each immersion or emersion step a monolayer is added to the surface, thus films with very accurate thickness can be formed. [60] Generally, a LB film contains one or more monolayers of an organic material. However, inorganic materials can also be sandwiched between such layers. The LB method is widely employed for the deposition of *amphiphilic molecules* or macro-molecules with *amphiphilic segments*.

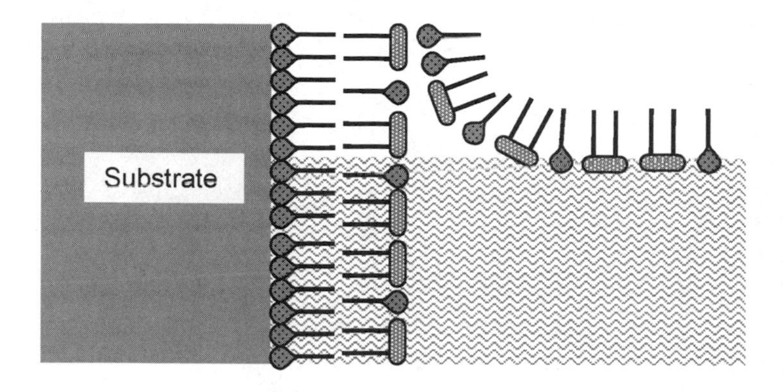

Fig. 4.33 Deposition of LB molecular assemblies of lipids onto solid substrates.

The monolayer is formed by spreading the organic molecules on the surface of water. These molecules are amphiphilic, usually having hydrophilic and hydrophobic parts. These molecules are generally free standing on the surface with their hydrophilic part anchoring to the substrate (surface-pressure diagrams **Fig. 4.34**).

Initially, the molecules are loosely packed and spread out on the water surface, forming a so called gas phase state (**Fig. 4.34**). The term is used because the area on the water surface available for each molecule is rather large, and the molecules do not interact with each other. In this state, the surface pressure is low. By using one or two sliding barriers in the fabrication apparatus (as shown in **Fig. 4.35**), the surface pressure can be increased.

At a certain point the surface pressure starts to increases rapidly, which results in a transition to the liquid phase (**Fig. 4.34**). As the barrier continues to close in, an even steeper rise in the surface pressure causes the onset of the solid phase. If we further increase the pressure, then ultimately, the monolayer will collapse (typically at a force of 30-60 mN/m). Orientation and local packing of molecular fragments can be controlled by the surface pressure during monolayer formation.

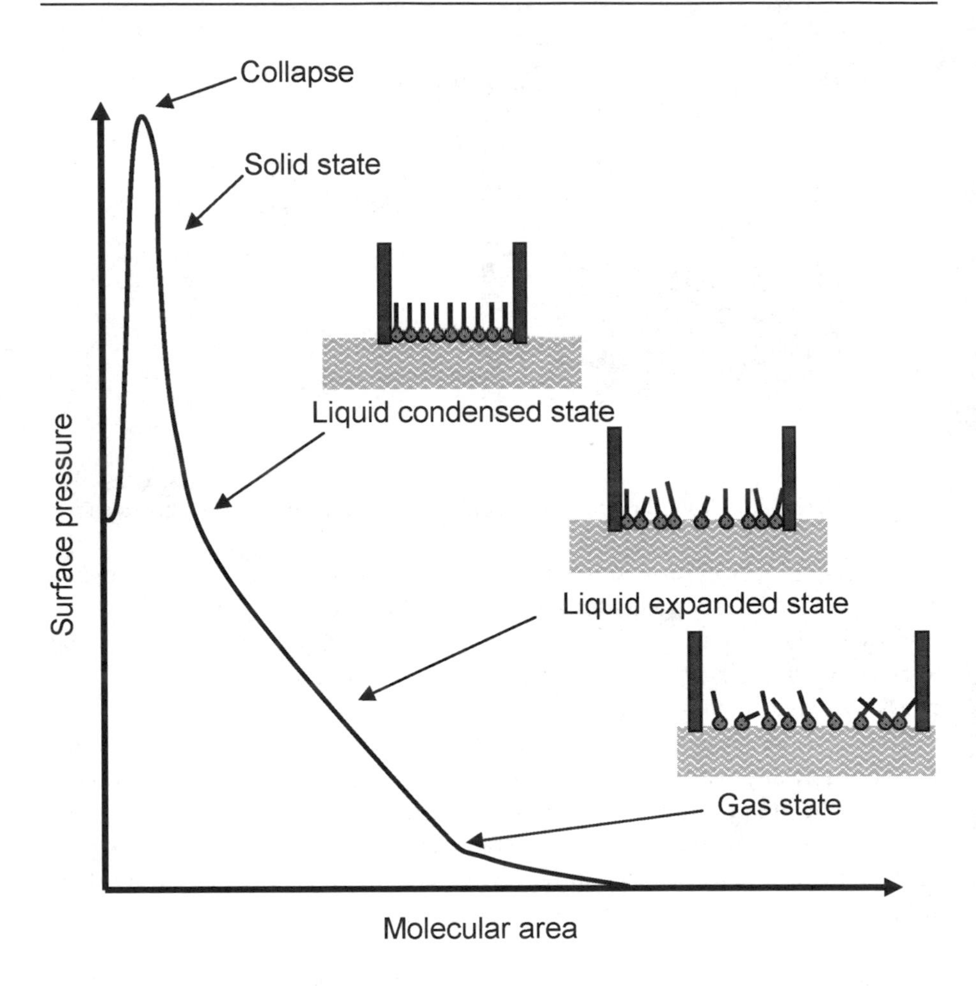

Fig. 4.34 Surface pressure - area isotherm – the process occurring in the LB method.

The LB technique can be used to produce well ordered multilayer structures which are expanded over the whole thickness of the film. The LB method can be used to produce accurate superlattices. However, if the number of layers is too large, then weak physical interactions result in low mechanical and thermal stability.

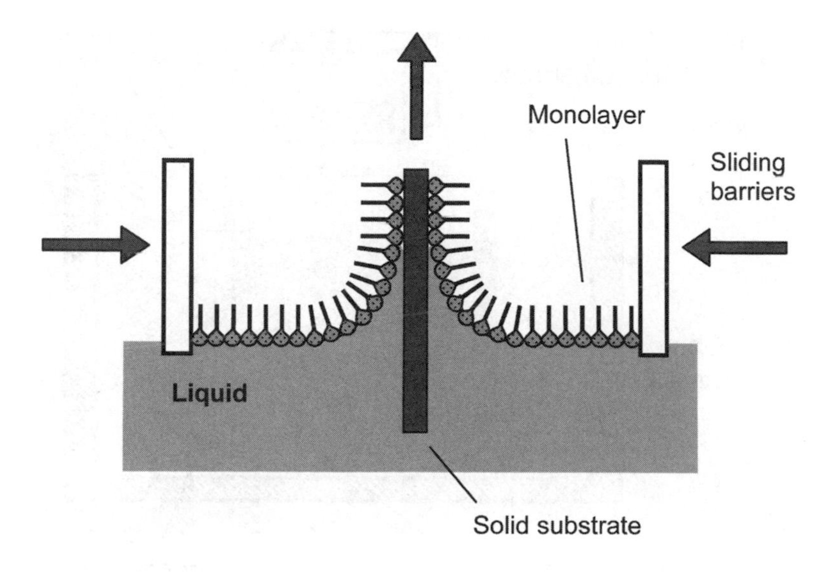

Fig. 4.35 Deposition of a monolayer on a solid substrate.

The thickness of the monolayers depends upon molecular length, orientation, rigidity and the type of packing. Their typical thickness is in the range 1-5 nm, while the roughness of LB films of intermediate thickness is in the range 0.1 to several nm.[61] LB films are widely applied to molecular electronic and bioelectronic devices because characteristics such as thickness and molecular arrangements are controlled at the molecular level.[62]

4.6.3 Electro-deposition

Also known as *electroplating*, *electro-deposition* is typically limited to electrically conductive materials. In this process, the substrate, which has a conductive surface, is placed in an electrolyte solution. A counter electrode is also placed in the electrolyte, and when a voltage is applied between the substrate and the counter electrode, a chemical redox process takes place on the substrate's surface (**Fig. 4.36**). This results in the deposition of a layer on the substrate.

The electro-deposition process is well suited for making thin films of metals such as gold, copper and nickel, whose thickness can be in the range from a few nm to well in the excess of 100 μm.

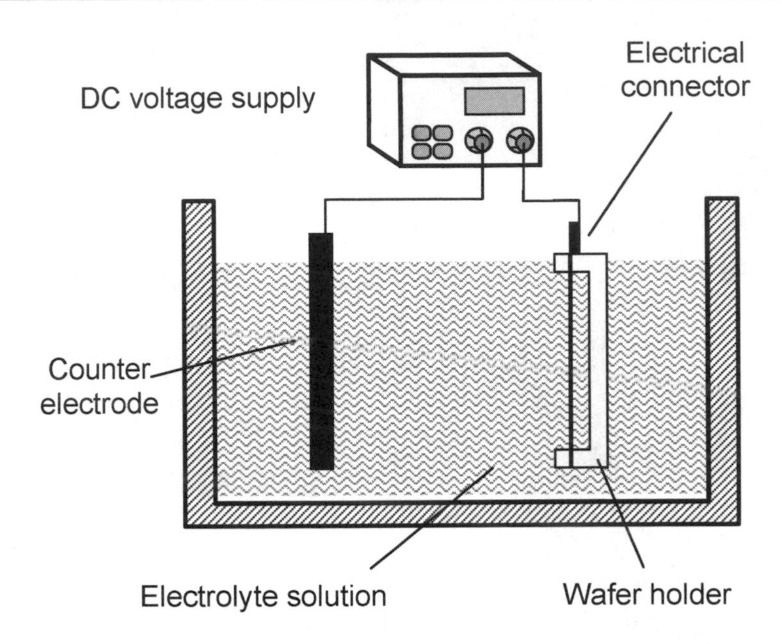

Fig. 4.36 A typical system used for electro-deposition.

Conventional electro-deposition methods can be employed for the fabrication of 1D materials. Nanopores membranes and substrates (**Fig. 4.37**) can be filled with metallic elements and their alloys via electrodeposition (**Fig. 4.38**). An auxiliary electrode can be used in the process to gain a larger control over the distribution of the electric field in the deposition media. Electro-deposition is quite a versatile and inexpensive technique to fabricate nanoscaled arrays with systematically reproducible properties.[63-66]

Electro-deposition of nanostructures can also be conducted without the assistance of nano-templates. As can be seen in **Fig. 4.39**, a nanostructured film, consisting of polyaniline nanofibers, has been deposited on a gold surface using an electrolyte containing aniline monomer.

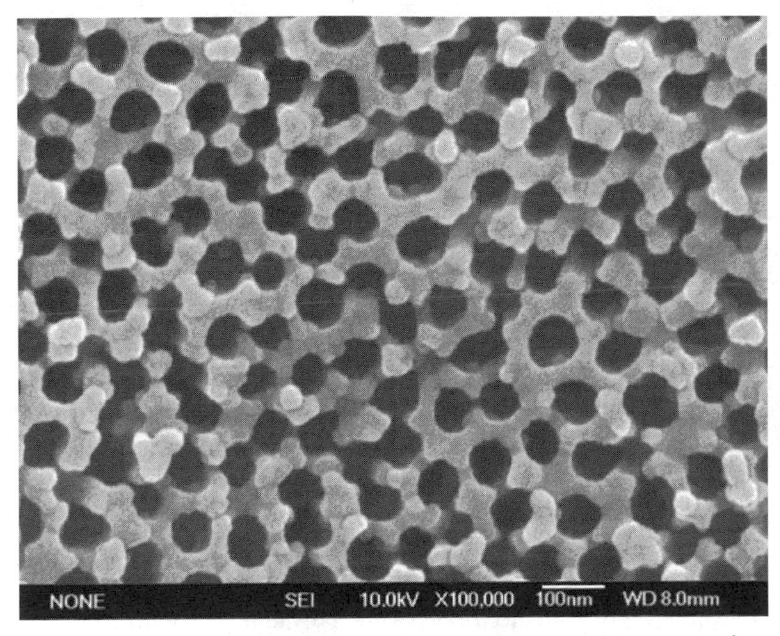

Fig. 4.37 SEM image of an anodized aluminum oxide (AAO) nano template.

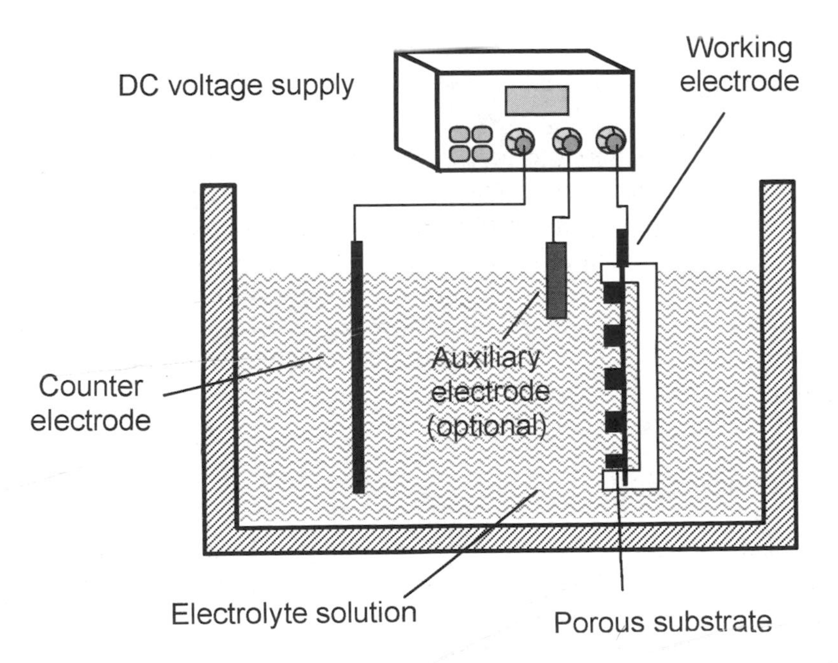

Fig. 4.38 A schematic electrode arrangement for the synthesis of nanowires though restrictive template-assisted electro-deposition.

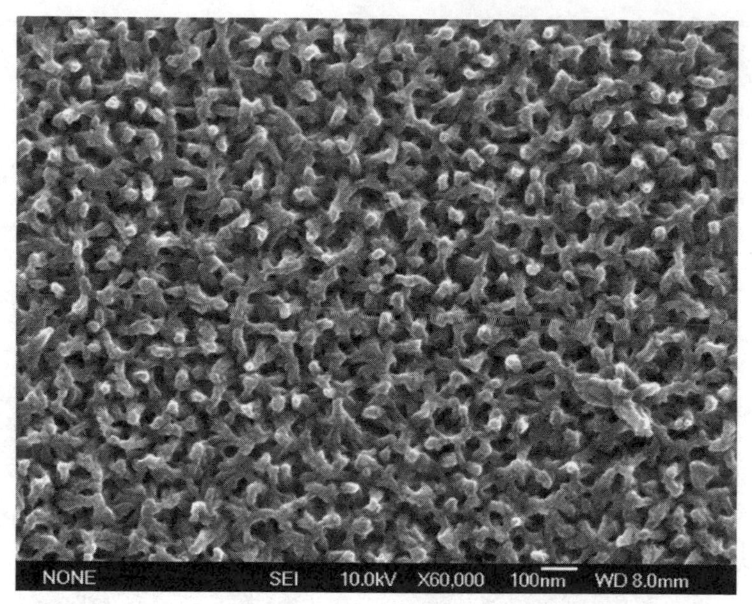

Fig. 4.39 An SEM image of electro-deposited polyaniline nano fibers.

4.7 Casting

In *casting*, the material to be deposited is in liquid state, generally in a solvent. This liquid is placed on the substrate by different means that include: *spin coating*, *drop casting*, *dip coating*, *spraying*. The solvent is then allowed to evaporate, leaving behind a thin film of the material on the substrate. This is particularly useful for organic and organometallic materials, which may be easily dissolved in organic solvents. The thicknesses of cast films can range from a single monolayer of molecules to tens of micrometers.

4.7.1 Spin Coating

Spin coating is a well-established technology, which is commonly used in the electronics industry.[67] Adjusting the rate and speed of spinning can control the uniformity and thickness of the film.[68]

Essentially, the spin coating process can be divided into the following four steps:[69]

(a) A small volume of liquid is placed on the surface.

(b) The substrate is initially spun-up and the liquid flows radially outwards, driven by the centrifugal force.
(c) While spinning, the excess liquid flows to the perimeter and leaves the surface as droplets. The thinner the film is, the larger the resistance of flow is.
(d) The final stage is evaporation, which is the primary mechanism of solidification.

A summary of the spin-coating process is seen in **Fig. 4.40**. The process of spinning is a reliable method for the deposition of thin films because the liquid tends towards a uniform thickness during spin-off. The film thickness tends to remain uniform, provided that the viscosity of the solvent is not shear dependent and does not vary across the substrate.[70]

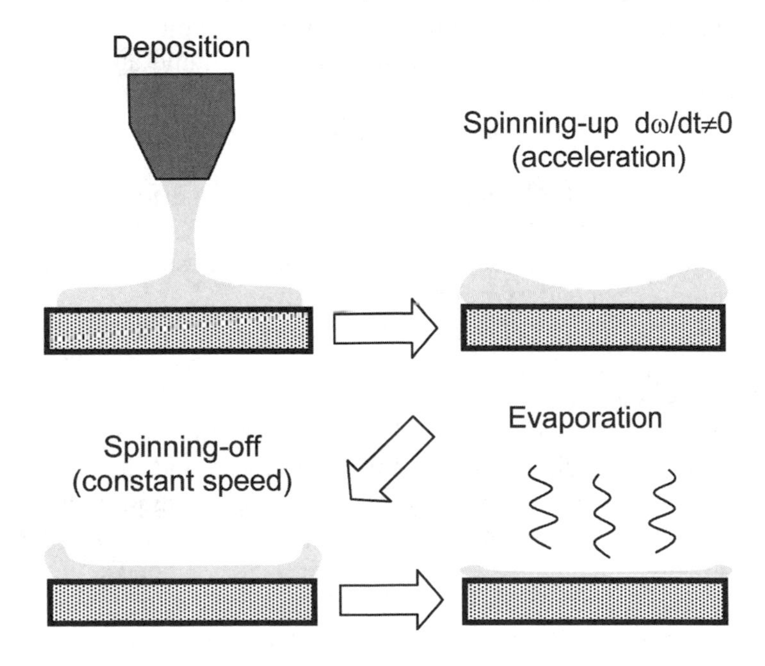

Fig. 4.40 The spin coating process.

To minimize contamination, the spin coating should be carried out in a clean environment, namely a *clean room*. The substrates are mounted on a spinning stage and held in place with suction.

Following the spin coating step, the substrates are placed in an annealing furnace to remove the solvent. Annealing is a heat treatment process in which a material is subjected to elevated temperature for an extended time and then slowly cooled down. Both the heating and cooling stages are controlled.

4.7.2 Drop Casting, Dip Coating and Spraying

Drop casting simply involves placing drops of the liquid containing the desired material onto the surface of the substrate, and subsequently allowing the solvent to evaporate.

Dip coating refers to the immersion of a substrate into a liquid coating the coating material, removing the substrate from the liquid, and allowing the excess liquid to drain off. The solvent can then be evaporated, leaving a solid thin film on the surface. Film thickness can be controlled with several parameters, including the rate at which the substrate is immersed and removed from the liquid, the immersion time, the liquid and substrate intrinsic properties (e.g. concentration, viscosity, rate of interaction between the surface and liquid etc.), and the number of times that the process is repeated.

In *spray coating,* the liquid containing the material to be deposited is sprayed in droplets from a nozzle onto the substrate's surface. Spray coating is commonly used nowadays for depositing nanostructured thin films as it can produce nanosized droplets of the solvent.

Thermal spray is one of the common spray coating techniques. It involves spraying melted (or heated) materials onto a surface. The thermal energy can be supplied by electrical (plasma or arc), or chemical means (combustion flame). *Cold spray* is a high kinetic energy coating process. In this method the particle spray velocity is increased and the particle temperature is reduced. The kinetic energy of the impinging particles is sufficient to produce considerable plastic deformation and high interfacial pressures and temperatures, can produce high energy bonds.[71] Despite its popularity, it does not result in uniform layers, making it impractical for producing thin films for certain applications. Other spray methods include *electrophoresis spray*, *thermophoresis spray* and *settling spray*.

In all of the casting techniques, the deposition strongly depends on the substrate's surface morphology and chemical composition. The degree of surface hydrophobicity is also a critical factor in film morphology.

4.8 Sol-gel

Sol-gel is a colloidal suspension that can be gelled to form a solid.[69] Although in the existence since the early 19th century, the sol-gel process gain popularity in the 1950s due to the work of Della and Rustrum Roy, who used the technique to prepare multi-component glasses.[72] Since then, the process has been employed to fabricate films for a vast number of industries, with applications in optics, electronics and of course sensors. Using the sol-gel process, it is possible to fabricate ceramic or glass materials in a wide variety of forms: ultra-fine or spherical shaped powders, thin film coatings, ceramic fibers, micro/nano-porous inorganic membranes, monolithic ceramics and glasses, or extremely porous aerogel materials.

In general, the sol-gel process involves the transition of a system from a liquid "sol" (mostly colloidal) into a solid "gel" phase.[69] The precursors for synthesizing these colloids consist of metal or metalloid elements surrounded by reactive ligands. Metal alkoxides (an alkoxide is the conjugate base of an alcohol, and therefore has an organic group bonded to a negatively charged oxygen atom) are popular precursors because they react readily with water. The most widely used metal alkoxides are the alkoxysilanes such as tetramethoxysilane (TMOS) and tetraethoxysilane (TEOS). However, other alkoxides such as aluminates, titanates, and borates are also commonly used in the sol-gel process.

The precursor is subjected to a series of hydrolysis and polymerisation reactions which lead to the formation of a new phase of *colloidal suspension*. A series of standard sol-gel reactions for organometallic precursors is shown below:[72]

$$\begin{aligned} &\text{M-O-R} + H_2O \rightarrow \text{M-OH} + \text{R-OH (hydrolysis)} \\ &\text{M-OH} + \text{HO-M} \rightarrow \text{M-O-M} + H_2O \text{ (condensation)} \\ &\text{M-O-R} + \text{HO-M} \rightarrow \text{M-O-M} + \text{R-OH (condensation)} \end{aligned} \tag{5.12}$$

where M is the metal, and R is an alkyl group. Once the sol is cast into a mould, the colloidal particles condense into a new phase, gel, in which a solid macromolecule is immersed in a liquid phase (solvent).

It is possible to produce thin films on a substrate with methods such as spin coating, dip-coating, and drop casting. With further drying and heat-treatment, the gel is converted into dense ceramic or glass. If the liquid in a wet gel is removed under a supercritical condition, a highly porous and extremely low-density material called an *aerogel* is obtained. The viscosity of a sol can be adjusted to a desired viscosity range to obtain ceramic fibers. Ultra-fine and uniform ceramic powders can also be formed by *precipitation*, *spray pyrolysis*, and *emulsion* techniques.

After the formation of the sol-gel film, it is generally annealed. In order to prepare high quality thin films, it is of paramount importance to control the annealing parameters. Annealing is carried out in order to convert the organometallic film to a metal oxide in addition to enabling and controlling crystallisation and grain growth. It also serves to remove organic compounds, which usually evaporate off the film at elevated temperatures. Furthermore, it promotes adhesion to the substrate.

Unlike many other fabrication methods, which require starting materials to have the same composition as the final product, sol-gel offers an economically feasible avenue to explore different ratios and combinations of the compounds.

The sol-gel method can be employed to form nanostructured thin films for sensing applications and the grain size of these thin films can be engineered for the right application. As can be seen in **Fig. 4.41**, nanostructured In_2O_3 was prepared using ethanolic solutions of an indium isopropoxide ($In(OC_3H_7)_3$) precursor. The solutions were spin coated on sapphire substrates and annealed at 500°C for 1 h in air. This resulted in a well-developed polycrystalline and highly porous microstructure composed of approximately spherical grains with an average size of approximately 20 nm.[73]

Fig. 4.41 SEM photomicrograph of an In_2O_3 film fabricated using the sol-gel method. Reprinted with the permission from the Elsevier publications.[73]

4.9 Nanolithography and Nano-Patterning

Nanolithography and *nano-patterning* are processes for making patterns on substrate surfaces with nanometer precision. The basis of lithography is very old. *Photolithography* has been used in microtechnology for more than 50 years. However, when we wish to position atoms or molecules precisely on surfaces many problems can occur. Many of such problems are due to the quantum nature of atoms and physical phenomena occurring at nanoscale dimensions. In this section we introduce some of the most widely utilized nano-patterning techniques.

4.9.1 Photolithography

This is a process that is extensively employed throughout the semiconductor industry to transfer a pattern from a photomask onto a substrate's surface. Photolithography coupled with the etching process (**Fig. 4.42**) involves the combination of the following steps which are described below:

1. *Thin film deposition*: The thin film material to be patterned is first deposited onto substrate. Any of the previously mentioned deposition techniques can be used.
2. *Application of photoresist*: A thin film of photoresist is deposited on the surface by spin-coating.
3. *Soft-baking*: the substrate is heated in an oven to evaporate the photoresist's solvent, solidifying and curing it.
4. *Exposure*: upon exposure to light (generally UV), the photoresist undergoes a chemical reaction. Depending on the photoresist's chemical composition, it can react in two different ways when the light impinges on it. A *positive photoresist*, is polymerized where it is exposed to light, however, the opposite is true for *negative photoresist*.
5. *Developing*: after exposure, the sample is soaked in a developer solution which removes the un-polymerized portion of the photoresist. This leaves a copy of the photomask's pattern on the surface.
6. *Etching:* the uncovered surface is etched. This typically involves immersion in a solution which dissolves the deposited thin film whilst not having any effect on the polymerized photoresist.

There is another type of lithography that utilizes the *lift-off* process (**Fig. 4.42**). In the lift-off process, a pattern is first defined on a substrate with standard photolithographic technique. Then a film is deposited over the substrate, covering the patterned photoresist and areas in which the photoresist has been cleared. During the actual lifting-off, the photoresist under the film is removed with a photoresist solvent taking the film with it. This leaves the substrate with only the film which was deposited directly onto it.

Limiting factors for achieving high resolution patterning are the wavelength of the light that is used in the photolithography process, and the diffraction caused by the mask and photolithography system. At present, *deep ultraviolet* (*DUV*) light with wavelengths of 250 and 190 nm are utilized. This allows feature sizes down to 50 nm. The current commercial standard for photolithographic system is 193 nm DUV.

The absolute resolution limit in a photolithographic system is given by a quarter-wavelength divided by the *numerical aperture* (*NA*).[74] The NA is equal to $n\sin\theta$, where n is the refractive index of the media in which the lens of the photolithographic mask is placed, and θ is the half-angle of the maximum cone of light that can enter or exit the lens. For a system with lenses operating in air, NA is equal to 1. This will limit the feature size to approximately 50 nm for a 200 nm wavelength.

It is possible, to achieve feature sizes less than 50 nm using DUV light together with *liquid immersion techniques*, which involve immersing the optics and samples in a liquid. This enables the use of optics with NAs exceeding 1. In this technique, deionised water is typically used as the photolithographic media, as it has a refractive index equal to 1.35, which is higher than that of air ($n = 1$). This will allow the effective NA to be increased which enhances the resolution without changing the light source. Features sizes less than 30 nm have been reported by researchers at IBM using this technique with a standard wavelength of 193 nm.[75]

Another new alternative is *extreme ultraviolet lithography* (*EUV*) . EUV lithography systems are currently under development.[74] They employ 13.5 nm wavelengths, which is approaching X-rays wavelengths of the electromagnetic spectrum. However, the biggest challenges in lens production for EUV remains in producing very flat mirrors (aberration free) and accurate lenses (or magnetic reflectors and lenses) with defect free coatings, and thermal management systems ensuring accurate imaging.

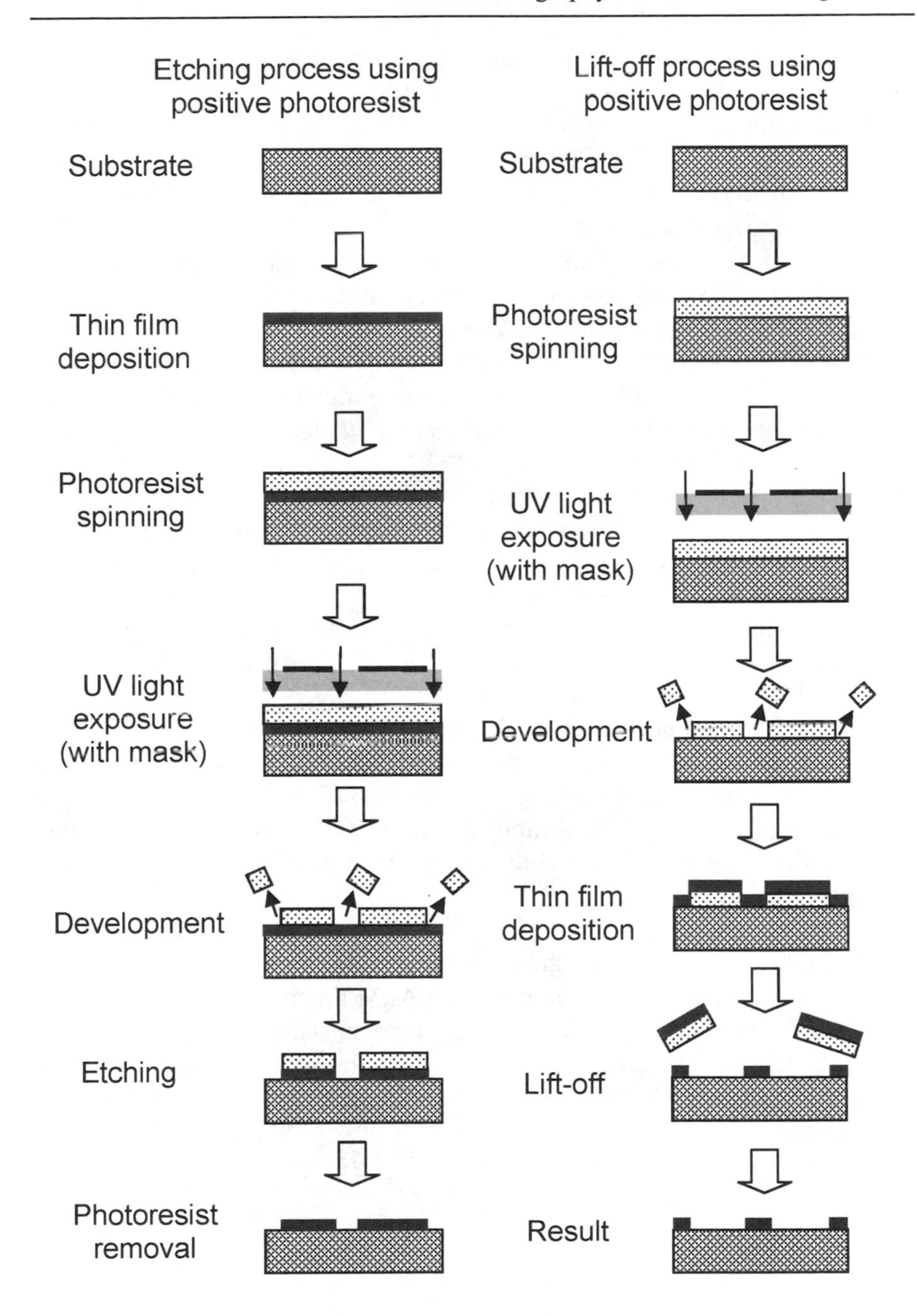

Fig. 4.42 The main processes in the photolithography.

4.9.2 Scanning Probe Nanolithography Techniques

A scanning probe tip, such as an *atomic force microscope* (*AFM*) tip, can be either to machine and manipulate surface molecules, or to place nano-dimensional particles on a surface. The detailed description of the scanning probe techniques is given in Chap. 5.

AFM can be employed to machine complex, high resolution patterns and to form free standing structural objects in thin layers.[76] The AFM tip can pattern lines that are less than several nanometers in resolution; and furthermore these patterns can be manipulated on the substrate surface with the AFM tip. These have applications in *nanometer-scale diffraction gratings*, *high-resolution lithography mask making*, and possibly the *assembly of nanostructures*.

In one of the pioneering examples, researchers at IBM used a *scanning tunneling microscope* (*STM*) at low temperatures (4 K) to position individual xenon atoms on a single-crystal nickel surface with atomic precision (carried out at the IBM laboratories).[77]

A tip of the AFM can also be used in a similar manner to a pen. The AFM tip can be coated with the material to be deposited (e.g. thiol molecules that are used to form self assembled monolayers). During the process of the AFM tip movement, the molecules migrate from tip to the substrate surface and as a consequence, a nanoscopic pattern is formed. This type of lithography is called *dip-pen nanolithography* (*DPN*) . DPN allows routine patterning with a lateral resolution of just a few nanometers.[78,79] At this stage, DPN is restricted to being a laboratory procedure, however it is nearing commercial availability.

A schematic representation of DPN for the deposition of molecules such as alkanethiols onto gold is shown in **Fig. 4.43**. The water meniscus (a crescent-shaped body) forms between the AFM tip coated with the alkanethiol and the gold substrate. The size of the meniscus, which is controlled by relative humidity, affects: (a) the rate of transport of the alkanethiol molecules, (b) the effective tip-substrate contact area, and (c) the DPN resolution.

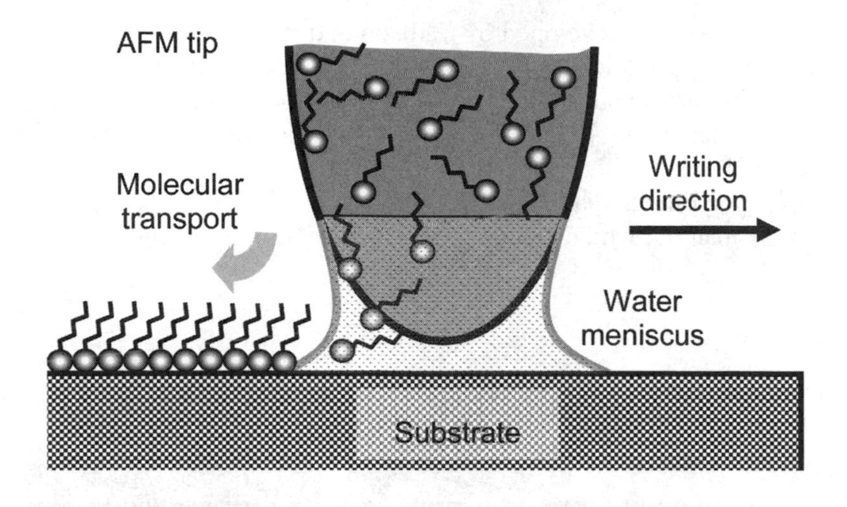

Fig. 4.43 The DPN process using an AFM.

4.9.3 Nanoimprinting

Another class of lithographic techniques can be termed as *embossing techniques* or *nanoimprinting*. The underlying principle is to imprint a pattern on the surface using a nanostructured stamp which is pressed against the surface, leaving a characteristic pattern behind. This technique is sometimes used in combination with UV exposure or thermal curing which stabilizes the pattern (generally a light sensitive film).

There are many different types of nanoimprint lithography methods. The two most important are: *thermoplastic nanoimprint lithography* (*T-NIL*) and *step and flash nanoimprint lithography* (*SF-NIL*) .

The earliest and most mature nanoimprint lithography is T-NIL which was developed in the mid 90s by SY Chou's group.[80] In this method, a thin layer of a thermal polymer is spin coated onto the substrate. A mold is then pressed against the spin coated polymer under a certain pressure. The polymer is cured by heating, and after cooling, the mold is separated from the sample and the patterned resist is left on the substrate. Chou et al.[81] fabricated arrays of 10 nm diameter and 40 nm periodic holes by nanoimprinting in polymethylmethacrylate (PMMA) thin films on both silicon and gold substrates. In this work, the smallest hole diameter successfully imprinted in PMMA was 6 nm.

SF-NIL was first developed by Willson and coworkers.[82] Instead of using a thermally curable resists (as in T-NIL), SF-NIL utilizes a UV curable resist on the sample substrate. The mold is comprised of a UV transparent material. After the mold and the substrate are pressed against each other, the resist is cured with UV light. An example shown in **Fig. 4.44**. In this example, polydimethylsiloxane (PDMS) photoresist has been used. PDMS is a "soft" material which is highly suited for nano-imprinting technology.

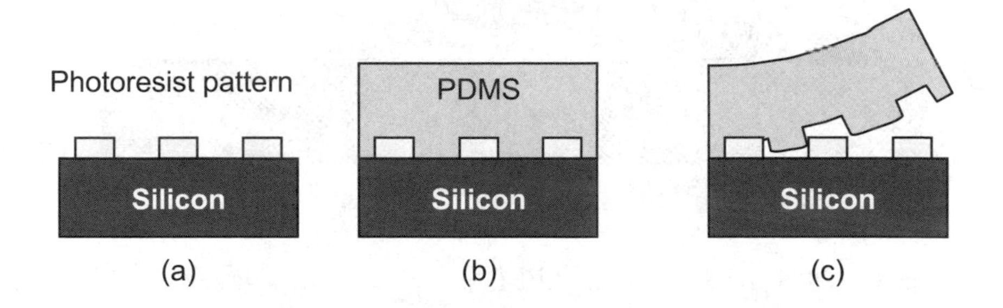

Fig. 4.44 An example of the SF-NIL procedure with PDMS as the UV curable resist.

Using this method, patterns of biological origin can be replicated. As can be seen in **Fig. 4.45** the wing of a cicada (an insect) has been molded in PDMS with the SF-NIL technique.

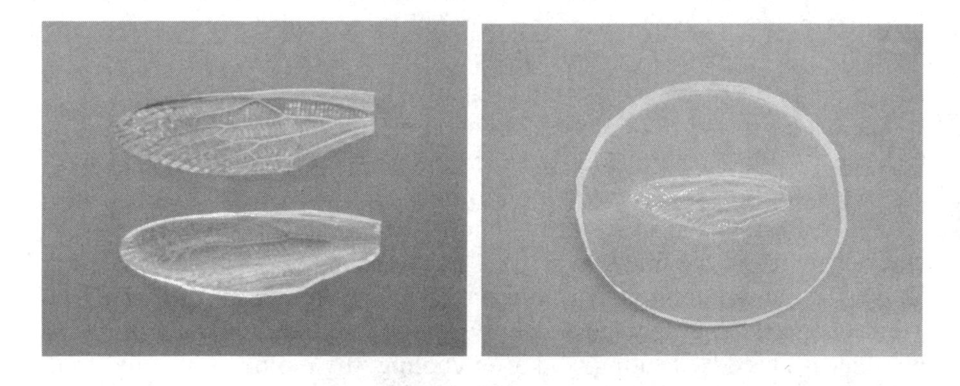

Fig. 4.45 Cicada wing (left) and its mold created by SF-NIL (right)

As shown in **Fig. 4.46**, the nanosized features in the cicada wigs surface have been successfully replicated with this technique.

Fig. 4.46 SEM images of (a) cicada wing's surface and (b) replicated in PDMS.

4.9.4 Patterning with Energetic Particles

Energetic particles such as electrons, ions, and electrically neutral metastable atoms may be used to form patterns with nanometer resolution in appropriate resist films. These particles have sufficiently short de Broglie wavelengths (<0.1 nm) which minimizes the diffraction effects that limits conventional photolithographic approaches as described earlier. Benefiting from small wavelengths, they can provide feature sizes smaller than that of the current standards achievable in the semiconductor industry. In these methods, a highly focused beam is scanned across the surface, leaving its mark and hence patterning the areas which it impinges on.

Electron beam lithography (EBL

EBL)is one of the most powerful methods for forming nano-patterns. In this technique, a beam of electrons is utilized to generate patterns on a substrate surface. It uses high voltage electrons in the range of 1 kV-100 kV to nano-pattern the deposited resist on the substrate by directly writing onto it. A very high resolution of several nm can be obtained with this method. Polymers such as poly(methyl methacrylate), or PMMA (that can also act as a resist), can be used.[83] The impinging electron beam causes local changes in the PMMA's solubility, and cause local chain scission and formation of nanopores, making the material become soluble in a developer solution.

The short wavelength associated with high-energy electron beams (for example, ~0.005 nm for 50 keV) gives EBL an extremely high resolution capability; approximately 0.5 nm focused spots were achievable very early in the development of this technique.[84,85] However, the resolution limit is determined not by the beam size but by forward scattering in the photoresist and secondary electron travel in the photoresist.[86] The effect of the forward scattering can be reduced with a higher energy electron beam or a thinner layer of photoresist. Unfortunately, the generation of secondary electrons is inevitable.

Despite the high resolution and ease of application, the commercial value of EBL is still limited. The main reason for this is speed. The beam must be scanned across the surface to be patterned as the pattern generation is produced in serial. This makes for the very slow pattern generation compared with techniques such as photolithography in which the entire surface is patterned at once.

An example of the EBL patterning which is used in conjunction with the lift-off process is shown in **Fig. 4.47**.

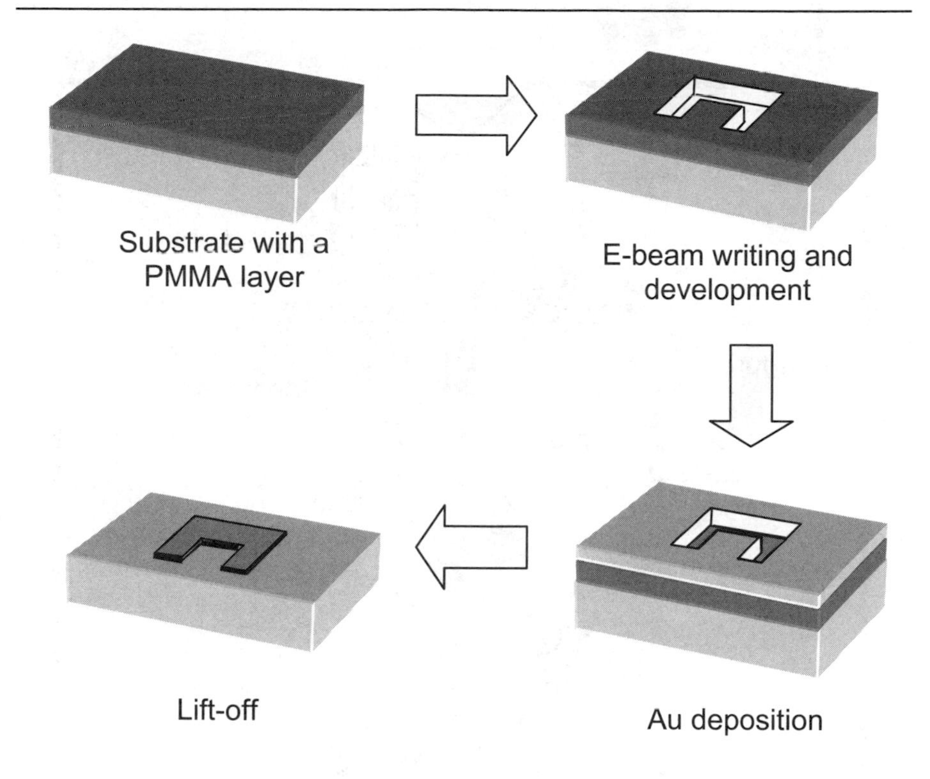

Fig. 4.47 Electron beam lithography and lift-off process for making Au patterns.

Focused Ion Beam (FIB)

FIB systems generally employ a highly focused beam of ions, such as Ga^+ (as gallium can be used to easily produce a liquid metal ion source), which is raster scanned over the sample surface in a similar manner to an electron beam in a scanning electron microscope.[87] However, it differs from an electron microscope in that the FIB is destructive to the specimen. Atoms on the sample surface are sputtered as the high-energy ions impinge on the sample. Furthermore, gallium atoms from the ion beam are also implanted into the top few nanometers of the sample surface.

As can be seen in **Fig. 4.48**, the primary Ga^+ ion beam strikes the sample surface. In doing so, a small amount of the material is sputtered and leaves the surface. These can be either neutral atoms (n^0) or secondary ions (i^+). Just as in sputtering, the incident ion beam also produces secondary electrons (e^-). Additionally, these secondary electrons, or sputtered ions, may also be collected by the detectors which convert the signal to an image of the surface.

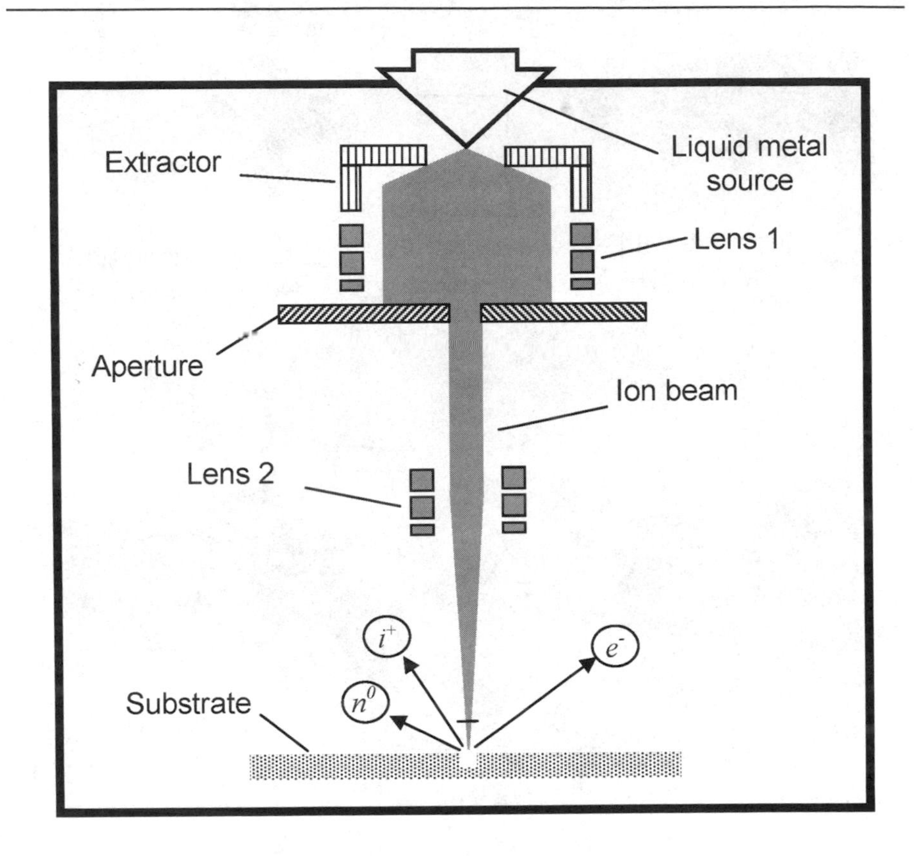

Fig. 4.48 A FIB system.

If the current in the primary ion beam is low, then only a very small amount of the material can be sputtered. However, at higher currents, significant amount of material can be removed due to sputtering. As a consequence, the specimen surface can be precisely milled with the ion beam.

In modern FIB systems, the resolution has been reduced to several nm. Ultrafine structures can also be fabricated with the FIB using direct deposition of metal (such as gold, iron, tungsten) or insulators (typically SiO_2). The applications of FIB not only include nano-sized-milling and patterning but also include: cross-sectional imaging through semiconductor devices, modification of the electrical routing on semiconductor devices, preparation for physico-chemical analysis, preparation of specimens for transmission electron microscopy (TEM), micro-machining, and production of photomasks with high resolutions. An example of patterning with FIB is shown in **Fig. 4.49**.

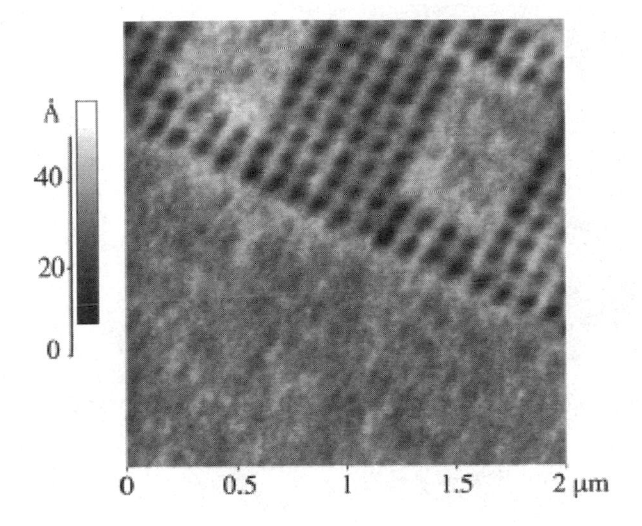

Fig. 4.49 An AFM image of nano-holes on an InP substrate surface created by FIB lithography using 30 keV Ga^{+} ions, current of 5– 6 pA and exposure time of 2 ms per point. Reprinted with the permission from the Elsevier publications.[88]

4.9.5 X-Ray Lithography (XRL) and LIGA

XRL is similar in principle to photolithography; however it uses collimated X-rays as the exposing energy. As X-rays have wavelengths that are much shorter in wavelength than visible or UV light, they can provide increased resolution. As early as 1972, Smith[89] proposed the use of X-rays in (semiconductor) lithography as a simple proximity imaging system. Although the idea itself is simple, the development of XRL systems has proved otherwise. These systems require a powerful and reliable X-rays source which can only be practically achieved using synchrotrons (although point source systems such as laser-driven plasma type are emerging as possible alternatives to synchrotrons). Other drawbacks include the requirement for a 1:1 mask and the small 10-20 μm gap between mask and substrate.[90]

In XRL, the primary beam consists of photons whose wavelengths are in the range between 0.5 and 4 nm. The most common substrate material used in this process is PMMA, as it exhibits high X-ray absorption (soft X-rays of 0.7–0.8 nm) and is sensitive to X-ray degradation.[91] An example of replication with XRL is shown in **Fig. 4.50**.

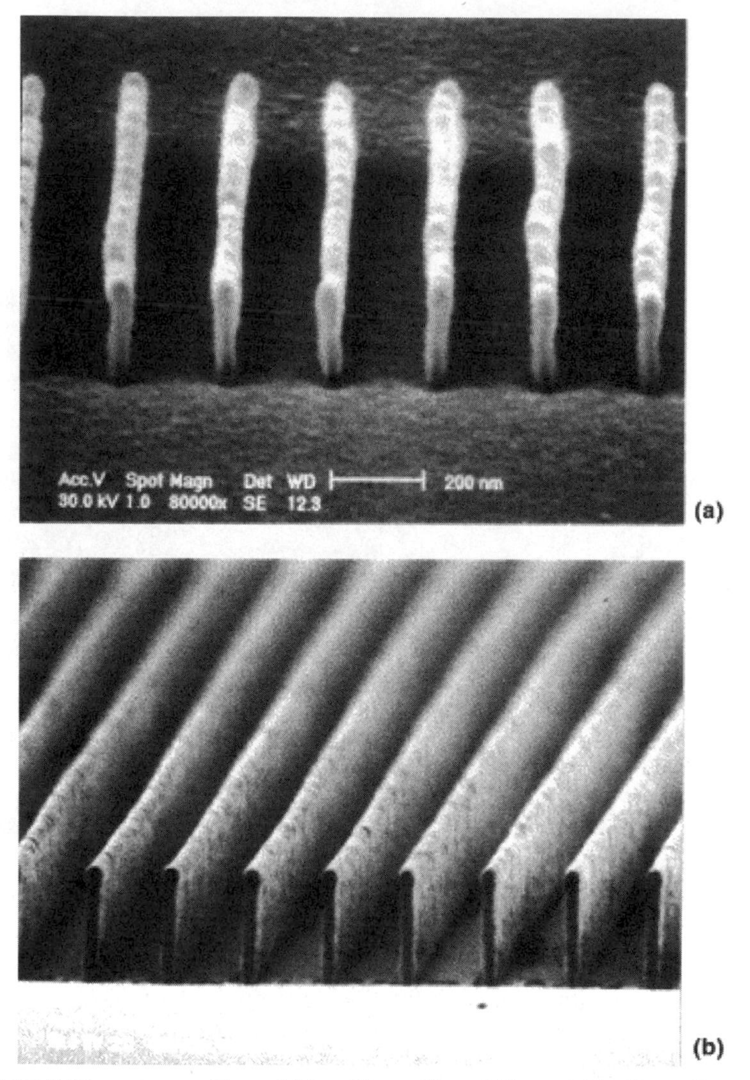

Fig. 4.50 PMMA resist profiles of (a) a 20 nm linewidth and (b) 60 nm linewidth (aspect ratio >10) developed by proximity X-Ray lithography. Reprinted with the permission from the WileyInterScience Publications.[90]

LIGA technology, in connection with X-ray lithography, can play an active part in emerging and competing micro and nano technologies.[92] LIGA is a German acronym for X-ray lithography (X-ray Lithographie), Electroplating (Galvanoformung), and Molding (Abformung). It is a process that was developed in the early 1980s by a team at the Institute for Nuclear Process Engineering (Institut für Kernverfahrenstechnik IKVT) at the Karlsruhe Nuclear Research Center.[93]

The LIGA technique allows high-aspect-ratio structures to be fabricated, whose lateral dimensions are below one micro-meter. For the fabrication of nano structures with high aspect ratios high-energy X-rays are required. Therefore, the X-ray source can be a synchrotron. Structures developed using the LIGA technique may be utilized in other fabrication technologies such as nanoimprint patterning (as a mold).

The advantages of using X-rays include its extreme precision, depth of field and very low intrinsic surface roughness. The quality of fabricated structures often depends on secondary effects during exposure and effects such as resist adhesion. UV-LIGA, rely on thick UV resists as an alternative for projects requiring less precision. The principle of X-ray lithography/LIGA is shown in **Fig. 4.51**.

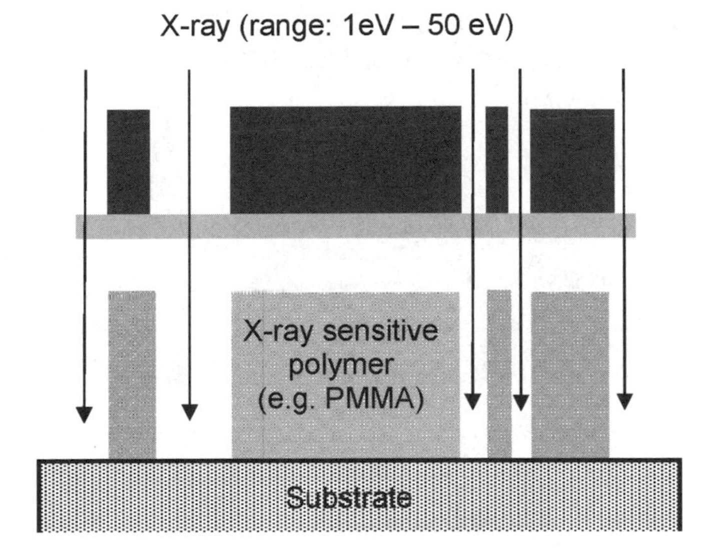

Fig. 4.51 Principle of X-ray lithography/LIGA.

In the LIGA process, X-rays pass through a lithographic mask, irradiating a polymer photoresist (usually PMMA, EPON). Conventional LIGA systems utilize x-ray energies in the range of 4–15 keV for the exposure of the polymer resist, whose maximum thickness is up to 1 mm. The lithographic patterning mask uses a 10–15 µm thick gold absorber (to prevent the X-rays from passing though it). The patterned photomask has structures on a suspended thin membrane (a few microns) made of materials such as quartz, diamond, or simply high-performance type glasses. Once exposed, the patterned resist is then developed with wet chemical processing.

4.9.6 Interference Lithography

An unconventional type of lithography is called *interference lithography* or *laser focusing lithography*.[94,95] In this technique, light waves, which are generated using different laser sources, interfere with each other to produce standing waves. These standing waves create high and low intensity regions on the surface of the substrate, that has been coated with a layer of photoresist. These standing waves act similar to of a photolithographic mask to produce bright and dark spots (**Fig. 4.52**).

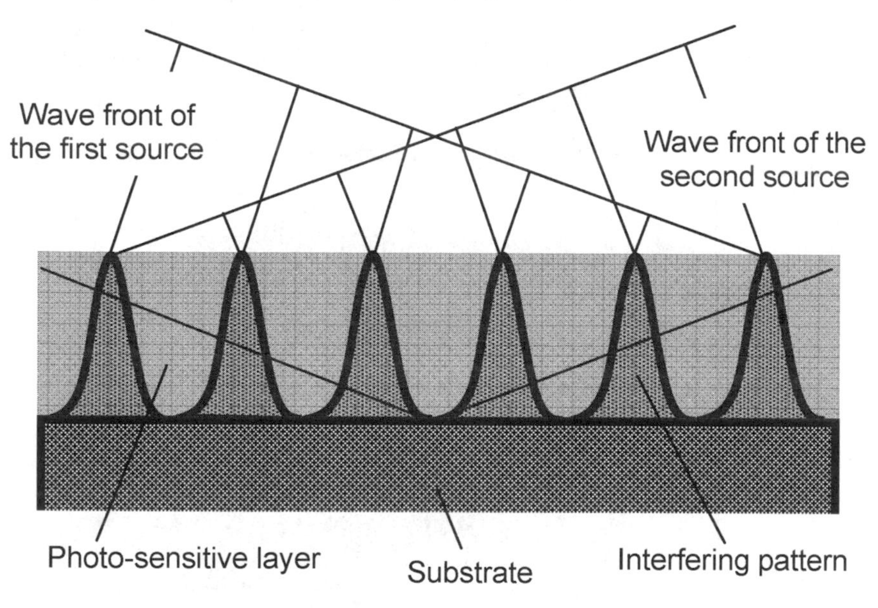

Fig. 4.52 The principle of the interference lithography.

Examples of patterns produced onto a commercially available photoresist are shown in **Fig. 4.53** and **Fig. 4.54**.

Fig. 4.53 Planar view of gratings produced by interference lithography.

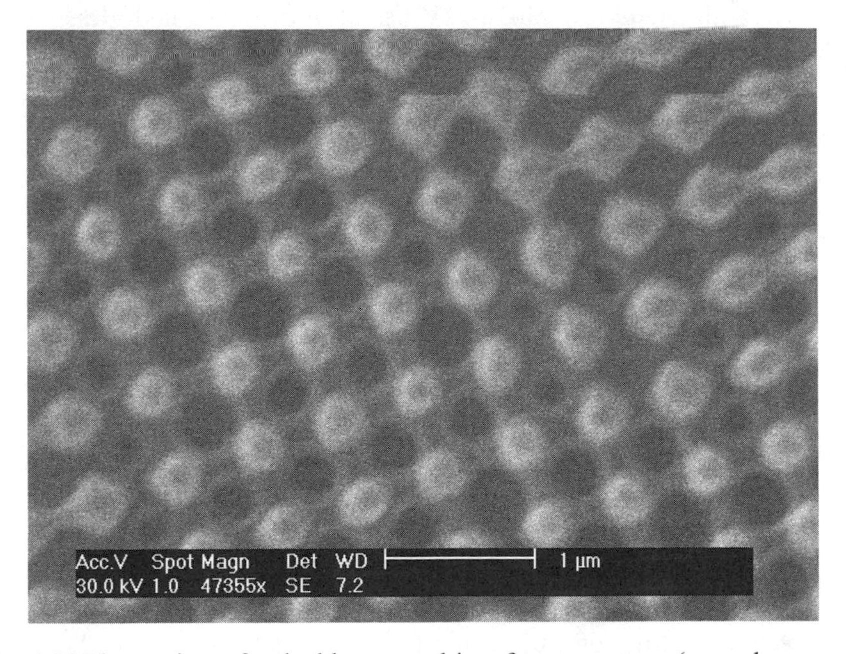

Fig. 4.54 Planar view of a doubly exposed interference pattern (second exposure after 90° rotation).

4.9.7 Ion Implantation

Ion implantation is a process by which ions of a material are implanted into another solid. Energies of the ions are typically in the range from 10 to 500 keV. Although energies in the range 1 to 10 keV may also be utilized, they have the drawback of lower penetration depths, only a few nanometers or less. Ion implantation is used in semiconductor device fabrication and in metal finishing, as well as various applications in materials science research.

As a result of the ion implanting, the physical properties of the solid changes. Additionally, the implanted ions can also introduce a chemical change in the sample, as the implanted ions may differ from the constituents of the substrate. Structural changes can also occur, as the implanted ions can alter or damage the sample's crystal structure. Additionally, ion implantation can be used for atomic mixing at the substrate's surface.

By ion implanting, surface morphology of nanostructured thin films can be precisely controlled and altered for sensing applications. In addition, using this method, single atoms of certain materials can be implanted onto the surface of the substrate which can be used as catalysts.

As an example, the controlled implantation of single ions into a silicon substrate with energy of sub-20-keV has been demonstrated[96], displaying potential application in the development of quantum computers.

4.9.8 Etching: Wet and Dry

In order to form the patterns, the electrodes and structures in transducers and sensors, it is necessary to etch the deposited thin films. In addition, the quality of sensitive layers can very much depend on the etching process. There are several important parameters in any etching process:

1. *Etching rate*: The rate of etching can greatly affect the quality of the produced structures. For the fabrication of electrode patterns generally the faster etch the better as it reduces the fabrication time, however, fast etching can also lead to deformities, particularly in the edges of the structures.
2. *Uniformity*: In fabrication of sensors and transducers, we need to obtain a consistent uniformity across a sample during the etching process. Some parts may be etched more than others for many reasons, such as non-uniform concentration of the etchant solution, heat, and contaminants in the sample. The etching environment should be properly controlled to circumvent such problems.

3. *Isotropy*: Some single crystal materials, such as silicon, exhibit anisotropic etching in certain etchants. Anisotropic etching, in contrast to isotropic etching, means different etch rates for different directions. As an example, different silicon crystal orientations are known to have different etch rates. A classic example is when potassium hydroxide (KOH) etches the silicon <100> plane, producing a characteristic anisotropic V-etch, in which sidewalls form a 54.7° angle with the surface. In contrast, for the <111> crystal plane, the etched sidewalls appear to be vertical.
4. *Selectivity*: It is a very important factor as we desire selective etch for the target materials and not the other materials exposed to the etchant.

The etching processes fall into two categories:

(a) *Wet etching:* In the wet etching process the material that we wish to etch is removed when immersed in a chemical solution. It works very well for etching thin films, and can also be utilized to etch the substrate itself to produce structural forms onto it.
(b) *Dry etching* is the removal of materials from a substrate surface through bombardment with energetic ions that carve out the surface. The dry etching technology can be divided into three major categories: *reactive ion etching* (*RIE*) , *sputter etching*, and *vapor phase etching*.

To conduct the RIE process, the substrate is placed inside a reactor. Consequently, several gases are introduced into the reactor chamber. The gas molecules are ionized using a RF power source which result in plasma generation. The ions are then accelerated towards the substrate that is to be etched. Upon arriving at the surface, they react with the material and etch it. Other gaseous materials are formed during this process, which are known as the *chemical part of reactive ion etching*. If the impinging ions have enough energy, then they can physically knock atoms out of the material without any chemical reactions occurring. Developing dry etching processes that balance chemical and physical etching is a delicate task, as there are many parameters that need fine tuning. Generally the chemical part causes an isotropic etch and the physical part produces anisotropic etch. By controlling these parameters, we can form features ranging from rounded to vertical, tailoring the anisotropy of the etch.

A subclass of RIE is *Deep RIE*, or *DRIE*. In DRIE, sidewalls are almost vertical and features with very large aspect ratios (as large as 100) can be obtained. This process is also termed the *Bosch process,* after the Robert

Bosch Company which first developed the method. The DRIE involves alternating different gas compositions during etching.

In the Bosch process initially a sample with a patterned surface is placed in the chamber. Once inside, the following steps are carried out: (a) A highly reactive gas is introduced to the chamber, which isotropically etches the substrate for just a few seconds, (b) a polymeric passivation layer is then conformally deposited over the whole surface which generally takes a few seconds, (c) through the sputtering process, ions remove the passivation layer from the bottom of the previously etched trench, but not from the sides, (d) the process is then iteratively repeated, starting with the introduction of the highly reactive gas. This time, the sidewalls are covered by the passivation layer, so introducing the reactive gas, for a short period, ensures that only the bottom of the trench will be etched. Continuing these steps several times allows high aspect ration structures to be obtained, and therefore etching is preferential in the vertical direction.[97,98] Such high aspect ratio structures are very useful in the fabrication of devices such as sensors, actuators, micro-fluidic systems, and scanning probe tips.

Another dry etching method is *sputter etching*. It is very similar in principle to the sputtering deposition process that was described previously in this chapter. However, the difference is that in this process, the substrate is now subjected to the ion bombardment instead of the target material.

Vapor phase etching is another dry etching method. The sample to be etched is placed inside a chamber in which gases are subsequently introduced and the gas molecules chemically react with the surface of the sample. The two most common vapor phase etching technologies are silicon dioxide etching using hydrogen fluoride (HF) and silicon etching using xenon diflouride (XeF_2). The byproducts condense on the surface and interfere with the etching process which have to be taken care of.[99]

4.10 Summary

Different methods for the synthesis of nano-materials in the form of nano-particles, one-dimensional nanostructures, and nanostructured thin films were presented in this chapter. In addition, the most common techniques for the development of nano-resolution patterns were described.

These methods and techniques are utilized for the fabrication of nanotechnology enabled sensors which will be presented in Chaps. 6 and 7, with Chap. 6 focusing on inorganic sensors. The complementary methods which can be used for the development of organic nanomaterials for sensing applications, as well as organic and hybrid sensors will be presented later in Chap. 7.

References

[1] J. I. Brauman and P. Szuromi, Science **273,** 855-855 (1996).

[2] A. P. Alivisatos, Science **271,** 933-937 (1996).

[3] Y. G. Sun and Y. N. Xia, Science **298,** 2176-2179 (2002).

[4] T. S. Ahmadi, Z. L. Wang, T. C. Green, A. Henglein, and M. A. El-Sayed, Science **272,** 1924-1926 (1996).

[5] M. C. Daniel and D. Astruc, Chemical Reviews **104,** 293-346 (2004).

[6] K. Grieve, P. Mulvaney, and F. Grieser, Current Opinion in Colloid & Interface Science **5,** 168-172 (2000).

[7] R. H. Kodama, J. Am. Chem. Soc., **122** 8581-8582 (2000).

[8] M. L. Steigerwald, A. P. Alivisatos, J. M. Gibson, T. D. Harris, R. Kortan, A. J. Muller, A. M. Thayer, T. M. Duncan, D. C. Douglass, and L. E. Brus, Journal of the American Chemical Society **110,** 3046-3050 (1988).

[9] G. Schmid, Journal of Materials Chemistry **12,** 1231-1238 (2002).

[10] F. E. Kruis, H. Fissan, and A. Peled, Journal of Aerosol Science **29,** 511-535 (1998).

[11] L. Madler, H. K. Kammler, R. Mueller, and S. E. Pratsinis, Journal of Aerosol Science **33,** 369-389 (2002).

[12] T. W. Smith, D. Wychick, Journal of Physical Chemistry B **84,** 1621-1629 (1980).

[13] K. S. Suslick, M. M. Fang, and T. Hyeon, Journal of the American Chemical Society **118** 11960-11961 (1996).

[14] S. H. Sun and C. B. Murray, Journal of Applied Physics **85** 4325-4330 (1999).

[15] C. Petit, A. Taleb, and M. P. Pileni, Journal of Physical Cheistry B **103** 1805-1810 (1999).

[16] G. Schmid, Chemical Reviews **92,** 1709-1727 (1992).

[17] M. Brust and C. J. Kiely, Colloids and Surfaces A-Physicochemical and Engineering Aspects **202,** 175-186 (2002).

[18] J. Turkevich, P. C. Stevenson, and J. Hillier, Discussions of the Faraday Society **11,** 55 - 75 (1951).

[19] G. Frens, Nature **241,** 20-22 (1973).

[20] T. Yonezawa and T. Kunitake, Colloids and Surfaces a-Physicochemical and Engineering Aspects **149,** 193-199 (1999).

[21] R. M. Crooks, M. Q. Zhao, L. Sun, V. Chechik, and L. K. Yeung, Accounts of Chemical Research **34,** 181-190 (2001).

[22] M. Q. Zhao and R. M. Crooks, Advanced Materials **11,** 217-+ (1999).

[23] D. A. Tomalia, H. Baker, J. Dewald, M. Hall, G. Kallos, S. Martin, J. Roeck, J. Ryder, and P. Smith, Polymer Journal **17,** 117-132 (1985).

[24] H. G. Lang, S. Maldonado, K. J. Stevenson, and B. D. Chandler, Journal of the American Chemical Society **126,** 12949-12956 (2004).

[25] M. Volmer and A. Weber, Zeitschrift Für Physikalische Chemie **119,** 277-301 (1926).

[26] I. N. Stranski and L. S. Kr'stanov, Akad. Wiss. Wien, Math.-naturw. Klasse, Abt. IIb **146,** 797-810 (1938).

[27] F. C. Frank and J. H. van der Merwe, Proceedings of the Royal Society of London: Series A, Mathematical and Physical Sciences **198,** 205-216 (1949).

[28] G. Guozhong, *Nanostructures & Nanomaterials, Synthesis, properties & Applications* (IImperial college press, London, 2004).

[29] P. P. Hartman and W. G. Perdok, Acta Crystallographica **8,** 525-529 (1955).

[30] E. I. Givargizov and D. Reidel, *Highly Anisotropic Crystals* (Dordrecht, The Netherlands 1987).

[31] W. Dittmar and K. Neumann, *Growth and Perfection of Crystals,* (New York, 1958).

[32] G. W. Sears, Acta Metallurgica **3,** 361-366 (1955).

[33] Z. W. Pan, Z. R. Dai, and Z. L. Wang, Science **291,** 1947-1949 (2001).

[34] Z. L. Wang, Annual Review of Physical Chemistry **55,** 159-196 (2004).

[35] R. S. Wagner and W. C. Ellis, Applied Physics Letters **233,** 1053-& (1964).

[36] M. S. Gudiksen, J. F. Wang, and C. M. Lieiber, Journal of Physical Chemistry B **105,** 4062-4064 (2001).

[37] M. S. Gudiksen, L. J. Lauhon, J. Wang, D. C. Smith, and C. M. Lieber, Nature **415,** 617-620 (2002).

[38] S. A. Campbell, *The Science and Engineering of Microelectronic Fabrication*, 2nd ed. (Oxford University Press, New York, USA, 2001).

[39] A. Cho and J. Arthur, Progress in Solid State Chemistry **10** (1975).

[40] A. Y. Cho, Journal of Vacuum Science & Technology **8,** S31-& (1971).

[41] W. S. Knodle and R. Chow in *Handbook of thin-film deposition processes and techniques: principles, methods, equipment, and applications*, 2nd ed. (Noyes Publications, Park Ridge, USA., 2002).

[42] A. Ichimiya and P. I. Cohen, *Reflection High-Energy Electron Diffraction* (Cambridge University Press, London, UK, 2004).

[43] H. S. Nalwa, *Experimental Methods in the Physics Sciences*, Vol. 38 Advances in Surface Science (Academic Press, New York, USA, 2001).

[44] K. Wasa, *Thin film materials technology: sputtering of compound materials* (Springer, Norwich, USA, 2004).

[45] R. A. Fischer, *Precursor Chemistry of Advanced Materials: CVD, ALD and Nanoparticles* (Springer, Berlin, Germany, 2005).

[46] C. A. Moore, Z.-q. Yu, L. R. Thompson, and G. J. Collins, in *Handbook of thin-film deposition processes and techniques: principles, methods, equipment, and applications*, 2nd ed., edited by K. Seshan (Noyes Publications, Park Ridge, USA, 2002).

[47] J. L. Zilko, in *Handbook of thin-film deposition processes and techniques: principles, methods, equipment, and applications*, 2nd ed. (Noyes Publications, Park Ridge, USA, 2002).

[48] M. J. Ludowise, Journal of Applied Physics **58,** R31-R55 (1985).

[49] M. S. Kabir, R. E. Morjan, O. A. Nerushev, P. Lundgren, S. Bengtsson, P. Enokson, and E. E. B. Campbell, Nanotechnology **16,** 458-466 (2005).

[50] K. B. K. Teo, M. Chhowalla, G. A. J. Amaratunga, W. I. Milne, G. Pirio, P. Legagneux, F. Wyczisk, J. Olivier, and D. Pribat, Journal of Vacuum Science & Technology B **20,** 116-121 (2002).

[51] K. B. K. Teo, S. B. Lee, M. Chhowalla, V. Semet, V. T. Binh, O. Groening, M. Castignolles, A. Loiseau, G. Pirio, P. Legagneux, D. Pribat, D. G. Hasko, H. Ahmed, G. A. J. Amaratunga, and W. I. Milne, Nanotechnology **14,** 204-211 (2003).

[52] H. Curtins, N. Wyrsch, and A. V. Shah, Electronics Letters **23,** 228-230 (1987).

[53] T. Sharda, T. Soga, T. Jimbo, and M. Umeno, Diamond and Related Materials **10,** 1592-1596 (2001).

[54] C. H. Lin, H. L. Chang, C. M. Hsu, A. Y. Lo, and C. T. Kuo, Diamond and Related Materials **12,** 1851-1857 (2003).

[55] T. P. Niesen and M. R. De Guire, Solid State Ionics **151,** 61-68 (2002).

[56] T. P. Niesen and M. R. De Guire, Journal of Electroceramics **6,** 169-207 (2001).

[57] A. Hishinuma, T. Goda, M. Kitaoka, S. Hayashi, and H. Kawahara, Applied Surface Science **48-9,** 405-408 (1991).

[58] T. Hamaguchi, N. Yabuki, M. Uno, S. Yamanaka, M. Egashira, Y. Shimizu, and T. Hyodo, Sensors and Actuators B-Chemical **113,** 852-856 (2006).

[59] R. B. Peterson, C. L. Fields, and B. A. Gregg, Langmuir **20,** 5114-5118 (2004).

[60] I. R. Peterson, Journal of Physics D-Applied Physics **23,** 379-395 (1990).
[61] V. V. Tsukruk, Progress in Polymer Science **22,** 247-311 (1997).
[62] A. N. Shipway, E. Katz, and I. Willner, Chemphyschem **1,** 18-52 (2000).
[63] Y. Gimeno, A. H. Creus, P. Carro, S. Gonzalez, R. C. Salvarezza, and A. J. Arvia, Journal of Physical Chemistry B **106,** 4232-4244 (2002).
[64] F. Favier, E. C. Walter, M. P. Zach, T. Benter, and R. M. Penner, Science **293,** 2227-2231 (2001).
[65] Y. Gimeno, A. H. Creus, S. Gonzalez, R. C. Salverezza, and A. J. Arvia, Chemistry of Materials **13,** 1857-1864 (2001).
[66] M. P. Zach, K. H. Ng, and R. M. Penner, Science **290,** 2120-2123 (2000).
[67] D. Meyerhofer, Journal of Applied Physics **49,** 3993-3997 (1978).
[68] B. D. Fabes, B. J. Zelinski, and D. R. Uhlmann, in *Ceramic films and Coatings*, edited by J. B. Wachtman and R. A. Haber (Noyes Publications, New Jersey, USA, 1993).
[69] C. J. Brinker and G. W. Scherer, *Sol-gel science - the physics and chemistry of sol-gel processing* (Academic Press, New York, USA, 1990).
[70] L. E. Scriven, Materials Research Society, 717-729 (1988).
[71] H. J. Kim, C. H. Lee, and S. Y. Hwang, Surface & Coatings Technology **191,** 335-340 (2005).
[72] I. M. Thomas, in *Sol-Gel Technology for thin films, fibres, preforms, electronics and specialty shapes*, edited by L. C. Klein (Noyes Publications, New Jersey, USA, 1988).
[73] C. Cantalini, W. Wlodarski, H. T. Sun, M. Z. Atashbar, M. Passacantando, and S. Santucci, Sensors and Actuators B **65,** 101-104 (2000).
[74] M. McCallum, G. Fuller, and S. Owa, Microelectronic Engineering **83,** 667-671 (2006).
[75] J. Markoff, New York Times **February 20** (2006).
[76] Y. Kim and C. M. Lieber, Science **257,** 375-377 (1992).
[77] D. M. Eigler and E. K. Schweizer, Nature **344,** 524-526 (1990).
[78] R. F. Service, Science **286,** 389-+ (1999).
[79] R. D. Piner, J. Zhu, F. Xu, S. H. Hong, and C. A. Mirkin, Science **283,** 661-663 (1999).
[80] S. Y. Chou, P. R. Krauss, and P. J. Renstrom, Applied Physics Letters **67,** 3114-3116 (1995).
[81] S. Y. Chou, P. R. Krauss, W. Zhang, L. J. Guo, and L. Zhuang, Journal of Vacuum Science & Technology B **15,** 2897-2904 (1997).

[82] B. D. Gates, Q. B. Xu, M. Stewart, D. Ryan, C. G. Willson, and G. M. Whitesides, Chemical Reviews **105,** 1171-1196 (2005).
[83] R. G. Jones and P. C. M. Tate, Advanced Materials for Optics and Electronics **4,** 139-153 (1994).
[84] A. V. Crewe, J. Wall, and J. Langmore, Science **168,** 1338-& (1970).
[85] A. V. Crewe and J. Wall, Journal of Molecular Biology **48,** 375-& (1970).
[86] A. N. Broers, A. C. F. Hoole, and J. M. Ryan, Microelectronic Engineering **32,** 131-142 (1996).
[87] S. Reyntjens and R. Puers, Journal of Micromechanics and Microengineering **11,** 287-300 (2001).
[88] J. Kapsa, Y. Robach, G. Hollinger, M. Gendry, J. Gierak, and D. Mailly, Applied Surface Science **226,** 31-35 (2004).
[89] D. L. Spears and H. I. Smith, Electronics Letters **8,** 102-& (1972).
[90] Y. Chen and A. Pepin, Electrophoresis **22,** 187-207 (2001).
[91] H. Becker and L. E. Locascio, Talanta **56,** 267-287 (2002).
[92] R. K. Kupka, F. Bouamrane, C. Cremers, and S. Megtert, Applied Surface Science **164,** 97-110 (2000).
[93] E. W. Becker, W. Ehrfeld, D. Munchmeyer, H. Betz, A. Heuberger, S. Pongratz, W. Glashauser, H. J. Michel, and R. Vonsiemens, Naturwissenschaften **69,** 520-523 (1982).
[94] H. H. Solak, D. He, W. Li, S. Singh-Gasson, F. Cerrina, B. H. Sohn, X. M. Yang, and P. Nealey, Applied Physics Letters **75,** 2328-2330 (1999).
[95] S. O. Kim, H. H. Solak, M. P. Stoykovich, N. J. Ferrier, J. J. de Pablo, and P. F. Nealey, Nature **424,** 411-414 (2003).
[96] D. N. Jamieson, C. Yang, T. Hopf, S. M. Hearne, C. I. Pakes, S. Prawer, M. Mitic, E. Gauja, S. E. Andresen, F. E. Hudson, A. S. Dzurak, and R. G. Clark, Applied Physics Letters **86** (2005).
[97] N. Maluf, *An Introduction to Microelectromechanical Systems Engineering* (Artech house publisher, Boston, USA,, 2004).
[98] M. Gad-el-Hak, *Mems: Design and Fabrication* (CRC Press Inc, London, UK, 2005).
[99] C. R. Helms and B. E. Deal, Journal of Vacuum Science & Technology A **10,** 806-811 (1992).

Chapter 5: Characterization Techniques for Nanomaterials

5.1 Introduction

In the previous chapter, methods for the synthesis of nanomaterials and the fabrication of the nanotechnology enabled sensors were presented. Along with the synthesis and fabrication processes, the nanomaterials utilized in the sensors need to be characterized to assess their physical and chemical properties.

When materials' dimensions are reduced to nanoscale they demonstrate unique properties which are different from those of their bulk counterparts. For example, their electronic and optical properties alter, their chemical activities can be increased or decreased and mechanical/structural stabilities are changed. Such features make nanomaterials attractive for unique sensing applications, and also at the same time, cause complications in their characterization processes. Therefore, the challenge lies in finding the right characterization techniques that have the optimum capabilities for studying the characteristics of nanomaterials.

A large number of techniques can be employed for nanomaterials characterization. Some of the most common characterization techniques in nanotechnology will be presented in this chapter. In addition, applicability of these techniques for investigating different types of nanomaterials and their relevance to sensor technology will also be described.

5.2 Electromagnetic Spectroscopy

Electromagnetic spectroscopy concerns the interaction of electromagnetic waves and matters, whether they are in the form of atoms, molecules, or larger assemblies.[1] These interactions involve the absorption and/or emission of electromagnetic radiation. In the electromagnetic spectroscopy, the intensity of the absorbed or emitted electromagnetic waves is plotted against wavelength or frequency. From this spectrum, materials can be identified and quantified.

The measured spectrum strongly depends on the ambient environment and the presence of other molecules nearby. Therefore, electromagnetic spectroscopy techniques can be directly employed for sensing applications, with the intensity of electromagnetic waves absorbed or emitted, or the shift in absorption/emission wavelengths as the output parameters. Electromagnetic spectroscopy also provides a contactless way to probe materials.

Electromagnetic radiation consists of mutually perpendicular propagating electric and magnetic waves, with particle-like (*photon*) properties. The relationships between frequency f, wavelength λ, and energy of the particle E, are given by:

$$c = \lambda f, \tag{5.1}$$

$$E = hf, \tag{5.2}$$

where h is Planck's constant (6.626×10^{-34} J·s), and c is the speed of light. In free space, c has the value of 2.998×10^{8} m/s, whereas in other media, its value is adjusted to c/n, where n is the *refractive index* of the medium.

The information that can be obtained from a material strongly depends on the wavelengths of electromagnetic wave that is applied. From **Table 5.1** it is observed that different wavelengths of incident waves are responsible for the observation of various phenomena. For example, microwaves stimulate rotations of molecules; infrared waves stimulates vibrations of higher energy orbitals; visible and ultraviolet waves promote electrons to higher energy orbitals; and X-rays and short wavelengths break chemical bonds and ionize molecules and can even damage living tissue.

When the near ultraviolet, visible, and near/mid infrared regions of the electromagnetic spectrum are utilized, the electromagnetic spectroscopy techniques are referred to as *spectrophotometry*. In these techniques, the wavelengths of the incident electromagnetic waves are generally scanned across a range to produce an absorption or emission spectrum. Analogous phenomena occur in the X-ray, microwave, radio, and other regions of the electromagnetic spectrum and these will be discussed later in this chapter.

Table 5.1 Types of interactions between electromagnetic waves of different frequencies with matter.

Region	Frequency (Hz)	Wavelength	Effect and Information	Energy, E (kJ/mol)
Radio waves	$< 3\times10^{8}$	Larger than 1 m	Nuclear and electron spin transitions	$E < 0.001$
Microwaves	$3\times10^{8} - 3\times10^{11}$	$1 - 10^{-3}$ m	Rotation	$0.001 < E < 0.12$
Infrared	$3\times10^{11} - 0.37\times10^{15}$	10^{-3} m – 800 nm	Vibration	$0.12 < E < 150$
Visible	$0.37\times10^{15} - 0.75\times10^{15}$	800 nm – 400 nm	Outer shell electron transitions	$150 < E < 310$
Ultraviolet	$0.75\times10^{15} - 3\times10^{16}$	800 nm – 10^{-8} m	Electronic excitation	$310 < E < 12000$
X- rays	$3\times10^{15} - 3\times10^{19}$	10^{-8} m – 10^{-11} m	Bond breaking and ionization	$12000 < E < 1.2\times10^{7}$
γ - rays	$3\times10^{19} - 3\times10^{20}$	10^{-11} m – 10^{-12} m	Nuclear	$1.2\times10^{7} < E < 1.2\times10^{8}$
Cosmic rays	$> 3\times10^{20}$	$< 10^{-12}$ m		$1.2\times10^{8} < E$

Electromagnetic spectroscopy techniques use light beams. The energy per unit area of the beam is the *irradiance*, P, and is related to the fraction of the light that passes through a sample, the *transmittance*, T, by:[2]

$$T = P_{out}/P_0 , \tag{5.3}$$

where P_0 is the impinging light on a sample and P_{out} is the light that emerges from the sample. The *absorbance*, A, is generally defined as:

$$A = \log(P_0/P_{out}) = -\log T. \tag{5.4}$$

The absorbance is directly proportional to the concentration, c (in mol/l), of the light absorbing species and can be obtained from the *Beer-Lambert law*:

$$A = \varepsilon bc, \tag{5.5}$$

where b (cm) is the path length and ε is the *molar absorptivity* or *molar extinction coefficient* (l/mol.cm). Molar absorptivity is characteristic of a substance and indicates the amount of light absorbed at a particular wavelength. The Beer-Lambert law applies to *monochromatic* (single frequency) radiation and is valid for dilute solutions of most substances; however, it fails at high concentrations as solute molecules influence one another as a result of their proximity.

A block diagram for a basic spectrometer is shown in **Fig. 5.**

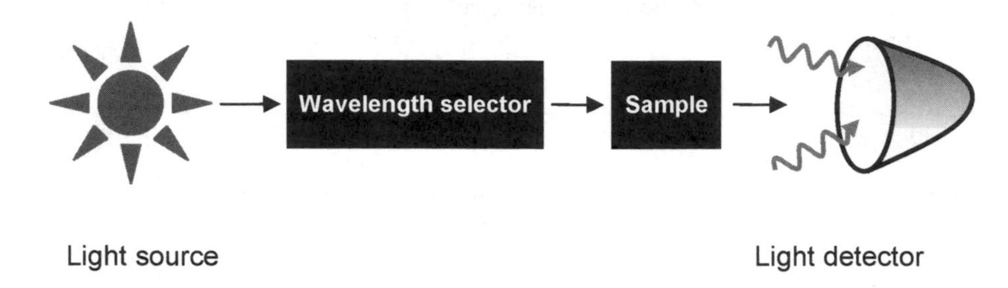

Fig. 5.1 Block diagram of a spectrophotometer.

In spectrophotometry, several processes occur after the material absorbs energy; these are changes in the *electronic energy*, *vibrational* and *rotational relaxations*, *intersystem crossings*, as well as *internal conversions* as shown in **Fig. 5.2**. Owing to the quantum nature of molecules, their energy distribution at any given moment can be defined as the sum of the contributing energy terms:[3]

$$E_{total} = E_{electronic} + E_{vibrational} + E_{rotational} + E_{translational}. \tag{5.6}$$

The electronic energy components correspond to all electrons energy transitions throughout the molecule and may be either localized, within specific bonds, or delocalized over structures such as aromatic rings.

Electronic transitions can occur when a molecule absorbs visible and ultraviolet electromagnetic radiations. Vibrational energy is an energy of less magnitude arising from the absorption in the infrared region, where the constituent atoms vibrate about the mean center of their chemical bonds.[3] Absorption of radiation in the microwave region gives rise to the rotational energy component, which is associated with tumbling motion of a molecule, and the translational energy is related to molecules being displaced as a function of the thermal motions of matter.

Having discussed some background of electromagnetic spectroscopy, the main techniques that are used in this field, namely *ultraviolet-visible spectroscopy*, *photoluminescence spectroscopy*, and *infrared spectroscopy*, will be presented.

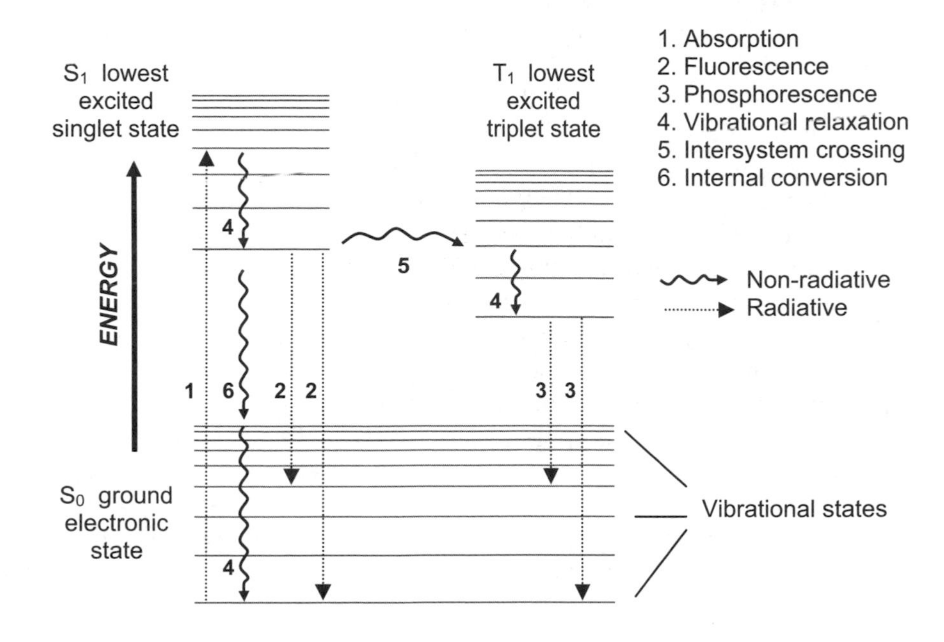

Fig. 5.2 Possible physical processes following absorption of a photon.

5.2.1 UV-Visible Spectroscopy

Ultraviolet-visible (*UV-vis*) spectroscopy is widely utilized to quantitatively characterize organic and inorganic nanosized molecules. A sample is irradiated with electromagnetic waves in the ultraviolet and visible ranges and the absorbed light is analyzed through the resulting sprectrum.[4,5] It can be employed to identify the constituents of a substance, determine their concentrations, and to identify functional groups in molecules. Consequently, UV-vis spectroscopy not only is used for characterization, but also for sensing applications. The samples can be either organic or inorganic, and may exist in gaseous, liquid or solid form. Different sized materials can be characterized, ranging from transition metal ions and small molecular weight organic molecules, whose diameters can be several Ångstroms, to polymers, supramolecular assemblies, nano-particles and bulk materials. Size dependant properties can also be observed in a UV-visible spectrum, particularly in the nano and atomic scales. These include peak broadening and shifts in the absorption wavelength. Many electronic properties, such as the band gap of a material, can also be determined by this technique.

The energies associated with UV-visible ranges are sufficient to excite molecular electrons to higher energy orbitals.[6,7] Photons in the visible range have wavelengths between 800-400 nm, which corresponds to energies between 36 and 72 kcal/mol. The near UV range includes wavelengths down to 200 nm, and has energies as high as 143 kcal/mol. UV radiation of lower wavelengths is difficult to handle for safety reasons, and is rarely used in routine UV-vis spectroscopy.

Fig. 5.3 shows a typical UV-vis absorption experiment for a liquid sample. A beam of monochromatic light is split into two beams, one of them is passed through the sample, and the other passes a reference (in this figure, a solvent in which the sample is dissolved).[8] After transmission through the sample and reference, the two beams are directed back to the detectors where they are compared. The difference between the signals is the basis of the measurement. Liquid samples are usually contained in a cell (called a cuvette) that has flat, fused quartz faces. Quartz is commonly used as it is transparent to both UV and visible lights.

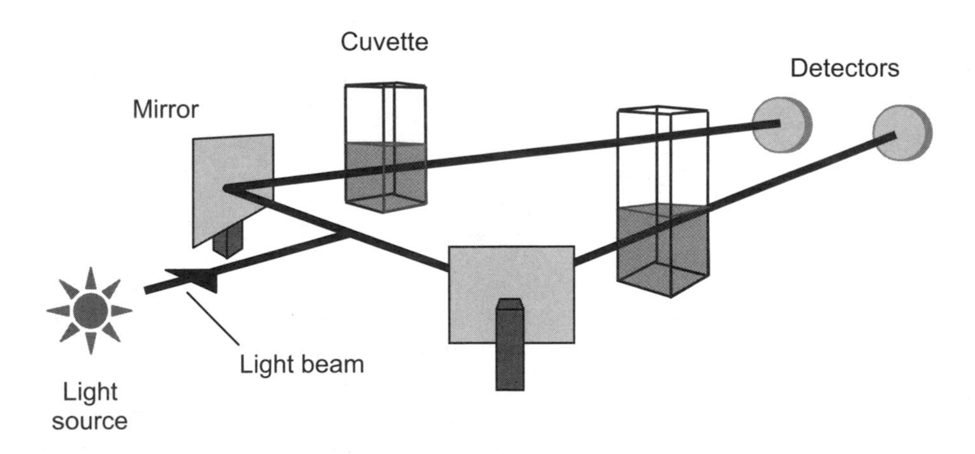

Fig. 5.3 A typical setup for a UV-vis spectrophotometer for liquid samples.

An example of a UV-vis absorption spectrum is shown in **Fig. 5.4**. The spectrum is of rhodamine B, which is a small organic molecule (molecular weight: 479.02 g/mol) that is commonly used as a dye and as a florescent label for larger molecules. The molecule has a maximum absorbance at around 550 nm. Provided that the path length and molar extinction coefficient for this molecule are known, the exact concentration of rhodamine B can be determined using the Beer-Lambert law.

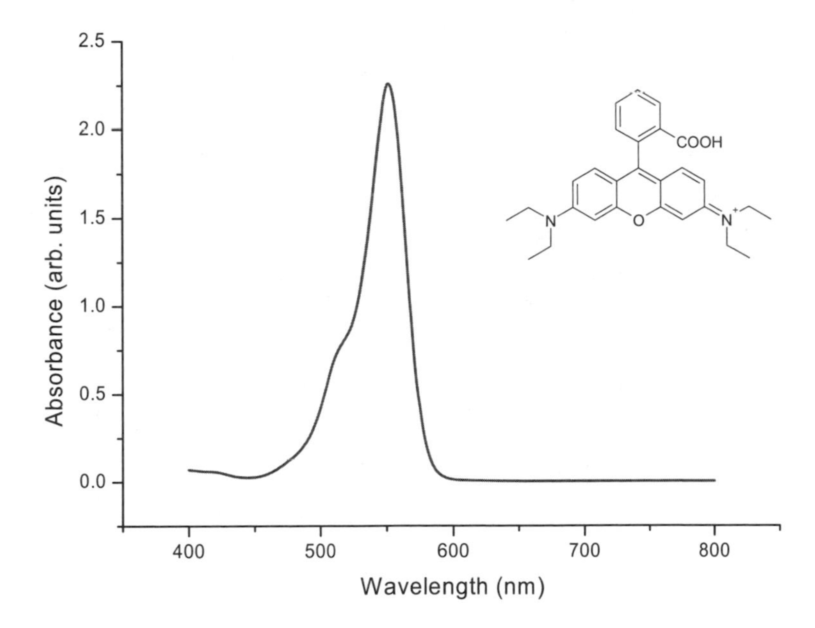

Fig. 5.4 UV-vis absorption spectrum of rhodamine B, show in inset, after having been dissolved in ethanol.

UV-vis spectroscopy offers a relatively straight forward and effective way for quantitatively characterizing both organic and inorganic nanomaterials. Furthermore, as it operates on the principle of absorption of photons that promotes the molecule to an excited state, it is an ideal technique for determining the electronic properties of nanomaterials.

In the spectrum of nanoparticles, the absorption peak's width strongly depends on the chemical composition and the particle size. As a result, their spectrum is different from their bulk counterparts. The detailed reasons for such distinct differences will be described in Chap. 6. For instance, for semiconductor nanocrystals, the absorption spectrum is broadened owing to quantum confinement effects,[9,10] and as their size reduces, there is no longer a distinct peak, rather there is a band. Furthermore, semiconductor nanoparticles' absorption peaks shift towards smaller wavelengths (higher energies) as their crystal size decreases.[10-13]

An important consequence of using the UV-vis spectroscopy is that the band gap of nanosized materials can be determined. As an example, the

UV-vis absorption spectra of different sized ZnO nanocrystals are shown in **Fig. 5.5.**[14] In this figure, there is no absorption peak; rather the absorption spectra of the ZnO nanocrystals resemble hyperbolic tangents. In this case, the band gap energy is the energy corresponds to the excitation wavelength at the variation point of each spectrum (indicated by the arrows). Due to quantum confinement effects, it is seen that the band gap is size dependent and it increases as the particle dimensions decrease.

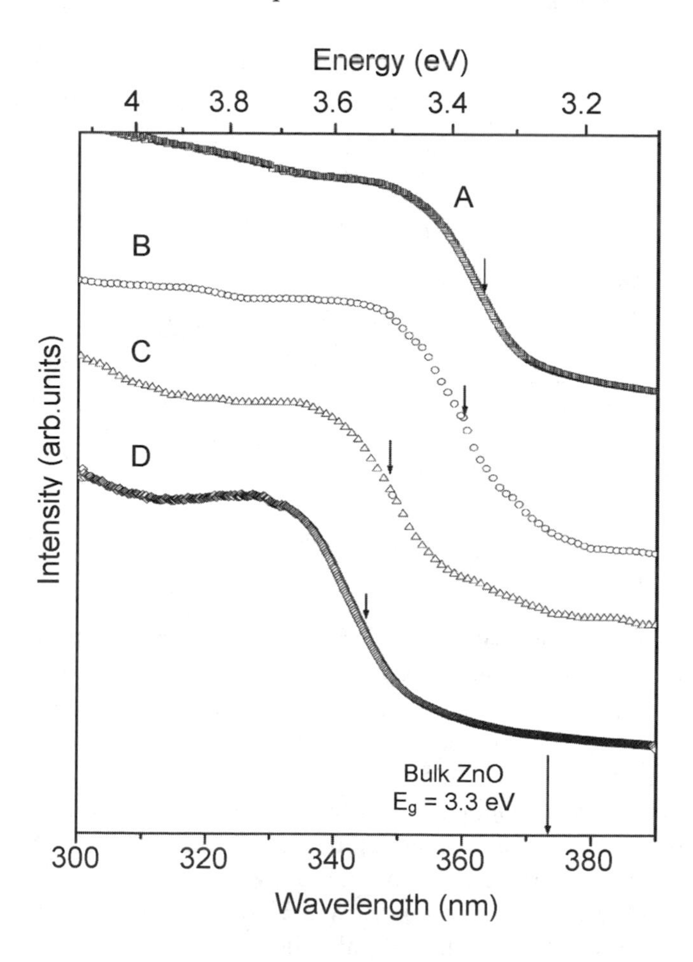

Fig. 5.5 UV-vis absorption spectra of various sizes of ZnO nanocrystals. The band gap of the bulk ZnO is 3.32 eV and is marked by an arrow. The absorption edges of the nanocrystals are marked by arrows and correspond to 3.42 eV (A), 3.44 eV (B), 3.55 eV (C) and 3.59 eV (D). Reprinted with permission from the Royal Society of Chemistry publications.[14]

5.2.2 Photoluminescence (PL) Spectroscopy

PL spectroscopy concerns monitoring the light emitted from atoms or molecules after they have absorbed photons.[15-17] It is suitable for materials that exhibit *photoluminescence*, which was described in Chap. 2. PL spectroscopy is suitable for the characterization of both organic and inorganic materials of virtually any size, and the samples can be in solid, liquid, or gaseous forms.

Electromagnetic radiation in the UV and visible ranges is utilized in PL spectroscopy. The sample's PL emission properties are characterized by four parameters: intensity, emission wavelength, bandwidth of the emission peak, and the emission stability.[18] The PL properties of a material can change in different ambient environments, or in the presence of other molecules. Many nanotechnology-enabled sensors are based on monitoring such changes. Furthermore, as dimensions are reduced to the nanoscale, PL emission properties can change, in particular a size dependent shift in the emission wavelength can be observed. Additionally, because the released photon corresponds to the energy difference between the states (see **Fig. 5.2**), PL spectroscopy can be utilized to study material properties such as band gap, recombination mechanisms, and impurity levels.

In a typical PL spectroscopy setup for liquid samples (**Fig. 5.6**), a solution containing the sample is placed in a quartz cuvette with a known path length. Double beam optics are generally employed. The first beam passes through an excitation filter or monochromator, then through the sample and onto a detector. This impinging light causes photoluminescence, which is emitted in all directions. A small portion of the emitted light arrives at the detector after passing through an optional emission filter or monochromator.[8] A second reference beam is attenuated and compared with the beam from the sample. Solid samples can also be analyzed, with the incident beam impinging on the material (thin film, powder etc.). Generally an emission spectrum is recorded, where the sample is irradiated with a single wavelength and the intensity of the luminescence emission is recorded as a function of wavelength. The fluorescence of a sample can also be monitored as a function of time, after excitation by a flash of light. This technique is called *time resolved fluorescence spectroscopy*.

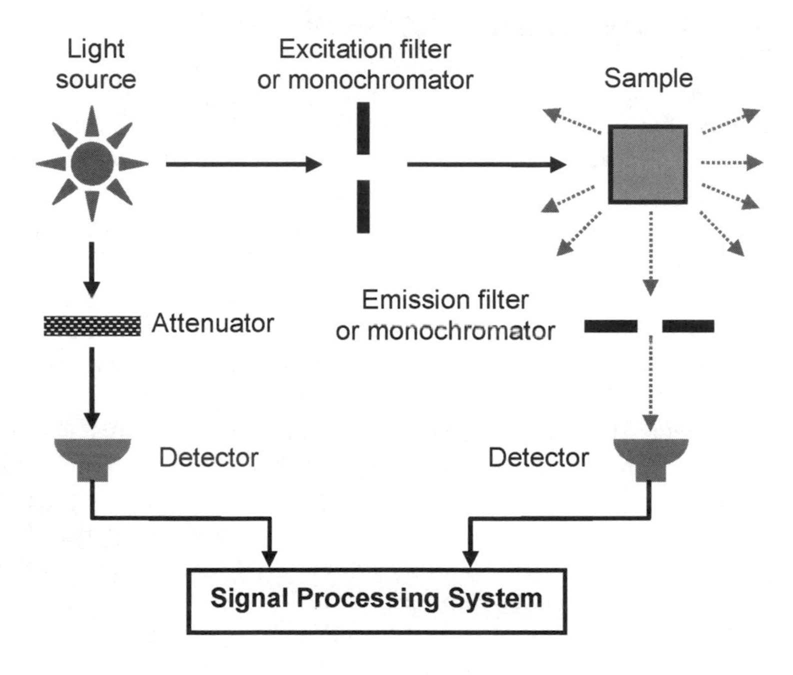

Fig. 5.6 A typical fluorescence spectrophotometer instrument setup.

Nanomaterials with PL effects, in particular nanocrystals, can reveal many interesting and improved optical properties. These include brighter emission, narrower emission band, and broad UV absorption.[19,20] For example, semiconductor nanocrystals produce narrower emission peaks than luminescent organic molecules, with bandwidth of around 30-40 nm.[9] Having a smaller bandwidth, it is much easier to discriminate individual wavelengths emanating from multiple sources, such as in an array of nanocrystals. The PL emission intensities and wavelengths are dependent on particle size. Hence, PL spectroscopy directly enables particle size effects, in particular those in the nanoscale, to be observed and quantified.[21,22] To illustrate this, the PL emission spectra of nanocrystals of different sizes are shown in **Fig. 5.7**,[19] where it is seen that the emission peaks shift to higher energy with decreasing dimensions.

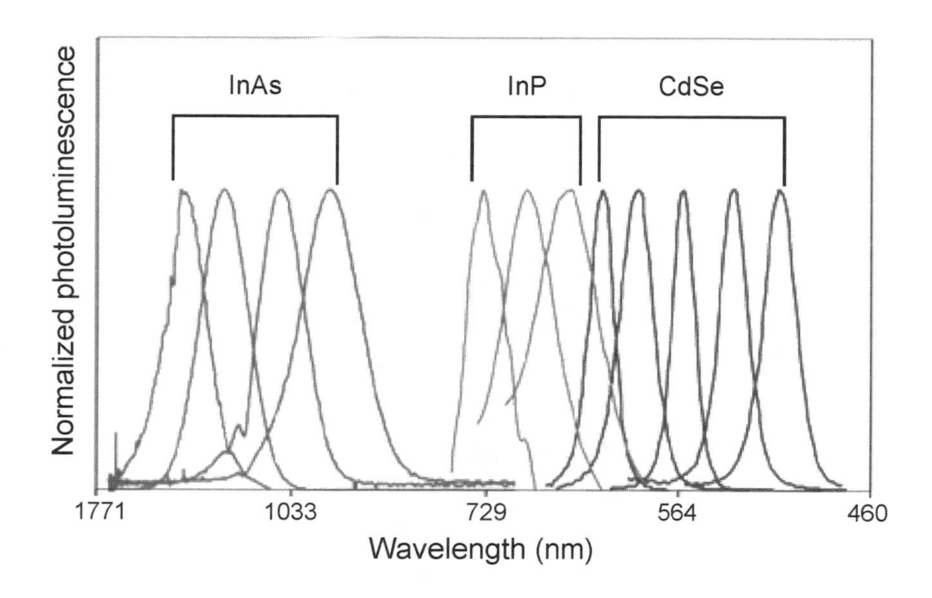

Fig. 5.7 Size and material dependent emission spectra of several surfactant-coated semiconductor nanocrystals having a variety of sizes. The peaks between 460 and 700 nm represent different sizes of CdSe nanocrystals with diameters of 2.1, 2.4, 3.1, 3.6, and 4.6 nm (from right to left). The peaks between 600 and 750 nm are of InP nanocrystals with diameters of 3.0, 3.5, and 4.6 nm. The peaks between 900 and 1700 nm are of InAs nanocrystals with diameters of 2.8, 3.6, 4.6, and 6.0 nm. Reprinted with permission from the Science Magazine publications.[19]

Changes in the PL spectra can be used in sensing applications. For instance, **Fig. 5.8** shows the PL emission spectra of an organic dye molecule used for monitoring the concentration of magnesium cations.[23] The PL properties of dye molecules in solution are modulated when complexed to the metal cations.[23,24] With no magnesium present, the uncomplexed molecule exhibits a relatively weak PL emission band at 540 nm. However, the addition of magnesium sees an increase in the emission intensity, which depends on the concentration of added ions (as seen in the inset of the figure) which allows for quantitative detection with high sensitivity.

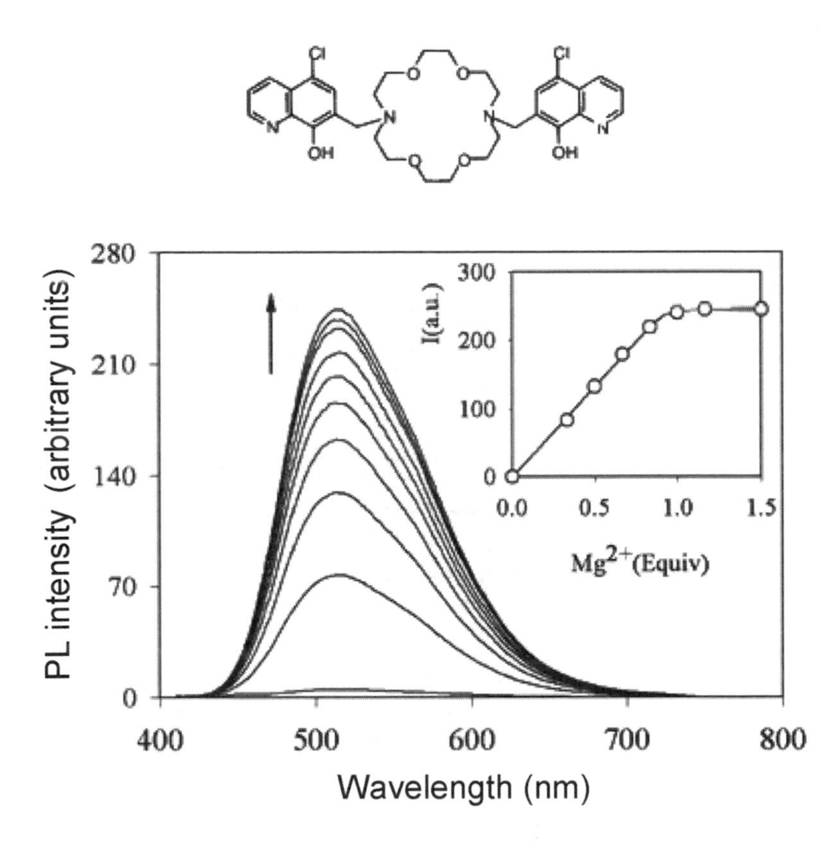

Fig. 5.8 *Top:* The organic dye molecule used in the sensing experiments. *Bottom:* PL emission spectra of the molecule (2.5×10^{-5} M, $\lambda_{ex} = 360$ nm) in methanol–water (11 v/v, pH 7.0) and upon addition of an increasing amount of Mg^{2+} ions. Inset: fluorescence intensity ($\lambda_{ex} = 360$ nm, $\lambda_{em} = 520$ nm) vs. equivalents of Mg^{2+} ions. Reprinted with permission from the Elsevier publications.[23]

PL spectroscopy is not limited to molecules dissolved in liquids. It may also reveal information regarding thin films and monolayers. For example, **Fig. 5.9** shows how the intensity of PL emission maxima of CdSe nanocrystals dispersed in films varies for different thicknesses,[25] and reveals a linear relationship with emission intensity and thickness.

The composition and dimensions of a material strongly influence its PL emission, and therefore these attributes need to be addressed when designing a PL based sensing system.[26] Organic dyes and semiconducting nanocrystals are amongst the most promising employed in nanotechnology enabled sensing applications. Numerous examples of the use of such materials in physical, chemical and bio-sensing applications will be presented in subsequent chapters.

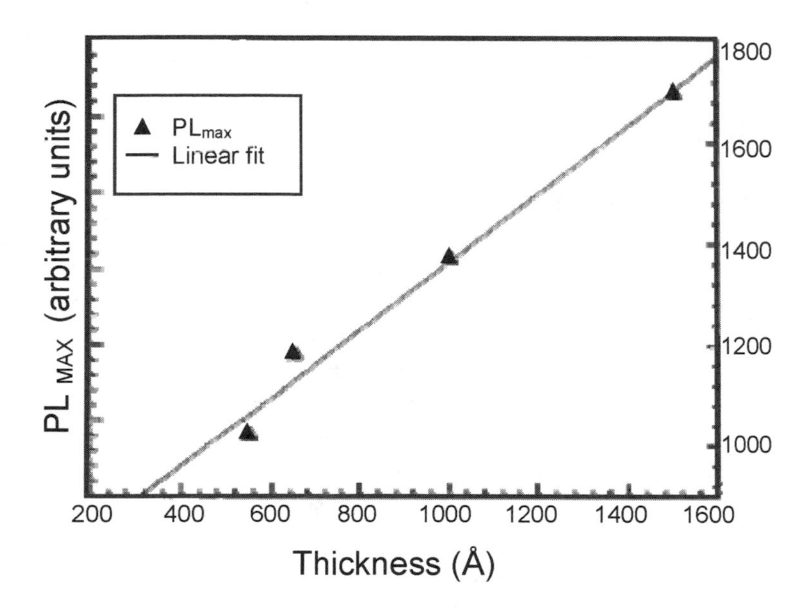

Fig. 5.9 Experimental measurements of the PL emission peak intensity as a function of film thickness of spin cast thin films containing CdSe nanosrytals. Reprinted with permission from the American Institute of Physics publications.[25]

5.2.3 Infrared Spectroscopy

Infrared (*IR*) *spectroscopy* is a popular characterization technique in which a sample is placed in the path of an IR radiation source and its absorption of different IR frequencies is measured.[27,28] Solid, liquid, and gaseous samples can all be characterized by this technique.

IR photons energies, in a range between 1 to 15 kcal/mol, are insufficient to excite electrons to higher electronic energy states, but transitions in vibrational energy states (see **Fig. 5.2**). These states are associated with a molecule's bonds, and consequently each molecule has its own unique signatures. Therefore, IR spectroscopy may be employed to identify the type of bond between two or more atoms and consequently identify functional groups. IR spectroscopy is also widely used to characterize the attachment of organic ligands to organic/inorganic nanoparticles and surfaces. Because IR spectroscopy is quantitative, the number of a type of bond may be determined.

Virtually all organic compounds absorb IR radiation, but inorganic materials are less commonly characterized, as heavy atoms show vibrational transitions in the far IR region, with some having extremely broad peaks that hamphers the identification of the functional groups. Furthermore, the peak intensities of some ionic inorganic compounds may be too weak to be measured.[29,30]

The covalent bonds that hold molecules together are neither stiff nor rigid, but rather they vibrate at specific frequencies corresponding to their vibrational energy levels. The vibration frequencies depend on several factors including bond strength and the atomic mass. The bonds can be modified in different ways, in a similar manner to a spring. As shown in **Fig. 5.10**, chemical bonds may be contorted in six different ways: stretching (both symmetrical and asymmetrical), scissoring, rocking, wagging, and twisting. Absorption of IR radiation causes the bond to move from the lowest vibrational state to the next highest, and the energy associated with absorbed IR radiation is converted into these types of motions.[31-33] Other rotational motions usually accompany these individual vibrational motions. These combinations lead to absorption bands, not discrete lines, which are commonly observed in the mid IR region.[33]

Weaker bonds require less energy to be absorbed and behave as though the bonds are springs that have different strengths. More complex molecules contain dozens or even hundreds of different possible bond stretches and bending motions, which implies the spectrum may contain dozens or hundreds of absorption lines. This means that the IR absorption spectrum can be its unique *fingerprint* for identification of a molecule.[34] The *fingerprint region* contains wavenumbers between 400 and 1500 cm^{-1}. A *diatomic molecule,* that has only one bond, can only vibrate in one direction. For a linear molecule (e.g. hydrocarbons) with n atoms, there are $3n$-5 vibrational modes. If the molecule is non-linear (such as methane, aromatics etc.), then there will be $3n$-6 modes.

Samples can be prepared in several ways for an IR measurement. For powders, a small amount of the sample is added to potassium bromide (KBr), after which this mixture is ground into a fine powder and subsequently compressed into a small, thin, quasi-transparent disc (**Fig. 5.11**). For liquids, a drop of sample may be sandwiched between two salt plates, such as NaCl. KBr and NaCl are chosen as neither compound shows an IR-active stretch in the region typically observed for organic and some inorganic molecules.

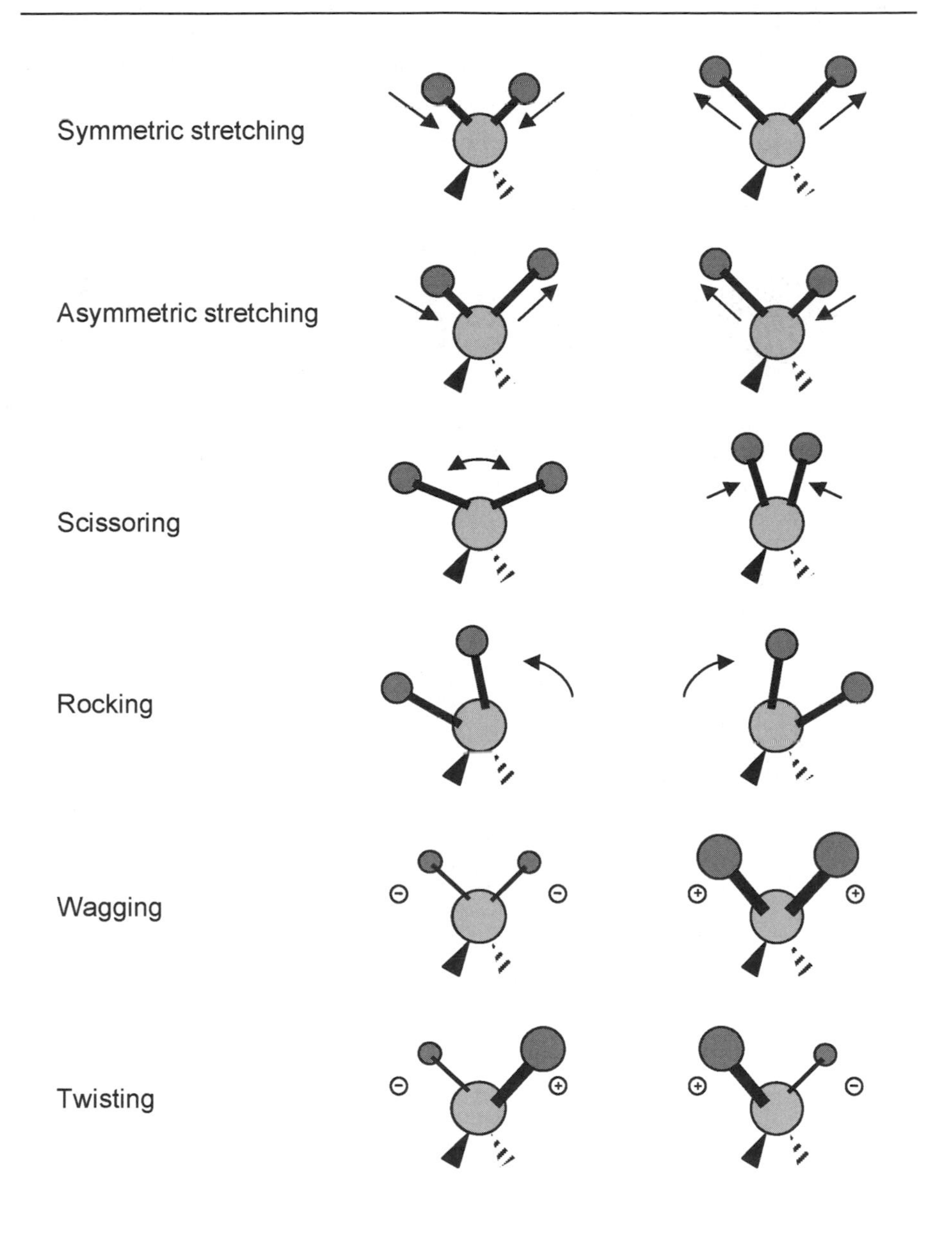

Fig. 5.10 Six different ways in which molecular bonds can vibrate.

Thin films can be investigated by either *reflection absorption infrared spectroscopy* (*RAIRS*), or using an *attenuated total reflectance* (*ATR*) crystal, or by measuring the *diffuse reflectance of the IR light* (*DRIFT*).[35]

Acquired IR spectra are subtracted from a background spectrum to remove unwanted signals, in particular any water present in the ambient.

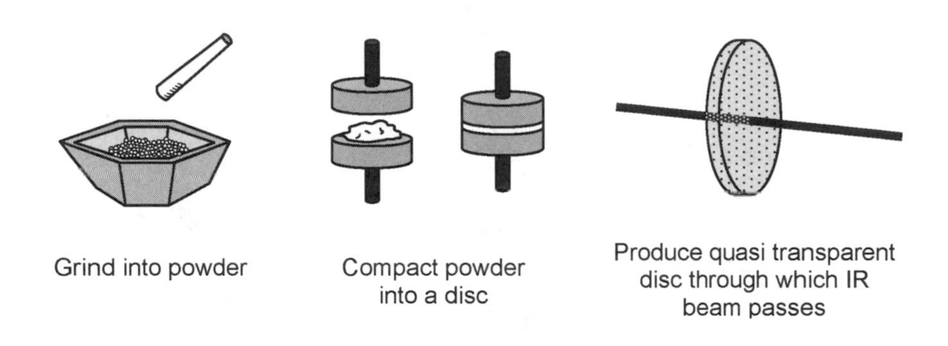

Fig. 5.11 The preparation of a KBr disc for IR spectroscopy.

The IR transmittance of the small organic dye molecule 9-anthracene carboxylic acid (molecular weight: 222.24 g/mol) is shown in **Fig. 5.12**. Most of the peaks are located within the fingerprint region. These peaks can be assigned to different vibrational motions of specific bonds. For example, aromatic compounds like anthracene show stretching of the C-H bonds at wavenumbers between 3130–3070 cm^{-1}, and several in-plane C-H bending peaks between 1225–950 cm^{-1}. Peaks observed between 1615 and 1450 cm^{-1} are attributed to both stretching in the aromatic ring, all of which can be seen in the spectrum.

IR spectroscopy is extremely useful for studying materials adsorbed on surfaces, and when they are arranged an ordered fashion as the characteristics peaks differ to those of molecules in solutions.

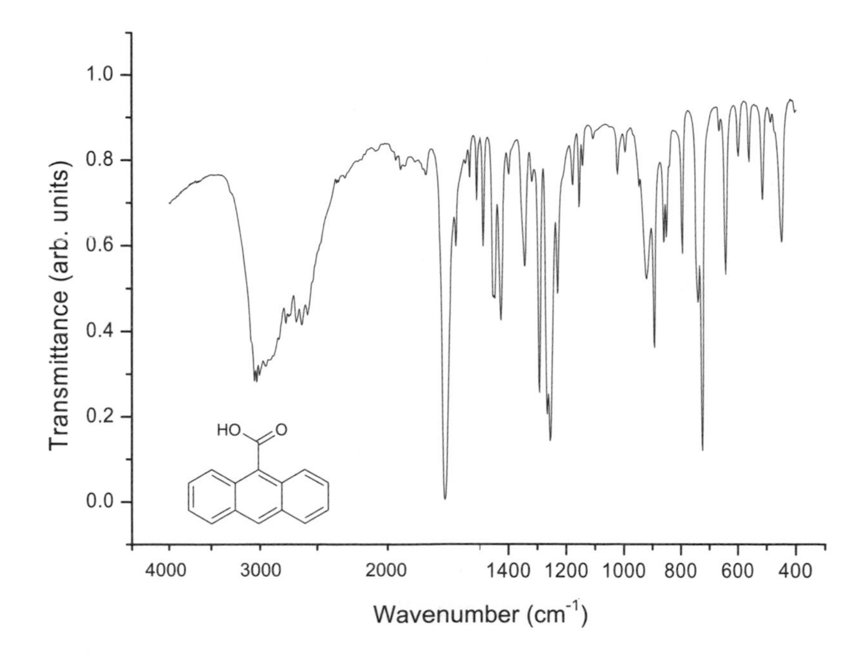

Fig. 5.12 An IR spectrum of an organic molecule (9-anthracene carboxylic acid, shown in the inset).

As previously mentioned, an application of IR spectroscopy, which is finding increased use, is monitoring the attachment of organic ligands to inorganic nanoparticles. It has been demonstrated that smaller particles exhibit a flatter and less noisy background spectrum, while lager particles can show absorption well into the near infrared region, which is a characteristic feature of conductive metal particles with larger nanodimensions.[38] Furthermore, the coupling of a molecule to a surface may result in byproducts and changes in the functional groups. As a result, the covalent attachment of nanomaterials to surfaces can be confirmed and quantified using IR spectroscopy, as illustrated in **Fig. 5.13**.[39]

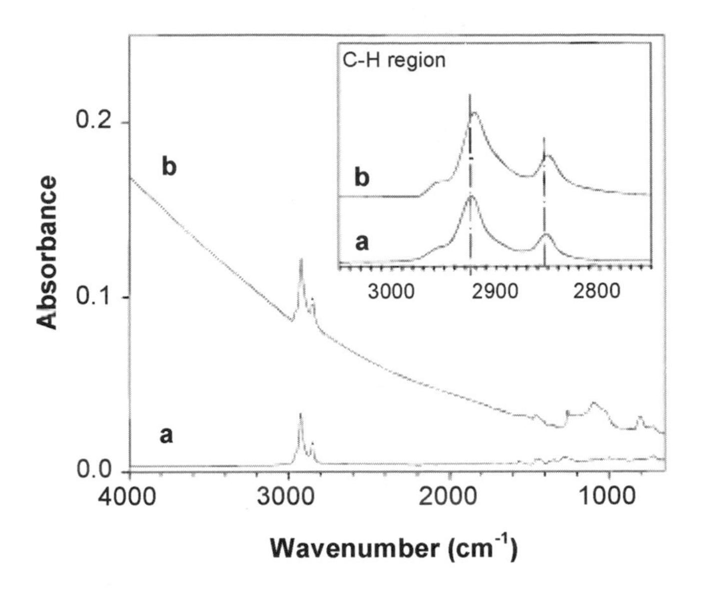

Fig. 5.13 IR spectra of thin films of the NDT-linked Au nanoparticles from 2- (a) and 5-nm (b) core sizes. As seen, the larger functionalized nanoparticles display a noisier background. The substrate was an Au film on glass slide. Reprinted with permission from the American Chemical Society publications.[39]

5.3 Nuclear Magnetic Resonance (NMR) Spectroscopy

NMR spectroscopy is a complex technique employed largely to study the chemical structure of organic and inorganic compounds, either in liquid or solid form.[40] Nuclei that contain odd numbers of protons or neutrons (such as hydrogen-1 and carbon-13) have an intrinsic magnetic moment. Such magnetic nuclei can be aligned with a very powerful external magnetic field. This alignment then can be perturbed with radio frequency (RF) electromagnetic field. The resulting response to the perturbing field is used in NMR spectroscopy.[41,42] NMR spectroscopy is extremely important in nanotechnology, particularly for the development of novel compounds and for determining the chemical structure of polymers and biomolecules such as polypeptides and proteins. Although NMR spectroscopy is mainly used in structural formation of synthesized materials, there are number of examples that highlight its application in sensing.

For a material to absorb electromagnetic radiation in the RF range, it must be NMR active.[15] If the electromagnetic radiation has the right frequency, the material's nucleus resonates, or flips, from one magnetic alignment to another, known as the resonance condition. Because the resonant frequencies depend not only on the nature of the atomic nuclei, but also on the atoms which are bound to that nucleus and chemical environment, functional groups and chemical bonds can be identified.

An example of a typical NMR spectroscopy setup for liquid samples is shown in **Fig. 5.14**. The dissolved sample is placed between the poles of a large magnet. This solution contains a small amount of material, such as tetramethylsilane (TMS), which produces a standard absorption line in the acquired spectrum. A pulse of RF radiation causes nuclei in the magnetic field to flip to a higher-energy alignment. The nuclei then re-emit RF radiation at their respective resonant frequencies, which creates an interference pattern in the resulting RF emission versus time. The emitted RF radiation is collected with a sensing coil and then amplified. The frequencies of the RF emissions are extracted through Fourier transformation.

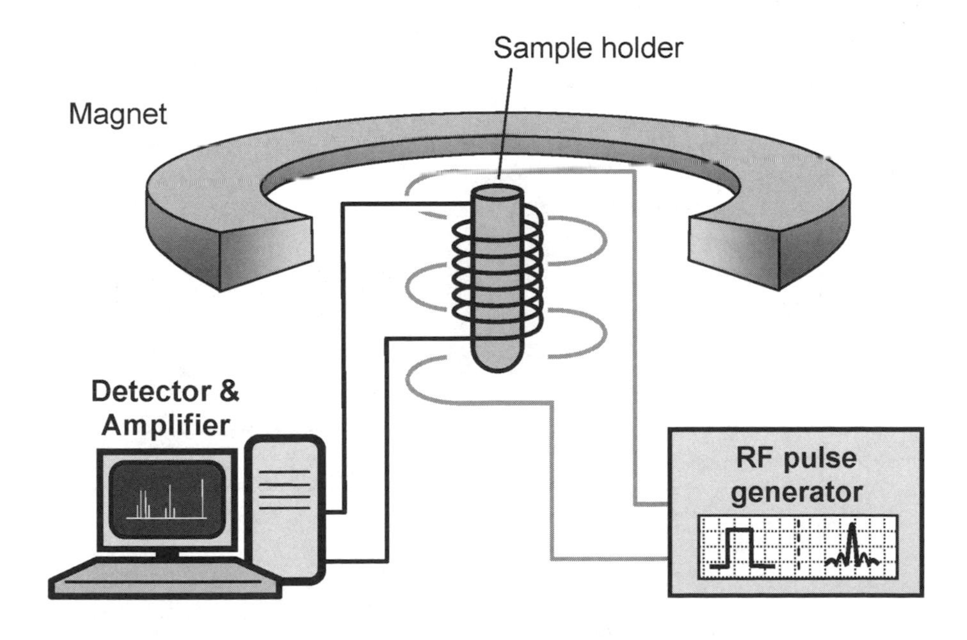

Fig. 5.14 Schematic diagram of an NMR spectroscopy set-up.

An NMR spectrum shows a plot of *chemical shift* against the intensity of absorption. The chemical shift corresponds to the exact position on the spectrum where a nucleus absorbs energy. NMR charts are calibrated using an arbitrary scale called the delta scale. Delta, δ, is equal to 1 ppm of the NMR's operating frequency, and is defined as:[8,34]

$$\delta = \frac{\text{Observed chemical shift from the TMS reference peak}}{\text{Spectrometer frequency}}. \quad (5.7)$$

To illustrate this, an NMR spectrum of chloroform ($CHCl_3$) tuned for the nucleus of hydrogen atoms, operating at 60 MHz is shown in **Fig. 5.15**. An NMR measurement tuned for the hydrogen nucleus is called proton, or ^1H, NMR. The $CHCl_3$ peak shown is measured relative to the TMS peak. If the RF frequency on the NMR is set to 60 MHz, then $CHCl_3$ will produce a peak which is at 437 Hz/60 MHz which is equal to a chemical shift of 7.28 ppm. For molecules with more intricate structures (or those not containing hydrogen at all) the ^1H-NMR may not provide sufficient information. In such cases, the NMR instrument may be tuned for the nuclei of isotopes such as ^{13}C, ^{15}N, ^{19}F, ^{29}Si, and ^{31}P.

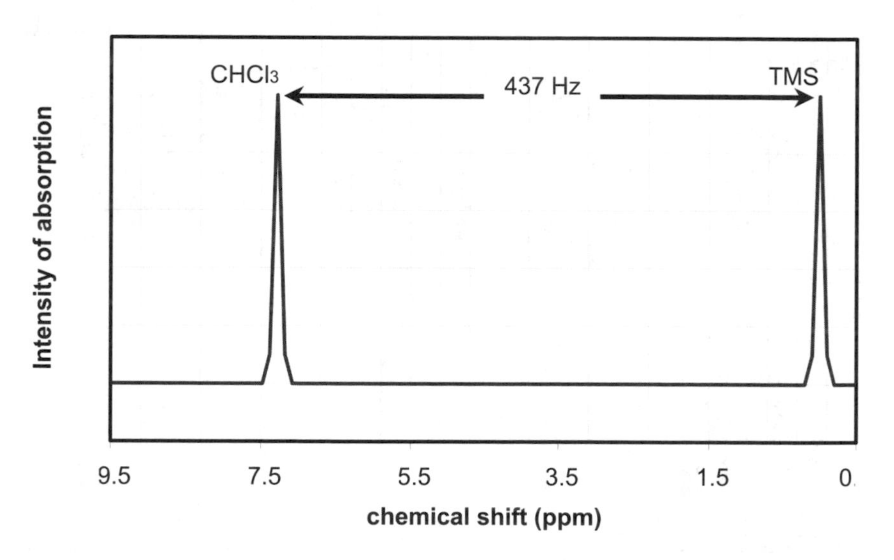

Fig. 5.15 Proton NMR spectrum of $CHCl_3$.

Fig. 5.16 shows an example of how NMR spectroscopy has been used in biosensing applications.[43] The molecule is designed to bind both xenon and protein with high affinity and specificity. Upon binding to the target protein a change in the xenon NMR spectrum occurs, and may be correlated to the amount of analyte with which the biosensor molecule has interacted.

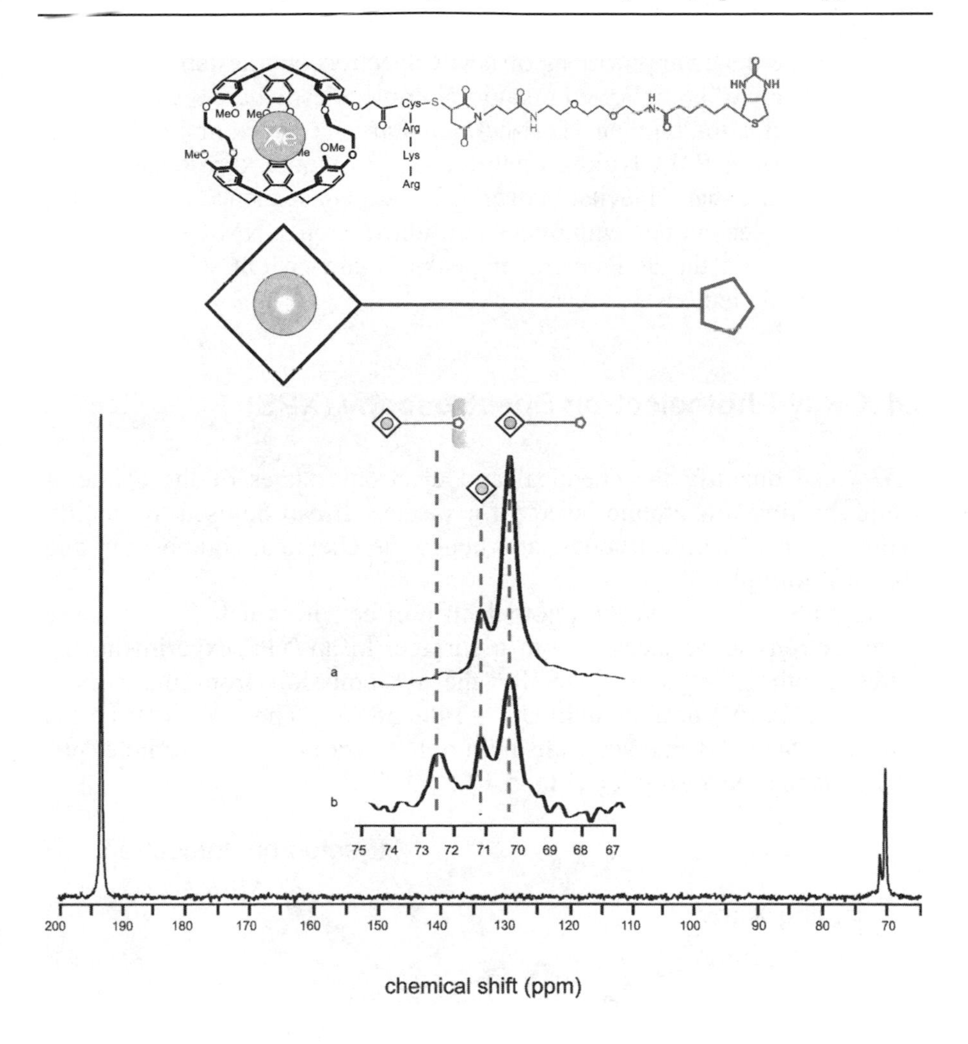

Fig. 5.16 *Top*: Structure of a biosensor molecule designed to bind xenon to a protein with high affinity and specificity. On either end are seen the cage binding xenon (cryptophane-A) and a biotin ligand, which are connected by a tethering molecule (the schematic representation of this structure is shown below it). *Bottom:* Xenon-129 NMR spectra monitoring the binding of biotin-functionalized xenon to avidin. (a) the functionalized xenon before the addition of avidin, with the more intense peak corresponding to functionalized xenon and the smaller peak corresponding to xenon in the cage without linker and ligand, serving as both a chemical shift and signal intensity reference. (b) the spectrum on the addition of approx 80 nmol of avidin monomer, which shown the decrease in existing peaks and the appearance of a new third peak. Reprinted with permission from the Highwire Press.[43]

One of the major applications of NMR spectroscopy systems is for the structural study of complicated organic biomolecules, where it can provide detailed information on their 3D structures. Large complex molecules such as proteins, DNA/RNA typically exhibit several thousand resonances and among them are many inevitable overlaps. As a consequence, complicated mathematical techniques employing multidimensional NMR experiments are required, and the assignment of peaks is carried out with advanced computational methods.

5.4 X-Ray Photoelectron Spectroscopy (XPS)

XPS can quantify the chemical and electronic states of the elements within the first few atomic layers of a surface. It can be used to identify elements, their chemical bonds, and hence the chemical composition and empirical formulae.[44-46]

The XPS is based on the photoelectric effect, in which X-rays cause photoelectrons to be ejected from a surface. In an XPS experiment, the source of the X-rays is generally the K_α emission from magnesium ($K_\alpha = 1253.56$ eV) or aluminum ($K_\alpha = 1486.58$ eV). The X-rays strike the sample surface and interact with the atomic electrons in the sample, primarily via photon absorption (**Fig. 5.17**).

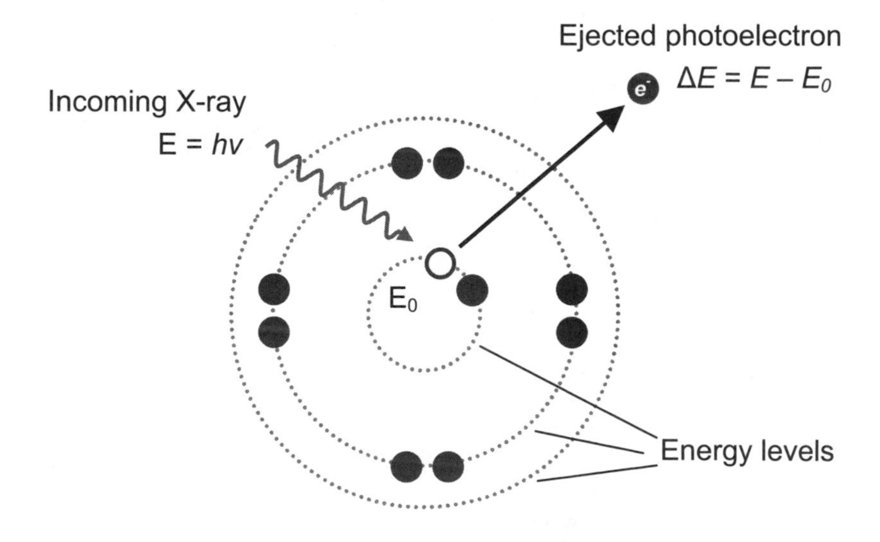

Fig. 5.17 Electronic orbits showing the ejection of a photoelectron after absorption of a photon.

The focused X-ray beam which impinges on the sample has energy of approximately 1.5 keV while the reflected photoelectrons have smaller energies. The reflected photoelectrons only escape from the very top surface of the sample (generally not more than 10 nm). The kinetic energy and number of the ejected photoelectrons is plotted as a spectrum of their binding energies. The acquired spectrum is compared with spectra from known databases. The peak positions and shapes correspond to the material's electronic configuration, and therefore elements and compounds show their own unique characteristic peaks. In sensing applications, XPS is used not only to study the chemical composition of a sensing layer, but also to investigate the interactions between the surface and the target molecules.

As XPS involves monitoring emitted photoelectrons, the experiments must be conducted under ultra high vacuum and therefore the sample should not outgas. Furthermore, exposure to the X-ray beam can damage certain materials, mainly organic molecules and polymers, and they may degrade during the measurement. XPS experiments are limited to just a few Ångstroms beneath the surface, despite the incoming X-rays being able to penetrate microns into the surface. This is because the ejected electrons must travel through the sample and yet retain enough energy to reach and excite the detector. Only electrons that are emitted by atoms near the surface have a chance to leave the sample.

When the XPS instrument is combined with ion beam sputtering, atomic layers can be continuously removed from the surface. After sputtering, the XPS may be performed once again on these layers, and as a result compositional depth profiles can be obtained down to a few micrometers.[47]

As the incident X-rays can penetrate deep into the atom, they can eject electrons from several energy levels. The incident X-rays have energy of $h\upsilon$ and can pass on energy to the ejected electron according to the following equation:

$$E_k = h\nu - E_b - \phi_S, \tag{5.8}$$

where E_k is the kinetic energy of the ejected electron, E_b is the binding energy of the ejected electron, h is Planck's constant (4.136×10^{-15} eV·s), υ is the frequency of the incident radiation, and ϕ_S is the work function of the material.

If an electron of an inner atomic shell is ejected from the atom, then an electron from the outer shell will fill the empty space it leaves behind. Two things may then happen, either a photon (whose energy is equal to the difference between the two energy levels) will be emitted, or the energy is transferred to another outer electron that is then emitted. This emitted outer electron is called an Auger electron, and consequently *Auger Electron*

Spectroscopy (*AES*) can be used to analyze these emitted electrons. Both these processes are schematically summarized in **Fig. 5.18**.[48]

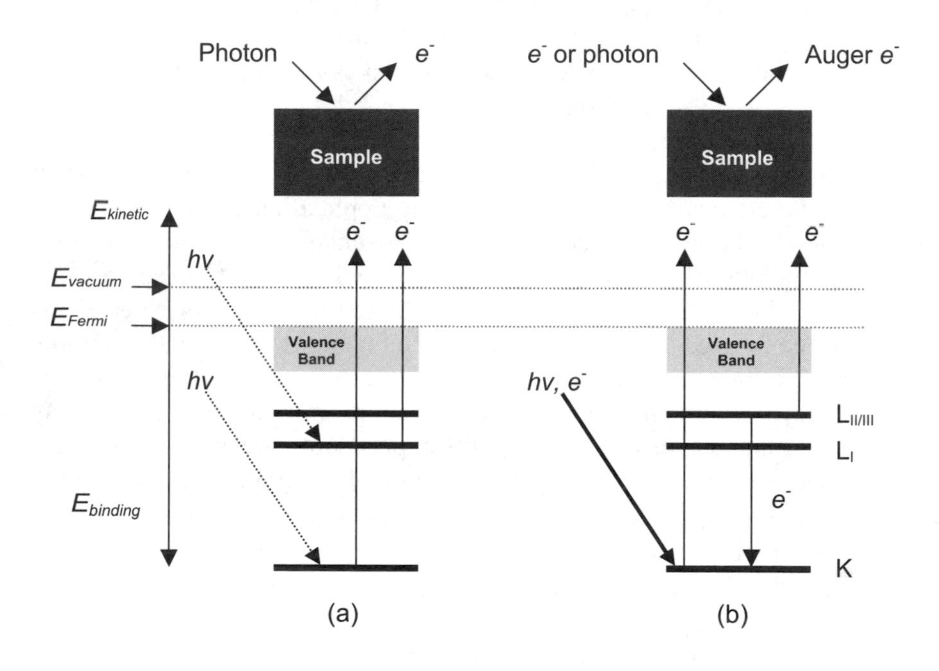

Fig. 5.18 The photoelectric and Auger processes in (a) XPS and (b) AES.

Because XPS is surface sensitive, it is ideal for characterizing sensing layers, as the interaction between the analyte and sensing layer occurs on the surface. By performing XPS measurements on the sensing surface, before and after interactions, an understanding of the sensing mechanisms can be obtained. For example, **Fig. 5.19** demonstrates how XPS was employed to study the interaction between NO_2 and O_2 gases with nanostructured In_2O_3 thin films.[49] The two peaks in the spectra correspond to the characteristic signatures used for indium metal and indium oxide that are determined from the In $3d_{5/2}$ and O $1s$ core level emissions, respectively. By observing the relative increase and decrease in these peaks with respect to each other, XPS was used to confirm the dissociative adsorption of NO_2 and the formation of adsorbed oxygen species on the In_2O_3 surface.

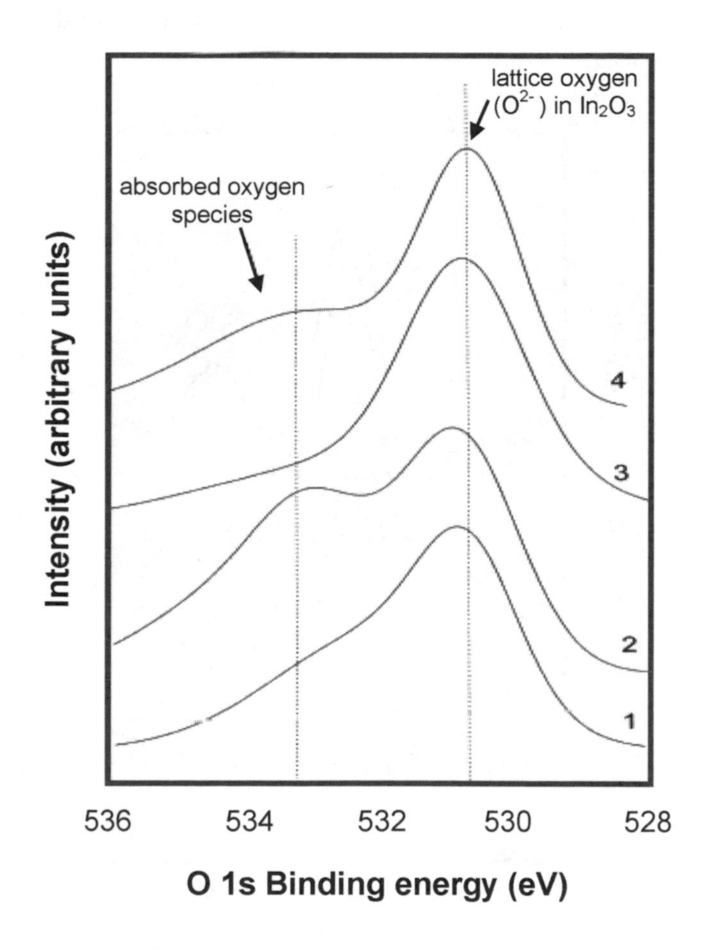

Fig. 5.19 O 1s core level XPS spectra of the In_2O_3 thin films at different conditions: **1.** After heating in air for 1 h at 700°C; **2.** Exposure to NO_2 1×10^2 Pa 200°C ½ h; **3.** Vacuum 1×10^{-4} Pa at 200°C for ½ h; **4.** Exposure to O_2 1×10^4 Pa at 200°C for ½ h. Reprinted with permission from the Elsevier publications.[49]

In another example, the XPS spectra of 40 Å CdSe semiconducting nanocrystals coated with trioctylphosphine oxide (TOPO) are seen in **Fig. 5.20**.[50] The coating prevents CdSe from oxidization, and in this example XPS was employed to confirm the presence of the coating. As seen in the figure, the Se peaks indicate its presence in the CdSe nanocrystals. An SeO_2 peak confirms the uncoated CdSe sample showing that nanocrystals oxidized in air. However, a smaller SeO_2 peak is present in one of the coated films (sample b), showing that the TOPO coating does not completely cover that CdSe nanocrystals.

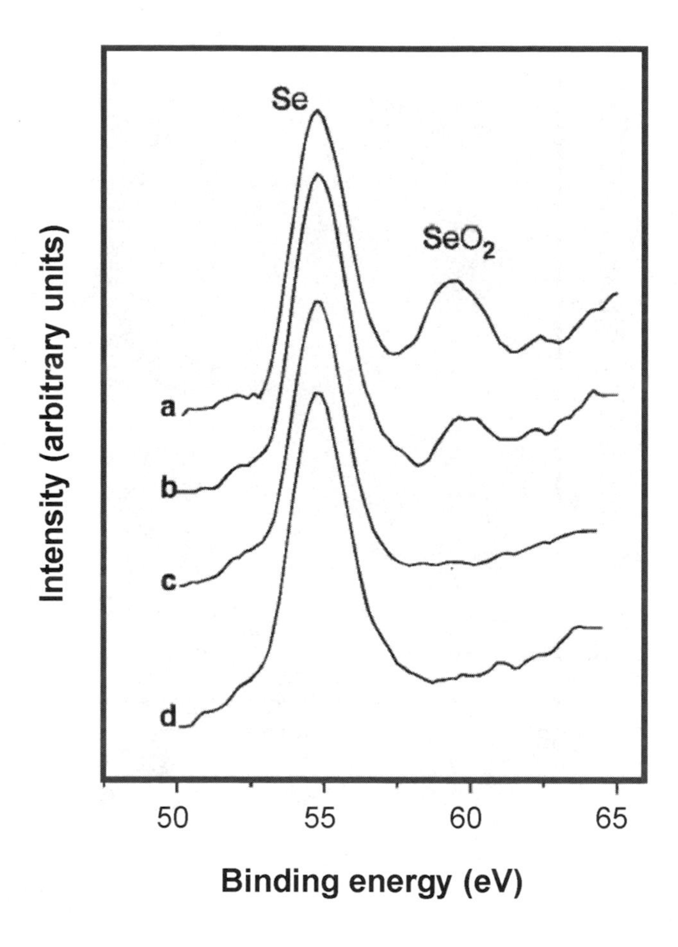

Fig. 5.20 X-ray photoelectron spectra highlighting the Se 3d core transitions from ~40 Å bare and ZnS overcoated CdSe dots: (a) bare CdSe, (b) 0.65 monolayers, (c) 1.3 monolayers, and (d) 2.6 monolayers of ZnS. The peak at 59 eV indicates the formation of selenium oxide upon exposure to air when surface selenium atoms are exposed. Reprinted with permission from the American Chemical Society publications.[50]

5.5 X-Ray Diffraction (XRD)

XRD involves monitoring the diffraction of X-rays after they interact with the sample. It is a crystallographic technique used for identifying and quantifying various crystalline phases present in solid materials and powders.[51,52] In XRD the crystal structure can be determined as well as the size of grains and nanoparticles. When X-rays are directed at a regular crystalline sample, a proportion of them are diffracted to produce a pattern. From such a pattern the crystal phases can be identified by comparison to those of internationally recognized databases (such as *International Center of Diffraction Data - ICDD*) that contain reference patterns. In sensing applications, XRD is generally used to correlate the properties of a material to its sensing performance.

XRD is one of the most utilized techniques for determining the structure of inorganic and organic[53] materials. It is also widely used for studying nanostructured thin films and nanoparticles. However, the materials must have ordered structure, and it cannot be used directly to study amorphous materials. Another inherent limitation of XRD is that mixtures of phases that have low symmetry are difficult to differentiate between because of the larger number of diffraction peaks. Furthermore, organic materials such as polymers are never completely crystalline, therefore XRD is primarily used to determine their crystallinity.

In crystallography, the solid to be characterized by XRD has a space lattice with an ordered three-dimensional distribution (cubic, rhombic, etc.) of atoms. These atoms form a series of parallel planes separated by a distance d, which varies according to the nature of the material. For any crystal, planes have their own specific d-spacing. When a monochromatic X-ray beam with wavelength λ is irradiated onto a crystalline material with spacing d, at an angle θ, diffraction occurs only when the distance traveled by the rays reflected from successive planes differs by an integer number n of wavelengths to produce constructive interference (**Fig. 5.21**). Such constructive interference patterns only occur when incident angles fulfill the Bragg condition such that:

$$n\lambda = 2d \sin\theta \, . \tag{5.9}$$

By varying the angle θ, the Bragg Law condition is satisfied for different d-spacings in polycrystalline materials. Plotting the angular positions versus intensities produces a diffraction pattern, which is characteristic of the sample. When a mixture of different phases is present, the resultant *diffractogram* is a superposition of the individual patterns.[54]

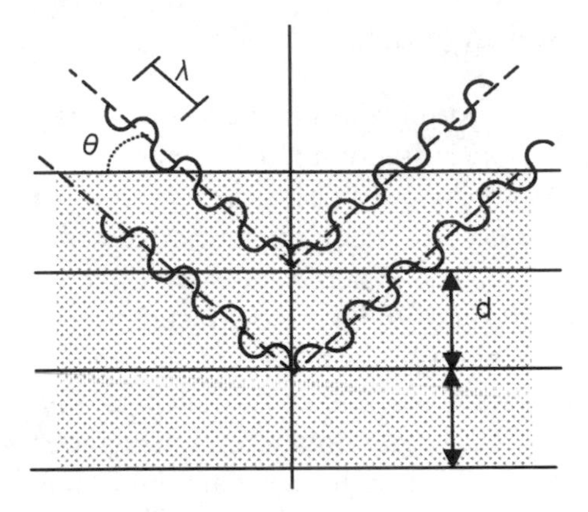

Fig. 5.21 Condition for constructive interference.

In a typical XRD pattern, the diffracted intensities are plotted versus the detector angle 2θ. Each peak is then assigned a label indicating the spacing of a crystal plane.

Bragg's law states the condition for sharp diffraction peaks arising from crystals which are perfectly ordered. Actual diffraction peaks have a finite width resulting from imperfections, either the irradiation source or the sample. A useful phenomenon is that as crystallite dimensions enter the nanoscale the peaks broaden with decreasing crystal size. It is known that the widths of the diffraction peaks allow the determination of crystallite size. Practically, the size of crystallites can be determined using variants of the Scherrer equation:[55,56]

$$t = \frac{K\lambda}{B\cos\theta}, \tag{5.10}$$

where t is the thickness of the crystal, K is a constant which depends on the crystallite shape, and B is the full width at half maximum of the broadened peak. If a Gaussian function is used to describe the broadened peak, then the constant K is equal to 0.89. The Scherrer equation is derived from Bragg's law and may be used to determine crystallite sizes if the crystals are smaller than 1000 Å.

XRD has many practical uses for nanotechnology enabled sensing applications. Not only does it allow for different phases to be identified, it can also be used to monitor the growth and formation of nanosized crystallites by examining the broadening of peaks in the XRD pattern. This is particularly important for studying sensing materials whose performance depends on the nanocrystals particle size. It is also valuable for determining the distribution of nanocrystals on the surface of a sensing layer.

In **Fig. 5.22**, XRD patterns for ZnO nanocrystals synthesized from zinc acetate and sodium hydroxide are shown.[14] As seen in the figure, peaks are broadened with respect to those of the bulk material, and the broadening increases as the crystallite size decreases. The crystal structure of the bulk (wurzite) has been preserved in the nanocrystals, despite being significantly broadened.

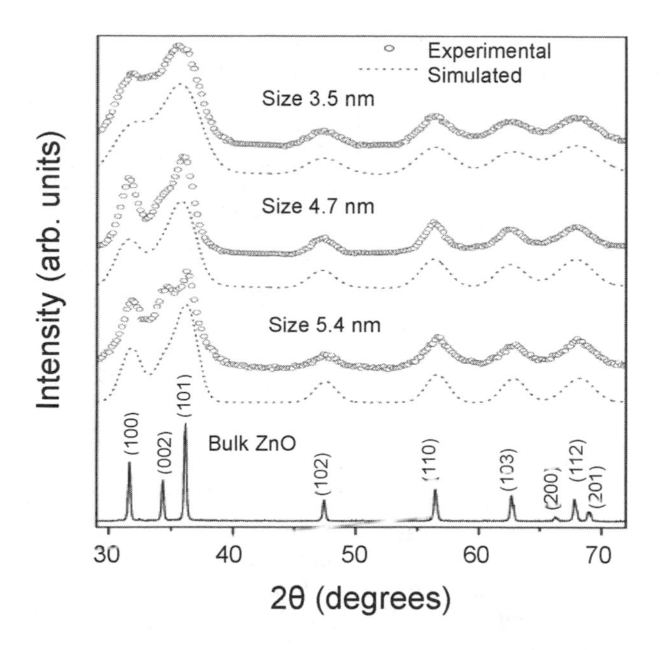

Fig. 5.22 Experimental XRD patterns of ZnO nanocrystals (circles) and the corresponding simulated patterns (dotted line) for different particle sizes. Reprinted with permission from the Royal Society of Chemistry publications.[14]

Performing XRD on powders made of nanocrystals can provide information about their structure and size. For instance, the powder diffraction patterns of InAs and InP nanocrystals of different sizes, are shown in **Fig. 5.23**.[22,57] It is seen that peak broadening occurs as the dimensions of the nanocrystals decrease. The peak positions for the nanocrystals shown agree with those of the bulk, demonstrating that the crystal structure hasn't changed as the particle sizes decrease, despite the substantial changes in the domain size. A small shift in peak positions can also be seen in **Fig. 5.23** patterns when crystallite size decreases which is due to lattice relaxation.

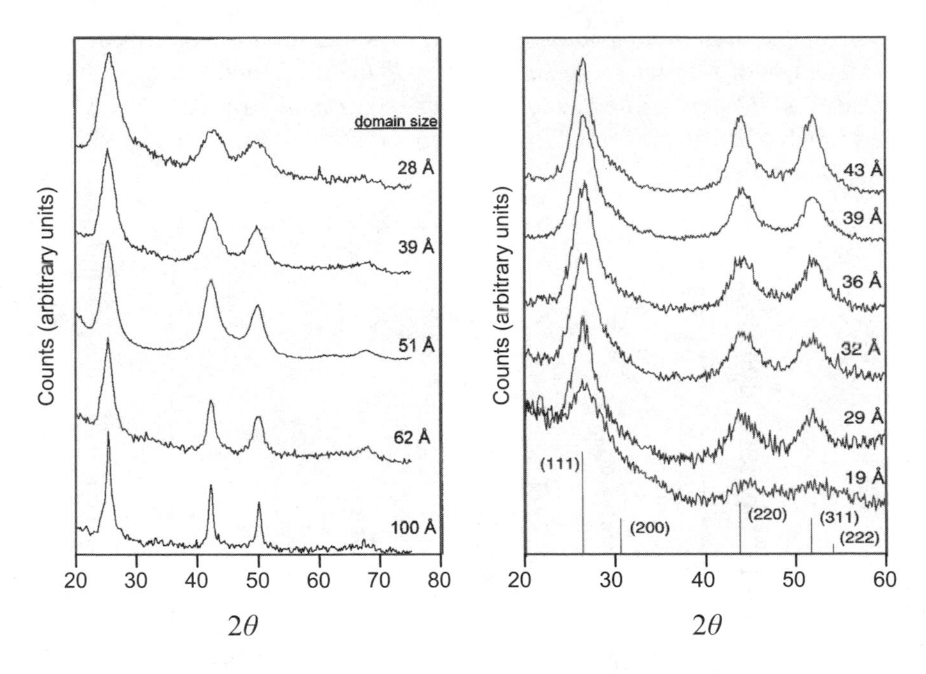

Fig. 5.23 *Left:* Powder XRD spectra of a series of InAs nanocrystal sizes. From the width of the reflections the crystalline domain size may be calculated.[22] *Right:* Powder X-ray diffraction of a series of InP nanocrystal sizes. The stick spectrum at the bottom gives the bulk reflection position with relative intensities. Reprinted with permission from the American Chemical Society publications.[57]

5.6 Light Scattering Techniques

These techniques involve analyzing the light that is scattered from materials. The types of light scattering include: *elastic*, where the wavelength of the scattered light remains unchanged from the incident light, and *inelastic*, where the scattered light's wavelength is different from that of the incident light. Depending on the frequency of light and the type of scattering, these techniques can provide information regarding the size, chemical composition and structure of nanomaterials. Changes in these material properties can be monitored with such techniques, and in some cases, the technique itself may be used for signal transduction in sensing applications.

Rayleigh scattering, which occurs when the particles are much smaller than the wavelength of the impinging light, is an example of elastic scattering. Conversely, Raman scattering is inelastic, where the scattering of a photon creates or annihilates a phonon, and the scattered light has a different wavelength than the incident light.

Two of the most relevant light scattering characterization techniques for nanotechnology enabled sensing are *Dynamic Light Scattering* and *Raman Spectroscopy*.

5.6.1 Dynamic Light Scattering (DLS)

DLS, also known as *quasi-elastic light scattering* and *photon correlation spectroscopy*, is commonly employed for studying colloidal systems as it is a relatively fast and straightforward technique in which a light beam is directed onto a sample that scatters the light elastically. This light is scattered over a period of time and then analyzed statistically.[47,58] DLS has two main applications: the study of system dynamics in real time; and the absolute determination of nanoparticle sizes,[59,60] with perhaps the latter being more important in nanotechnology. In fact, it is well suited for examining the monodispersity of synthesized nanoparticles, and for determining small changes in particles' mean diameter after the adsorption of molecules/layers, such as when forming a coating.

DLS has several advantages over other techniques for particle size determination; while scanning electron microcopy techniques are very accurate a DLS measurement uses non-ionizing lower energy light sources and is carried out at room pressure. Also time-dependent DLS is ideal for studying the growth of nanocrystals in solution, unlike other techniques that require the samples be dried.[61]

DLS does have some limitations. It is only suitable for particles that exhibit Rayleigh scattering. For particles whose size is larger than about a tenth of the illuminating wavelength, the intensity is angle dependent and the scattering is explained by *Mie theory*.[58] DLS is utilized to measure particle sizes in the range from a few nanometers to a few microns. Other limitations include: the requirement (in most cases) for dilute suspensions in order to minimize multiple scattering; the difficulty in differentiating between tiny fluctuations and noise;[62] and it is impossible to discriminate between light scattered from a single primary particle agglomerate.

A typical setup is shown in **Fig. 5.24**. A beam of monochromatic light passes through a solution containing the particles, which are generally colloids or micelles and scattering occurs. The intensity of the scattered light is uniform in all directions with the amount of Rayleigh scattering depending on the size of the particles and the wavelength of the incident light. Whilst in solution, the particles move about in small random patterns

(Brownian motion). At constant temperature, larger particles move more slowly than the smaller ones, and hence the distance between the particles is constantly varying. Therefore time-dependent fluctuations in the scattering intensity can be observed by counting the number of photons returning to the detector after scattering.[63,64] These time dependent fluctuations can then be related to particle speed by autocorrelation of the average of the product of the photon count with a delayed version as a function of the delay time. The autocorrelation function is analyzed by numerically fitting the data with calculations based on assumed particle size distributions. Analysis of the time dependent fluctuation gives the diffusion coefficient of the particles, D.

If the viscosity η, of the solution is known, then the radius or a spherical particle, r, can be obtained from the *Stokes-Einstein* relation:[65]

$$r = \frac{kT}{6\pi\eta D}, \tag{5.11}$$

where k is Boltzmann's constant (1.3806505×10^{-23} J/K) and T is the absolute temperature.

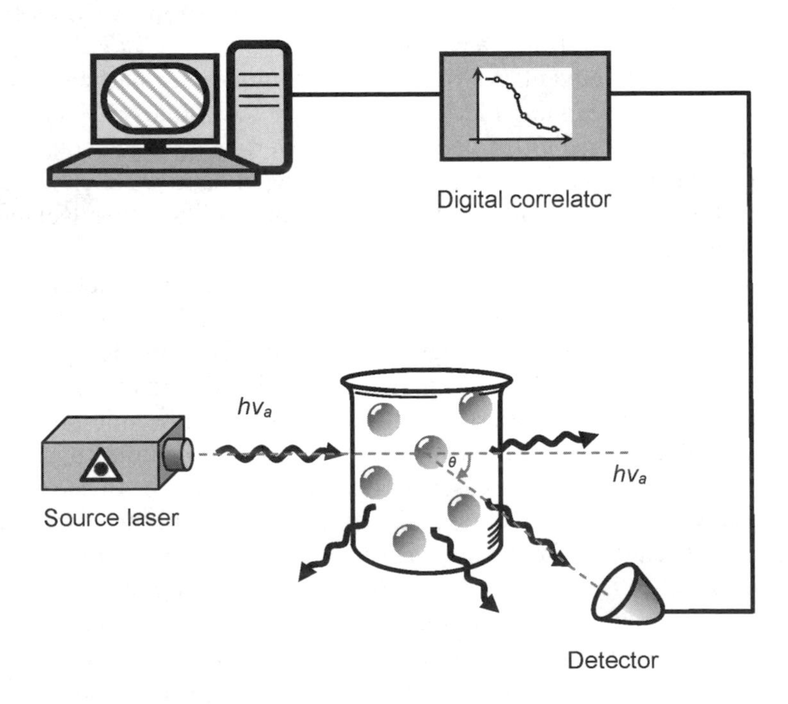

Fig. 5.24 An illustration of a DLS experiment.

Fig. 5.25 shows an example where DLS has been employed to determine the size of helical rosette nanotubes,[66] built from multiple stacks of the rosette shaped compound shown. The average size of the nanotubes (32.5 nm) derived from the dynamic DLS was used to reveal that the nanotubes are composed on average of a stack of approximately 140 rosette molecules.[66] However, this example highlights one of thc key assumptions in DLS calculations, which is that the molecules must be spherical. Therefore, thc actual length is most likely larger.

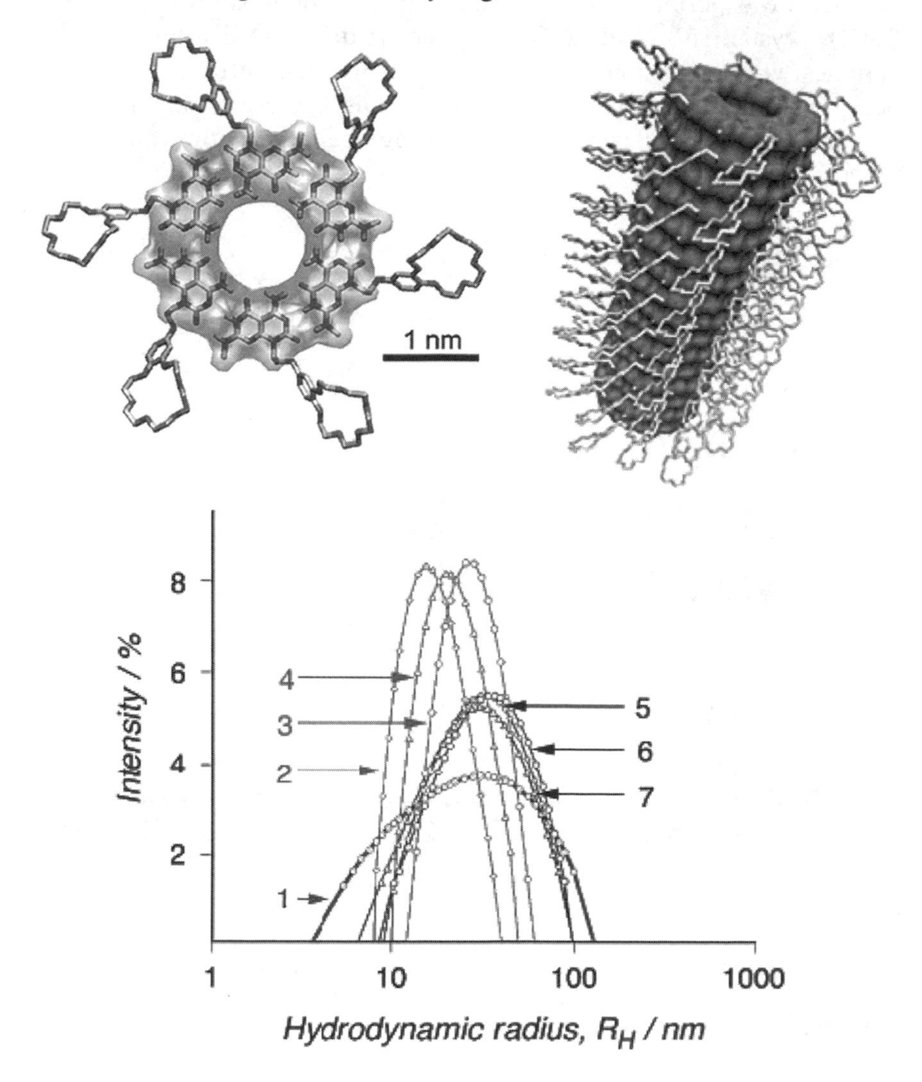

Fig. 5.25 *Top:* The hierarchical self-assembled six-membered ring supermacrocycle (rosette) on the left and the resulting nanotube stack on the right. *Bottom:* DLS regularization diagrams of the nanotube, measured in solution at different concentrations. Reprinted with permission from the American Chemical Society publications.[66]

In another example shown in **Fig. 5.26**, the advantage of DLS over other particle size determination techniques is highlighted.[61] Here, time-dependent DLS was performed on a crystallizing solution to study the growth of $CaCO_3$ crystallites. After 20 minutes the particle size distribution shows particles with dimensions well in excess of the micron range. With time, peak broadening becomes prevalent, as this represents the increasing polydispersity of the particle sizes during their nucleation and growth processes.[61]

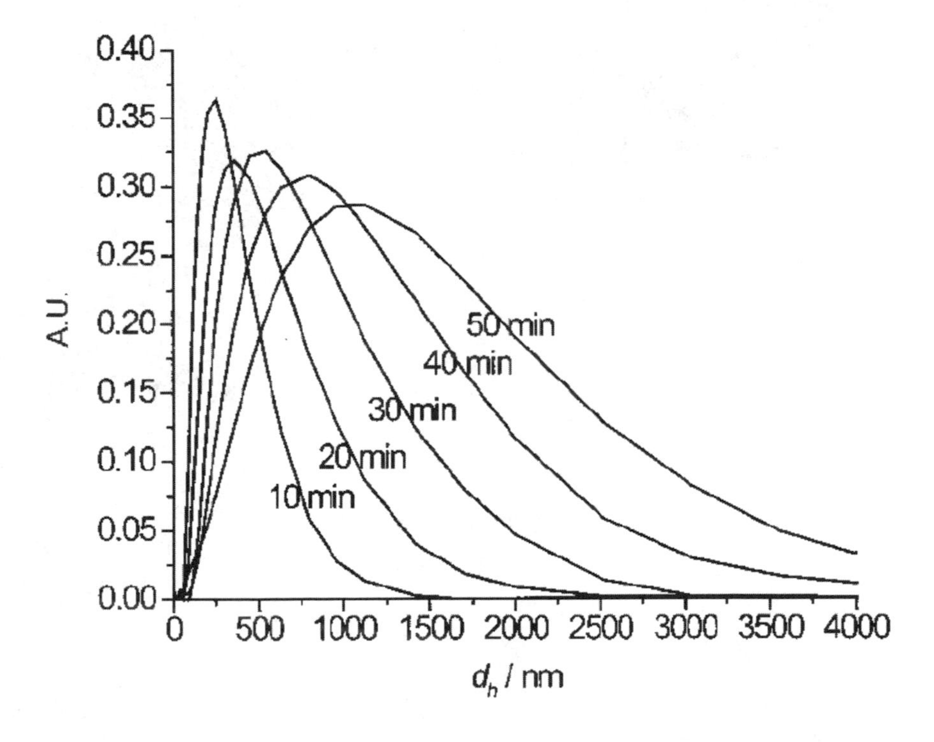

Fig. 5.26 Time dependence particle size distributions of $CaCO_3$ crystals from dynamic light scattering measurement. Reprinted with permission from the Wiley InterScience publications.[61]

5.6.2 Raman Spectroscopy

Raman spectroscopy is based on monitoring the intensity and wavelength of light that is scattered inelastically from molecules or crystals. It is suitable for characterizing organic and inorganic samples. In a Raman experiment, a sample is irradiated with light of known polarization and wavelength (generally in the visible or infrared ranges). Inelastic (or Raman) scattering occurs and the scattered light is wavelength-shifted with respect to the incident light (**Fig. 5.27**). The spectrum of the scattered light is then analyzed to determine the changes in its wavelength. Raman spectroscopy is a powerful analytical tool for qualitatively and quantitatively investigating the composition of materials.[67,68]

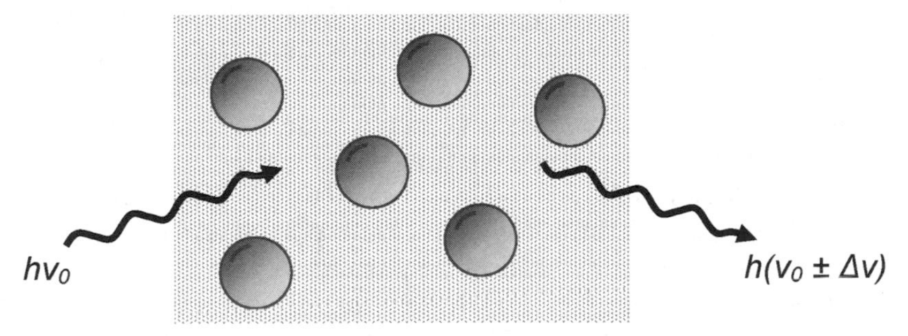

Fig. 5.27 Representation of Raman scattering from particles.

In a Raman spectrum, the wavenumbers of the Raman shifts are plotted against their respective intensities, which originate from the interaction photons with molecular vibrations (phonons in a crystal).[69] When irradiation from a laser source interacts with phonons in the sample, there can be an exchange of energy between them. The phonons may either gain or lose energy. The phonon modes are intrinsic properties related to chemical bonding.[35,70] Therefore, the information contained within a Raman spectrum may provide a "fingerprint" by which molecules can be identified. This not only serves for characterization purposes, but also in sensing applications. When molecules bind together, new bonds may appear, the number of existing types of bonds may increase/decrease, functional groups may change. In sensing applications, monitoring the peak intensities and positions can provide quantitative information on the number of analyte molecules that have taken part in the interaction, such as in affinity based sensing.

If the Raman scattered photon has a lower energy than the incident light, its frequency is shifted down and is referred to as a *Stokes emission*. On the other hand, if the scattered photon has higher energy than the incident light, the frequency is shifted up and it is referred to as an *anti-Stokes* emission.[71-73] The energy of the scattered photon, E, is related to the energy of the incident photon, $E_0 = h\upsilon_0$, by:

$$E = h\upsilon_0 \pm \Delta E_\upsilon , \tag{5.12}$$

$$\text{Stokes:} \quad \upsilon = \upsilon_0 - \Delta\upsilon, \tag{5.13}$$

$$\text{anti-Stokes:} \quad \upsilon = \upsilon_0 + \Delta\upsilon. \tag{5.14}$$

where ΔE_υ is the change in energy and $\Delta\upsilon$ is the change in frequency. Although there is a change in wavelength, no energy is lost during the interaction. This is because the electron cloud in a bond becomes deformed during interaction with the impinging light. The amount of deformation is the polarizability, and this determines the shift Raman of intensity and frequency.[74]

Raman scattering comprises a very small fraction of the scattered light, only one Raman scattered photon from 10^6 to 10^8 excitation photons. Therefore, the main limitation of Raman scattering is discerning the weak inelastically scattered light from the intense Rayleigh scattered light. Furthermore, depending on the incident light energy, photoluminescence may occur which can obscure the Raman spectrum. The Raman scattering intensity is inversely proportional to the fourth power of the emission wavelength.[75] So decreasing the wavelength of the light source should result in an increase in the Raman signal. However, decreasing the wavelength increases the likelihood of observing photoluminescence.[76] Stokes shifts are susceptible to photoluminescence interference, but not anti-Stokes. If the sample is highly fluorescent then a different excitation, one that will not cause electronic excitation, can be employed to circumvent this problem. Unlike IR spectroscopy, where water contamination can block out entire regions of the spectrum, a Raman spectrum is less susceptible to the presence of water.

The Raman signal is inherently weak, which prevents achieving low detection limits with normal Raman spectroscopy.[77] The Raman signal can be enhanced if molecules are adsorbed on roughened metal surfaces, typically gold or silver. This is called *Surface Enhanced Raman Spectroscopy*[78] (*SERS*), [79] and exploits changes in analyte polarizability perpendicular to the surface, which enhances scattering by factors of more than a million-fold.[77] The metal surface must have a plasmon in the frequency region close to that of the excitation laser. Surface roughness or curvature is required for the scattering of light by surface plasmons, however, it is not the defining factor, as an intrinsic surface enhancement effect plays a fundamental role.[80]

SERS is an ideal technique in nanotechnology enabled sensing, particularly for monitoring very small traces of an analyte. In order for SERS to occur, the particles or features must be small when compared the wavelength of the incident light. SERS-active systems should possess structures typically in the 5–100 nm range.[81]

The following examples illustrate the applicability of Raman spectroscopy for sensing as well as in compound identification. **Fig. 5.28** shows an example of SERS in chemical sensing,[82] where the Langmuir-Blodgett technique was used to assemble monolayers of aligned, thiol capped silver nanowires that are ~50 nm in diameter and 2–3 μm in length.[82] The fluorescent molecule Rhodamine 6G (R6G), which produces a distinct Raman spectrum when excited by light of wavelength 532 nm, was adsorbed onto the films. The inset of the figure shows that there is a linear relationship between the R6G concentration and the intensity of the of the Raman peak at 1650 cm^{-1}. This system was capable of detecting 0.7 pg of the analyte.

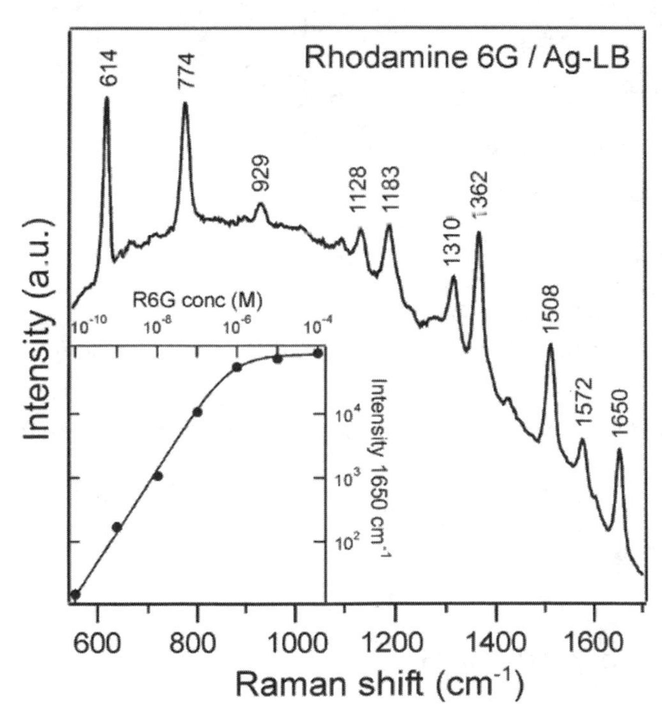

Fig. 5.28 SERS spectrum of R6G on the thiol-capped Ag-LB film (incident light: 532 nm, 25 mW) after 10 min incubation in a 10^{-9} M R6G solution. The inset shows the linear relationship between the Raman intensity at 1650 cm^{-1} and the R6G concentration. Reprinted with permission from the American Chemical Society publications.[82]

Fig. 5.29 shows the spectra obtained from a fiber optic based SERS system employed for chemical sensing.[83] The fiber end incorporate alumina nanoparticles and a silver coating, and the SERS effect is induced through laser light propagating down the fiber. For sensing experiments the coated fiber tip is immersed in solutions of different concentrations of cresyl fast violet (CFV) in groundwater. As seen in the figure, the CFV molecules produce a distinct SERS spectrum when compare to that of the groundwater. By monitoring the intensity of the main SERS peak a quantitative calibration curve for CFV in groundwater could be obtained, with a detection limit approximately 50 ppb.

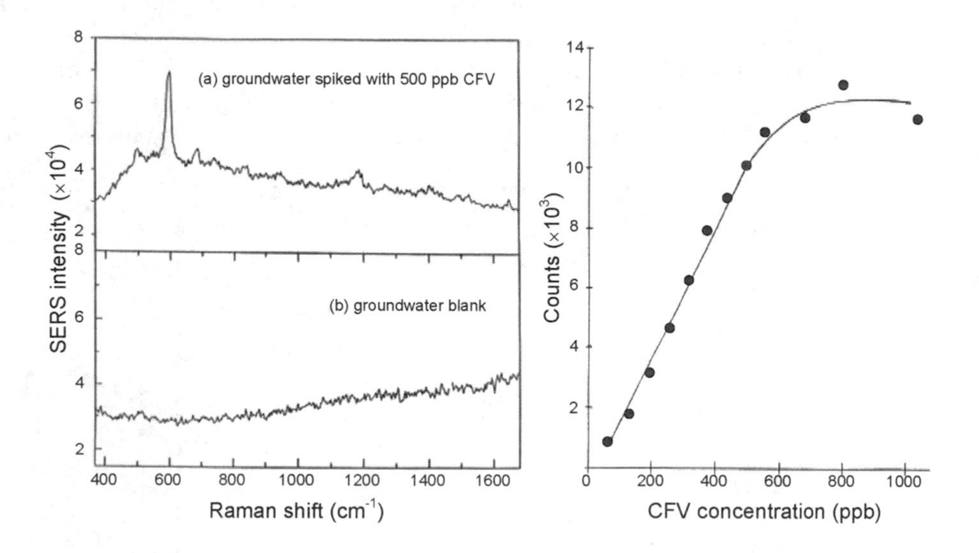

Fig. 5.29 *Left:* In situ SERS spectra of (a) groundwater containing 500 ppb of CFV, and (b) groundwater blank using the fiberoptic sensor. *Right:* Calibration curve for CFV measured in situ with the fiber optic sensor. Reprinted with permission from the Elsevier publications.[83]

5.7 Electron Microscopy

Electron microscopes have played an integral role in the rapid advancements of nanotechnology and nanoscience. In 1929 Ernst Ruska completed his PhD thesis on magnetic lenses and he and Max Knoll consequently developed the first electron microscope. In that microscope, the sample surface was placed normal to the viewing direction, and was illuminated by an electron beam at grazing incidence angle to the surface. He obtained images of copper and gold surfaces at a magnification of only 10 times! Later in 1938, Albert Prebus and James Hillier at the University of Toronto developed the first practical transmission electron microscope. Von Borries (1940) was much more successful with his grazing incidence method for the development of the transmission electron microscope (TEM). He placed the sample surface at a few degrees both to the viewing direction and to the illuminating beam.

Unlike the characterization techniques discussed so far, electron microscopy concerns the interaction between electrons and matter. Electron microscopes utilize a highly energetic beam of electrons that interacts with a material. From the interaction, information regarding topography, chemical composition, morphology and crystallographic structure can be obtained. Such instruments are capable of examining features in the nanoscale and are perhaps the most routinely employed characterization instruments in nanotechnology.

Electron microscopes are suitable for the characterization of both organic and inorganic materials. However, as they employ high-energy electron beams, prolonged exposure to the beam may cause certain materials, particularly organic polymers and biological materials, to be damaged, deformed, or destroyed. Furthermore, electron microscopes generally operate at high vacuum and consequently not suitable for samples that contain a liquid component or outgas.

Electron microscopes operate similarly to optical microscopes, as they both have an illumination source and magnifying lenses. However, the illumination source is a high-energy electron beam. Optical microscopes use glass lenses, whereas the electron beam is deflected and focused by electromagnetic fields, with what are called electron optics or electromagnetic lenses. This is because it not practically possible to fabricate materials-based lenses for electrons. The beam is then focused onto the sample, where electro-matter interactions give rise to measurable signals.

Electron microscopes have a much better resolution than their optical counterparts because of the interaction of an electron's matter wave with the sample. From Bragg's law, the minimum separation, $d_{\min}$, which can be resolved by any microscope, is given by:

$$d_{\min} = \frac{\lambda}{2\sin\theta}. \tag{5.15}$$

The resolution can be improved by using shorter wavelengths. The wavelength associated with an electron is given by the de Broglie relation:[48]

$$\lambda = \frac{h}{p} = \frac{h}{\sqrt{2m_e E_k}}, \tag{5.16}$$

where, m_e the electron mass, E_k is its kinetic energy, and h is Planck's constant. The electrons in the microscope obtain their kinetic energy by an applied electric potential in the order of kilo or even mega volts. Depending on the kinetic energy, a typical range of wavelengths can be between 1–0.005 Å, which is much smaller than that of visible light (400-700 nm) used in optical microscopes, resulting in far better resolution. As a consequence, nanoscale features, not observable with optical microscopes, can be observed.

The incident electron beam causes secondary electrons to be emitted from a surface, which can be monitored to produce a topological image. Using secondary electrons to image surfaces by rastering electron beam was realized by Manfred von Ardenne and Max Knoll in 1933 and 1934 respectively.[84] At that time, Knoll employed this method to study the targets of television camera tubes. These experiments paved the way for the construction of the first *transmission electron microscope* by von Ardenne, and subsequently a *scanning electron microscope*.[84]

5.7.1 Scanning Electron Microscope (SEM)

The *SEM* is perhaps the most routinely utilized instruments for the characterization of nanomaterials. With an SEM it is possible to obtain secondary electron images of organic and inorganic materials with nanoscale resolution, allowing topographical and morphological studies to be carried out, by scanning an electron probe across a surface and monitoring the secondary electrons emitted. Compositional analysis of a material may also be obtained by monitoring X-rays produced by the electron-specimen interaction. Thus detailed maps of elemental distribution can be produced. In sensor technology, this is predominantly used to study surfaces of thin films and sensing layers.

A schematic diagram of an SEM is shown in **Fig. 5.30**.[85] The electron beam is emitted from a heated filament, which is commonly made from lanthanum hexaboride (LaB_6) or tungsten. The filament is heated by applying a voltage, which causes electrons to be emitted. Alternatively, electrons can be emitted via *field emission* (*FE*).

The electrons are accelerated towards the sample by applying an electric potential. This resulting electron beam is focused by a condenser lens, which projects the image of the source onto the condenser aperture. It is then focused by an objective lens and raster-scanned over the sample by scanning coils. This is achieved by varying the voltage produced by the scan generator on the scan coils that are energized, creating a magnetic field, which deflects the beam back and forth in a controlled pattern.

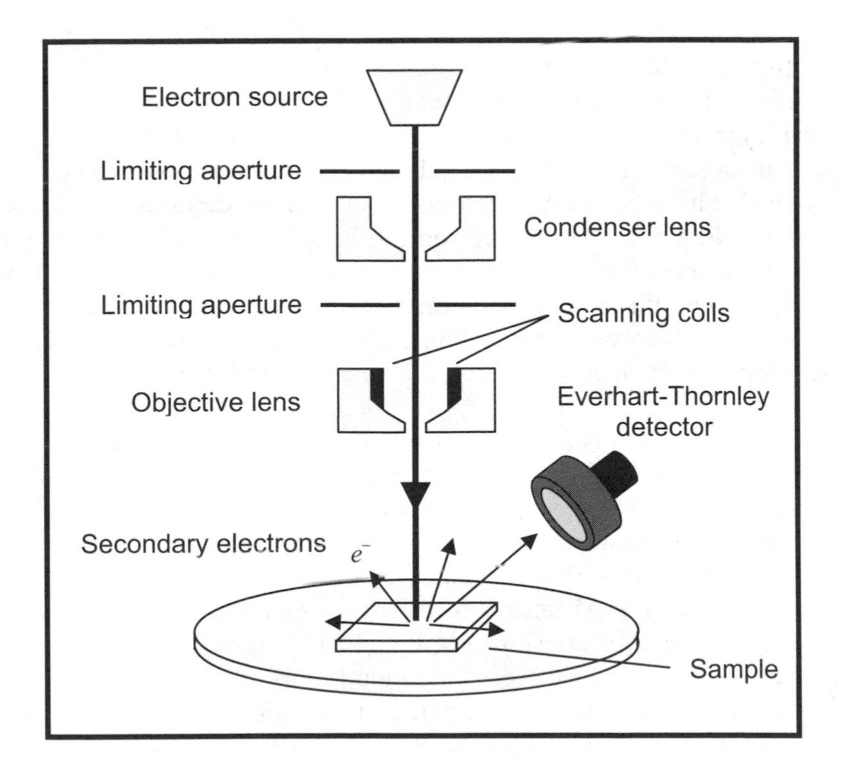

Fig. 5.30 Schematic diagram of an SEM set-up.

When the primary electrons hit the sample, they give part of their energy to electrons in the sample, resulting in the emission of secondary electrons. These secondary electrons have lower energies (around 20 eV). These secondary electrons are collected by an Everhart-Thornley detector, converted to a voltage, amplified and build the image. Their intensity is displayed versus the position of the primary beam on the sample. The samples placed in the SEM must be either conducting or covered with a thin metal layer in order to avoid electric charging. Scanning takes place at low pressures, so that the electrons are not scattered by gas molecules inside the chamber.[86] Furthermore, with an SEM it is possible to obtain images from comparatively large area of the sample.

In addition to secondary electrons there are also high-energy electrons, originating in the electron beam (producing X-rays), that are back-scattered from the specimen interaction volume. These electrons may be used to detect contrast between areas with different chemical compositions.

SEM can monitor the formation and growth of thin films and nanostructures. In nanotechnology enabled sensing applications, the SEM plays a vital role in helping researchers understand the interaction between the sensing layer/media and the analyte. This is because sensitivity is strongly dependant on surface morphology and topography. An SEM image of ZnO nanorods that have been fabricated via a liquid phase deposition technique is shown in **Fig. 5.31**. The figure shows highly cylindrical nanorods protruding from the surface, whose diameter can be estimated at ~100 nm. Since the advent of high resolution electron microscopes, features as small as 1 nm can be resolved. This is illustrated in **Fig. 5.32** which shows the distribution of FePt nanoparticles that have been stabilized on Si substrates with amino-silanes.[87]

For many thin film based gas sensors, the surface to volume ratio of a sensing layer film can influence its sensitivity. High resolution SEM images of carbon nanotubes, with large surface to volume ratio, employed for nitrous oxide gas sensing are shown in **Fig. 5.33**.[88] Multilayered materials and nanostructures can also be studied by examining their cross sections. The cross sectional SEM images of different organic multi-layers used in biosensing applications are shown in **Fig. 5.34**,[89] From these SEM images it is possible to observe their conformation to the substrate as well as the thickness of each layers. All information (topological and morphological) obtained using SEM can be correlated to the response of the sensors, and enable researchers to optimize sensor performance.

Fig. 5.31 SEM image of ZnO nanorods fabricated via liquid phase deposition.

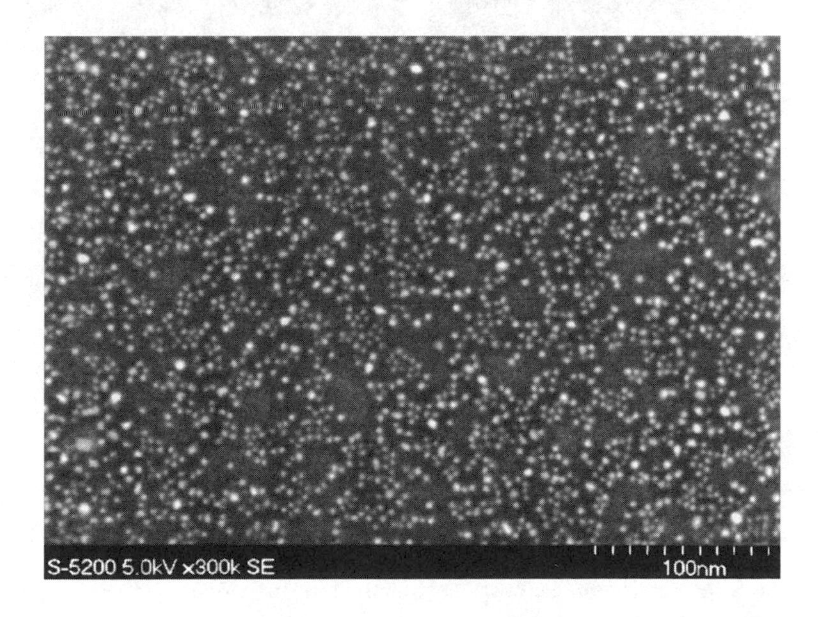

Fig. 5.32 High-resolution SEM image of a monolayer FePt nanoparticle film. Reprinted with permission from the American Institute of Physics publications.[87]

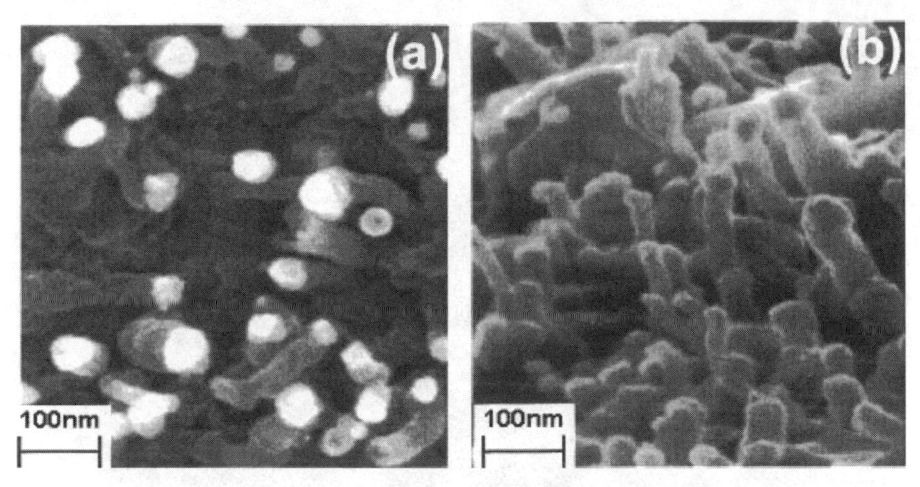

Fig. 5.33 High-resolution SEM images of carbon nanotubes grown on a 5 nm Ni catalyst layer thickness; (a) top view and (b) side view. Reprinted with permission from the American Institute of Physics publications.[88]

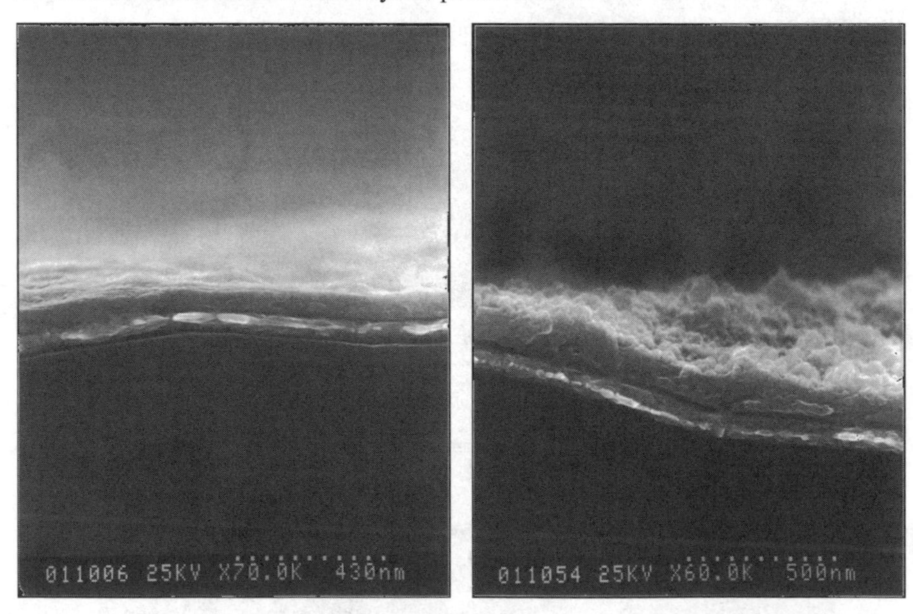

Fig. 5.34 SEM micrographs (cross-sectional view) of organic layers deposited onto monolayers of mercaptopropionic acid (MPA) on gold coated quartz crystal microbalance (QCM) transducers for biosensing applications. *Left:* $(PAH/PSS)_{12}$/anti-IgG multilayer film on a MPA-treated gold QCM electrode; *Right:* $(PAH/PSS)_{12}$/$(anti\text{-}IgG/PSS)_4$/anti-IgG multilayer film on a MPA-treated gold QCM. Reprinted with permission from the American Chemical Society publications.[89]

Nowadays SEMs are designed for specific purposes ranging from routine morphological studies, cryogenic studies, to high-speed compositional analyses or for the study of environment-sensitive materials.

5.7.2 Transmission Electron Microscope (TEM)

In a *TEM*, a beam of focused high energy electrons is transmitted through a thin sample to reveal information about its morphology, crystallography, particle size distribution, and its elemental composition. It is capable of providing atomic-resolution lattice images, as well as giving chemical information at a spatial resolution of 1 nm or better. Because the unique physical and chemical properties of nanomaterials not only depend on their composition, but also on their structures, TEM provides a means for characterizing and understanding such structures. TEM is unique as it can be used to focus on a single nanoparticle in a sample, and directly identify and quantify its chemical and electronic structure. Perhaps the most important application of TEM is the atomic-resolution real-space imaging of nanoparticles.[90]

A TEM operates in a similar manner to a slide projector. As the electron beam passes through the sample, only certain parts of it are transmitted, making an amplitude contract image. The image passes through a magnifying lens and is then projected onto a phosphor screen or a charge coupled device (CCD), which allows for quantitative data processing. Information may also be obtained from backscattered and secondary electrons, as well as emitted photons. A wide variety of materials can be characterized with a TEM, including metals, minerals, ceramics, semiconductors, and polymers. Because electrons must be transmitted through the material this requires that the sample be appropriately thin.

A simplified TEM setup is shown in **Fig. 5.35**. The electron gun is a pin shaped cathode that is typically made from materials such as LaB_6. Heating this cathode by applying a large current produces a stream of almost monochromatic electrons that travel down a long column after being accelerated by a large voltage. Increasing this voltage increases the kinetic energy of the electrons and hence decreases their wavelength. The smaller the electron beam wavelength the higher the resolution, although the quality of the lens-systems is the limiting factor. The condenser lenses focus the beam to a small and coherent cylinder while the condenser aperture removes electrons scattered at large angles. The beam strikes the specimen on the sample holder, and the majority is transmitted, focused by the objective lens, after which it passes through the intermediate and projector lenses and enlarged. Eventually by striking a phosphor or CCD surface it forms an image.

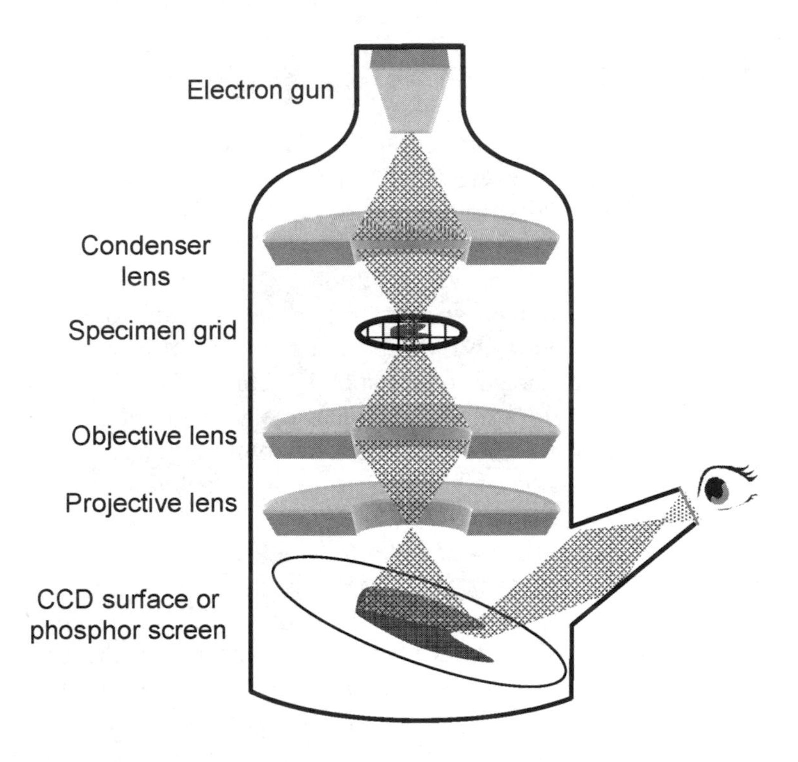

Fig. 5.35 Schematic diagram of a TEM set-up.

The TEM has the capability to create both electron microscope images (information in real space) and diffraction patterns (information in reciprocal space) for the same region by adjusting the strength of the magnetic lenses.[91] By inserting a selected area aperture and using parallel incident beam illumination, a selected area electron diffraction pattern from an area as small as several hundreds to a few nm in diameter is obtained.

Crystal structures can also be investigated by *high resolution transmission electron microscopy* (*HRTEM*) where the images are formed due to differences in phase of electron waves scattered through a thin specimen. The emergence of *HRTEM* has allowed the direct reconstruction of Bragg differential electron beams to create interference patters, which in favorable cases of simple projected electrons, give a representation of the underlying crystallographic diffraction grating.

Fig. 5.36 shows near atomic resolution images of SnO_2 nanowires that were synthesized at elevated temperatures.[92] Such nanowires are of particular interest because of their potential in the fabrication of nanoscale devices and sensors. SnO_2 is a well-known functional material used particularly in optoelectronic devices and gas sensors. In this example, the TEM has been employed to carry out detailed characterization of the crystal structure. From the figure, the nanowires are flat and straight in the form of nanobelts, and their widths are several tens of nanometers. More interestingly, from the high resolution TEM images and selected area electron diffraction, the crystallographic configuration of the SnO_2 nanowires could also be determined.

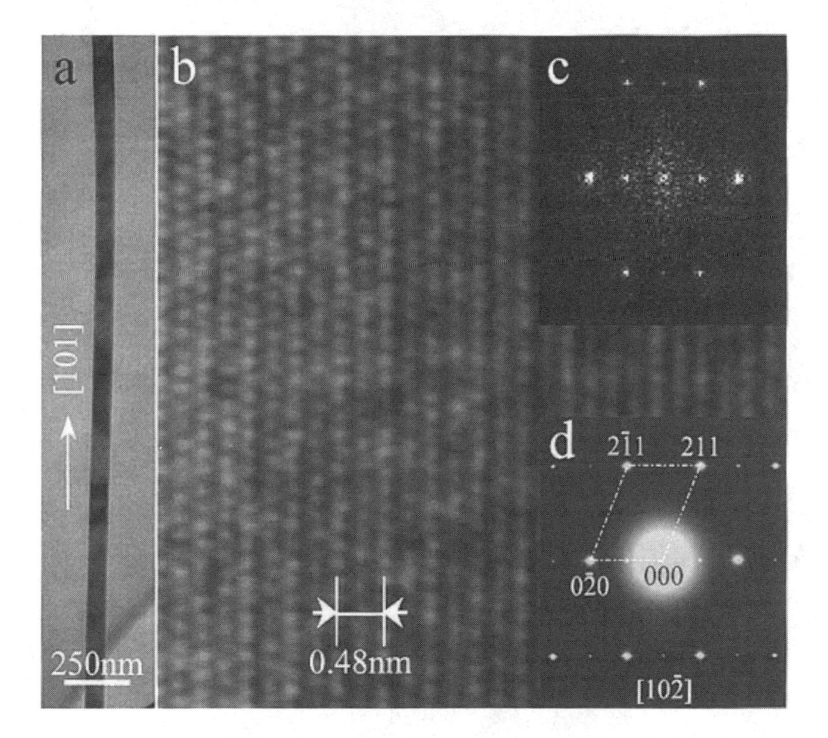

Fig. 5.36 (a) TEM image of a rutile structured SnO_2 nanowire. (b) High-resolution TEM image of the nanowire. (c) Corresponding FFT of the image, and (d) selected area electron diffraction pattern from the nanowire. Reprinted with permission from the American Chemical Society publications.[92]

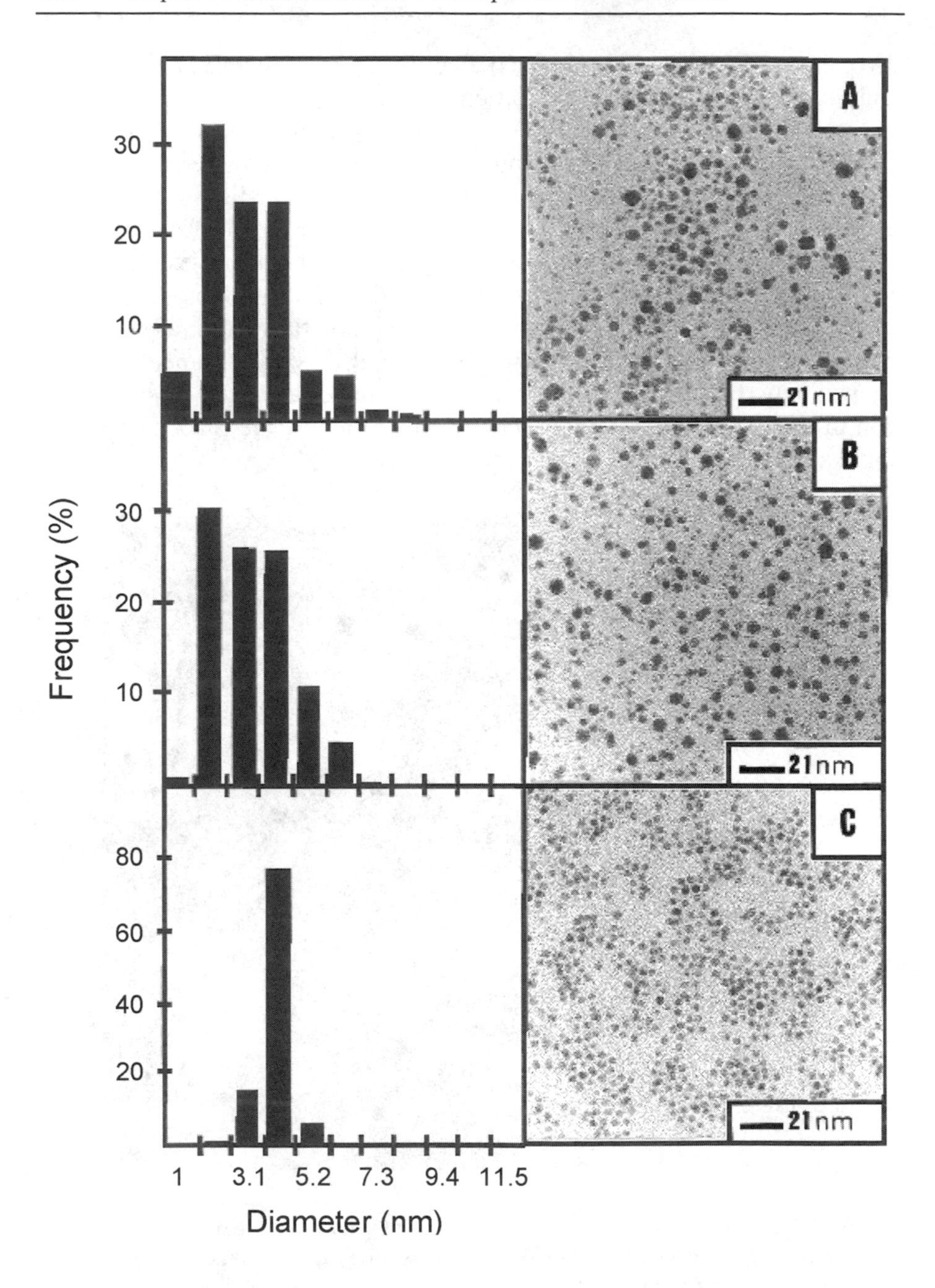

Fig. 5.37 TEM micrograph and histograms of size of silver particles: in reverse micelles (A), after extraction (B), and at the end of the size-selected precipitation process (C). Reprinted with permission from the American Chemical Society publications.[93]

Another feature of TEM is highlighted in **Fig. 5.37**, which shows how amplitude contrast can be used to investigate size distributions of silver nanoparticles.[93] In this work, the colloidal silver particles were prepared from reverse micelle solutions, with images obtained at different stages of the growth process. The particle size profiles were obtained directly from the images by measuring the particles diameters from several TEM images on different parts of the grid at a specified magnification. The histograms are used to illustrate the polydispersity of the particles, and from these, the average particle size and the standard deviation can be calculated.

Clearly, the TEM can to provide a wide range of information from nanostructured materials, including diffraction patterns, lattice constants and particle sizes. Furthermore, the fact that an individual nanoparticle can be selected and characterized individually makes TEM one of the most powerful characterization techniques in nanotechnology.

5.8 Rutherford Backscattering Spectrometry (RBS)

The *RBS* is a quantitative technique that utilizes elastic scattering of highly energized ions with low mass, such as He^{2+} alpha particles, to analyze the surface and the outer few micrometers of solids. These particles collide with a surface and by measuring the energy spectrum of the back-scattered ions, information such as elemental composition, stoichiometry, quantitative measurements of film impurities, film thickness and the depth distribution of elements can be measured.[94] It is an important tool for studying thin films, and in particular sensing layers, both prior and subsequent to interaction with an analyte.

The RBS instrument generally consists of a helium gas source and a chamber of several meters long in which He ions are accelerated. Collisions of He molecules with N_2 gas molecules also take place in this chamber to produce He ions. The high-energy ion beam (> ~MeV) is then focused and directed onto the sample. The ions hit the sample, reflected and consequently detected. The detector is generally placed such that the scattered particles at close to a 180° angle can be collected (around 165°). An RBS experiment is carried out under high vacuum conditions and an area of a few square millimeters is analyzed. The schematic of an elastic collision in an RBS experiment is shown in **Fig. 5.38**.

If the positively charged ion comes close to the positively charged nucleus of an atom, it is repelled. Obviously, the repulsion force increases with the mass of the target atom in the sample. The collision energy is transferred from the incident ions to the stationary atoms, while kinetic energy and momentum are conserved[48] and the scattered ions have less energy. The reduction in energy is a function of the masses of both the incident ion and target atom. The ratio of this backscattered ion's energy to its initial energy is called the *kinematic factor*, $K = E_1/E_0$ (**Fig. 5.38**). The kinematic factors can be calculated for any atoms and detector angles. The amount of energy which is transferred to the sample atom depends on the kinematic factor. As a result, the energy of scattered ions is an indicator for the chemical composition of the sample. The resulting RBS spectrum is then plotted against a numerical fit from which various properties are determined.

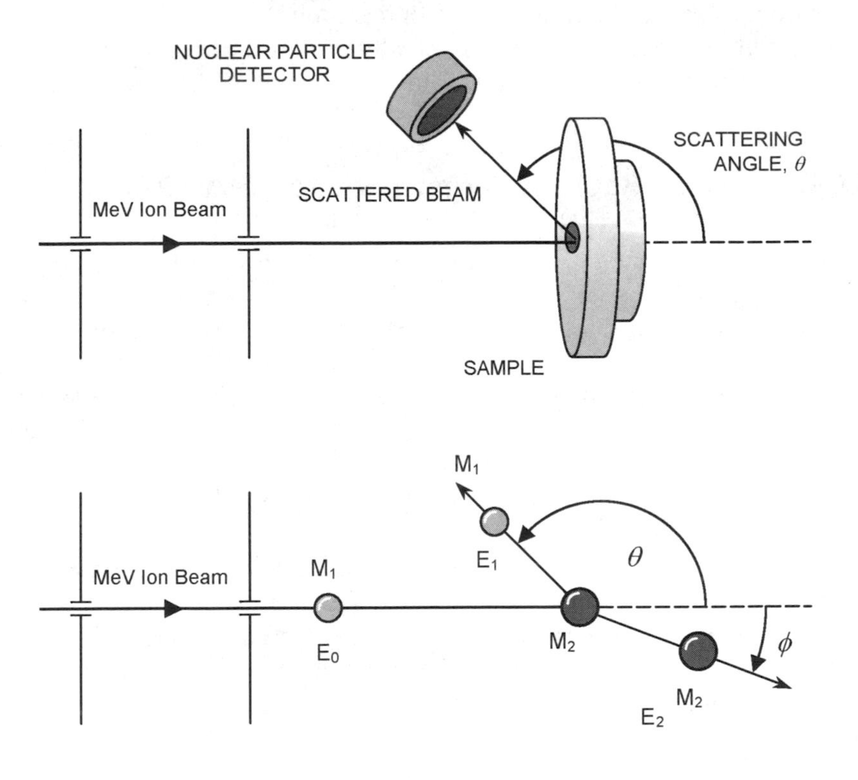

Fig. 5.38 A typical RBS experimental set-up and ion-atom collision.

In addition to the compositional analysis, information about the sample's crystal structure and impurities along the crystallographic directions can also be obtained by RBS. If the incident ion doesn't hit any of the atoms near the surface, but instead hits an atom deeper in, the incident ion loses energy gradually as it passes through the solid before leaving the solid (**Fig. 5.39**). This passing through the crystal lattice is sensitive to very small displacements of atoms from crystalline lattice sites, and is strongly influenced by the crystal disorder and crystal orientation. Therefore, a depth profile of the composition of a sample can also be obtained. As seen from, channeling information can be acquired by aligning the sample's crystallographic axes with the incoming alpha particles (**Fig. 5.39**). RBS can also cause ion-implantation damage in single-crystal films and wafers under analysis which should be taken into consideration.

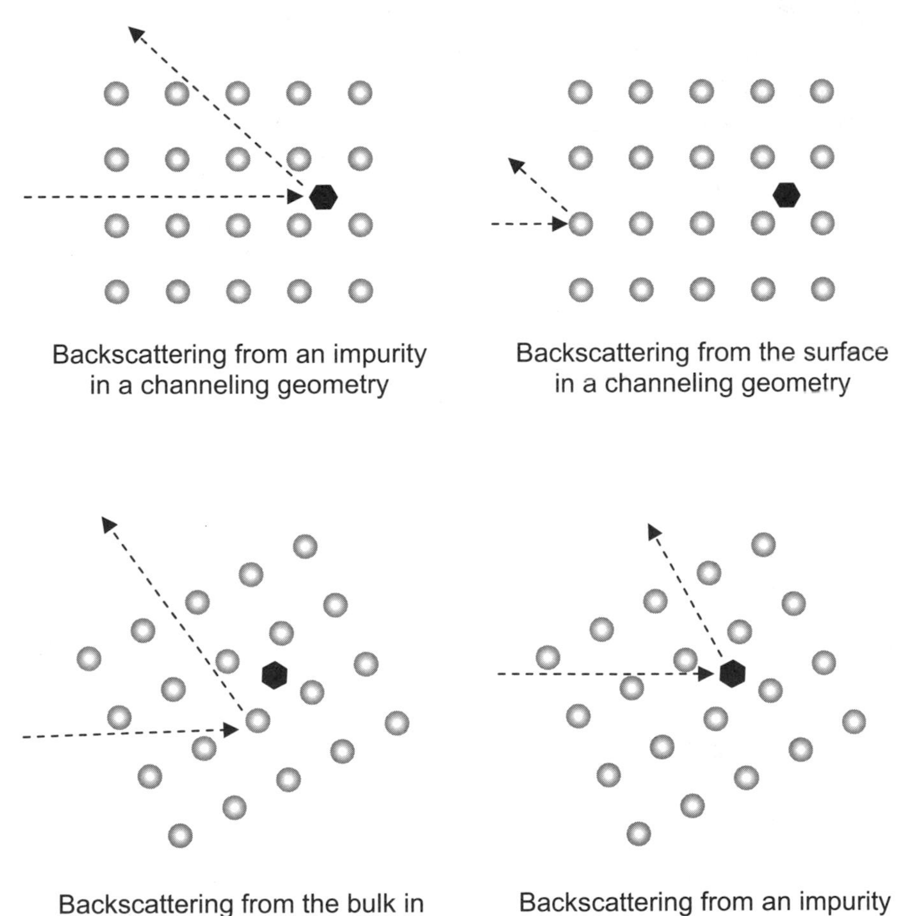

Fig. 5.39 Channeling in a crystal structure.

The RBS spectra of several Ga_2O_3 based films deposited on SiC substrates and annealed in air at 300 and 900°C are shown in **Fig. 5.40**. The spectra have been normalized with respect to each other so that valid comparisons can be made. With the aid of data fitting software, RBS analyses the film thickness to be approximately 100 nm. Numerical fitting also provides information on film stoichiometry and reveals the ratio of Ga to O to be approximately 1:1.51. From the spectra it is also seen that the height of the SiC substrate signal for the film annealed at 900°C appears to be lower than that of the film annealed at 300°C. This can be attributed to channeling in the film, due to its improved crystallinity after annealing, which has resulted in reduced backscattering. **Fig. 5.41** shows a section of the RBS spectrum of a binary metal oxide film of Ga-Zn oxide. The measured data shows a peak with a distinct "shoulder" occurring at around 1616 keV. As both Ga and Zn produce peaks quite close to each other, the shoulder is characteristic of the superposition of both signals, which can be resolved by numerically fitting peaks corresponding to the two individual elements.

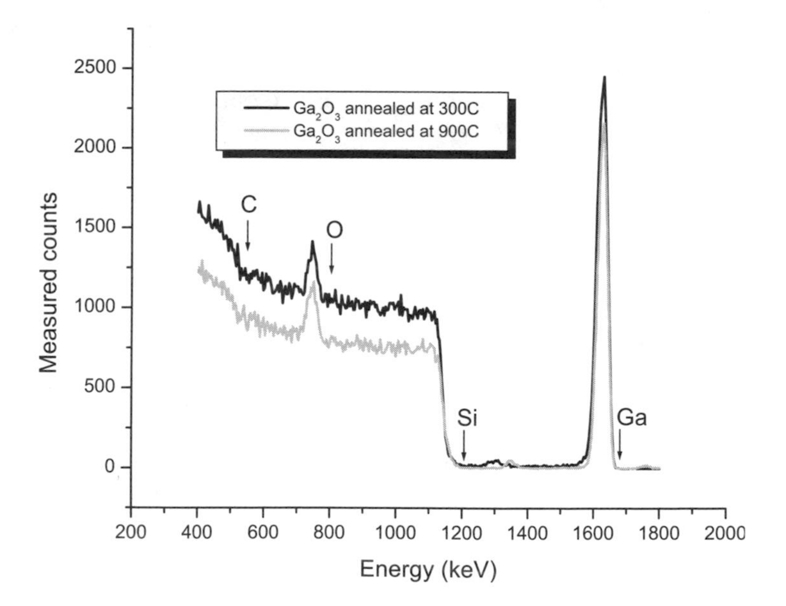

Fig. 5.40 RBS spectra of Ga_2O_3 based films annealed at two different temperatures.

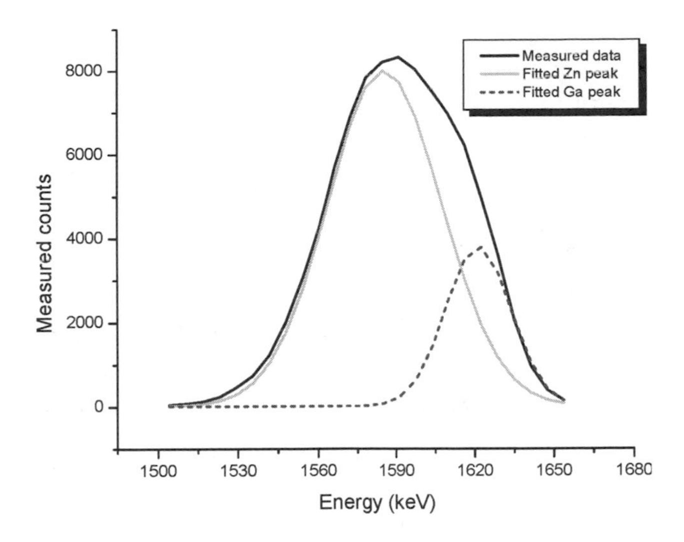

Fig. 5.41 Section of the RBS spectrum of Ga-Zn binary oxide.

5.9 Scanning Probe Microscopy (SPM)

SPM emerged with the invention of the scanning tunneling microscope in 1981 by Heinrich Rohrer and Gerd Binning and since it has enabled a myriad of achievements in nanotechnology.

SPM techniques utilize a physical probe to image and measure surfaces at the level of molecules and groups of atoms. They are often employed to study forces, electrical properties, crystal structure, and interactions on surfaces. Typically, an ultra fine probe tip is mechanically raster-scanned over a sample surface. Interactions between the probe tip and material can be measured as a function of probe position, with the type of interaction defining the type of scanning probe microscopy employed. Several interactions can also be measured simultaneously. Scanning probe techniques can provide atomic-resolution images of crystals. At this stage they are also under investigation to quantitatively resolve the atomic lattices of nanoparticles at near zero Kelvin where wobbling and random noise of the structures are minimum.[90]

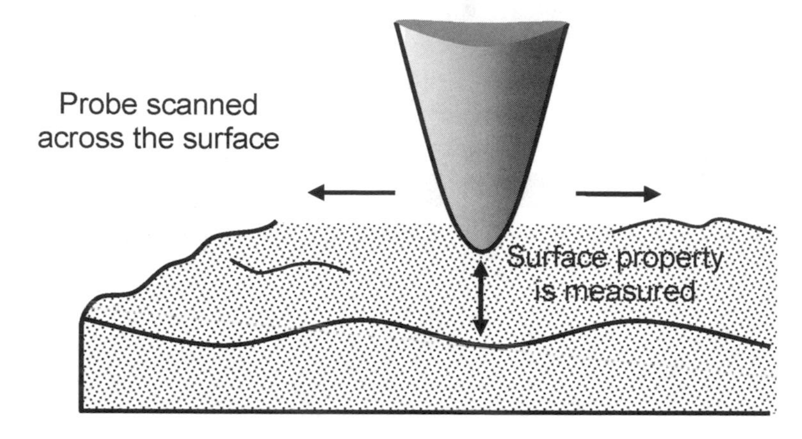

Fig. 5.42 Illustration of the basic concept behind scanning probe techniques.

SPM techniques not only allow materials to be imaged and characterized, they even allow for the manipulation of structures and features at atomic proportions. The types of materials that can be studied range from inorganic to biological samples.

There are quite a few established SPM techniques available, however, two of the most widely used that have relevance for sensing are *scanning tunneling microscopy* (*STM*) and *atomic force microscopy* (*AFM*), which are described in more detail below.

5.9.1 Scanning Tunneling Microscope (STM)

An STM functions by measuring the small current between the probe tip and the sample surface. From it, a three-dimensional profile of the surface can be obtained, which allows surface roughness, size and conformation of atoms and molecules, crystal orientations and defects to be characterized.[95] For sensing applications, STM is particularly useful to study the surface of sensing molecules and layers. Furthermore, as it involves precise current and voltage measurements across the surface, it can be used to determine the density of electronic states at localized areas. STM can also be used to activate chemical reactions, or reversibly produce ions by removing or adding individual electrons from atoms or molecules.[96]

A typical STM setup is shown in **Fig. 5.43**. A voltage is applied between an atomically sharp probe tip and the sample. The tip is generally made from tungsten, and as it approaches the conducting surface at a distance in the nanometer range, a tunneling current begins to flow. The magnitude of this current alters exponentially with distance between probe tip and surface. The tip is mounted on a piezoelectric stage, which accurately control its movement. When the appropriate voltage is applied to this stage, it can move the tip several Ångstroms or nanometers laterally or longitudinally. The stage movement is controlled by a feedback system, which is configured to keep the distance between the sample surface and tip constant, and maintains constant tunneling current. If the tip encounters a surface protrusion, such as a step, the tunneling current briefly changes. The feedback system will then automatically adjust the current, restoring it to its preset value. These changes in the tunneling current are recorded as a function of the tip position create an image, or map, of the surface. In ideal circumstances, individual surface atoms can be resolved.[97] Since the STM requires measuring a current between the tip and sample, only conducting or semiconducting materials can be characterized.

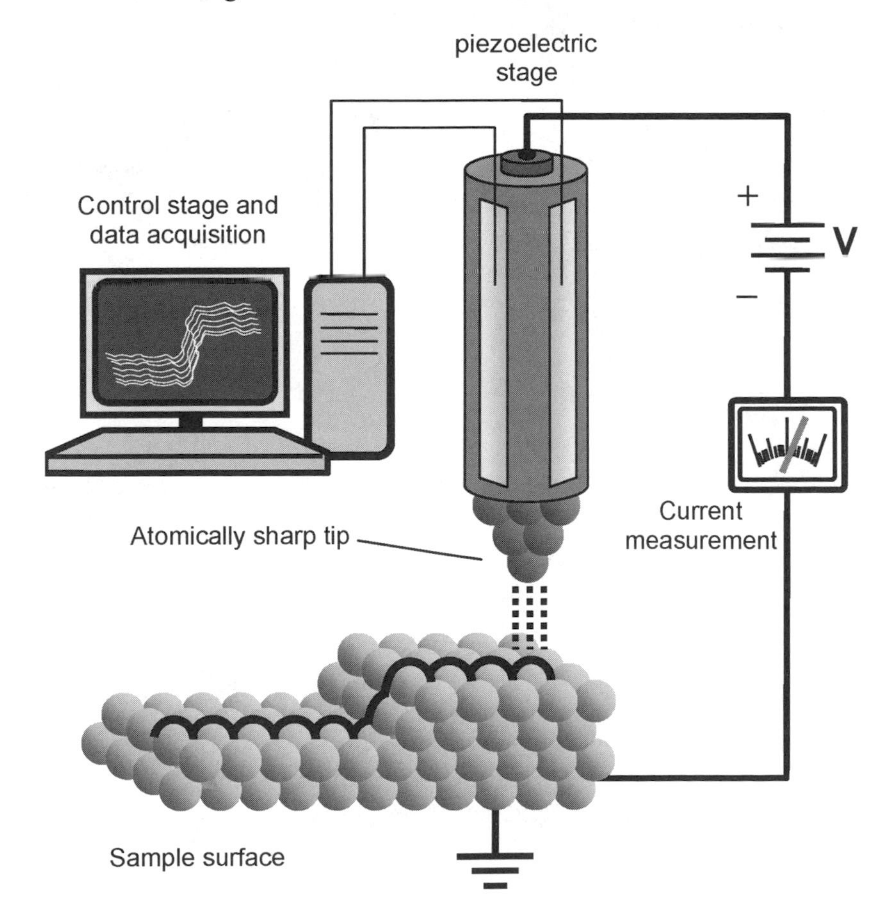

Fig. 5.43 Example of an STM setup.

Fig. 5.44 shows STM images of single-walled carbon nanotubes (SWCNTs) measured in ultra-high vacuum at 77 K,[98] with bias voltages of 50 and 150 mV, respectively, and a tunneling current of 150 pA. It is seen that the STM analysis was able to resolve the hexagonal-ring structure of the walls of the nanotubes, with the size of each hexagons being just a few Ångstroms. A small portion of a two-dimensional expected honeycombe lattice is overlayed in the figure to highlight the atomic structure of the nanotubes.

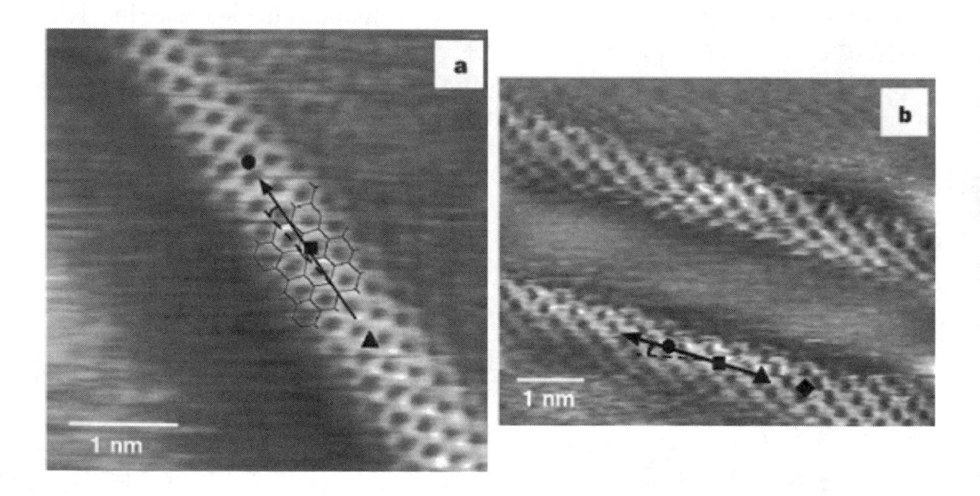

Fig. 5.44 STM images of (**a**) a SWCNT exposed at the surface of a rope and (**b**) isolated SWCNTs on a Au(111) substrate. The tube axes in both images are indicated with solid, black arrows, and the zigzag direction are highlighted by dashed lines. A portion of a two-dimensional graphene layer is overlaid in a to highlight the atomic structure. Reprinted with permission from the Nature publications.[98]

With the STM molecules adsorbed on surfaces and the strength of their bonds can also be studied. Obviously, whether such molecules are adsorbed during the growth process of thin films or nanostructures, or whilst being immobilized onto the surface of another material, or whether they are gas molecules being adsorbed during a chemical reaction, studying such adsorptions may have strong relevance for sensing applications. This issue will be discussed more in Chap. 7.

5.9.2 Atomic Force Microscope (AFM)

An AFM is utilized to measure attractive or repulsive forces between the scanning probe tip and the sample surface, as well as to acquire atomic-scale images of surfaces. It was developed in 1986 by Binnig, Quate and Gerber,[101] and since then has had a major impact on science and technology.

When the AFM tip and the sample surface are brought within a few nanometers of each other, forces between atoms in the tip and in the sample cause it to deflect. The amount of deflection is then measured and correlated to the force. The forces generally measured are: Van der Waals, electrostatic, magnetic, capillary, Casimir, and solvation forces.

AFM is not only used as a characterization tool for obtaining force and topographical maps of surfaces (in particular sensing surfaces), but it is also capable of functioning as a current, chemical, physical and biosensor.[102-106] With the AFM virtually any surface, whether it is an insulator, conductor, organic or biological can be imaged.[47] The materials can also be analyzed in different environments, such in liquid media, under vacuum, and at low temperatures. Unlike the STM, it does not require a current between the sample surface and the tip. This means that that it can move into potential regions that are inaccessible to the STM, and also take images of delicate samples that would be damaged irreparably by the STM tunneling current.

AFM does have drawbacks such as only being able to image a maximum height in the order of micrometres, as well as having a limited scan area. Furthermore, the tip's shape may become deformed as it moves across a surface, particularly if it encounters a substantial shear force.

An AFM setup is shown in **Fig. 1.1**. The AFM has a microscale cantilever, usually made from silicon or silicon nitride. At one end of the cantilever is a sharp tip that has a radius of curvature in the order of nanometers. By bringing the tip into close proximity of the sample surface, forces acting between the tip and the sample surface cause the cantilever to bend and deflect in a similar manner to a tiny diving board. The forces are not measured directly. Rather, they are calculated by measuring the cantilever's deflection. If the stiffness of the cantilever is known, then the force is measured using *Hooke's law*:

$$F = -kz \,, \tag{5.17}$$

where F is the force, k is the stiffness of the cantilever, and z is the distance the lever is bent. The sample is mounted on a piezoelectric stage whose position can be precisely controlled by applying a voltage. The sample is laterally scanned relative to the tip, and the cantilever's deflection is measured as a function of position. The cantilever deflection can be detected in several different ways, with the most common to measure laser light that is reflected from the top of the cantilever onto a position-sensitive detector made from an array of photodiodes.

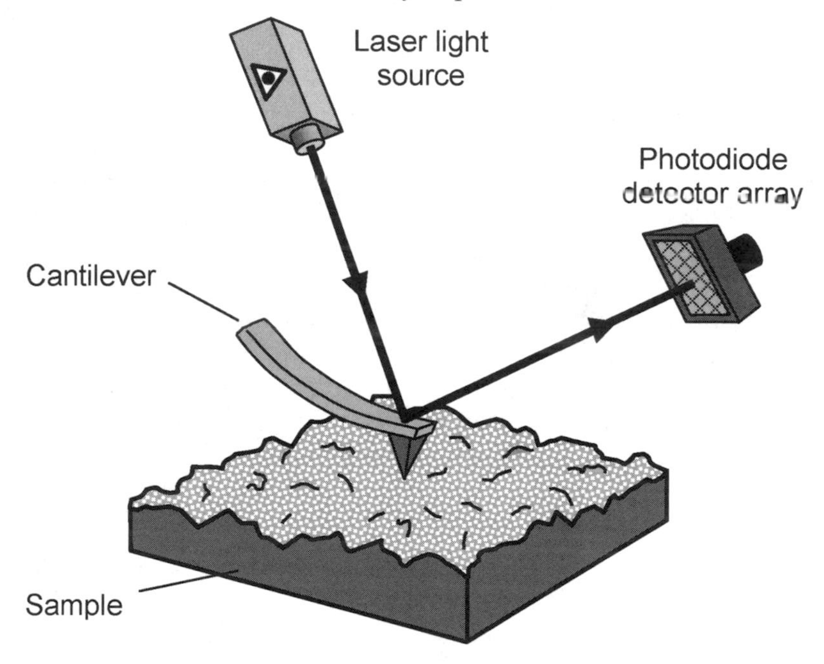

Fig. 5.45 An AFM probe raster-scan over a sample surface.

The AFM can be operated in several different modes. The most common is *contact mode*, where the tip is in intimate contact with the surface. As it is raster-scanned across the surface, it is deflected as it encounters surface corrugations. Another way is *constant force mode*, where the distance between the tip and surface is continually adjusted so as to maintain a constant deflection, and therefore constant height above the surface. *Tapping mode* or *dynamic force mode* (*DFM*) is another approach in which a stiff cantilever is brought within close proximity of the surface. It is then oscillates and changes in the resonant frequency or amplitude of the cantilever are measured during a scan. During the oscillation, part of the tip intermittently touches or taps the surface. Tapping mode requires very stiff cantilevers to avoid them becoming stuck in the sample surface. The choice of operation modes depends on the sample and its environment. Tapping mode offers improved lateral resolution on soft samples, as lateral forces such as drag, which is common in contact mode, are virtually eliminated. It is also advantageous for materials that are poorly adsorbed on a substrate. In general, for delicate samples, constant force or tapping modes are utilized.

The sensitivity of the force measurements is directly related to the size and stiffness of the cantilever, which can range from 1 pN to 1 nN per nm, and the deflection sensor measures motions ranging from several microns to even 0.01 nm. The minimum force that can be measured with the AFM is typically 5 pN.

Fig. 5.46 illustrates how the AFM can be utilized to characterize the surface of a sensor. These AFM images are of WO_3 crystals, which were deposited onto a multilayered $ZnO/LiTaO_3$ SAW transducer (described in Chap. 3). WO_3 is known for its sensitivity towards different gas species. On this device, some of the deposited WO_3 covers the regions where the metallic interdigital electrodes are located beneath. From **Fig. 5.46** it is clear that the presence of the underlying metal results in the deposited WO_3 having vastly different surface morphology to that on the areas without underlying metal. Such observations are important because in gas sensing applications the surface morphology and surface-to-volume ratio can have a strong influence on its sensitivity. From the figure the average grain size for WO_3 over the non-metalized to metalized regions are 132 nm and 156 nm, respectively. Furthermore rms surface roughness also changes for both of the regions, being ~12 nm on the non-metalized and ~16 nm on the metalized regions.

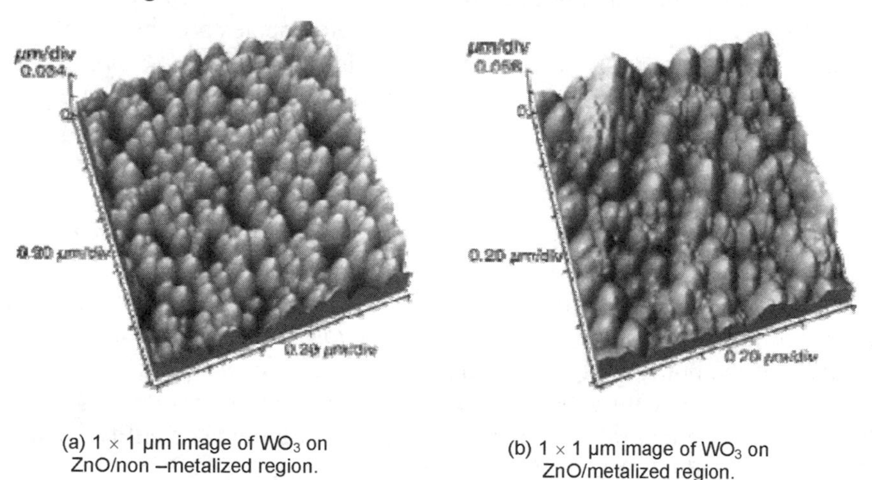

(a) 1 × 1 µm image of WO_3 on ZnO/non –metalized region.

(b) 1 × 1 µm image of WO_3 on ZnO/metalized region.

Fig. 5.46 AFM surface images of 150 nm thick WO_3 sensing layer deposited on a ZnO/36° YX $LiTaO_3$ SAW device: (a) WO_3 deposited on the non-metalized region; (b) WO_3 deposited on the metalized region. Reprinted with permission from the Elsevier publications.[107]

5.10 Mass Spectrometry

Mass spectrometry is an analytical technique for measuring the charge to mass ratio of ions. It is an invaluable tool in nanotechnology that allows unknown compounds, to be identified by determining their mass. Furthermore, the structure of a compound can be analyzed through observing the mass to charge ratio of its fragments. It is also used for determining the isotopic composition of elements in a compound, and for studying the fundamental properties of gas phase ion chemistry.

Mass spectrometry is extremely sensitive, as the molecular masses can be measured within an accuracy of 0.01%. This is enough to allow minor mass changes in biomolecules to be detected, e.g. the substitution of one amino acid for another.[108] However, it is generally not a quantitative technique, as it can only reveal which fragments of a compound exist, and not their abundance. Mass spectrometry is suitable for both organic and inorganic compounds. It has become a fundamental tool in nanotechnology, particularly for determining the structure of organic compounds and biomolecules such as peptides and sequencing of oligonucleotides. It is extensively used in drug testing and drug discovery, pharmacokinetics, and in monitoring water quality and sensing food contaminations.[108,109]

Mass spectrometry may also be employed to study surfaces, by bombarding the sample's surface with a stream of high energy ions. Consequently both neutral and charged species (atoms, molecular fragments) are ejected from the surface. The ejected secondary species are then analyzed by mass spectrometry. This technique is called *secondary ion mass spectrometry* (*SIMS*) and is one of the most sensitive techniques for studying surfaces.[110]

In a mass spectrometer set-up, shown in **Fig. 5.47**, high-energy electrons, released from a heated filament, collide with a sample molecule causing it to become ionized. This is carried out under a vacuum in a range of 10^{-5} to 10^{-8} torr, as the ions are extremely reactive and short-lived. Residual energy from the collision may cause the molecular ion to fragment into neutral pieces and/or smaller *fragment ions*. The molecular ion is a *radical cation*, but the fragment ions may either be *radical cations* or *carbocations* (an ion with a positively-charged carbon atom), depending on the nature of the neutral fragment. A radical is a molecular entity such as $^{\bullet}CH_3$, $^{\bullet}SnH_3$, and $Cl^{\bullet}$ that possess an unpaired electron (the dot symbolizes the unpaired electron), and if it carries a charge, it is called a radical ion.[111] The anions are accelerated towards the repellor plate whilst the cations are accelerated away from it to the accelerating electrode. The cations pass through the accelerating electrode slits as an ion beam. When this ion beam experiences a strong magnetic field perpendicular to its direction, the ions are deflected in an arc whose radius is inversely proportional to the mass of the ion. Lighter ions are deflected more than heavier ions. By varying the strength of the magnetic field in time, ions of different masses impinge on a detector that is placed at the end of a curved tube at different times.[15]

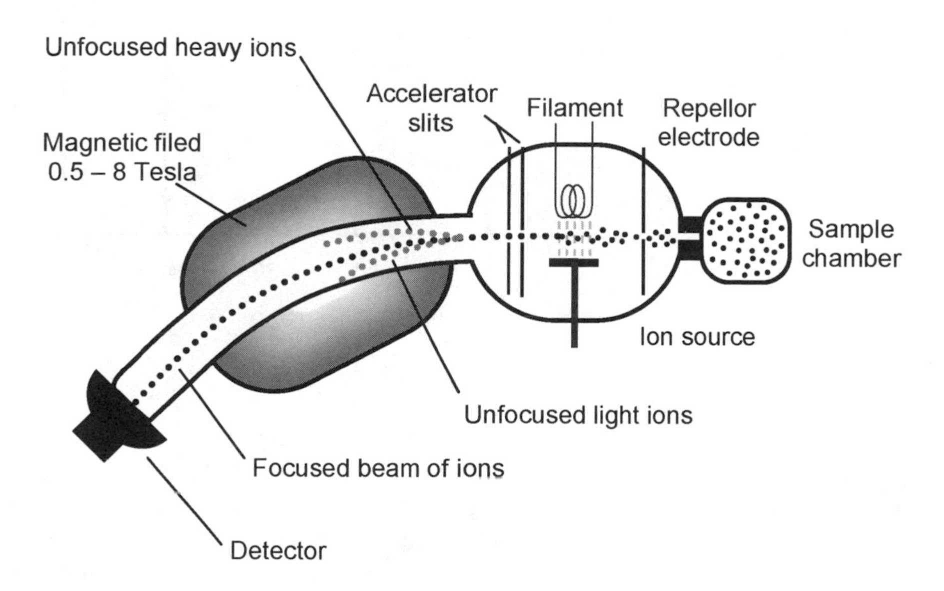

Fig. 5.47 The Mass Spectrometry setup.

Fig. 5.48 shows an example of a mass spectrum of a relatively simple carbon dioxide molecule, CO_2. Mass spectra are generally presented as a graph having vertical bars, where each bar represents an ion having a specific *mass-to-charge ratio* (*m/z*) and the length of the bar indicates its relative abundance. The ion with the highest abundance is assigned to 100, and it is referred to as the base peak. CO_2 has a nominal mass of 44 amu. As can be seen in **Fig. 5.48**, the molecular ion is the strongest ion in the CO_2 mass spectrum. Carbon dioxide's mass spectrum is very simple, as its molecule is comprised of only three atoms. In this graph, the molecular ion is the base peak, and the only fragment ions are CO (m/z=28) and O (m/z=16).

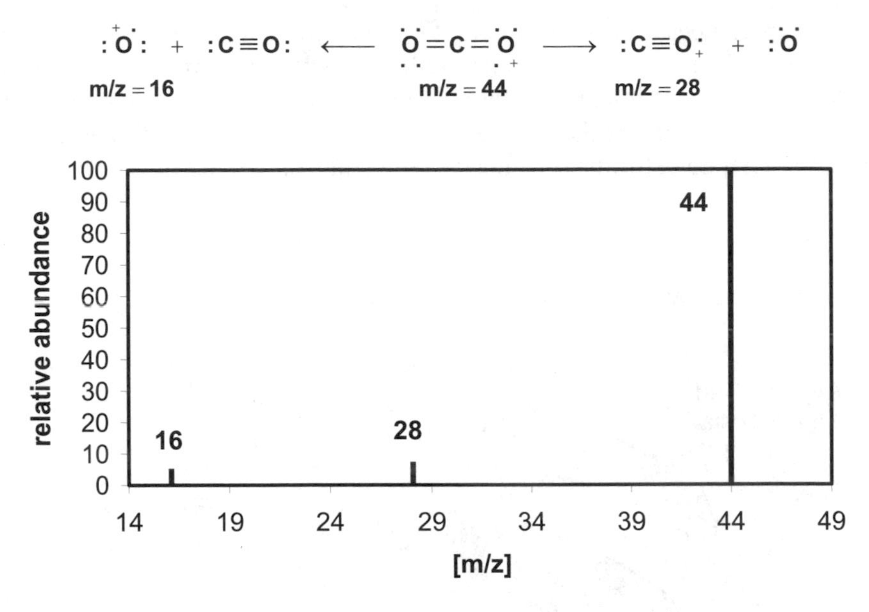

Fig. 5.48 Carbon dioxide mass spectrum.

Mass spectrometry is also an important method for the characterization and analysis of proteins and peptides, in which it is called peptide mass fingerprinting. It involves generating peptides from proteins using residue-specific enzymes, after which mass spectrometry is performed to determine the masses of the peptides. The acquired mass spectrum is compared and matched against peptide libraries generated from protein sequence databases of predicted masses.

There are many types of mass spectrometers, yet two of the most prominent are *matrix-assisted laser desorption/ionisation*, and *time of flight*. However, hybrid instruments utilizing various mass spectrometry techniques are widely used to provide high resolution quantification.

5.10.1 Matrix-Assisted Laser Desorption/Ionisation (MALDI) Mass Spectrometer

In the early 1960's, it was shown that the irradiation of low-mass organic molecules with a high-intensity laser pulse leads to the formation of ions that could be successfully mass analyzed. This technique was utilized to transform biopolymers and organic macromolecules from non-volatile form to volatile. However, the technique could not be employed for molecules with masses less than 5-10 kDa, because ions were created in bursts during laser desorption, the system couldn't be coupled to scanning mass analyzers. Michael Karas and Franz Hillenkamp[112] overcome this limitation by demonstrating that a small organic molecule matrix could be mixed with a small amount of sample (hence called *matrix-assisted laser desorption/ionization - MALDI*) and probed as shown in **Fig. 5.49**. A pulse of laser beam impinged on the sample to produce ions needed for the analysis. The matrix also allows for the laser incidence spot to transfer the heat to the back metal plate. Koichi Tanaka[113] later demonstrated the application of MALDI to a range of biological macromolecules, who together with John Fenn received the 2002 Noble Prize for chemistry.

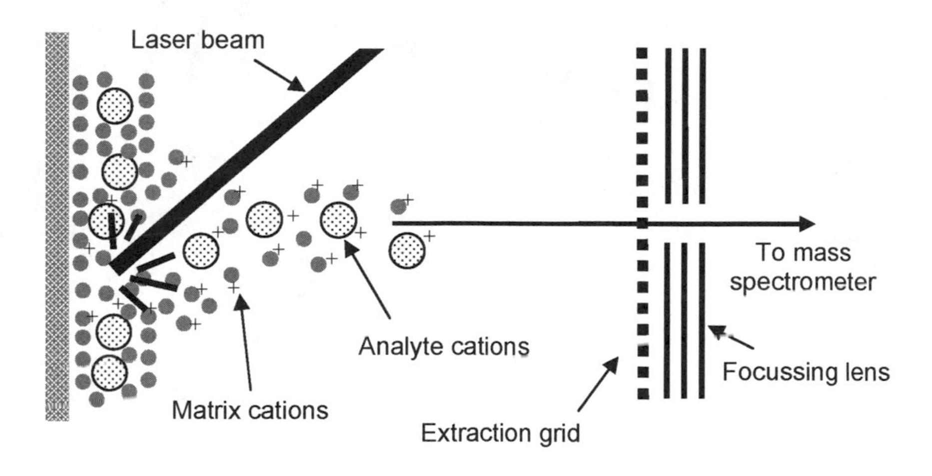

Fig. 5.49 A schematic diagram of the mechanism of MALDI.

5.10.2 Time of Flight (TOF) Mass Spectrometer

The *TOF mass spectrometer* was first developed in the late 1950's by the Bendix corporation, based on the work of Wiley and McLaren.[114] TOF mass spectrometers, and their hybrids with MALDI, are known for having high sensitivity. A representation of the TOF system is shown in **Fig. 5.50**. In the TOF spectrometry, a pulse of ions with the same initial kinetic energy is introduced into the flight tube. However, ion velocity depends on the mass-to-charge ratio, with lighter particles moving faster (and having shorter flight times) than heavier ones. The field pushes the ions along until they encounter the reflectron field, which directs the ions back towards a reflectron detector. When the ions arrive at the detector, the travel time from source to detector is transformed to the m/z value, and a spectrum is obtained. TOF mass spectrometers are capable of providing mass data for very high-mass molecules, in particular biomolecules.

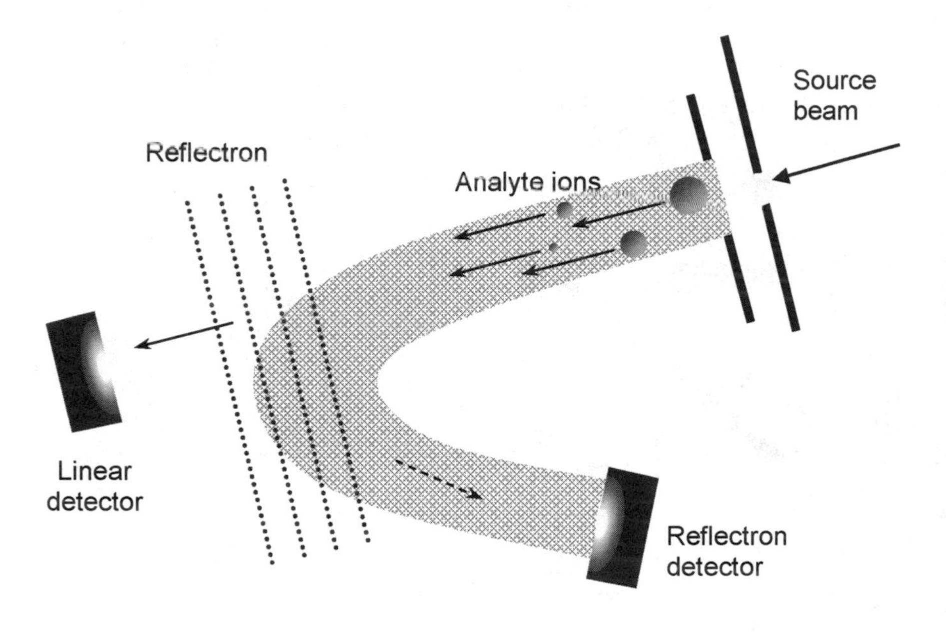

Fig. 5.50 A schematic representation of a Time-of-Flight mass spectrometer.

5.11 Summary

In this chapter the most common techniques for characterizing nanoparticles and nanostructured materials were presented, many of which can also be used directly in sensing applications. It is important to note that the properties of the sample such as the material type, composition and dimensions as well as the environment in which the study is conducted, place limitations on the characterization technique. A brief comparison of the characterization techniques, presented in this chapter, can be seen in **Table 5.2**.

Table 5.2 A brief comparison of the characterization techniques.

	Characterization environment		Type of material
Surface Imaging		**Resolution**	
SEM	Vacuum	~ nm	*s
TEM	Vacuum	~ Å	*s
AFM	-	~ Å	*s,l
STM	-	~ Å	*s,l
Crystallinity of material		****Qual./Quant.**	
XRD	-	**Quant.	*s
TEM	Vacuum	**Quant.	S
STM		**Qual.	s,l
Particle size determination		**Resolution**	
DLS	Liquid	~ nm	*l
XRD	-	~ Å	*s
TEM	Vacuum	~ Å	*s
Chemical ccomposition			
RBS	Vacuum		*s
XPS	Vacuum		S
Chemical bonding			
XPS	Vacuum		*s
IR	-		*s,l,g
Raman	-		*s,l,g
NMR	-		*s,l
Compound identification		****Qual./Quant.**	
Mass spec	Vacuum	**Qual.	*s,l
UV-vis	-	**Quant.	*s,l
PL	-	**Quant.	*s,l
IR	-	**Quant.	*s,l,g
Raman	-	**Quant.	*s,l,g

***s – solid, l – liquid, g – gas**
**** Qual. –qualitative, Quant.- quantitative**

Although some of the major material characterization techniques have been presented in this chapter, there are many others that can be used for examining nanomaterials. Some of the techniques used for chemical analysis include *low energy electron diffraction* (*LEED*), *thermal neutron analysis* (*TNA*), *fast neutron analysis* (*FNA*), *nuclear reaction analysis* (*NRA*) to name a few. Furthermore, there are also other techniques which are almost exclusively utilized by biochemists and chemists for structural analysis of organic and biological materials. Such techniques include *atomic absorption spectroscopy* (*AAS*), *small angle X-ray scattering* (*SAXS*), *circular dichroism* (*CD*), and *turbidity measurements* to name a few.

Research into novel materials has also resulted in material-specific characterization set-ups, which are versions of the techniques already described. In addition, for some characterization techniques, simply changing the radiation source, such as electrons or ion beams, may reveal different information and different aspects of materials.

References

[1] S. J. Strickler, in *Encyclopedia of Chemical Physics and Physical Chemistry*; *Vol. 2*, edited by J. H. Moore and N. D. Spencer (Inst of Physics Pub Inc, Bristol, UK, 2002).

[2] D. A. Skoog, D. M. West, and F. J. Holler, *Fundamentals of Analytical Chemistry*, 5th ed. (Saunders College Publishing, New York, USA, 1988).

[3] J. Coates, in *Encyclopedia of Analytical Chemistry*, edited by R. A. Meyers (John Wiley & Sons Ltd., Chichester, UK, 2000), p. 10815-10837.

[4] B. J. Clark, T. Frost, and M. A. Russell, *UV spectroscopy: techniques, instrumentation, data handling* (Chapman & Hall, London, UK, 1993).

[5] H.-H. Perkampus, *UV-VIS spectroscopy and its applications* (Springer-Verlag, Berlin, Germany, 1992).

[6] P. W. Atkins and J. de Paula, *Atkins' Physical Chemistry*, 7th ed. (Oxford University Press, New York, USA, 2002).

[7] P. W. Atkins and R. S. Friedman, *Molecular Quantum Mechanics*, 3rd ed. (Oxford University Press, New York, USA, 1997).

[8] C. K. Mann, T. J. Vickers, and W. M. Gulick, *Instrumental analysis* (Harper & Row, New York, USA, 1974).

[9] X. Michalet, F. Pinaud, T. D. Lacoste, M. Dahan, M. P. Bruchez, A. P. Alivisatos, and S. Weiss, Single Molecules **2,** 261-276 (2001).

[10] A. D. Yoffe, Advances in Physics **50,** 1-208 (2001).

[11] A. P. Alivisatos, Science **271,** 933-937 (1996).

[12] J. H. Park, J. Y. Kim, B. D. Chin, Y. C. Kim, J. K. Kim, and O. O. Park, Nanotechnology **15,** 1217-1220 (2004).

[13] N. Venkatram, D. N. Rao, and M. A. Akundi, Optics Express **13,** 867-872 (2005).

[14] R. Viswanatha, S. Sapra, B. Satpati, P. V. Satyam, B. N. Dev, and D. D. Sarma, Journal of Materials Chemistry **14,** 661-668 (2004).

[15] D. A. Skoog and J. J. Leary, *Principles of Instrumental Analysis*, 4th ed. (Saunders College Publishing, Orlando, USA, 1992).

[16] T. H. Gfroerer, in *Encyclopedia of Analytical Chemistry*, edited by R. A. Meyers (John Wiley & Sons Ltd., Chichester, UK, 2000), p. 9209-9231.

[17] E. N. Harvey, *A History of Luminescence* (American Philosophical Society, Philadelphia, USA, 1957).

[18] L. H. Qu and X. G. Peng, Journal of the American Chemical Society **124,** 2049-2055 (2002).

[19] M. Bruchez, M. Moronne, P. Gin, S. Weiss, and A. P. Alivisatos, Science **281,** 2013-2016 (1998).
[20] W. C. W. Chan and S. M. Nie, Science **281,** 2016-2018 (1998).
[21] M. A. Hines and P. Guyot-Sionnest, Journal of Physical Chemistry B **102,** 3655-3657 (1998).
[22] A. A. Guzelian, U. Banin, A. V. Kadavanich, X. Peng, and A. P. Alivisatos, Applied Physics Letters **69,** 1432-1434 (1996).
[23] L. Prodi, F. Bolletta, M. Montalti, N. Zaccheroni, P. B. Savage, J. S. Bradshaw, and R. M. Izatt, Tetrahedron Letters **39,** 5451-5454 (1998).
[24] L. Prodi, New Journal of Chemistry **29,** 20-31 (2005).
[25] H. Mattoussi, L. H. Radzilowski, B. O. Dabbousi, E. L. Thomas, M. G. Bawendi, and M. F. Rubner, Journal of Applied Physics **83,** 7965-7974 (1998).
[26] L. Prodi, F. Bolletta, M. Montalti, and N. Zaccheroni, Coordination Chemistry Reviews **205,** 59-83 (2000).
[27] D. N. Kendall, *Applied infrared spectroscopy* (Reinhold Pub. Corp., New York, USA, 1966).
[28] H. W. Siesler and K. Holland-Moritz, *Infrared and Raman spectroscopy of polymers* (M. Dekker, New York, USA, 1980).
[29] R. A. Shaw and H. H. Mantsch, in *Encyclopedia of Analytical Chemistry*, edited by R. A. Meyers (John Wiley & Sons Ltd., Chichester, UK, 2000).
[30] J. R. Ferraro and K. Krishnan, *Practical Fourier transform infrared spectroscopy: industrial and laboratory chemical analysis* (Academic Press, San Diego, USA, 1990).
[31] H. H. Hausdorff, *Analysis of polymers by infrared spectroscopy* (Perkin-Elmer Corporation, Norwalk, USA, 1951).
[32] J. H. van der Maas, *Basic infrared spectroscopy*, 2nd ed. (Heyden & Son, London, UK, 1972).
[33] C.-P. Sherman Hsu, in *Handbook of Instrumental Techniques for Analytical Chemistry*, edited by F. Settle (Prentice-Hall, 1997).
[34] J. McMurry, *Organic Chemistry*, 2nd ed. (Brooks/Cole, Pacific Grove, USA, 1988).
[35] B. Schrader, *Infrared and Raman spectroscopy: methods and applications* (VCH, Weinheim, German, 1994).
[36] M. D. Porter, T. B. Bright, D. L. Allara, and C. E. D. Chidsey, Journal of the American Chemical Society **109,** 3559-3568 (1987).
[37] H. Hoffmann and T. Leitner, in *Encyclopedia of Analytical Science*, edited by C. F. Poole, A. Townshend, and P. J. Worsfold (Academic Press, New York, USA, 2004).

[38] F. Brouers, J. P. Clerc, G. Giraud, J. M. Laugier, and Z. A. Randriamantany, Physical Review B **47,** 666-673 (1993).

[39] F. L. Leibowitz, W. X. Zheng, M. M. Maye, and C. J. Zhong, Analytical Chemistry **71,** 5076-5083 (1999).

[40] R. W. Darbeau, Applied Spectroscopy Reviews **41,** 401-425 (2006).

[41] A. Abragam, *The principles of nuclear magnetism* (Clarendon Press, Oxford, UK, 1978).

[42] J. B. Grutzner, in *Encyclopedia of Analytical Science*, edited by C. F. Poole, A. Townshend, and P. J. Worsfold (Academic Press, New York, USA, 2004).

[43] M. M. Spence, S. M. Rubin, I. E. Dimitrov, E. J. Ruiz, D. E. Wemmer, A. Pines, S. Q. Yao, F. Tian, and P. G. Schultz, Proceedings of the National Academy of Sciences of the United States of America **98,** 10654-10657 (2001).

[44] D. Briggs, *Handbook of x-ray and ultraviolet photoelectron spectroscopy* (Heyden, London, UK, 1977).

[45] T. L. Barr *Modern Esca: The Principles and Practice of X-Ray Photoelectron Spectroscopy* (CRC Press Inc, Boca Raton, USA, 1994).

[46] D. Briggs and J. T. Grant, *Surface analysis by Auger and x-ray photoelectron spectroscopy* (SurfaceSpectra Limited, Chichester, UK, 2003).

[47] A. J. Milling, *Surface characterization methods: principles, techniques, and applications* (Marcel Dekker, New York, USA, 1999).

[48] L. C. Feldman and J. W. Mayer, *Fundamentals of surface and thin film analysis* (North-Holland Publishing, New York, USA, 1986).

[49] A. Gurlo, N. Barsan, M. Ivanovskaya, U. Weimar, and W. Göpel, Sensors and Actuators B **47,** 92-99 (1998).

[50] B. O. Dabbousi, J. RodriguezViejo, F. V. Mikulec, J. R. Heine, H. Mattoussi, R. Ober, K. F. Jensen, and M. G. Bawendi, Journal of Physical Chemistry B **101,** 9463-9475 (1997).

[51] B. E. Warren, *X-ray diffraction* (Addison-Wesley Pub. Co., Reading, USA, 1969).

[52] D. Keith Bowen and B. K. Tanner, *High Resolution X-Ray Diffraction and Topography* (Taylor & Francis, London, UK, 1998).

[53] J. Drenth, *Principles of Protein X-Ray Crystallography* (Springer Verlag, New York, USA, 1999).

[54] B. D. Cullity, *Elements of x-ray diffraction* (Addison-Wesley Pub. Co., Reading, USA, 1978).

[55] P. Scherrer, in *Göttinger Nachrichten* (1918).

[56] A. L. Patterson, Physical Review **56,** 978-982 (1939).

[57] A. A. Guzelian, J. E. B. Katari, A. V. Kadavanich, U. Banin, K. Hamad, E. Juban, A. P. Alivisatos, R. H. Wolters, C. C. Arnold,

and J. R. Heath, Journal of Physical Chemistry **100,** 7212-7219 (1996).

[58] C. F. Bohren and D. R. Huffman, *Absorption and scattering of light by small particles* (Wiley, New York, USA, 1983).

[59] G. Bryant and J. C. Thomas, Langmuir **11,** 2480-2485 (1995).

[60] G. Bryant, C. Abeynayake, and J. C. Thomas, Langmuir **12,** 6224-6228 (1996).

[61] H. Cölfen and L. M. Qi, Chemistry-A European Journal **7,** 106-116 (2001).

[62] W. van Criekinge, P. van der Meeren, J. vanderdeelen, and L. Baert, Particle & Particle Systems Characterization **12,** 279-283 (1995).

[63] R. Pecora, *Dynamic Light Scattering, Applications of Photon Correlation Spectroscopy* (Springer, New York USA, 1985).

[64] B. J. Berne and R. Pecore, *Dynamic light scattering with applications to chemistry, biology and physics* (Wiley-Interscience, New York, USA, 1976).

[65] P. C. Hiemenz and R. Rajagopalan, *Principles of colloid and surface chemistry* (Marcel Dekker, New York, USA, 1997).

[66] H. Fenniri, B. L. Deng, and A. E. Ribbe, Journal of the American Chemical Society **124,** 11064-11072 (2002).

[67] C. Ross and K. T. Carron, in *Encyclopedia of Analytical Science*, edited by C. F. Poole, A. Townshend, and P. J. Worsfold (Academic Press, New York, USA, 2004).

[68] J. R. Ferraro and K. Nakamoto, *Introductory Raman Spectroscopy* (Academic Press, San Diego, USA, 1994).

[69] I. De Wolf, C. Jian, and W. M. van Spengen, Optics and Lasers in Engineering **36,** 213-223 (2001).

[70] N. B. Colthup, L. H. Daly, and S. E. Wiberley, *Introduction to infrared and Raman spectroscopy*, 2nd ed. (Academic Press, New York, USA, 1975).

[71] D. A. Long, *Raman spectroscopy* (McGraw-Hill, New York, USA, 1977).

[72] R. L. McCreery, *Raman Spectroscopy for Chemical Analysis* (Wiley, New York, USA, 2000).

[73] A. Szymanski, *Raman spectroscopy: theory and practice* (Plenum Press, New York, USA, 1967).

[74] J. A. Koningstein, *Introduction to the Theory of the Raman Effect* (D. Reidel, 1972).

[75] M. J. Pelletier and C. C. Pelletier, in *Encyclopedia of Analytical Science*, edited by C. F. Poole, A. Townshend, and P. J. Worsfold (Academic Press, New York, USA, 2004).

[76] M. J. Pelletier, *Analytical applications of Raman spectroscopy* (Blackwell Science, Malden, USA, 1999).

[77] T. Vo-Dinh, Trac-Trends in Analytical Chemistry **17,** 557-582 (1998).

[78] R. E. Littleford, D. Graham, W. E. Smith, and I. Khan, in *Encyclopedia of Analytical Science*, edited by C. F. Poole, A. Townshend, and P. J. Worsfold (Academic Press, New York, USA, 2004).

[79] M. Fleischmann, P. J. Hendra, and McQuilla.Aj, Chemical Physics Letters **26,** 163-166 (1974).

[80] D. L. Jeanmaire and R. P. van Duyne, Journal of Electroanalytical Chemistry **84,** 1-20 (1977).

[81] M. Moskovits, in *Surface-Enhanced Raman Scattering: Physics and Applications*; *Vol. 103* (Springer-Verlag, Berlin, 2006), p. 1-17.

[82] A. Tao, F. Kim, C. Hess, J. Goldberger, R. R. He, Y. G. Sun, Y. N. Xia, and P. D. Yang, Nano Letters **3,** 1229-1233 (2003).

[83] D. L. Stokes and T. Vo-Dinh, Sensors and Actuators B-Chemical **69,** 28-36 (2000).

[84] M. von Ardenne, in *Advances in electronics and electron physics. Supplement 16: The beginnings of electron microscopy*, edited by P. W. Hawkes (Academic Press, Orlando, USA, 1984).

[85] H.-J. Butt, K. Graf, and M. Kappl, *Physics and chemistry of interfaces* (Wiley-VCH, Weinheim, Germany, 2003).

[86] S. Amelinckx, D. van Dyck, J. van Landuyt, and G. van Tandeloo, *Electron microscopy: principles and fundamentals* (Wiley-VCH, Weinheim, Germany, 2003).

[87] A. C. C. Yu, M. Mizuno, Y. Sasaki, M. Inoue, H. Kondo, I. Ohta, D. Djayaprawira, and M. Takahashi, Applied Physics Letters **82,** 4352-4354 (2003).

[88] L. Valentini, I. Armentano, J. M. Kenny, C. Cantalini, L. Lozzi, and S. Santucci, Applied Physics Letters **82,** 961-963 (2003).

[89] F. Caruso, D. N. Furlong, K. Ariga, I. Ichinose, and T. Kunitake, Langmuir **14,** 4559-4565 (1998).

[90] Z. L. Wang, Journal of Physical Chemistry B **104,** 1153-1175 (2000).

[91] L. A. Bendersky and F. W. Gayle, Journal of Research of the National Institute of Standards and Technology **106,** 997-1012 (2001).

[92] Z. R. Dai, J. L. Gole, J. D. Stout, and Z. L. Wang, Journal of Physical Chemistry B **106,** 1274-1279 (2002).

[93] A. Taleb, C. Petit, and M. P. Pileni, Chemistry of Materials **9,** 950-959 (1997).

[94] J. R. Tesmer and M. Nastasi, *Handbook of modern ion beam materials analysis* (MRS, Pittsburgh, USA, 1995).

[95] G. Binnig and H. Rohrer, Reviews of Modern Physics **59,** 615-625 (1987).

[96] J. Tersoff and D. R. Hamann, Physical Review Letters **50,** 1998-2001 (1983).
[97] J. Golovchenko, Science **232,** 48-53 (1986).
[98] T. W. Odom, J. L. Huang, P. Kim, and C. M. Lieber, Nature **391,** 62-64 (1998).
[99] J. J. Davis, C. M. Halliwell, H. A. O. Hill, G. W. Canters, M. C. van Amsterdam, and M. P. Verbeet, New Journal of Chemistry **22,** 1119-1123 (1998).
[100] D. Losic, J. G. Shapter, and J. J. Gooding, Langmuir **18,** 5422-5428 (2002).
[101] G. Binnig, C. F. Quate, and C. Gerber, Physical Review Letters **56,** 930-933 (1986).
[102] G. U. Lee, D. A. Kidwell, and R. J. Colton, Langmuir **10,** 354-357 (1994).
[103] T. Nakagawa, K. Ogawa, and T. Kurumizawa, Journal of Vacuum Science & Technology B **12,** 2215-2218 (1994).
[104] E. L. Florin, M. Rief, H. Lehmann, M. Ludwig, C. Dornmair, V. T. Moy, and H. E. Gaub, Biosensors & Bioelectronics **10,** 895-901 (1995).
[105] T. Nakagawa, Japanese Journal of Applied Physics Part 2-Letters **36,** L162-L165 (1997).
[106] J. W. Zhao and K. Uosaki, Langmuir **17,** 7784-7788 (2001).
[107] S. J. Ippolito, A. Ponzoni, K. Kalantar-Zadeh, W. Wlodarski, E. Comini, G. Faglia, and G. Sberveglieri, Sensors and Actuators B-Chemical **117,** 442-450 (2006).
[108] G. Siuzdak, *The Expanding Role of Mass Spectrometry in Biotechnology* (MCC Press, 2003).
[109] E. De Hoffmann and V. Stroobant, *Mass Spectrometry: Principles and Applications*, 2nd ed. (John Wiley & Sons, New York, USA, 2001).
[110] D. Briggs, Surface and Interface Analysis **9,** 391-404 (1986).
[111] A. D. McNaught and A. Wilkinson, *IUPAC compendium of chemical terminology*, 2nd ed. (Blackwell Science, Boston, USA, 1997).
[112] M. Karas, D. Bachmann, U. Bahr, and F. Hillenkamp, International Journal of Mass Spectrometry and Ion Processes **78,** 53-68 (1987).
[113] K. Tanaka, H. Waki, Y. Ido, S. Akita, Y. Yoshida, T. Yoshida, and T. Matsuo, Rapid Communications in Mass Spectrometry **2,** 151-153 (1988).
[114] W. C. Wiley and I. H. McLaren, Review of Scientific Instruments **26,** 1150-1157 (1955).

Chapter 6: Inorganic Nanotechnology Enabled Sensors

6.1 Introduction

When the sizes of materials are reduced in one or more dimensions their physical and chemical properties can change dramatically. These changes affect their electromagnetic (electronic, magnetic, dielectric, etc), mechanical (lattice dynamics, mechanical strength, etc.), thermal (Seebeck coefficient, thermal resistance, etc.), optical (Stokes shift, resonance, etc.) and chemical properties (chemoluminescence, surface funtionalization, etc.). Furthermore, in materials with one or more nanoscale dimensions, these properties can be purposefully engineered, enhancing and tailoring the performance of sensors developed with these novel materials.

This chapter presents the theoretical behavior of materials with nanoscale dimensions and how they are utilized in sensing applications. These materials can be confined in one dimension, creating a two-dimensional layer with a thickness on the order of nanometers (2D); can be confined in two dimensions, leaving a one-dimensional object with a nanoscale cross-section (1D); or can be confined in three dimensions, which effectively results in a zero-dimensional object (0D). This chapter will focus mainly on inorganic nanostructures for sensing applications. Nanotechnology enabled organic sensors will be left for Chap. 7.

6.2 Density and Number of States

Two-dimensional structures such as *quantum wells* are particularly important in the development of the next generation of the semiconductor devices (e.g. giant magnetoresistors, lasers and ultra-high-speed transistors). In such structures, a thin semiconductor region is sandwiched between different materials (**Fig. 6.**). Other important nano-materials are described as *nanowires*, *nanobelts* and *nanorods*, which are one-dimensional, and *nanoparticles* and *quantum dots*, which are considered zero-dimensional (**Fig. 6.**).

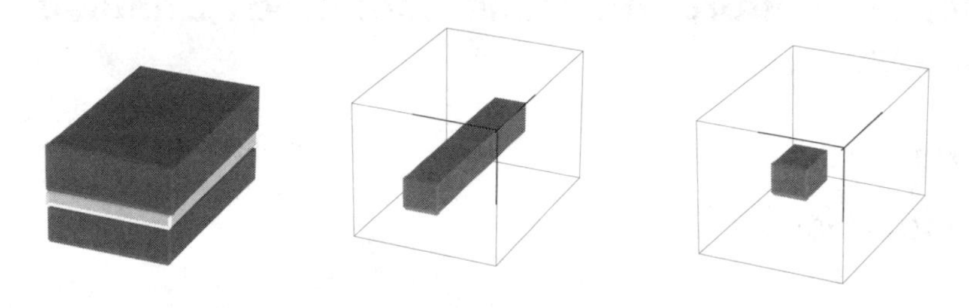

Fig. 6.1 Two, one and zero-dimensional structures.

One of the ways to achieve greater efficiency for sensors is to reduce the dimensions of the functional elements of the transducer and/or sensitive layers. Devices with nanometer dimensions exhibit quantum mechanical effects.[1] This causes an alteration of the *number of states* (*NOS*) and *density of states* (*DOS*), which are fundamental to the operation of nanosensors. The NOS, $N(E)$, and the DOS, $D(E)$, which can be defined as a function of energy E, affect the physical and chemical properties such as resistance, signal to noise ratio, optical properties and thermal characteristics.

The NOS and DOS can be calculated using quantum mechanical analytical and computational evaluations. The 2D, 1D, and 0D approximations can be utilized to analyze structures such as quantum wells, nanowires and nanoparticles, where quantum effects are observed in dimensions of less than a few nanometers. For materials with larger dimensions, (although the bulk of the material is treated classically) the phenomena occurring at the boundaries and surfaces are best treated quantum mechanically.

In this section, the numerical calculations for the NOS and DOS will be presented. These results will be used to describe the performance of sensors and their electronic, optical, mechanical and thermal behavior.

6.2.1 Confinement in Quantum Dimensions

When treated classically, a free particle displays a continuous energy spectrum. However, this assumption is not valid for particles in confined systems, which are described using quantum mechanics. Under such conditions, particles exist in discrete energy levels that form distinct bands. Energy levels outside these bands are forbidden.

The energy band model is crucial for detailed studies of nanoscale semiconductor devices, magnetic nanoparticles, optical sensors and nanodimensional mechanical structures, as it provides the necessary tools to understand their behaviour and describe their properties.

6.2.2 Momentum and Energy of Particles

Free particles, such as molecules in an ideal gas, do not interact significantly with each other. For example, in a host material the valence electrons detach themselves from the host atoms and move freely in a state much like molecules in an ideal gas. The collective state of these electrons is called *Fermi gas*.

For a system of free particles, the potential energy is zero, and the total energy of the system is comprised of the kinetic energy of the constituent particles. Kinetic energy is classically defined as:

$$E = \tfrac{1}{2}mv^2 = p^2/2m, \tag{6.1}$$

where p is the momentum, m is the particle's mass and h is the Plank's constant.

Louis De-Broglie was the first to hypothesize that particles could exhibit wave-like properties. Quantum mechanics treats particles in this fashion. In quantum mechanics, the total energy of a particle is given by the Hamiltonian of the system, which is comprised of kinetic and potential energy components.

To determine the momentum of a free particle using quantum mechanics, the wave-like nature of the particle must be considered. The relationship between frequency f and wavelength λ of a wave is:

$$\lambda f = v, \tag{6.2}$$

where v is the velocity of the wave. The wavevector (or wavenumber), k, is defined as:

$$k = 2\pi/\lambda. \tag{6.3}$$

In classical physics, momentum p is defined as:

$$p = mv. \tag{6.4}$$

However, in quantum mechanics, matter-waves are responsible for the propagation of momentum, which is thereby defined as:

$$p = \left(\frac{h}{2\pi}\right)k = \hbar k\,, \tag{6.5}$$

In many expressions, the reduced Planck's constant $\hbar$ is used for convenience.

6.2.3 Reciprocal Space

From Eqs. (6.3) and (6.5) it can be seen that wavenumber k is inversely proportional to wavelength and proportional to momentum. As a result, it is useful to define *reciprocal space*, also commonly known as *k space* or the *Brillouin zone*, for describing the motion and energy of electrons. As we shall see, k space is also critical for determining NOS and DOS.

A two-dimensional infinite potential well can be useful scenario to conceptualize k space. Imagine an electron confined to one dimension, in the region 0 to L. For this system, the solution to the time-independent Schrödinger equation is (**Fig. 6.2**):

$$\psi_n(x) = \sqrt{\frac{2}{L}} \sin(\frac{n\pi}{L} x), \tag{6.6}$$

where n is an integer corresponding to the nth state. *Eigenvalues* of this equation represent the discrete energy states in which the particle may exist. These are given by:

$$E_n = \frac{\hbar^2 \pi^2}{2mL^2} n^2 = \frac{h^2}{8mL^2} n^2, \tag{6.7}$$

where E_n is the energy of the nth state. These states correspond to electrons of wavelength:

$$\lambda_n = 2L/n. \tag{6.8}$$

The wavenumber, k_n, is consequently given by:

$$k_n = n\pi/L. \tag{6.9}$$

The wavelengths of the first few excited states, expressed as $2L$, L and $2L/3$ correspond to reciprocal space values of π/L, $2\pi/L$ and $3\pi/L$. As a result, the smallest value of k is π/L which corresponds to the largest wavelength and the largest value of k is $3\pi/L$ corresponding to the smallest wavelength, thus demonstrating the reciprocal nature of k space.

It can be seen that increasing the energy state results in a smaller wavelength, greater momentum, a larger wavenumber, and hence a larger vector in k space, thus demonstrating the proportionality between energy levels and wavenumbers. Reciprocal space in two and three dimensions can be defined and visualized similarly.

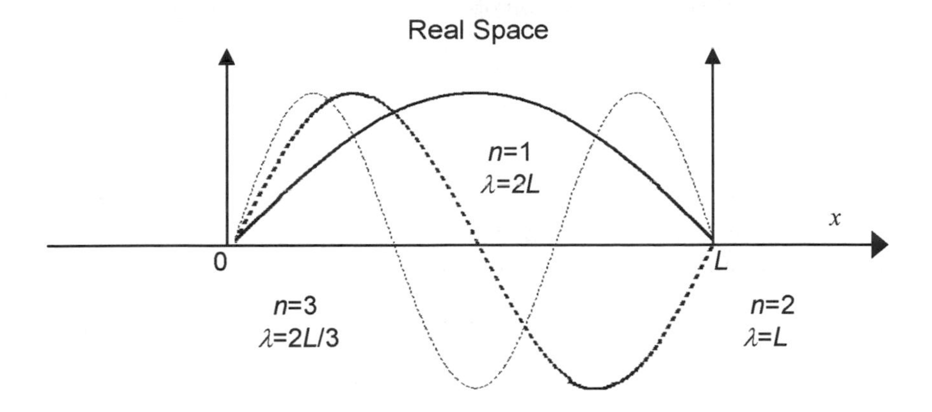

Fig. 6.2 The time-independent Schrödinger equation solutions for a two-dimensional infinite potential well.

6.2.4 Definition of Density of States

The electrical, mechanical, optical and magnetic properties of materials depend strongly on how the energies are distributed. To assess these properties, we need to consider the NOS and DOS.

We define the number of free particles having an energy of E as $N(E)$. The density of states is then defined as $D(E) = dN(E)/dE$. DOS is then expressed as $D(E)dE$, which is the number of allowed energy levels per unit volume of the material, within the energy range E to $E + dE$.

6.2.5 DOS in Three-Dimensional Materials

In three-dimensional k space, the points of vectors of equal magnitude in all directions form a spherical shell. These vectors represent equal energy states, proportional to $k^2 = k_x^2 + k_y^2 + k_z^2$, given by:

$$E = \frac{\hbar^2}{2m}\left(k_x^2 + k_y^2 + k_z^2\right) = \frac{\hbar^2 k^2}{2m}. \tag{6.10}$$

If we consider the free particles being contained in a cube of length L, the total volume will be equal to $V = L^3$, while the difference between k-space points will be equal to $2\pi/L$.

In three-dimensional reciprocal space, the volume of a cubic unit cell is defined as:

$$V_{3D} = \left(\frac{2\pi}{L}\right)^3, \tag{6.11}$$

which is the volume of a single energy state.

The vector k encompasses a volume of $4/3\ \pi k^3$. As a result, with reference to (**Fig. 6.3**), the volume between the two shells which are infinitely close to each other, at a distance of k from the centre is found to be:

$$d\left(\frac{4}{3}\pi k^3\right) = 4\pi k^2 dk\ . \tag{6.12}$$

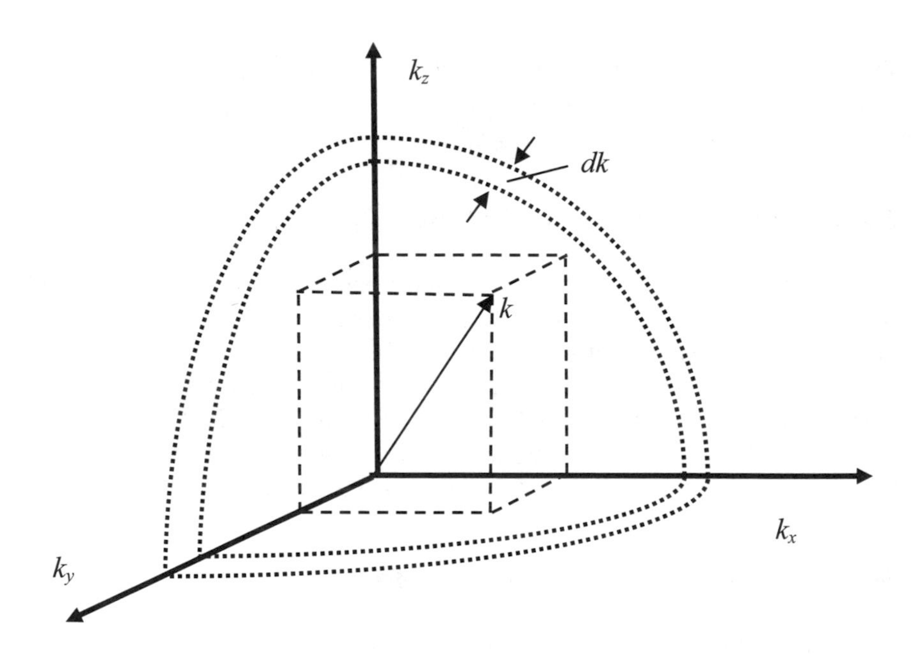

Fig. 6.3 The k-space (reciprocal space) in 3D. Calculation of the NOS for a wavenumber between k and $k+dk$.

By dividing this volume by the volume of a single energy state (Eq. (6.11)), the density of states, $D(k)$, in reciprocal space can be obtained as:

$$D(k)dk = 2 \times \frac{4\pi k^2 dk}{V_{3D}} = \frac{k^2 L^3 dk}{\pi^2}\ . \tag{6.13}$$

In this equation, to account for the intrinsic angular momentum of the electrons (or spin states), a factor of two has been used. For phonons this factor is 3.

In order to calculate the DOS as a function of energy, by referring to Eq. (6.10) we can see that:

$$dk = \frac{1}{\sqrt{2mE\hbar^2}} mdE\,, \tag{6.14}$$

and the following equation can be obtained:

$$D(E)dE = \frac{L^3 k^2 dk}{\pi^2} = \frac{2L^3 mE}{\pi^2 \hbar^2} \frac{1}{\sqrt{2mE\hbar^2}} mdE = \frac{L^3}{2\pi^2}\left(\frac{2m}{\hbar^2}\right)^{3/2} E^{1/2} dE\,. \tag{6.15}$$

This shows that for a three-dimensional material the DOS is proportional to $E^{1/2}$ and continuous in the E space.

6.2.6 DOS in Two-Dimensional Materials

The DOS can also be calculated for 2D structures, such as superlattices, where the structure is confined to quantum states in one dimension and effectively infinite in the other two.

The approach is similar to the three-dimensional calculations. In this case, particles which have the same energy, occupying the same points in the reciprocal space, are have their k space points located in a ring of radius k and infinitely small width dk (**Fig. 6.4**).

The area of this ring in two-dimensional reciprocal space is:

$$d\left(\pi k^2\right) = 2\pi|k|dk\,, \tag{6.16}$$

while the area occupied by a single energy state is:

$$V_{2D} = \left(\frac{2\pi}{L}\right)^2. \tag{6.17}$$

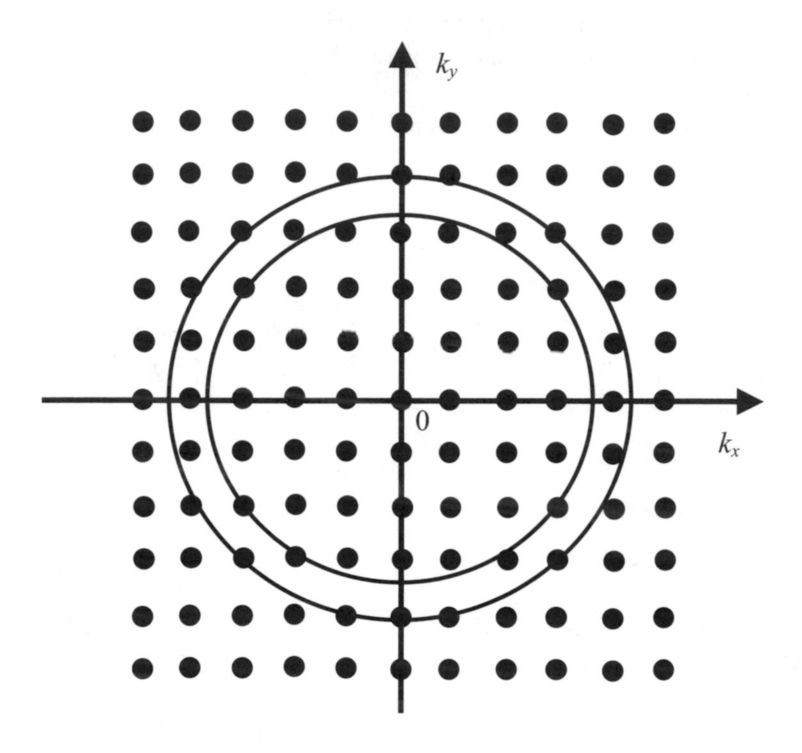

Fig. 6.4 *k*-space in 2D.

The DOS in reciprocal space can therefore be calculated as:

$$D(k)dk = 2 \times \frac{2\pi |k| dk}{\left(\frac{2\pi}{L}\right)^2} = \frac{|k| L^2 dk}{\pi}, \tag{6.18}$$

where again the multiplication by 2 accounts for the intrinsic angular momentum of the electrons (or 3 for phonons). Consequently, from (6.14), the 2D DOS in E space is:

$$D(E)dE = \frac{L^2 k dk}{\pi} = \frac{L^2 m}{\pi \hbar^2} dE . \tag{6.19}$$

It can be seen that the 2D DOS does not depend on the energy value unlike the 3D DOS. This time, the DOS takes on a staircase-function as follows:

$$D(E)dE = \frac{L^2 m}{\pi \hbar^2} \sum_i H(E - E_i) dE, \tag{6.20}$$

where $H(E\text{-}E_i)$ is the *Heaviside function* and *i* an integer.

6.2.7 DOS in One-Dimensional Materials

In one-dimensional structures, the material is confined in two dimensions with only one *k* component. The space area is a line and the area in reciprocal space is defined as:

$$V_{1D} = \frac{2\pi}{L}. \tag{6.21}$$

2dk is used for the calculation of DOS where factor 2 accounts for the spin degeneracy of electrons (and 3 for phonons). The DOS per unit length in 1D is:

$$D(k)dk = 2 \times \frac{2dk}{\frac{2\pi}{L}} = 2\frac{L}{\pi} dk, \tag{6.22}$$

Transforming to *E* space, the DOS per unit volume at energy value of *E* is:

$$D(E)_{1D}\, dE = \frac{2L}{\pi} dk = \frac{L}{\pi} \left(\frac{m}{\hbar^2} \right)^{1/2} \frac{1}{E^{1/2}} dE. \tag{6.23}$$

As a result, the DOS function is:

$$D(E)_{1D}\, dE = \frac{L}{\pi} \left(\frac{m}{\hbar^2} \right)^{1/2} \sum_i \left(\frac{H(E - E_i)}{(E - E_i)^{1/2}} \right) dE. \tag{6.24}$$

where $H(E\text{-}E_i)$ is the Heaviside function.

6.2.8 DOS in Zero-Dimensional Materials

A system of free particles fully confined in all dimensions is a 0D structure. This structure has discrete charge and electron states. In this case, the NOS is a step function of *E*. The calculation of NOS can be found in quantum mechanics textbooks which is described as:[2]

$$N(E) = K_0 \sum_i d_i H(E - E_i) . \tag{6.25}$$

where $H(E\text{-}E_i)$ is the Heaviside function and K_o and d_i are the coefficients that can be calculated depending on the shape of the unit cell.

From Eq. (6.25) the DOS for a quantum box can be calculated as:

$$D(E) = K_0 \sum_i d_i \delta(E - E_i) . \tag{6.26}$$

where $\delta(E\text{-}E_i)$ is the *Kronecker delta* function.

6.2.9 Discussions on the DOS

The NOS and DOS for 0D, 1D, 2D and 3D materials as a function of energy, E, are presented in **Table 6.1**.

Table 6.1 NOS and DOS for 0D, 1D, 2D and 3D materials.

Type	NOS – $N(E)$	DOS – $D(E)$
0D	$K_0 \sum_i d_{0i} H(E - E_i)$	$K_0 \sum_i d_{0i} \delta(E - E_i)$
1D	$K_1 \sum_i d_{1i} (E - E_i)^{1/2} H(E - E_i)$	$\frac{1}{2} K_1 \sum_i d_{1i} (E - E_i)^{-1/2} H(E - E_i)$
2D	$K_2 \sum_i d_{2i} (E - E_i) H(E - E_i)$	$K_2 \sum_i d_{2i} H(E - E_i)$
3D	$K_3 (E)^{3/2}$	$\frac{3}{2} K_3 (E)^{1/2}$

The graphs for NOS and DOS of 0D, 1D, 2D, and 3D structures as a function of E have been presented in **Fig. 6.5**.

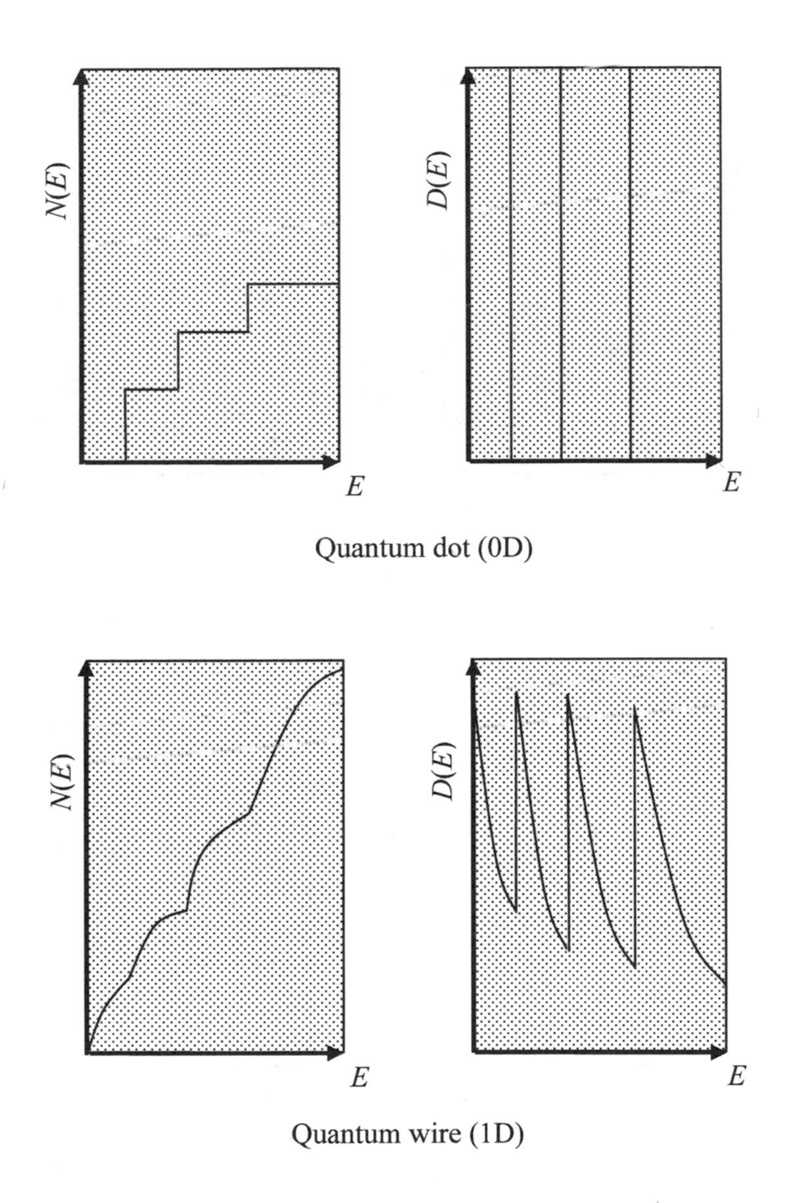

Fig. 6.5 NOS and DOS of 0D, 1D, 2D, and 3D structures as a function of E.

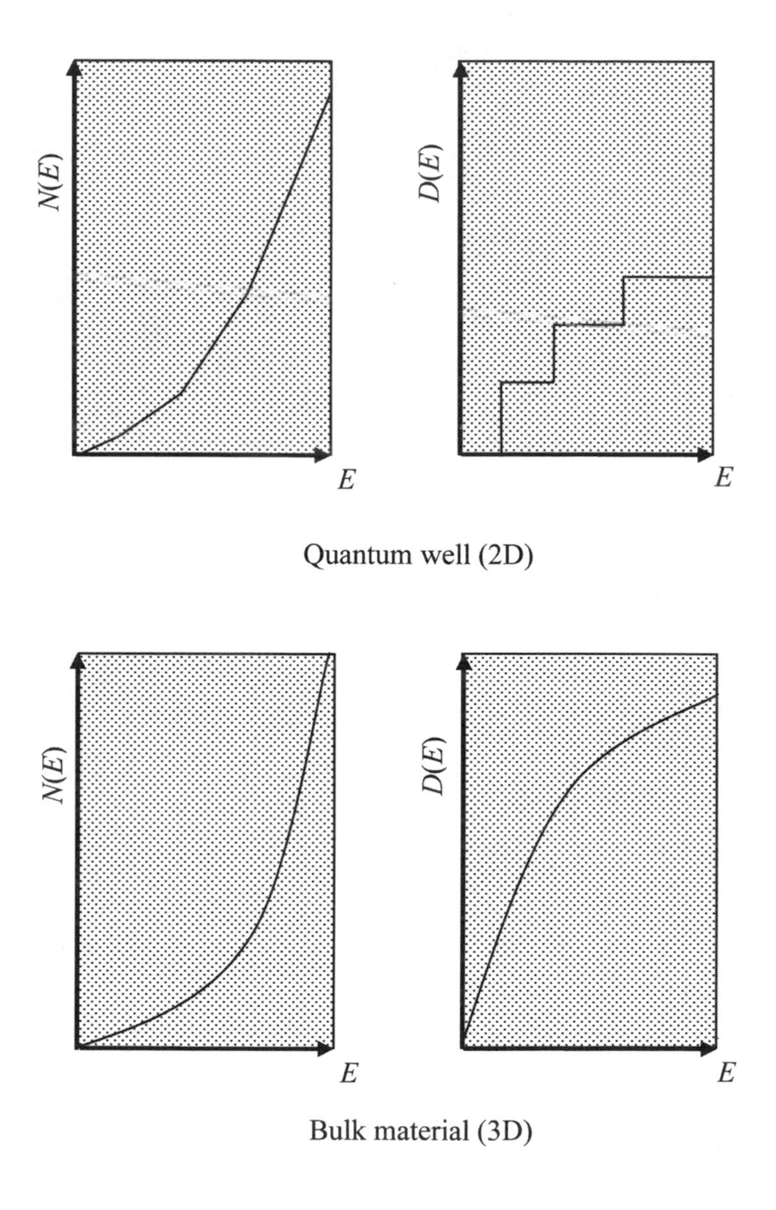

Fig. 6.5 (cont) NOS and DOS of 0D, 1D, 2D, and 3D structures as a function of E.

Obviously, the **Fig. 6.5** graphs can be utilized to visualize how the properties of nanomaterials can be quantized when the dimensions of a material are reduced to quantum sizes.

For instance, the conductance of a 1D structure can be derived from the DOS. In a 1D structure (such as a nanowire) with a length of L, if a potential equal to V is applied, for a mobile electron the energy difference (ΔE) between two sides of this 1D structure is equal to:

$$\Delta E = eV/L. \tag{6.27}$$

where e is an electron charge. Now, if q is charge (equal to $-e$ for an electron), and v is the velocity, the net current, I, flowing through the channel will be:

$$I = N(E)qv = D(E)\Delta E\, qv, \tag{6.28}$$

where $D(E)$ and $N(E)$ are the density of states and the number of states in a 1D case, respectively, for which $D(E)$ was presented by Eq. (6.24). Replacing v_i which is the velocity of an electron from Eq. (6.1) in sub-band i the DOS can be obtained as:[3]

$$D(E)_{1D}\, dE = \frac{L}{\pi}\left(\frac{m}{\hbar^2}\right)^{1/2}\left(\frac{1}{(E-E_i)^{1/2}}\right) = \frac{4L}{hv_i}, \tag{6.29}$$

where E_i is the energy in each sub-band and m is the mass for an electron. From Eqs. (6.28) and (6.29) we can obtain the equation defining current, I, as:

$$I = 2e^2 V/h. \tag{6.30}$$

Interestingly, Eq. (6.30) shows that in a 1D structure, the current only depends on the voltage across it via fundamental constants. Consequently, the two-terminal resistance is calculated to be 12.906 kΩ for such a 1D structure.

As can be observed, an ideal 1D structure has a finite resistance. This is called a *resistance quantum* (or the *inverse conductance quantum*). Such a resistance can also be experimentally measured. At the end of the 1980s, the University of Cambridge and Delft groups reported the measurements of steps in the conductance of a quasi 1D configuration in a *field effect transistor* structure by means of the voltage applied at the gate as shown in **Fig. 6.6.**[4,5]

All the above assumptions are for near zero degree Kelvin temperatures. At higher temperatures a thermal effect also appears which smoothes out the step shape.

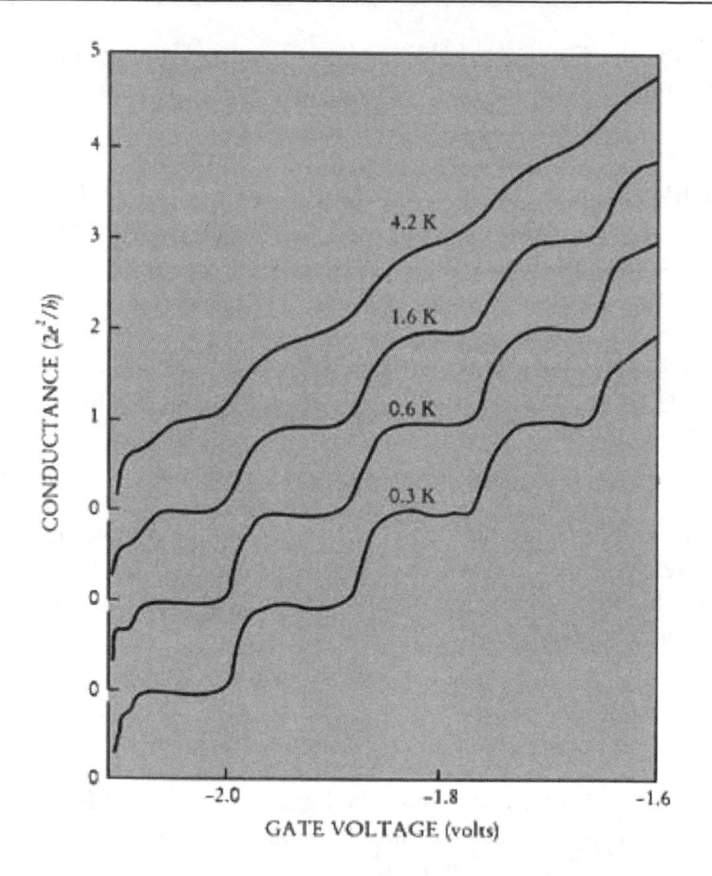

Fig. 6.6 Conductance quantization of a quantum point contact in units of $2e^2/h$. As the gate voltage defining the constriction is made less negative, the width of the point contact increases continuously, but the number of propagating modes at the Fermi level increases stepwise. The resulting conductance steps are smeared out when the thermal energy becomes comparable to the energy separation of the modes. Reprinted with permission from the Institute of Physics publications.[4]

If there are multiple 1D structures (i.e. a few of them in parallel or in a bundle), using the *Landauer equation*, we can extend Eq. (6.30) to calculate the general Ohm's law.[3]

6.2.10 Theoretical and Computational Methods

Theoretical and computational methods are becoming ever more popular for realizing the performance of different sensing systems. Such methods take the advantage of the ever increasing calculation power of computers. They help to understand the quantitative and qualitative properties of the systems. Depending on the dimensions of the structure, a variety of methods ranging from the classical calculations to quantum mechanical calculations and semi empirical techniques can be used.[6]

Several electronic structure methods can be utilized to determine the total energies of a particular system. The determination of parameters such as the total energy is a prerequisite for the theoretical treatment of any particular nanoparticles, sensitive layers or layered systems which are used in nanotechnology enabled sensing applications.

Some of the major techniques in such calculations are *electron density* based focusing on *quantum chemistry* and *solid state physics* fundamental. These electronic structure calculations are generally dominated by *density functional theory* (*DFT*) and wave methods.[6]

The results of such calculations define properties such as energy of the structure, work function charges, binding energies, geometries of the stable structures, as well as chemisorption and physisorption properties of the surfaces. The approach suggested by Du and Leeuw is a good example such calculations.[7]

6.2.11 One-Dimensional Transducers

In this section, transducers based on single 1D structures will be presented and it will be shown how their conductivity can be modulated in response to target gases. Applications of such structures as mechanical resonators and photodiodes will be presented later in this chapter. The biosensing application will be described in Chap. 7.

There are plenty of examples in literature about developing sensors with one-dimensional nanostructures. The simplest structure is a single nanorod or nanowire connected between two electrical contact pads as shown in **Fig. 6.7**.

At near zero Kelvin temperatures, the resistance measured between two electrical pads is almost equal to 12.9 kΩ as previously discussed. Obviously, this resistance alters at higher temperatures depending on whether the material is an insulator, semiconductor or conductor.

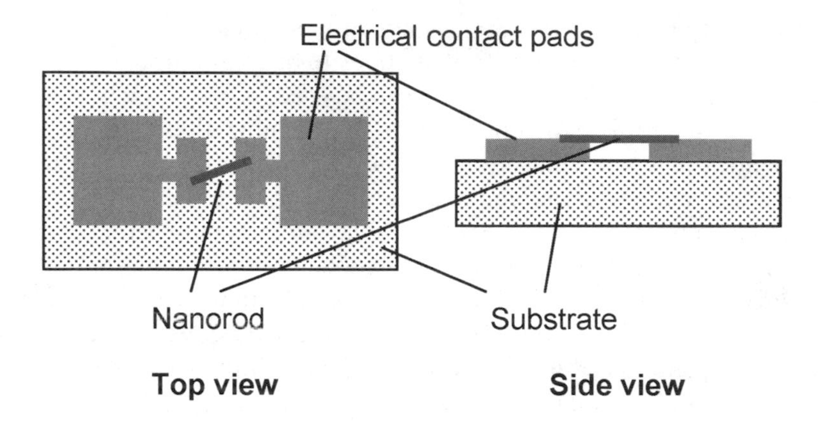

Fig. 6.7 A sensor based on a single one-dimensional nanostructure.

When this one-dimensional nanostructure is exposed to target molecules or environmental fluctuations, its physical and chemical properties change producing a segmented nanorods or nanowire (**Fig. 6.8**).

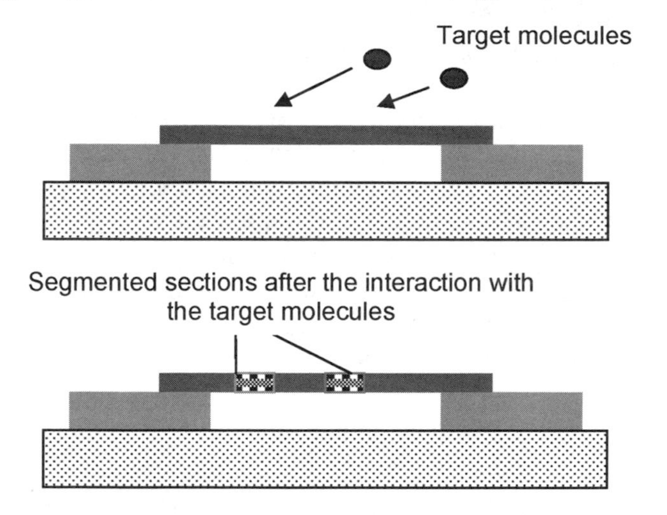

Fig. 6.8 Exposure of a nanorod to target analytes. A segmented nanorod is produced after the exposure.

The segmented structure consists of many almost-zero-dimensional sections which are connected in series with quantized resistance within each segment approximating a hetrostructure. As a result, the electrical, mechanical, optical and chemical properties of the 1D structure changes.

1D materials including carbon nanotubes, metals such as Au, Ag, Pt, and metal oxides such as ZnO, SnO_2, In_2O_3, Ga_2O_3 and CdO in the form of nanorods, nanowires and nanobelts have been extensively studied and employed for their applications in the fabrication of electronic devices. Not only 1D structures can be individually used for the fabrication of transducers and sensors, but the incorporation of a large number of such structures together can also be utilized to form nanostructured thin films for the fabrication of transducers and sensing elements.

Field effect transistors based on 1D structures are now being widely investigated in the electronic industry as an alternative for current silicon transistor technologies. 1D structures can be directly used in the fabrication of the mechanical transducing platforms such as resonators. They are also employed as efficient photo sensitive elements for the fabrication of photodiodes and sensors. These 1D nanostructures can exhibit well engineered chemical compositions and crystallographic formation.[8,9] Metal oxides can interact with different gas species as sensitive elements at elevated temperatures. Metals such as Pt and Au can function as catalytic elements for gas species. Carbon nanotubes can interact with gases such as hydrogen and store them in their cages at room temperature. In addition, these materials can be functionalized as 1D surfaces for chemical and bio sensing applications.

Field-effect transistors and conductometric sensors can be fabricated using individual 1D materials such as nanobelts and nanorods.[10,11] There are many methods that can be used for this task. A few examples are as follows:

(*a*) Long bundles of metals oxides nanobelts can be dispersed in a liquid which does not react with them.[8] Ultrasonication is used until most of the individual nanobelts are isolated. The liquid is then dispersed onto substrates with prefabricated metallic pads. *Non-contact-mode AFM* (AFM tip hovers electrostatically over the surface) is then used for imaging and to locate the metallic pads and the bundles. Field-effect transistors can then be formed using the AFM, electron or ion beam lithographic methods ensuring a selected number of bundles connect to the metal pads. Heat treatment is generally needed to make sure that the connections between the metal pads and the one-dimensional nanostructures are correctly formed. It is also possible to apply the metallic contacts after the dispersion of bundles. In this case, electron beam lithography has to be applied. An AFM image of the FET and the schematic diagram are shown in Fig. 6.9.[8] After forming the electrical contacts a FET is produced.[12]

A typical ZnO FET is depicted in Fig. 6.9 with a gate threshold voltage of -15 V, a switching ratio of nearly 100, and a peak conductivity of 1.25×10^{-3} $(\Omega\text{-cm})^{-1}$. Similar characteristics have been observed in using carbon nanotubes instead of metal oxide nanobelts.[13]

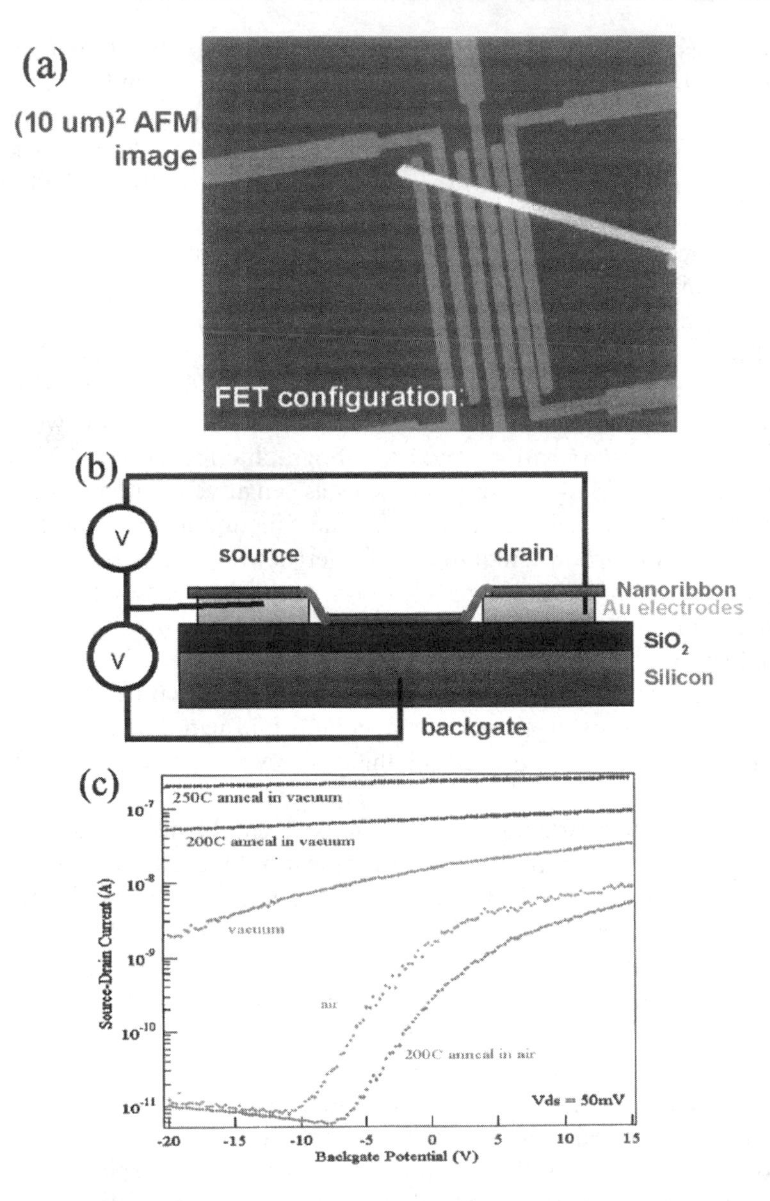

Fig. 6.9 Atomic force microscopy image of a field-effect transistor (FET) device made using a single nanobelt. (b) Schematic working diagram of the device, (c) SnO_2 nanobelt FET oxygen sensitivity. Source-drain current versus gate bias for a SnO_2 FET after various treatments was measured in this order: air, vacuum, 200°C vacuum anneal, 250°C vacuum anneal, 200°C air anneal. Reprinted with permission from the Annual Reviews publications.[8]

(*b*) One-dimensional nanowires can also be attached onto or grown from a tip (such as a scanning probe microscopy tip) to form a 1D electronic device as shown in **Fig. 6.10**.[14] In this example, heterojunctions of carbon nanotubes (CNT) and silicon nanowires (SiNW) were attached on a tip. An Fe catalyst was used to grow CNTs. And to create NT/SiNW junctions, SiNWs were grown from the CNT tip ends by using silane. CNTs were then attached to scanning probe microscopy tips using a simple method reported by Smalley et al.[15] The tip was coated with an acrylic adhesive and then brought in contact with a bundle of CNTs under the direct view of a dark-field optical microscope. The tip was withdrawn from the bundle as soon as one CNT was attached to the end of the tip. The CNT/SiNW was then attached to Ga–In liquid metal at the SiNW end.

The current–voltage (IV) measurements on NT/SiNW junctions with only the SiNW in contact with the Ga–In liquid exhibit reproducible rectifying behavior of a diode (**Fig. 6.10 - b**). When the tips are advanced past the SiNW so that only the CNT is in contact with Ga–In liquid, symmetrical, non-rectifying IV curves were observed.

In addition, the IV measurements made on SiNWs grown directly from the Pt–Ir tips (the same conditions as for CNT/SiNW growth) are symmetrical, exhibiting no evidence of rectifying behavior but have characteristics of a tunnel junction.

(*c*) Nanowires can also be fabricated by drawing out a metal wire. They can also be formed by indenting a contact tip into a metallic surface, when the tip is retracted, a Ångstrom dimension nanowire is formed.[16]

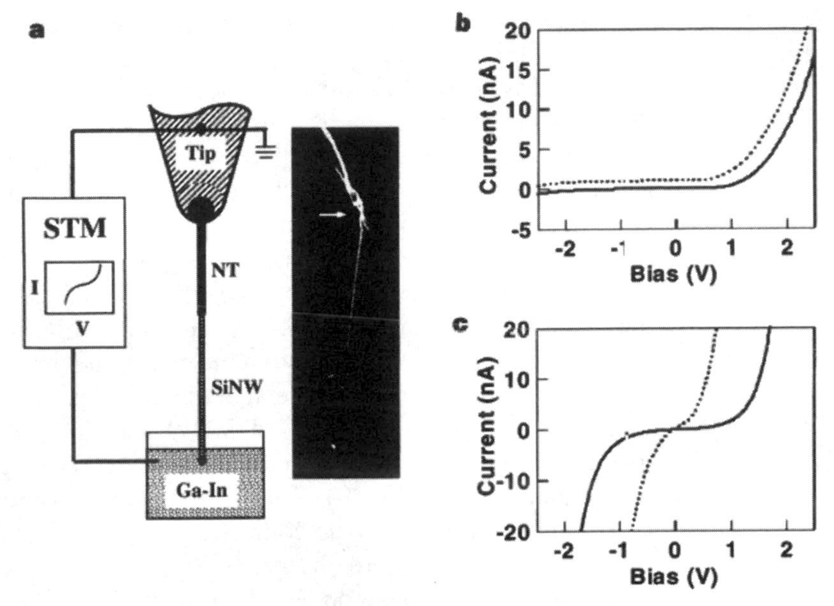

Fig. 6.10 (**a**) Diagram of the set-up used to measure the electrical properties of nanojunctions. Inset, another NT/SiNW junction grown. The arrow indicates the position of the junction with the NT and SiNW above and below this point, respectively. (**b**), $I-V$ curves of two independent NT/SiNW junctions (solid and dotted curves). The reverse bias region of the dotted curve overlaps the solid curve and was shifted for clarity. (**c**), $I-V$ curves of a NT (dotted), coated with a thin layer of amorphous silicon, and a SiNW (solid) connected directly to the Pt–Ir tip. Reprinted with permission from the Nature publications.[14]

6.2.12 Example: One-Dimensional Gas Sensors

As described in the previous section, 1D transducers can be efficiently used for gas sensing applications. Li et al demonstrated the capability of such sensors using individual ZnO-nanowire-based transistors as shown in **Fig. 6.11**.[17] The transistors show a carrier density of 2300 μm^{-1} and mobility up to 6.4 cm^2/Vs, which are derived from their IV curves. The threshold voltage shifts in the positive direction and the source-drain current decreases as ambient oxygen concentration increases. Surface adsorbents on the ZnO nanowires affect both the mobility and the carrier density.

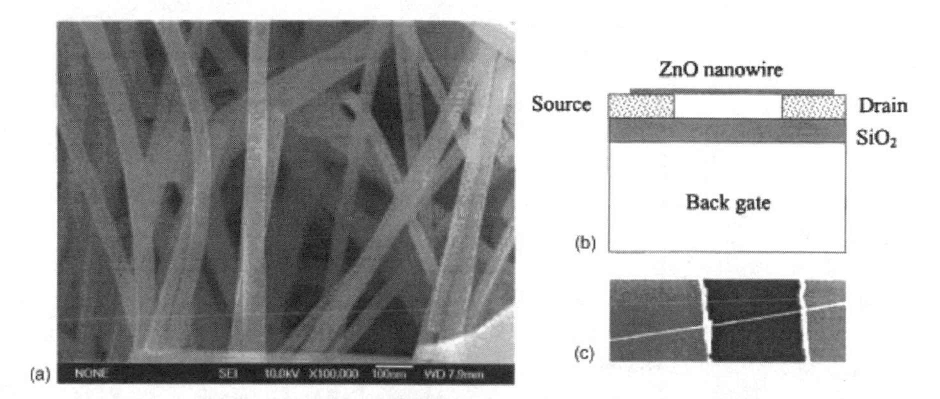

Fig. 6.11 (a) An SEM image of the ZnO nanowires. (b) Schematic illustration of the ZnO nanowire transistor, a single ZnO nanowire bridges the two electrodes (source and drain). The substrate is used as the back gate. (c) The SEM image of the nanowire transistor, the separation between the two electrodes is about 1 μm. Reprinted with permission from the Applied Physics Letters publications.[17]

To show that the adsorbed species affect the conductance of the individual ZnO nanowire transistors, a measurement was first performed in a vacuum of 10^{-4} Pa. Oxygen was then gradually added. The transfer characteristics from – 40 V to 10 V at $V_{SD} = 2$ V were obtained under different oxygen pressures, as shown in **Fig. 6.12**. The current decreased and $V_{\rm th}$ was positively shifted as more oxygen was added. It was clearly demonstrated that the slope of the four curves in the linear region decreased with increasing oxygen pressure. The adsorbed oxygen molecules depleted the ZnO nanowire of electrons and formed oxygen ions (O^-, O^{2-} or O_2^-), which led to surface depletion in the nanowire's conductance channel. As the oxygen pressure was raised, more electrons were captured by the oxygen. As a result, the carrier density in the ZnO nanowire decreased and the depletion layer widened. Both gave rise to a decrease of conductance and a positive shift of threshold voltage.

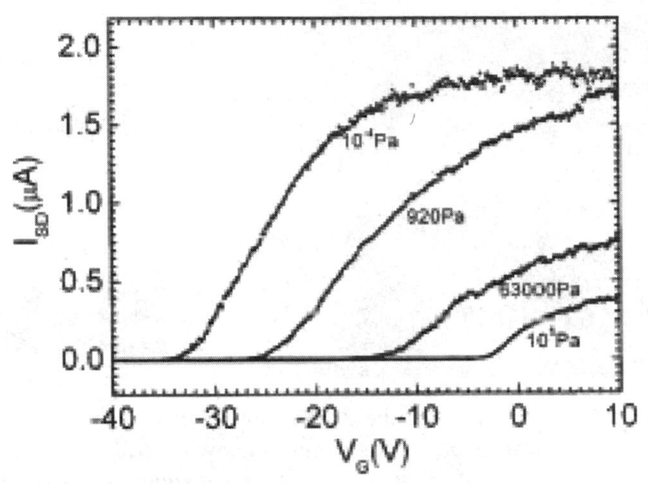

Fig. 6.12 The I_{SD}–V_G curves at V_{SD} = 2 V as V_G varies from –40 V to 10 V at different oxygen pressures: 10^{-4}, 920, 6.3×10^4, and 10^5 Pa, respectively. Reprinted with permission from the Applied Physics Letters publications.[17]

6.3 Gas Sensing with Nanostructured Thin Films

Gas sensing is a field in which the use of innovative nanomaterials is expected to result in sensor products with substantially enhanced sensitivity, improved selectivity, decreased power consumption, and excellent reproducibility. The unique properties of nanomaterials can play a significant role in the development of such sensors, due to the fact that they often show physical and chemical characteristics which differ from those of the bulk counterparts.[18,19]

Metal oxides (such as SnO_2, WO_3, TiO_2) are well known for their high sensitivity to changes in their surrounding gas atmosphere at elevated temperatures.[20] Nanostructures of semiconducting oxides have been shown to be very promising due to their large surface area-to-volume ratio, single crystallinity, and possible complete depletion of carriers within the nanostructure when exposed to a target gas.[19,21,22]

Despite the aforementioned advantages, sensors based on nanostructured thin films still face problems of high cross-sensitivity to different gas species (i.e. low selectivity), high sensitivity to humidity of ambient atmosphere, and long-term instabilities.[20,23,24] Current efforts to resolve these issues tend to implement the results obtained mainly from empirical approaches and include development of new materials, streamlining of fabrication processes, and the elaboration of smart systems (arrays) with signal conditioning and filtration algorithms.[25-27]

The *structural engineering of* metal oxide nanostructured thin films is an effective method which can be employed for the optimization of these types of gas sensors. In this approach, operating parameters such as response time, output signal, selectivity and stability can be improved and tuned through the optimization of the structure itself.[28]

Using a structural engineering method, the following geometric parameters of a metal oxide gas sensing matrix can be controlled: grain size; agglomeration; area of inter-grain and inter-agglomerate contacts; film thickness; porosity, and dominant orientation and faceting of crystallite, forming gas sensing surface.[28] However, these parameters are only a part of the metal oxide matrix. We should also take into account the physical–chemical parameters of a gas sensing matrix, such as: chemical and phase composition; type of additives; the bulk and surface oxygen vacancy concentrations (stoichiometry); size and density of both metal catalyst particles and single atoms on the surface of metal oxides; surface architecture; type and concentration of uncontrolled impurities, etc.

6.3.1 Adsorption on Surfaces

The study of adsorption is of fundamental importance in the field of sensors. Adsorption depends on the *potential energy surface* (*PES*) of the sensing system. The PES corresponds to the energy plane over the configuration space of the atomic coordinates of the involved atoms.[29] It gives information about the adsorption sites, energies, vibrational frequencies of *adsorbate species* and the existence adsorption barriers.

The potential energy curves for adsorption dubbed Lennard-Jones potentials are shown in **Fig. 6.13**.[29] In this figure, AB+S represents the potential energy of a molecule, AB, approaching a surface, S. The other curve is A+B+S corresponds to the interaction of two widely separated atoms, A and B, with the surface. Such potential graphs can be used for determining the type of adsorption on a surface.

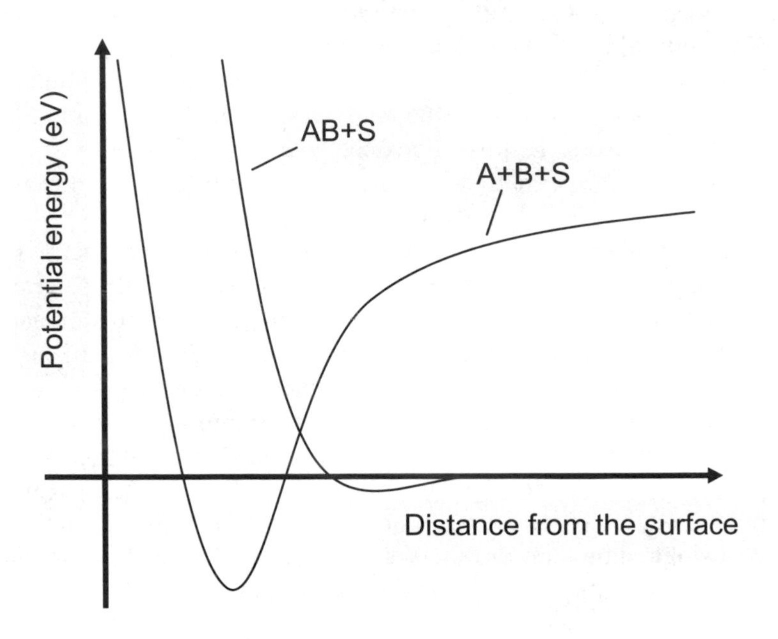

Fig. 6.13 The Lennard-Jones potentials.

In the weakest form of adsorption (*physisorption*) no true chemical bond between the surface and the adsorbate is established. This bonding can be due to the induced dipole moment of a nonpolar adsorbate interacting with its own image charges on the polarized surface. This bonding is rather weak in the order of 0.1 eV. However, it is observed in a large range of applications. For instance many enzymatic interactions (Chap. 7) consist of a large number of physical adsorptions. Physisorptive forces such as van der Waals are purely attractive. However, many physical forces can be repulsive. The orbitals of the approaching atom have to be orthogonal to the substrate wave function which increases their kinetic energy. This generates a strong repulsive force. As a result, there is a balance between the short range Pauli repulsion and the long range attractions.

Adsorption can also be the result of a chemical reaction. *Chemisorption* corresponds to the creation of chemical bonds between adsorbate and substrate, which causes the resulting electronic structure perturbation. In gas sensors, the target gas may be chemisorbed or physisorbed on the surface. The interaction depends on the type of PES, gas molecules and environmental parameters. When they adsorb on surface, molecules can be dissociated on or diffused in the sensitive layer.[29]

Currently the theoretical calculations do not yield quantitative results but they are ideal for illustrating qualitative trends. Williams and Lang were the first who used the *atom-jellum model* to study the chemisorption of an atom bonded on a metallic surface.[30] They solved the jellium simulation along with the Kohn-Sham equations to show the behavior of the chemisorpted species (**Fig. 6.14**).

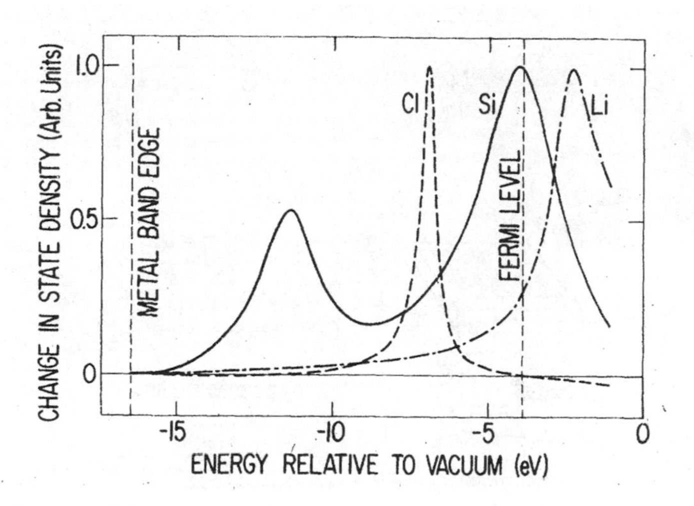

Fig. 6.14 Change of the density of states upon the adsorption of Si, Li, and CL on Al as calculated by William and Lang. Reprinted with permission from the American Physical Society publications.[30]

Changes of the DOS upon the adsorption of Si, Li and Cl on an Al surface have been shown in **Fig. 6.14**. The location of the resonance relative to the Fermi level determines the degree of occupation. The difference in eigenstate density can also be observed. The DOS in Si show two prominent peaks which are associated with the Si 3*s* and 3*p* atomic levels. The 2*s* resonance of Li lies above and the 3*p* resonance of Cl below the E_f, which are clear examples of positive and negative chemisorptions.

6.3.2 Conductometric transducers Suitable for Gas Sensing

The proper choice of contact geometry is very important in the fabrication of reliable nanostructured thin film/thick film sensors. The transducer should be compatible with the sensitive thin film deposition method, and it is desirable to have a heater integrated in the design. A heater can also be needed as most of the metal oxide based sensors operate at elevated temperatures. Some of these contact geometries are shown in **Fig. 6.15**.[31]

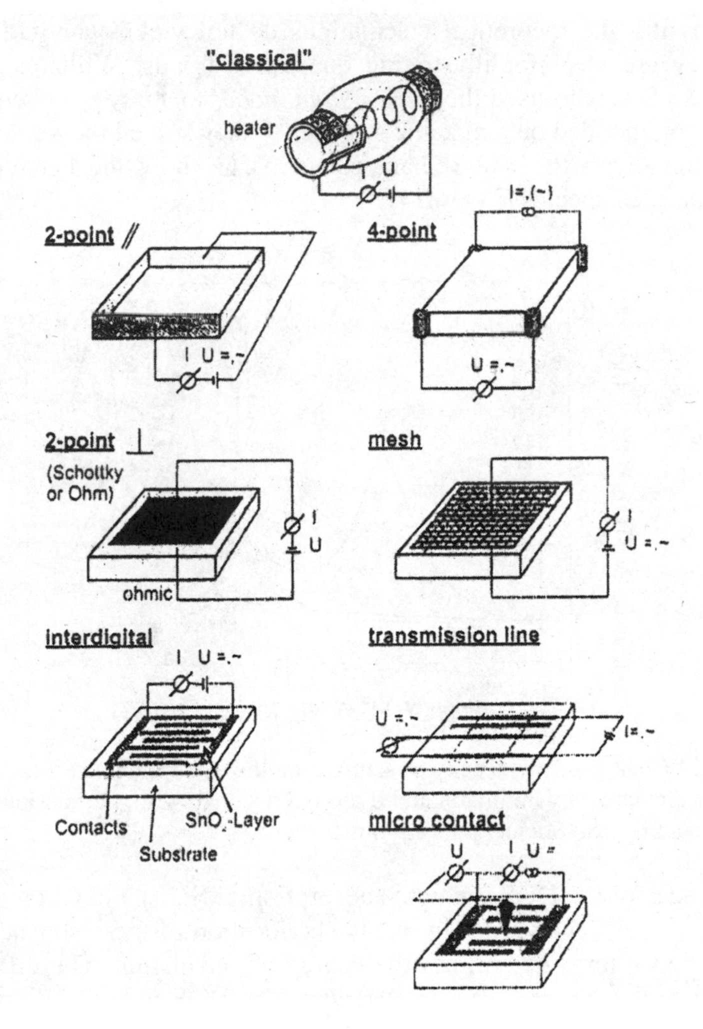

Fig. 6.15 Typical contact geometries for semiconductor gas sensors. Reprinted with permission from the Elsevier publications.[31]

Such sensors are generally either based on ohmic or Schottky contacts which are formed after the deposition of the sensitive layer on the electric pads. For a given contact geometry and material, different operation modes can be used. Changes in the operation mode correlate with the target gas concentration. An overview of the IV characteristics and responses is shown in **Fig. 6.16**.[31]

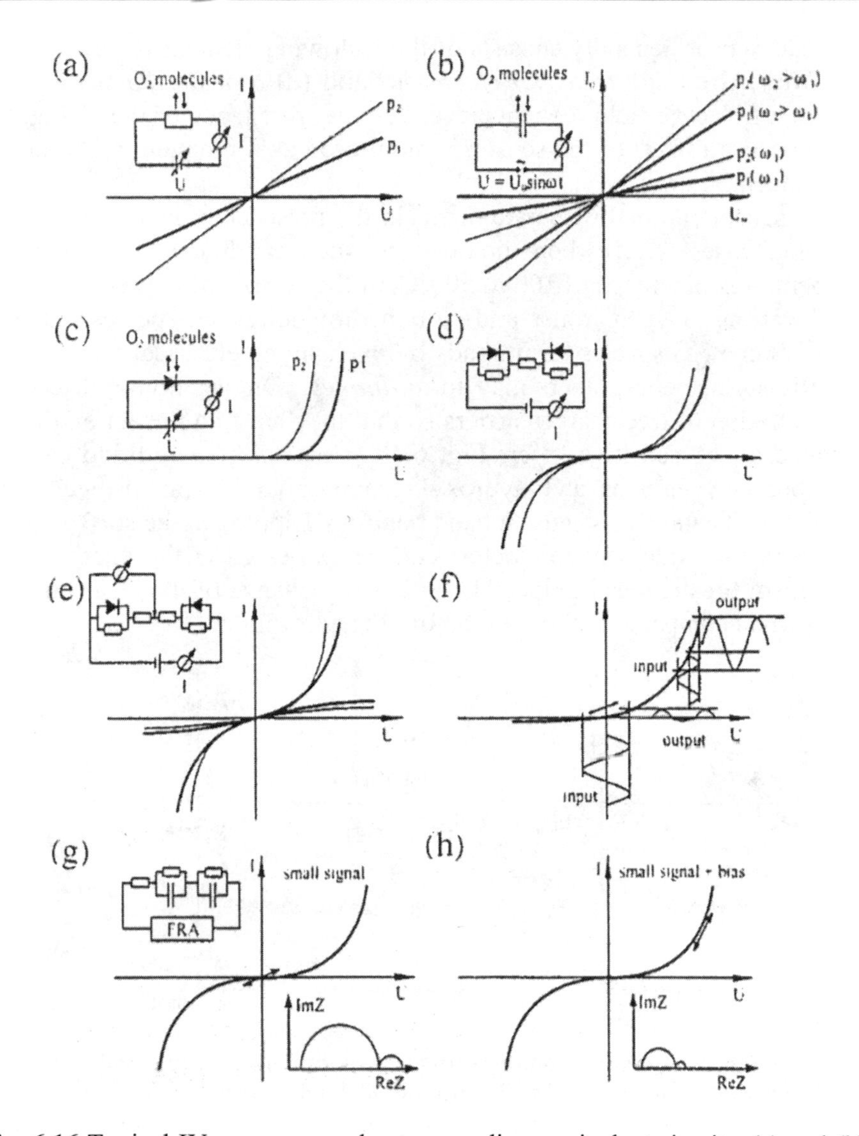

Fig. 6.16 Typical IV responses and corresponding equivalent circuits: (a) and (b) characterize ohmic contact, (c) to (h) characterize nonohmic barrier contacts, (c) with one Schottky barrier and one ohmic contact, (d) identical barriers at both contacts monitored in a two-point arrangement, (e) a point contact in a three-point arrangement, (f) large amplitude modulation of one Scottky barrier at one electrode with characteristics non-linear response, (g) small amplitude ac response monitored without bias voltage, (h) small amplitude ac response monitored with bias voltage. Reprinted with permission from the Elsevier publications.[31]

A gas sensor generally consists of the following elements:[32] (a) *a sensitive layer,* (b) a *substrate,* (c) *electrodes* and (d) *a heater.* Generally the change of electric field (conductance, voltage, resistance or the change of piezoelectric effect) of the sensor is monitored as a function of the target gas concentration.[33]

Gas sensors normally operate in air, in the presence of humidity and interfering gases (e.g. carbon dioxide). In such conditions, for operating temperatures in a range of 100 to 500°C, at the surface of the sensitive material various oxygen, water and carbon dioxide-related species are present.[34] Some gas species form bonds by exchanging electrical charge with specific surface sites, others may form *dipoles.* Dipoles do not affect the concentration of free charge carriers so that they have no impact on the resistance of the sensitive layer. **Fig. 6.17** presents the simplified case of adsorbed oxygen ions and hydroxyl groups bound to an *n*-type semiconductor. Their effects cause a band bending. Dipoles, at the surface of an *n*-type metal oxide semiconductor; can be expressed in the energy band formalism for the metal oxide. The effect is a change of the electronic affinity when compared to the state before the adsorption.[33]

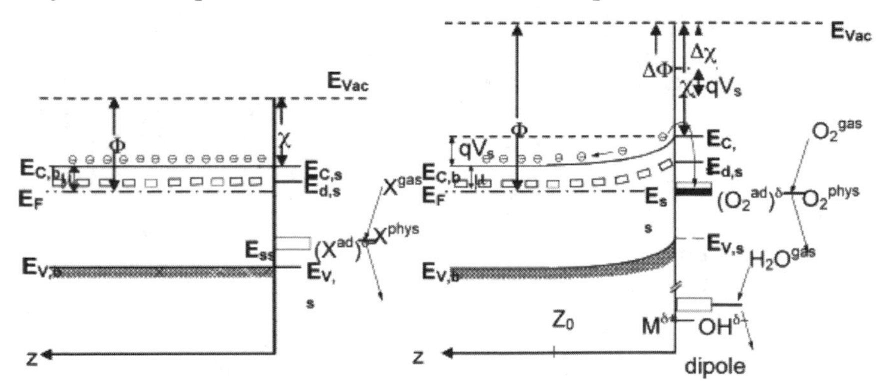

Fig. 6.17 Schematic representation of the flat band in *n*-type semiconductor (left) and band banding model (right) illustrating adsorption at the surface of a *n*-type semiconductor. The changes of the work function ($\Delta\Phi$) are determined by band bending (qV_S—due to ionosorption) and changes the electron affinity ($\Delta\chi$) due to building of dipoles at the surface ($M^{\delta+}$– $OH^{\delta-}$). Reprinted with permission from the Elsevier publications.[33]

The sensitive layer can be made of bulk materials and/or small grains. The grains can be large microsize, or nanosize (**Fig. 6.18**) or a combination of both. These sensitive layers are in contact with electrodes. These electrodes are generally made of metals. However, they can also be fabricated from materials such as conductive polymers or conductive metal oxides.

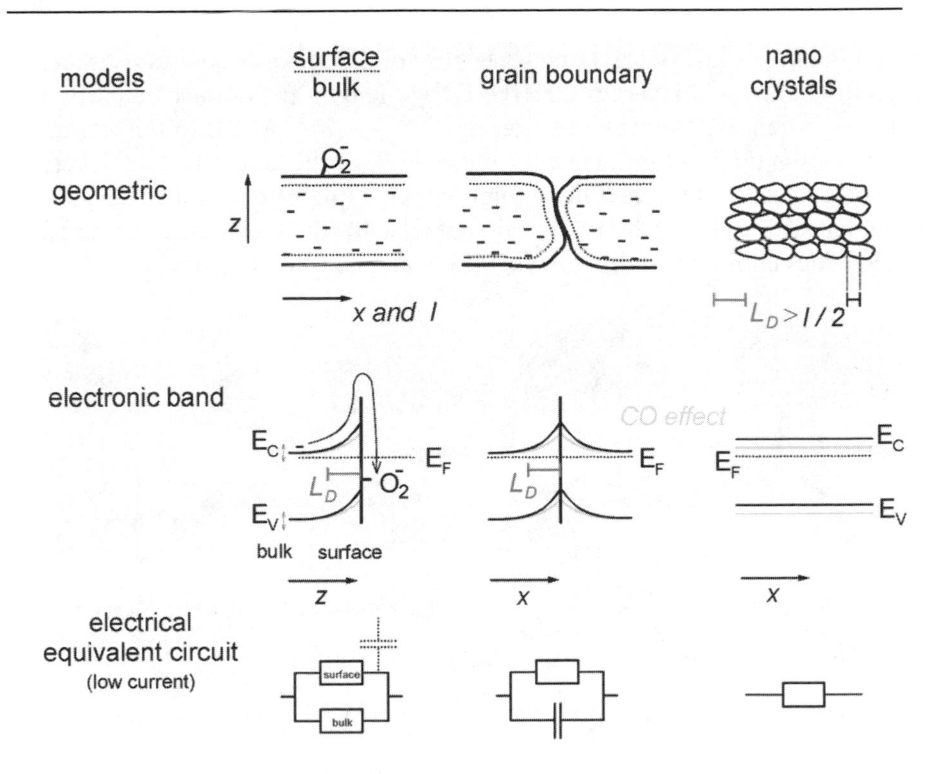

Fig. 6.18 Different conduction mechanisms and changes upon exposure of a sensing layer to O_2 and CO. This survey shows geometries, electronic band pictures and equivalent circuits. E_C minimum of the conduction band, E_V maximum of the valence band, E_F Fermi level, and λ_D Debye length.[32]

The material can be completely or partly depleted which depends on the thickness of the depleted layer that is formed on its surface after exposure to a target gas and the Debye length λ_D *of the material.*[32]

Two types of sensitive layers (compact and porous) are presented in **Fig. 6.19**.[32] The contributions to the overall resistance corresponding to the changes in the band bending at the material/grain surface and the potential barriers that appear due to the metal/metal oxide contact are shown **Fig. 6.19**.

Although, the idea that only the resistance in the sensitive material is altered when exposed to a target gas, is widely accepted; however, it is viewed as an over-simplification. It should be remembered that the overall resistance of the sensor depends not only on the gas sensing material properties but also on other sensor parameters (transducer morphology, electrode, etc.).

When the sensitive layer consists only of a compact continuous material and the thickness is larger than the Debye length, it can only be partly depleted when exposed to the target gas, $z_g > z_0$.(**Fig. 6.19**). In this case, the interaction does not influence the entire bulk of the material. Two levels of resistance are established in parallel, which explains the limited sensitivity as only the underneath layer is in contact with the electrodes. As a result, the better choice is a thinner layer which can be fully depleted.

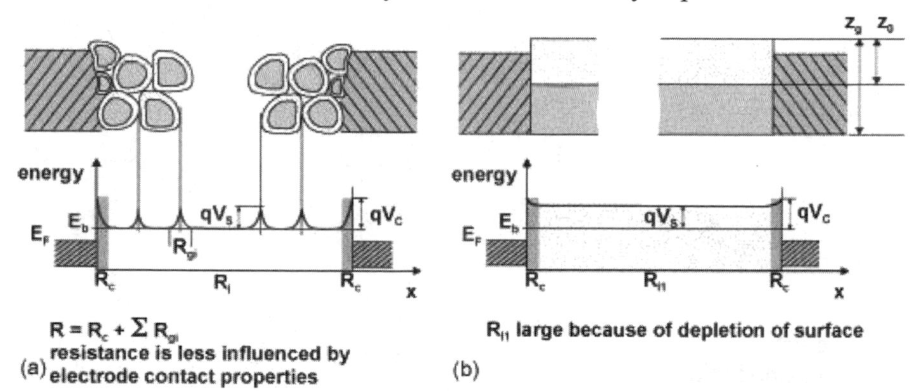

Fig. 6.19 The schematic representation of a porous (a) and a compact (b) sensing layer and their energy bands. The representations shows the influence of electrode-sensing layer contacts. R_C is resistance of the electrode–metal oxide contact, R_{l1} is the resistance of the depleted region of the compact layer, R_l is the equivalent of series resistance of R_{l1} and R_C, and the equivalent series resistance of ΣR_{gi} and R_C, in the porous and compact situations, respectively. R_{gi} is the average inter-grain resistance in the case of porous layer, E_b minimum of the conduction band in the bulk, qV_S band bending associated with surface phenomena on the layer, and qV_C also contains the band bending induced at the electrode–metal oxide contact. Reprinted with permission from the Elsevier publications.[33]

Contacts in gas sensors can have both electrical and electro/chemical roles. A useful tool for determining the electrical contribution of contact resistance is the use of transmission line measurements.[35] In this method, the dependence of the sensor resistance is plotted as a function of the spacing between electrodes at different ambient conditions. A fit of the experimental data can be made using ohms law. For thin compact films, contact resistance plays an important role as a dominant factor in resistance.[36] The contribution of contact resistance is also extremely important for the case in which individual nanorods, nanowires or nanobelts are to be used as sensing layers.[37] In addition to direct current measurement methods for measuring the resistance, ac impedance spectroscopy can also be very useful in identifying contact-related elements, and the presence of surface depletion regions. This is possible because the depletion region behaves like a charged capacitor. **Fig. 6.18 (bottom)** presents the *equivalent circuits* corresponding to different types of contributions found in a sensing layer. Attention should be paid to the fact that finding parallel RC elements in the equivalent circuit of a sensor does not automatically indicate the presence of a contact; these types of measurements are very sensitive to errors. The correct interpretation of error-free experimental results requires critical analysis of the experimental data in conjunction with information on the morphology and microscopic characteristics of both sensing layer and sensor.[33]

A simple calculation, applied in the case of the *Schottky approximation* for contact allows for observation of the main trends. It is based on the fact that the total charge Q_S, trapped on the surface level (E_S in **Fig. 6.17**) associated with gas adsorption (oxygen ionosorption in **Fig. 6.17**) can be written as:[33]

$$Q_S = qn_b s z_0, \tag{6.31}$$

where q is the charge of an electron, n_b is the electron concentration, s is the surface on which the adsorption takes place, and z_0 is the depth of the depletion region. Just as a reminder, in our Schottky approximation, it is considered that all the electrons in the conduction band from the depletion layer are captured on the surface trap levels. The relationship between the band bending (V_S) and Q_S, can be written as:[32]

$$Q_S = s(2q\varepsilon\varepsilon_0 n_b)^{1/2}(V_S)^{1/2}, \tag{6.32}$$

where ε_0 is the permittivity of air and ε is the relative permittivity of the media. The use of the general definition for the capacitance of the depletion region is:

$$C_S = \frac{1}{s}\left(\frac{\partial Q_S}{\partial V_S}\right) = \left(\frac{q\varepsilon\varepsilon_0 n_b}{2}\right)^{1/2} V_S^{\ 1/2}. \tag{6.33}$$

This equation allows for the elimination of experimental error and providing an accurate interpretation of the gas sensing results.

6.3.3 Gas Reaction on the Surface - Concentration of Free Charge Carriers

In this section, the charge carrier concentration, n_S, will be described as a function of parameters of a gas sensitive layer. The conductance and the electric field in a sensitive layer depend on the charge carrier concentration.

We will only present the analysis of the oxygen response on the surface of SnO_2. Similar analysis for other gas species can be found in the review paper by Barsan et al.[32]

An example: oxygen interaction with a metal oxide surface

Interaction with the metal oxide layer causes adsorption of oxygen gas in molecular (O_2^-) and atomic (O^-, O^{2-}) forms (**Fig. 6.20**).[32] Such an interaction generally occurs in a temperature range between 100 and 500°C.

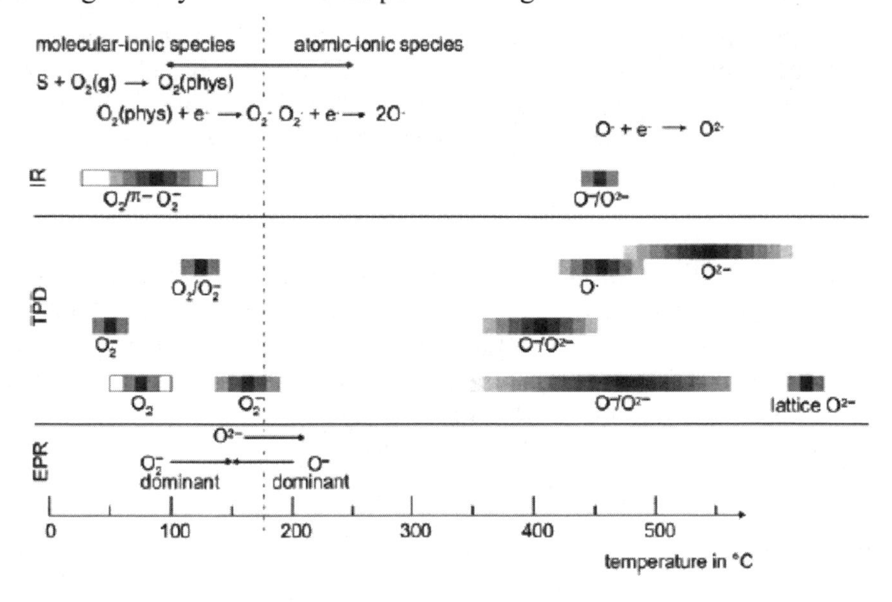

Fig. 6.20 Oxygen species detected at different temperatures at SnO_2 surfaces with infrared analysis, temperature programmed desorption, electron paramagnetic resonance.[32]

Investigations using *infrared analysis*, *temperature programmed desorption*, *electron paramagnetic resonance* show that below 150°C the molecular form dominates interactions. However, at higher temperatures the ionic species participate in the interactions. These interactions produce a depletion layer at the metal oxide surface.[32] After surface charge accumulation is completed, a strong potential barrier at the surface has to be overcome by the electrons wanting to reach the surface.

The oxygen chemisorption can be described as:[32]

$$\frac{\beta}{2}O_2^{gas} + \alpha.e^- + S \leftrightarrow O_{\beta S}^{-\alpha}, \tag{6.34}$$

where O_2^{gas} is the oxygen molecule in the ambient atmosphere, e^- is the electron that reaches the surface when it has sufficient energy and S is an unoccupied chemisorption site for oxygen. The number of electrons is denoted as n_S ($n_S = [e^-]$).

$O_{\beta S}^{-\alpha}$ is a chemisorbed oxygen species with:

$\alpha = 1$ for singly ionised form
$\alpha = 2$ for doubly ionised form
$\beta = 1$ for atomic form
$\beta = 2$ for molecular form

Using mass action law, and defining the activation energies for adsorption and desorption as the reaction constants k_{ads} and k_{des}, respectively, Eq. **(2.1)** can be rewritten as:[32]

$$k_{ads} \cdot [S].n_S^{\alpha} \cdot p_{O_2}^{\beta/2} = k_{des} \cdot [O_{\beta S}^{-\alpha}]. \tag{6.35}$$

If we assume that $[S_t]$ is the total number of available surface sites, the surface coverage with chemisorbed oxygen, θ, can be defined as:

$$\theta = \frac{[O_{\beta S}^{-\alpha}]}{[S_t]}, \tag{6.36}$$

and using the conservation of surface sites:

$$[S] + [O_{\beta S}^{-\alpha}] = [S_t], \tag{6.37}$$

it can be written as:

$$(1-\theta).k_{ads}.n_S^{\alpha}.p_{O_2}^{\frac{\beta}{2}} = k_{des}.\theta, \tag{6.38}$$

where p_{O_2} is the concentration of oxygen in the gaseous phase. This equation defines the relationship between the surface coverage of adsorbed oxygen and the number of electrons with enough energy to reach the surface.

Eq. (6.38) is not sufficient for extracting the relationship between n_S and p_{O_2}, as the surface coverage, θ, and n_S are related. The *electroneutrality condition* (this condition states that the charge in the depletion layer is equal to the charge captured at the surface) together with the *Poissons equation* are used to overcome this problem. It is also assumed that the temperature is high enough to ionise all donors (concentration of ionised donors equals the bulk electron density n_b), and all the electrons in the depletion layer are captured on surface levels.

For obtaining the second relation, two different cases will be considered:[32]

Case 1. Large grains/crystallites (d>>λ_D comparable to the Debye length)

In this case, only the surface, and not the bulk region, is affected. It is possible to treat the situation for large grains as dealing with a planar or quasi-infinite object where qV_S is the band bending, z_0 is the depth of the depleted region and A the covered area. In this case, the electroneutrality and the Poisson equations for energy (E) can be written as (Q_{ss} is the surface charge):[32]

$$\alpha\theta[S_t]A = n_b z_0 A = Q_{SS}\,, \tag{6.39}$$

$$\frac{d^2E(z)}{dz^2} = \frac{q^2.n_b}{\varepsilon\varepsilon_0}. \tag{6.40}$$

The boundary conditions for the Poisson equation are:

$$\left.\frac{dE(z)}{dz}\right|_{z=z_0} = 0\,, \tag{6.41}$$

$$E(z)\big|_{z=z_0} = E_C\,. \tag{6.42}$$

It can be shown that:

$$E(z) = E_C + \frac{q^2.n_b}{2.\varepsilon.\varepsilon_0}.(z-z_0)^2. \tag{6.43}$$

As $V = E/q$, the potential is found to be:

$$V(z) = \frac{q^2 n_b}{2\varepsilon\varepsilon_0}.(z-z_0)^2, \tag{6.44}$$

and the surface band bending is:

$$V_S = \frac{q^2 n_b}{2\varepsilon\varepsilon_0}.z_0^{\,2}. \tag{6.45}$$

Combining Eqs. (6.39) and (6.45) and using $n_S = n_b \exp\left(-\frac{qV_S}{k_B T}\right)$, it is found that:

$$\theta = \sqrt{\frac{2.\varepsilon.\varepsilon_0.n_b.k_B.T}{\alpha^2.[S_t]^2.q^2}.\ln\frac{n_b}{n_s}}, \tag{6.46}$$

which together with Eq. (6.38) can be employed for the calculations of n_S and θ as a function of partial pressures (p_{O_2}), temperature T in Kelvin, ionisation and chemical state of oxygen α, β, reaction constants k_{ads}, k_{des}, material constants ε, n_b, $[S_t]$ and fundamental constants, k_B and ε_0.

Case 2. Grains/crystallites smaller than or comparable to the Debye length ($d \leq \lambda_D$)

Many gas sensor thin films consist of cylindrical grains. As a result, a good approximation is to limit the calculations to the case where the conduction takes place in cylindrical objects with radius R. With this assumption, the Poisson equation in cylindrical coordinates for energy, E, can be obtained as:[32]

$$\left(\frac{1}{r}\frac{d}{dr} + \frac{d^2}{dr^2}\right)E(r) = \frac{q^2 n_b}{\varepsilon\varepsilon_0}, \tag{6.47}$$

with the boundary conditions:

$$\left.\frac{dE(r)}{dr}\right|_{r=0} = 0, \tag{6.48}$$

$$\left.E(r)\right|_{r=0} = E_0. \tag{6.49}$$

As a result the $\Delta E = E(R) - E_0$ can be obtained as:

$$\Delta E = \frac{q^2 n_b}{4\varepsilon\varepsilon_0} R^2. \tag{6.50}$$

Using the Debye length obtained in the Schottky approximation:

$$\lambda_D = \sqrt{\frac{\varepsilon\varepsilon_0 k_B T}{q^2 n_b}}, \tag{6.51}$$

it can be eventually obtained:[32]

$$\Delta E \sim k_B.T.\left(\frac{R}{2.\lambda_D}\right). \tag{6.52}$$

If ΔE is comparable with the thermal energy, a homogeneous electron concentration in the grain result which in turn produces a flat band of energy.

It can be shown that for SnO_2 using data found in literature for grain sizes lower than 50 nm, complete grain depletion and a flat band energy condition is obtained for practical temperatures in gas sensing (e.g. 700 K is not acceptable since the value of ΔE is larger than k_BT).

The electroneutrality condition now takes the form of (in flat band energy condition):

$$\alpha\theta[S_t]A + n_S V = n_b V, \tag{6.53}$$

where n_S is the concentration of electrons in the bulk of the metal oxide. Assuming that the cylinder length is L, the surface of A is:

$$A = 2\pi R(R + L), \tag{6.54}$$

and the volume V:

$$V = \pi R^2 L, \tag{6.55}$$

using Eqs. (6.53),(6.54) and (6.55), it can be rewritten:

$$\theta = \frac{n_b R}{2\alpha[S_t]\left(1+\dfrac{R}{L}\right)}.\left(1-\frac{n_S}{n_b}\right). \tag{6.56}$$

With the approximation of R/L much smaller than one (correct for high aspect ratio crystallites), it can be obtained that:

$$\theta = \frac{n_b R}{2\alpha[S_t]}.\left(1-\frac{n_S}{n_b}\right). \tag{6.57}$$

Again this equation together with Eq. (6.38) allows the determination of n_S and θ as functions of only partial pressures (p_{O_2}), temperature T in Kelvin, ionisation and chemical state of oxygen α and β, reaction constants k_{ads} and k_{des}, material constants n_b and $[S_t]$ and the fundamental constant k_B.

The aforementioned model of a thin film gas sensor, based on the *chemisorptional theory of Volkenshtein*, theory can be used to predict the response of sensors.[38] This model can be utilized to find the optimum operating conditions. The gas sensitivity, gas response peaks, and the rate of response and associated recovery processes can be calculated.

For example, **Fig. 6.21** shows gas sensitivity and temperature of maximum gas response as functions of adsorption/desorption parameters such as coefficients of O_2 and CO adsorption (α_O, α_{CO}), coefficients of O_2 and CO desorption (β_O, β_{CO}), total number of adsorption sites (N), and the activation energy of oxygen and CO adsorption ($E_{act}(O_2)$, $E_{act}(CO)$).[28]

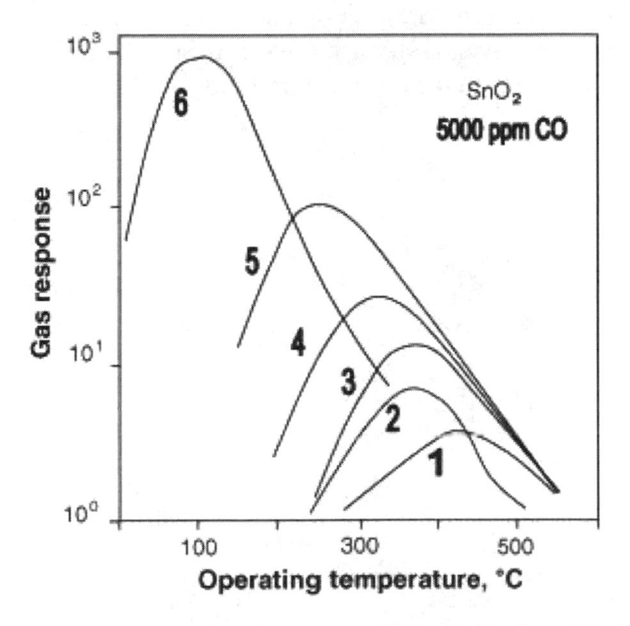

Fig. 6.21 Theoretically calculated temperature dependencies of SnO_2 gas response to CO at various CO adsorption parameters on the SnO_2 surface. $N_d = 5.10^{18}$ cm^{-3}; $d = 10$ nm; $\alpha_O = 0.05$ N exp($-0.5/kT$); $\beta_O = 7.10^{14}$ N exp($-2.1/kT$); $\alpha_{CO} = (2–4) \times 10^{14}$ exp($-E_{act}(CO)/kT$); 1–3: $E_{act}(CO) = 0.9$ eV; 4: 0.85 eV; 5: 0.8 eV; 6: 0.6eV. Reprinted with permission from the Elsevier publications.[28]

6.3.4 Effect of Gas Sensitive Structures and Thin Films

The adsorption and desorption processes, which determine the gas sensitivity, depend on the crystal orientation and morphology. The shape and size of nano-crystals intensely influences the type of adsorbate bonding to the surface. The reduction of dimensions to the nm range generally augments the crystallite shape influence on the adsorption properties.[28]

Different types of bonding can be generated on the edge/corner sites or on the flat planes of crystallites in the thin film which act as the sensitive layer. These sites have significant influences on gas sensing performance and film functionality depends on them. For instance during the deposition process, SnO_2 nano-crystallites may adopt different crystallographic planes (**Fig. 6.22**). In addition to the most stable SnO_2 plane [1 1 0], the external facets of SnO_2 nanocrystals can be formed as crystallographic planes with crystallographic planes such as (1 1 1), (2 0 0), (1 0 1), (0 1 1), (−1 −1 2), and (2 1 0). The combination of these planes depends on the grain size. As can be seen **Fig. 6.22**, even a slight increase in SnO_2 grain size can produce a significant change in the crystal morphology.[28]

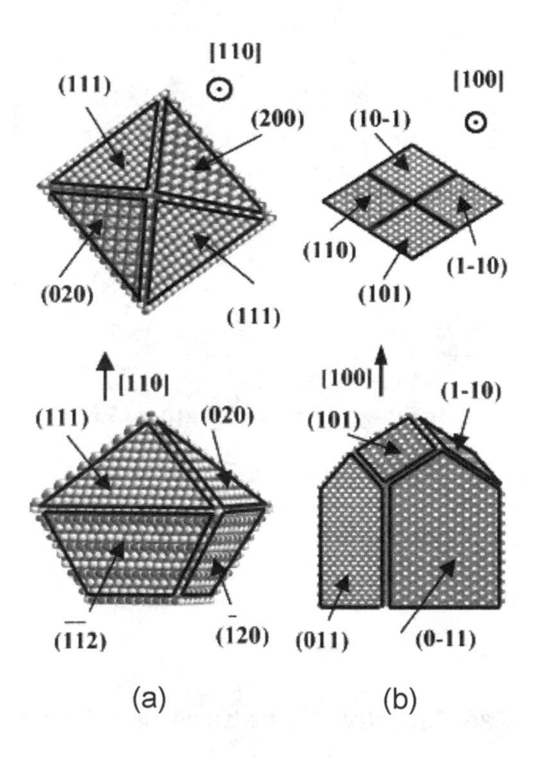

Fig. 6.22 Crystallographic models of SnO_2 nanocrystals deposited by spray pyrolysis at a temperature range from 400 to 500°C: (a) *d*~40–80 nm; (b) *d*~300 nm. Reprinted with permission from the Elsevier publications.[39]

In addition to grain size and shape, film thickness and agglomeration also play important roles in gas sensor response.[40,41] It is suggested that the gas sensing matrix of polycrystalline films can be schematically presented using an equivalent circuit shown in **Fig. 6.23**.[28,42]

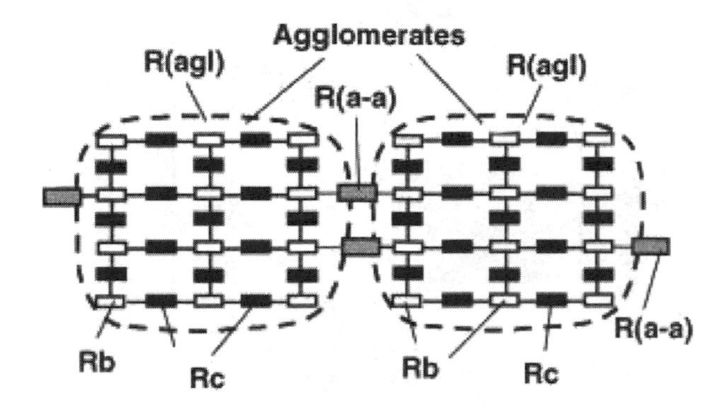

Fig. 6.23 Principle electrical circuit of agglomerated polycrystalline film. Reprinted with permission from the Elsevier publications.[42]

This model consists of four gas sensitive elements: $R_{(a\text{–}a)}$ the representation of the resistance of inter-agglomerate contacts (an integral resistance, representing a three dimensional grain network), R_c the resistance of inter-crystallite (inter-grain) contacts, R_b the bulk resistance of crystallites (grains), and R_{agl} the resistance of the agglomerate.

The small size of crystallites is an essential but not sufficient condition for obtaining both maximum gas sensitivity, and fast response.[43] Manipulating the porosity of both agglomerates and the gas sensing matrix can also affect the sensor. Low gas permeability of agglomerates increases the influence of inter-agglomerate contacts. The agglomeration of small crystallites into large masses is one of the key phenomena, varying the response characteristics.[28] Generally, the gas sensing matrix should be highly porous, at the same time the dimensions of pores should be less than a few nm to allow the tunneling of electrons. However, for strongly agglomerated structures, the small grain size is not advantageous. Agglomerates, which are made of smaller grains, are more densely packed, which may lead to a smaller gas penetrability.[28]

It is necessary to highlight the effect of ambient interferences in gas measurements such as carbon and its compounds, humidity, and other airborne chemisorbed species. For example, high concentrations of carbon can block surface sites of adsorption on a metal oxide. Fortunately, the effect of carbon and many other contaminants on the sensitive layer can be removed with targeted treatments (**Fig. 6.24**).[28]

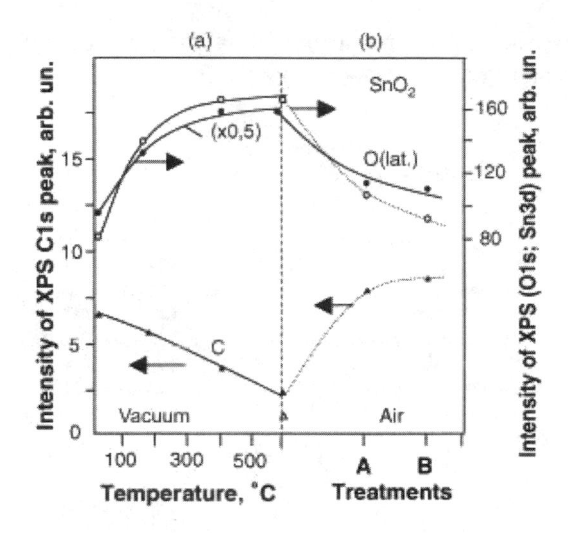

Fig. 6.24. Influence of various thermal treatments on the intensities of main peaks of SnO_2 X-ray photoelectron spectra ($\varphi = 90°$): (a) treatments in vacuum; (b) treatments in air, $T_{an} = 300°C$, $t_{an} = 10$ min. (A) Dry air; (B) humid air. Reprinted with permission from the Elsevier publications.[28]

6.3.5 Effects of Deposition Parameters and Substrates

The structure of metal oxide thin films largely depends on the deposition technique used. The thin film can be mostly epitaxial such as CVD deposited films or composed of nanocrystals which are deposited by methods such as thermal evaporation, sputtering and PLD. In addition, parameters such as substrate temperature, precursor concentration, gas flow, oxygen concentration in the deposition chamber and rate of deposition also influence the structure of the layer. During the deposition of a non-epitaxial thin film, generally an initial layer (*transition layer*) is formed from which the crystallites (*textured layered*) are seeded and grown (**Fig. 6.25**). Generally the transition layer is less ordered than the textured layer.

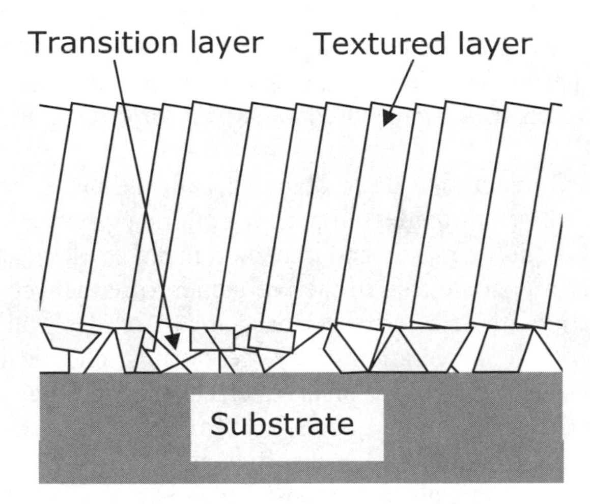

Fig. 6.25 Simplified cross section of a metal oxide film containing a transition layer.

A large number of substrates can be employed for the deposition of gas sensitive metal oxide films. These include: various dielectric polycrystalline Al_2O_3-based substrates; Al_2O_3 single crystals with specific crystallographic orientation; SiO_2 deposited on Si substrate; MgO substrate; SiC substrate; glass ceramics; etc.

The structure of nanocrystals can greatly depend on the substrate's material and lattice structure. An example of ZnO thin films deposited on four different substrates and consequently taking different forms is shown in Chap. 2 in the sputtering deposition method section.

The substrate can also have a significant influence on the mechanical structure of the deposited thin films. Metal oxide thin films can be mechanically strained by the substrate.[39] By selecting substrates with a favourable lattice parameters and thermal expansion coefficients, it is also possible to minimize the mechanical strain.[28]

6.3.6 Metal Oxides Modification by Additives

Mixing metal oxides with metals that function as catalysts, the use of binary compounds and complex multi-component materials, as well as doping, are among the most common methods utilized to enhance or tailor the gas sensing performance of metal oxide thin films.[28,44-52] These changes can be introduced to the base oxide matrix during the deposition process or incorporated post deposition by annealing at elevated temperatures. These additives can be used for modifying the catalytic activity of the base oxide, stabilizing a particular valence state, favoring formation of active phases, stabilizing the catalyst against reduction, or improving the electron exchange rate.

Well-defined materials can be obtained with the incorporation of additional phases (different oxides) in nanocrystalline systems.[53] Also catalytic active additives (noble metals and transition metal oxides), and inert impurities can play a positive role in the production of enhanced sensitive layers. Introduction of additives into base metal oxides can change such parameters as the concentration of charge carriers, chemical and physical properties of the metal oxide matrix, electronic and physical–chemical properties of the surface (energetic spectra of surface states, energy of adsorption and desorption, adhesion coefficients, etc.), surface potential and inter-crystallite barriers, phase composition and crystallite size.[31]

It is known that the incorporation of a second phase, even in small quantities, can change the conditions of base oxide growth. For example, doping of SnO_2 with Nb (0.1–4 mol%) causes a decrease in crystallite size from 220 nm for pure SnO_2 to about 30 nm for Nb (0.1 mol%) doped samples (calcination temperature approximately 900°C).[54,55] Phase modification opens up exciting possibilities for the manipulation of the chemical and physical structures as well as the gas sensing properties of metal oxides. Experimental realities, pertaining to influence of individual additives on the gas sensing properties of various metal oxides has been reviewed elsewhere.[28,56]

The additional influence observed on gas sensing characteristics can be a consequence of grain size change, and the change in concentration of free charge carriers.[57] For example, SnO_2 doping by Nb and Sb in the range of 0.01 and 1.0 mol% during sol–gel preparation which is annealed at $T = 900$°C leads to a film resistance decrease of 10^2 and 10^4 times, respectively, while doping with In resulted in a rise in film resistance, approximately by a factor of 10^2.[54,55]

The effect of doping on gas sensing properties of metal oxide gas sensors can be different from the catalytic activity of these additives.[28] For example, this can be observed using transition metals (Cr, Mn, Fe, Ni, Co, Cu, Zn, and Ga), which are widely used in catalysis.[28] The improvement of gas response of In_2O_3-based sensors, containing a second oxide at a level 2–10 wt.%, was observed for elements with minimal catalytic activity.

6.3.7 Surface Modification

The change in the concentration of conduction band electrons in metal oxides is a direct result of surface chemical reactions, including: chemisorption, reduction/oxidation and/or catalysis. The results of experimental and theoretical study demonstrate that different surface sites participate in electrical response, interaction with water vapors, and adsorbed species.[28] Different test gases interact differently at various surface sites as well. For example, SnO_2 response to methane can be due to four different types of reactive surface sites found on polycrystalline SnO_2.[58]

Nanoscale particles of noble metals (Pd, Rh, Pt, Au, Ta), and oxides of other elements (such as Co, Fe, Zn, Se, Cu), deposited on the surface of metal oxides can act as surface sites for adsorbates and promoters for surface catalysis.[28] They could produce additional adsorption sites and surface electronic states, which mediate in electronic transfer processes.[59] As a result, parameters such as gas sensitivity, rate of response, and selectivity can be significantly altered.[60,61] High catalytic activity of additives is an essential but insufficient requirement. For achieving high gas response, the noble metal should create optimal conditions for both electron and ion exchange between surface nanoclusters and metal oxide support (**Fig. 6.26**).[62]

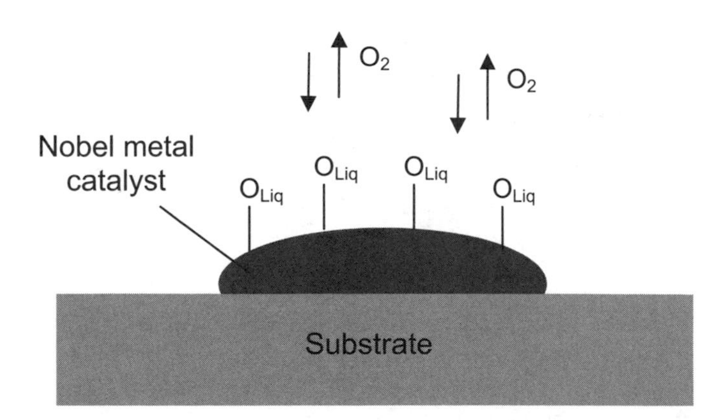

Fig. 6.26 Illustration of Pd influence on adsorbed species on the SnO_2 surface.

The effect of Au catalyst on the sputtered TiO_2 gas sensitive layer can be seen in the following example. The SEM image of such a thin film is shown in **Fig. 6.27**.

Fig. 6.27 SEM image of annealed TiO_2 film.

The sputtered TiO_2 film has a thickness of approximately 100 nm. The catalyst is a thin layer of Au which was sputtered for 10 s in a low power DC sputterer. **Fig. 6.28** shows the response and recovery times of both pure and Au-coated TiO_2 sensors to 510 ppb of NO_2 at a temperature range between 230°C and 275°C. **Fig. 6.29** shows the Au-coated TiO_2 sensor exhibited the largest response to 510 ppb of NO_2 at 230°C. The Au-doped TiO_2 sensor response was approximately 2 times larger than that of the pure TiO_2 sensor, with the ratio varying with temperature. At higher operating temperatures, the sensitivity of both pure TiO_2 and Au-coated TiO_2 sensors are reduced.

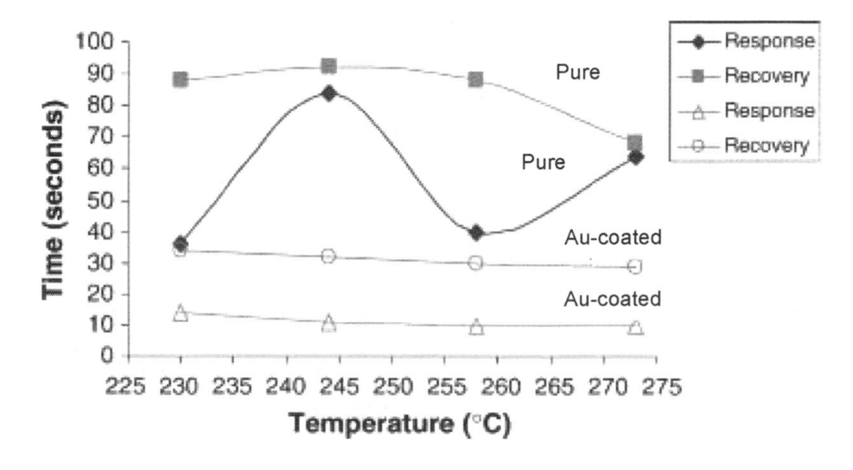

Fig. 6.28 Response and recovery times for pure and Au-coated TiO_2 sensors to 510 ppb of NO_2 at a temperature range between 230°C and 275°C.

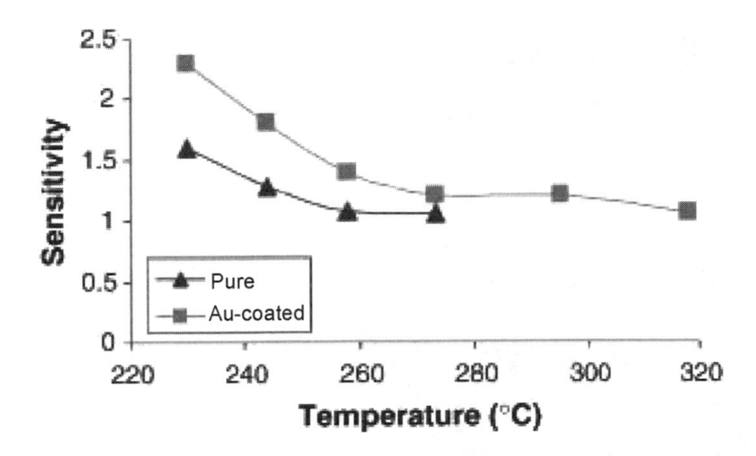

Fig. 6.29 Sensitivity of pure TiO_2 and Au-coated TiO_2 sensor to 510 ppb of NO_2 at a temperature range between 220°C and 320°C.

The nature of the noble metals, their oxidation state, and their distribution both over the oxide surface and into the film, are determining factors in gas sensor sensitivity and selectivity.[28]

While physical methods of noble metal deposition are the most common approach, it is very difficult to obtain homogeneous distribution. Thin films generally need to be annealed following deposition of a noble metal. Annealing promotes the formation of metallic clusters, improves homogeneity of their distribution by layer thickness, and stabilizes the properties of the gas sensing matrix.[28]

The distribution of catalysts both throughout the thin film and in terms of catalyst particle size, have an important influence on sensor response. For example, McAleer et al[63] found that optimal responses occur when the cluster size of the catalysts exceeds 1–5 nm on a sensing surface. An optimal distance between clusters can be estimated as equal to an oxygen surface diffusion length at operating temperature.[61]

Surface morphology has a significant effect on the shape and distribution of catalysts. Noble metal clusters generally accumulate at step edges and kinks of metal oxides during their deposition.[64] Accumulation at the corners strongly affects the size of clusters (**Fig. 6.30**). The size of clusters also strongly depends on the degree of the wettability of the surface.

Fig. 6.30 Underlying surface morphology significantly influences the shape and distribution of the catalyst clusters.

6.3.8 Filtering

Most filters, which are used in commercial gas sensors, are passive membranes with differing diffusion parameters and pore–molecule dimensions.

For example, a simple filter that resists permeation of unwanted molecules can be fabricated with several protective coating such as SiO_2, Al_2O_3, zeolites, ZrO_2, Pd, Pt, or with various polymeric membranes. This type of adsorbent filter may become saturated for large interfering gas concentrations.[28] Catalytically active layers, such as Pt and Pd on the top of a passive membrane (Al_2O_3 or SiO_2), utilize catalytic activity for the conversion of certain gases. Such technology promotes even higher selectivity of gas response. It is also possible to burn reactive gases in the catalytic layer while the less reactive ones pass through.[59]

Catalysts, based on noble metals, can be poisoned by many organic and inorganic chemicals that contain sulphur (H_2S, SO_2, thiols, etc.) and phosphorous.[65]

6.3.9 Post Deposition Treatments

Following deposition, thin films can undergo annealing steps and various radiation treatments. Generally during the annealing process the smallest grains are joined. As a result, thermal treatments can be used for stabilization of thin film properties.[28] The stages that may occur in the annealing process are as follows: structural stability of the film (small grain joining); the coalescence of grains forming agglomerates; local structural reconstruction; and global (comprehensive) structural reconstruction (**Fig. 6.31** and **Fig. 6.32**).[66]

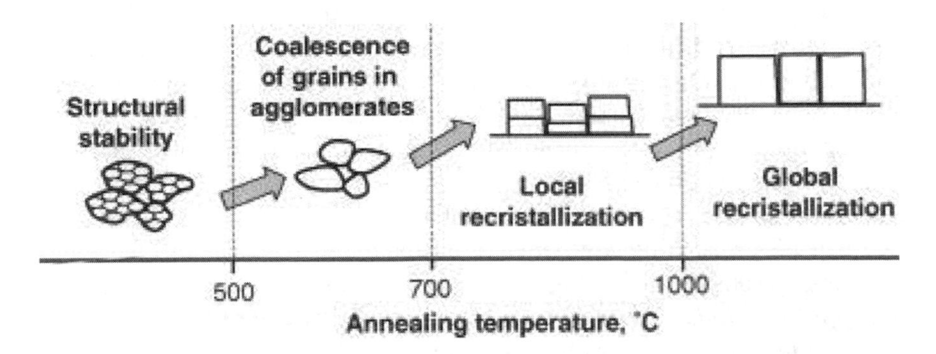

Fig. 6.31 Stages of structure transformation of In_2O_3 films during thermal treatment. Reprinted with permission from the Elsevier publications.[66]

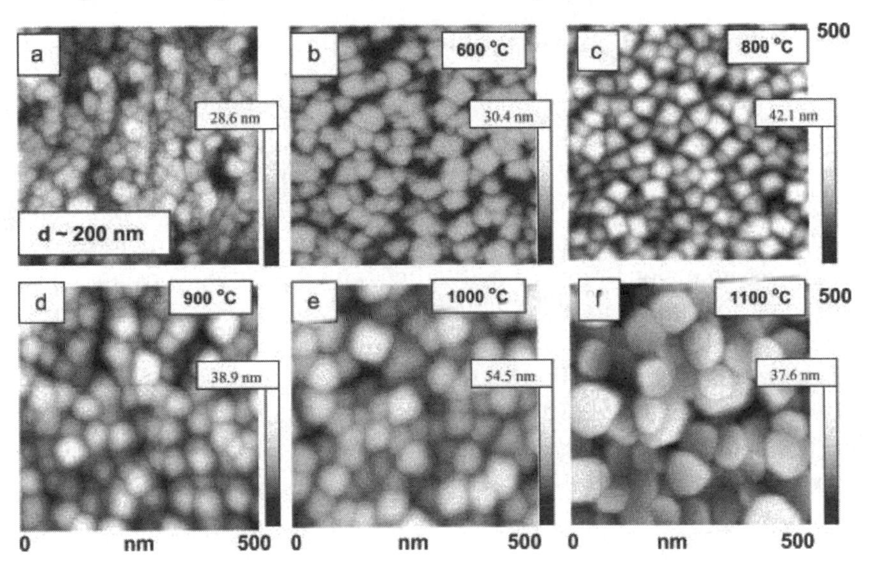

Fig. 6.32 Influence of annealing on AFM images of In_2O_3 films: (a) 500°C, (b) 600°C, (c) 800°C, (d) 900°C, (e) 1000°C and (f) 1100°C. Reprinted with permission from the Elsevier publications.[66]

6.4 Phonons in Low Dimensional Structures

A *phonon* is a quantized mode of vibration occurring in a rigid crystal lattice. Phonons can scatter electrons and interact with photons. They play an important role in the electrical and thermal conductivity of micro/nano devices. Similar to electrons, phonons can be confined within the low dimensions of nanostructures. The dynamics of phonons confined in quantum dimensions differ from their non-confined counterparts. Phonons are responsible for transferring mechanical energy. Mechanical sensors, thermoelectric devices and many optical measurement systems are based on the function of phonons. Phonon vibrations affect atomic structures and electrons. Interactions of phonons with photons, atomic structures and electrons can be used in sensing applications and also they can be effectively used as sensing elements. For instance, a direct application of phonons in sensing technologies appears in *acoustic phonon pulse spectroscopy* and *Raman spectroscopy*. However, the effects of phonons can be unfavourable for in dimensional structures as they can broaden the spectrum of electromagnetic measurements and add electronic noise.

6.4.1 Phonons in One-Dimensional Structures

The simplest one-dimensional structure is made of similar atoms (*monoatomic*) connected in series as shown in **Fig. 6.33**. Despite the simplicity, such a structure is a good example to understand the behavior of phonons, propagation of waves and dispersion relationships. This understanding can then be expanded to visualize more complex structures.

Let's assume a one-dimensional lattice with the lattice constant equal to a. When a force is exerted on the lattice the whole system moves with the same frequency. Now, let's consider that the system consists of N atoms and the force applied to the nth atom is only due to neighboring atoms. If *harmonic approximation* (it assumes that force is proportional to relative displacement), Hooke's law and the inter-atomic force constant α are used we will obtain:[2]

$$M\frac{d^2u_n}{dt^2} = -\alpha(2u_n - u_{n+1} - u_{n-1}), \tag{6.58}$$

where u_n is the displacement of the nth particle, M is the mass of the atom and t is time. To solve the Eq. (6.64), n coupled equations for the N atoms in the system would have to be solved simultaneously. With a solution in the form of $Ae^{i(kx_n - \omega t)}$ (x_n the location of the nth atom) the result can be written as:[2]

$$\omega = \left(\frac{4\alpha}{M}\right)^{1/2} \left|\sin(ka/2)\right|. \tag{6.59}$$

This is the dispersion relation for a one-dimensional lattice as shown in **Fig. 6.33**.

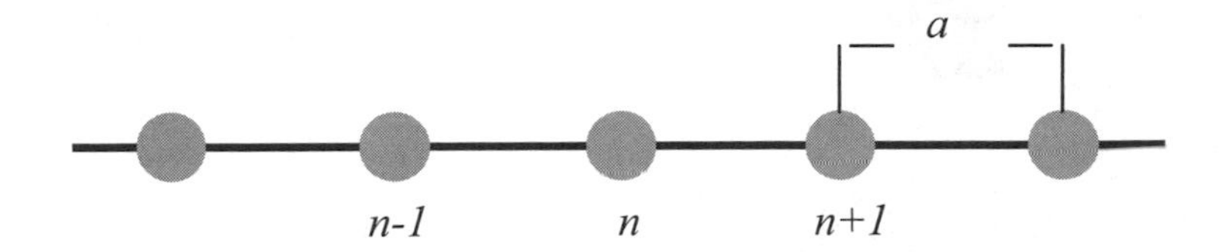

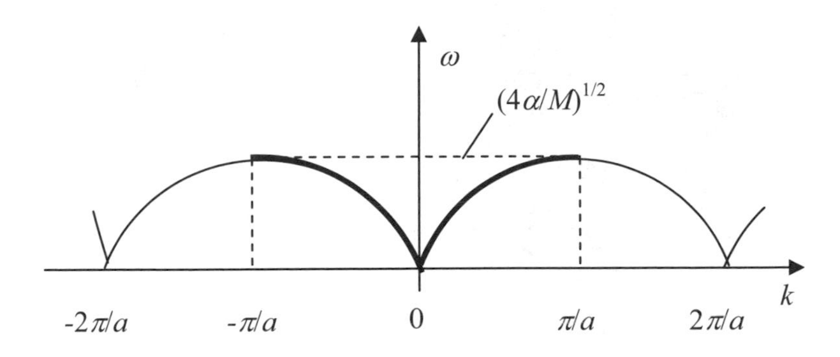

Fig. 6.33 Segment of a one-dimensional lattice (top) and the dispersion curve for a monoatomic one-dimensional lattice (bottom). The curve is periodic.

The phase velocity ($v_p = \omega/k$) and group velocity ($v_p = d\omega/dk$) can be obtained from the dispersion-relation Eq. (6.59). As the wavelength decreases and k increases the quantized property of the lattice becomes more prevalent. Atoms begin to scatter waves. This scattering of waves results in a decrease in the velocity of the waves.

This dispersion equation is periodic. If the frequencies are covered in the range of $0 < \omega < (4\alpha/M)^{1/2}$, they are transmitted while the others are highly attenuated.

The dispersion curve is periodic in k space. By solving the equation for a harmonic function, it can be demonstrated that the number of permissible k points is equal to the number of unit cells in the lattice.

The discussion above concerned monoatomic lattices with similar atoms. The monoatomic lattice only displays acoustic dispersion. *Acoustic phonons*, described above, have frequencies that become small at long wavelengths, and correspond to longitudinal and transverse mechanical waves in the lattice.

The dispersion equation of a *diatomic lattice* has two roots (**Fig. 6.34**). It has an acoustic branch and an optical branch. The frequency gap between the acoustic branch and the optical branch is forbidden as the lattice cannot transmit them. Therefore, a diatomic lattice can operate as a *band-pass mechanical filter*.

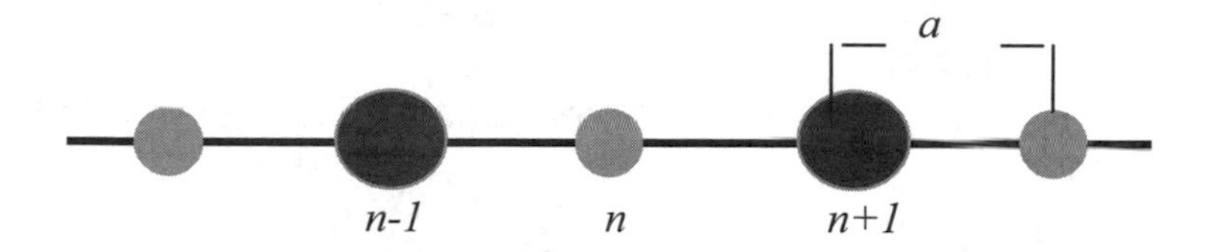

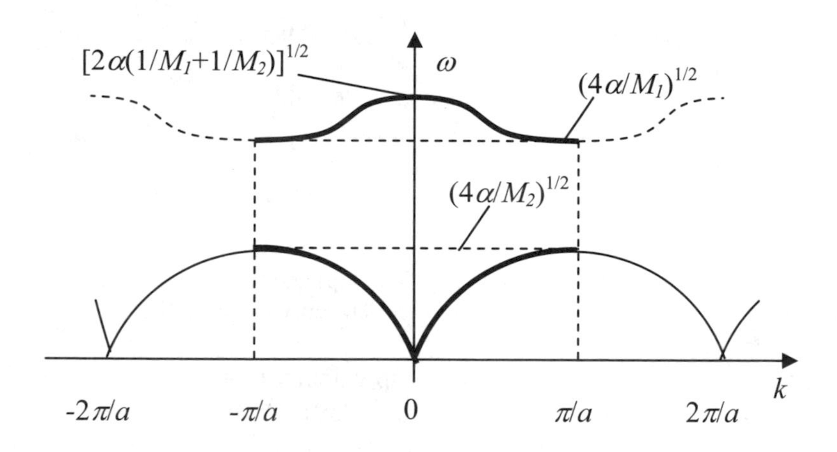

Fig. 6.34 Segment of a one-dimensional diatomic lattice (top) and the dispersion curve for a diatomic one-dimensional lattice (bottom). The curve is periodic. Mass of one atom is different from the other ($M_1 < M_2$).

This upper branch in the dispersion curve is classified optical as the frequencies of this branch can be approximated by $(2\alpha/M)^{1/2}$, which is approximately 3×10^{13} s^{-1} which falls in infrared range.

Optical phonons occur in crystals with more than one type of atom in the unit cell and are excited by infrared radiation. Optical phonons that interact in this way with light are called infrared active. Optical phonons which are Raman active can also interact indirectly with light, through Raman scattering.

The mean free path of a phonon (before being scattered) is determined by several interaction classes. The collisions of phonons with other phonons results in scattering due to *anharmonic interactions* between them. This kind of *phonon-phonon interaction* is important at high temperatures where atomic displacements are large. The collision of phonons with imperfections (*phonon-defect interaction*) of a crystal destroys the perfect periodicity of the structure which is the very basis of propagating lattice waves. Phonons may also collide with external boundaries. At low temperatures phonon-phonon and phonon-defect collisions become less effective. In the former case, the number of phonons at such temperatures is low and in the latter case, the wavelengths at low temperatures are so long that the effect of impurities cannot be sensed. As a result, the main reason for the scattering mechanism is the geometrical and boundary effects of a crystal as the wavelengths are large and comparable with the effects of surfaces and boundaries.

Raman spectroscopy can be used for determining the dimensions and structure of nanomaterials. In pioneering work by Rao et al[67], Raman spectroscopy was used to determine the diameters and structural formation of *single wall carbon nanotubes* (*SWNTs*) with laser excitation wavelengths in the range from 510 to 1320 nm. Numerous Raman peaks were observed and corresponding with vibrational modes of armchair symmetries (n,n) SWNTs (**Fig. 6.35**) In a SWNT the wrapping of sheets is represented by (n,m) which is called the chiral vector. The integers n and m denote the number of unit vectors along two directions in the honeycomb lattice. Theoretical calculations have shown that carbon nanotubes may be metallic or semiconducting, depending on the nanotube symmetry and diameter. The electronic and magnetic properties of metallic tubules have been shown theoretically to be diameter-dependent.

Fig. 6.35 shows the Raman spectrum (top) of SWNTs. The spectra were collected in a backscattering measurement conducted at room temperature. The spectrum in **Fig. 6.35** was collected using 514.5 nm excitation and a spectral slitwidth of approximately 2 cm^{-1}; the spectrum contains light scattered with the electric vector both parallel and perpendicular to the horizontal plane of incidence. The bottom of **Fig. 6.35** shows the results of a theoretical calculation for the frequency and scattering intensity of the seven strongest Raman-active normal modes of armchair (n,n) carbon nanotubes, with $8<n<11$. Raman-allowed $(q=0)$ phonons were calculated by using C-C force constants optimized to fit the experimental phonon dispersion for a flat graphene sheet. The downward-pointing arrows in **Fig. 6.35** indicate the calculated frequencies of the much weaker Raman modes. The calculated frequencies are in good agreement with those obtained previously by a *zone folding* (*ZF*) *model* for a flat graphene sheet in which the same force constants were used.[68] The predicted diameter dependence of the SWNT mode frequencies can be easily observed **Fig. 6.35** particularly in the low-frequency region ($\omega \leq 500$ cm^{-1}).

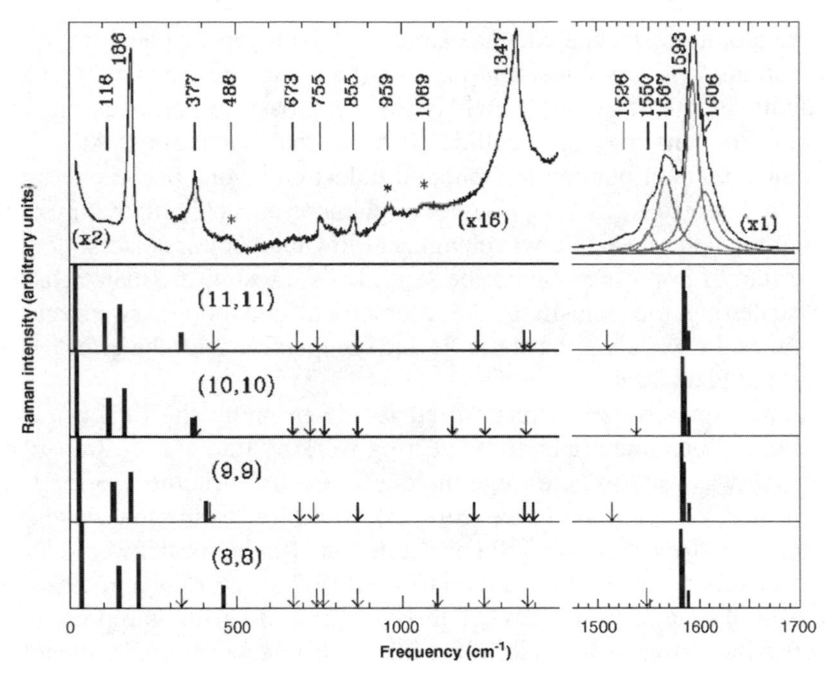

Fig. 6.35 Raman spectrum (top) of SWNT samples taken with 514.5 nm excitation at approximately 2 W/cm^2. The * in the spectrum indicates features that are tentatively assigned to second-order Raman scattering. The four bottom panels are the calculated Raman spectra for armchair (n,n) nanotubes, $n = 8$ to 11. The downward pointing arrows in the lower panels indicate the positions of the remaining weak, Raman-active modes. Reprinted with permission from the Science Magazine publications.[67]

The phonon behavior in two and three dimensional materials can be described similarly. However, the degree of complexity is much greater.

6.4.2 Electron-Phonon Interactions in Low Dimensional Materials

Understanding the underlying physics of phonon-electron interactions is important, as it describes the mechanism by which many electronic and thermal devices operate. Let's assume that $\psi_a(\mathbf{r})$ is the energy eigenstate of an electron confined within an ideal heterostructure potential and that the vibrational degrees of freedom of the heterostructure in the quantum state is defined by N_α where the number of phonons in each state is α.

In the case of a quantum well (2D structure) with the thickness of d and infinite well-plane dimensions, the *electron energy* eigensates and eigenvalues are:

$$\psi_k(\mathbf{k}) = Ce^{i(k_x x + k_y y)} \sin(\pi z / d), \tag{6.60}$$

and:

$$E_k = \frac{\hbar^2}{2m^*}\left(k_x^{\ 2} + k_y^{\ 2}\right), \tag{6.61}$$

where C is the normalization constant and m^* is the effective electron mass.

For a quantum wire of d and w dimensions (1D structure) they are:

$$\psi_k(\mathbf{k}) = Ce^{ik_x x} \sin(\pi y / w)\sin(\pi z / d), \tag{6.62}$$

and:

$$E_k = \frac{\hbar^2 k_x^{\ 2}}{2m^*}. \tag{6.63}$$

If the potential experienced by the electron in the $\psi_a(\mathbf{r})$ eigenstate in time t with the frequency of ω_α due to mode α is:[69]

$$V_\alpha(\mathbf{r},t) = V_{+\alpha}(\mathbf{r})e^{i\omega_\alpha t} + V_{-\alpha}(\mathbf{r})e^{-i\omega_\alpha t}, \tag{6.64}$$

the time dependant perturbation theory can be used for the calculation of *the scattering probability per unit time*: [69]

$$\frac{1}{\tau_a} = \frac{2\pi}{\hbar}\sum_b\sum_a\left[\left|\left\langle\psi_b\right|\hat{V}_{+\alpha}\left|\psi_a\right\rangle\right|^2 (N_\alpha + 1)\delta(E_b - E_a + \hbar\omega_\alpha) + \left|\left\langle\psi_b\right|\hat{V}_{-\alpha}\left|\psi_a\right\rangle\right|^2 N_\alpha \delta(E_b - E_a - \hbar\omega_\alpha)\right]. \tag{6.65}$$

where $\psi_b(\mathbf{r})$ is the eigenfunction after perturbation and δ is the *Kronecker delta* function. The two terms on the right hand side of Eq. (6.65) are responsible for phonon emission and phonon absorption, respectively.

Acoustic phonons generate a perturbation which changes the properties of the material at low temperatures. Acoustic phonons; however, can generate a perturbing potential in two different ways: (a) a small change in the relative position of atoms perturbs the electrostatic potential experienced by each electron and the perturbation changes the electron energies (this perturbation is known as *deformation*) or (b) it can also produce changes in relative positions of oppositely charged ions which produces electric polarization and therefore long-range electric fields which affect the electron energies (this perturbation is known as *piezoelectricity*) . The deformation and piezoelectric potentials take the following forms:

$$V_{+q}^{DP}(\mathbf{r}) \approx C^{DP} k^{1/2} e^{-i\mathbf{q}.\mathbf{r}} . \tag{6.66}$$

$$V_{+q}^{PE}(\mathbf{r}) \approx C^{PE} k^{-1/2} e^{-i\mathbf{q}.\mathbf{r}} . \tag{6.67}$$

where C^{DP} and C^{DP} are the coupling strength constants for deformation and piezoelectric potentials, respectively, and q is the phonon wavenumber.

For a quantum well with the thickness of d the matrix element is:[69]

$$\left|\left\langle \psi_{k'} \middle| \hat{V}_{\pm}(q) \middle| \psi_{k} \right\rangle\right|^2 \approx \frac{\left(\left|C^{DP}\right|^2 q + \left|C^{PE}\right|^2 q^{-1}\right)}{q_y^2\left[(2\pi/w)^2 - q_y^2\right]^2} \times \sin^2(q_z d/2)\delta_{k_x,k'_x \pm q_x}\delta_{k_y,k'_y \pm q_y}, \tag{6.68}$$

where q is the phonon wave vector, k and k' are the electron wavenumbers before and after scattering, respectively. For a quantum wire element the matrix element is:[69]

$$\left|\left\langle \psi_{k'} \middle| \hat{V}_{\pm}(q) \middle| \psi_{k} \right\rangle\right|^2 \approx \frac{\left(\left|C^{DP}\right|^2 q + \left|C^{PE}\right|^2 q^{-1}\right)}{q_y^2\left[(2\pi/w)^2 - q_y^2\right]^2 q_z^2\left[(2\pi/d)^2 - q_z^2\right]^2} \times \sin^2(q_y w/2)\sin^2(q_z d/2)\delta_{k,k' \pm q_x}, \tag{6.69}$$

where d and w are the dimensions of the base of the wire. From Eqs. (6.68) and (6.69) it can be seen that momentum components parallel to the quantum well plane and wire length are conserved.

The research on the behavior of phonons in 1D structures is becoming increasingly popular in recent years.[70,71] In such structures the effects of phonons have been studied in thermal transport, Raman scattering, and electrical transport.[71] For example, it has been shown that the electron-acoustic phonon scattering in metallic single-walled carbon nanotubes (SWCNTs) contributes to the resistance at room temperature. It is also shown that at low bias voltages, the scattering is weak, resulting in long mean-free paths at room temperature. In this case, both measurements and calculations confirm that the mean-free path is in the range of a few hundred nanometers to several micrometers. At high bias voltages, electrons gain enough energy to emit optical phonons. Yao et al showed that this scattering leads to a saturation of the current at ~20 μA for such high bias voltages.[72] Park et al investigated the scaling of the resistance in such SWCNTs for lengths (L) ranging from 50 nm to 10 μm using the tip of an atomic force microscope (AFM) as a movable electrode.[70] For low bias voltages, they find a length-independent resistance for $L < 200$ nm, indicating ballistic transport of electrons in the SWCNTs. At longer lengths, the scaling of the resistance with length corresponded to a low-bias mean-free path equal to 1.6 μm.

The above mentioned calculations can also be extended to 0D structures such as quantum dots (Q-dots) . In Q-dots, it appears that the electronic performance is limited by decoherence effects arising from acoustic–phonon interactions. Stroscio et al theoretically calculated a variety of mechanisms that can shift or broaden the frequencies of acoustic phonons in quantum dots.[73]

6.4.3 Phonons in Sensing Applications

Phonons with wavelengths in the order of several tens of nanometers can travel relatively large distances in semiconductors without being severely scattered. This phenomenon can be used in *phonon pulse spectroscopy*.[74] By detecting the flux of acoustic phonons emitted from free electrons heated above the lattice temperature or the interaction of free electrons with acoustic phonons, information about free electrons and their distributions can be obtained. For instance, the two-dimensional distribution of electrons that is generated by the Hall effect can be investigated.[75] Also this method can be utilized for measuring geometrical properties of semiconductors such as surface roughnesses.[76]

In *Raman spectroscopy* a photon which absorbs or emits a phonon experiences a change in frequency. The frequency shift is called a Raman shift, as was explained in Chap. 5. Raman shift as a function of the intensity of the frequencies of the allowed modes can be measured. The phonon frequencies are determined by the inter-atomic forces. As was described previously, Raman spectroscopy can be widely used in the study of the surfaces. In addition, it can also be utilized for the study of heterostructures.[77] Raman measurement systems are gradually becoming more available and smaller in size. For instance, DeltaNu Ltd. recently presented a *hand-held Raman sensor* with a 120 mW - 785 nm laser, which provides the user with the resolution of 8 cm^{-1} and spectral range of 100 – 2000cm^{-1}.

Phonons play a crucial role in the performance of thermoelectric materials. For the fabrication of efficient Peltier coolers, it is desired to have substances which are highly electrically conductive and thermally isolative. In such materials, phonons have to be scattered but electrons should be able to move without any impediments. The use of nanosized heterostrucures or arrays of nanowires (Chap. 4) makes such structures possible.

6.4.4 One-Dimensional Piezoelectric Sensors

The piezoelectric effect was explained in Chap. 2 and several widely used piezoelectric transducers were introduced in Chap. 3. As we already learnt, piezoelectricity is due to the displacement of ionic charges within a crystal and is generally very small for most crystals. A field of about 1 kV/cm can only produce a strain of 10^{-7} in quartz.

As seen in Chap. 3, piezoelectric transducers can be used as sensing platforms. They are ideal transducers for the detection of mass in nano and pico gram ranges. Reducing the dimensions of the resonator based piezoelectric transducers generally improves their operational frequency which increases the mass sensitivity.

Piezoelectric Resonators

The Young's modulus and the piezoelectricity of nanobundles can be very different from their bulk counterparts. Based on electric field-induced resonant excitation, the mechanical properties of individual nanowire-like structures can be measured in situ by TEM.[78] The observations on ZnO cantilevers are shown in **Fig. 6.36**.[8] It shows the harmonic resonance, with the vibration planes, respectively, near perpendicular and parallel to the viewing direction. In calculating the bending modulus, the fundamental resonance frequency (v_1) and the dimensional sizes of the investigated ZnO nanobelts were measured. To determine v_1, one end of the nanobelt was tightly fixed and the resonant excitation was measured at around half the value of the resonance frequency. The nanobelt was aligned perpendicular to the electron beam so the real length of the nanobelt was measured. The projection direction along the beam was determined by electron diffraction; so that the true thickness and width were determined (the normal direction of the nanobelt is [2110]). Based on experimental data, the Young's modulus of ZnO nanobelts were calculated to be approximately 50 GPa. Experiments show that nanobelts can be effective nanoresonators exhibiting two orthogonal resonance modes, which can be used as probes for scanning probe microscopy operated in tapping and scanning modes or as mass sensor cantilevers.[8]

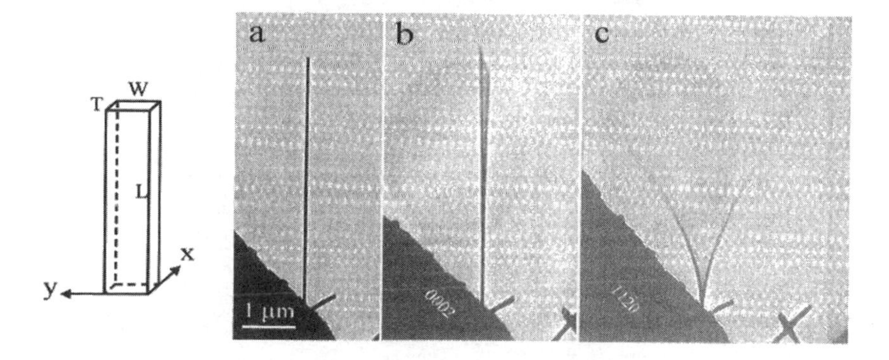

Fig. 6.36 Nanoresonators made of nanobelts. Measuring the Young's modulus of a ZnO nanobelt by electric field-induced mechanical resonance in TEM. (a) Geometrical shape of a nanobelt. (b,c) Mechanical resonance of a nanobelt along the two orthogonal directions closely perpendicular to the viewing direction ($v_x = 622$ kHz) and nearly parallel to the viewing direction ($v_y = 691$ kHz), respectively. Reprinted with permission from the Annual Reviews publications.[8]

Piezoelectric Filed Effect Transistors

By connecting a ZnO nanowire across two electrodes that are used for applying a bending force to the nanowire a *piezoelectric field effect transistor* (*piezo FET*) can be fabricated.[79] This piezo FET can be considered as a new type of transistor that operates by applying a mechanical force. It can act as a force sensor capable of forces in the nano Newton ranges or less. It can obviously be used as a mass sensor that can operate in gas media as well. It is a non-resonating device with excellent performance for biosensing applications which have to be conducted in liquid media as acoustic dampening does not occur during measurements.

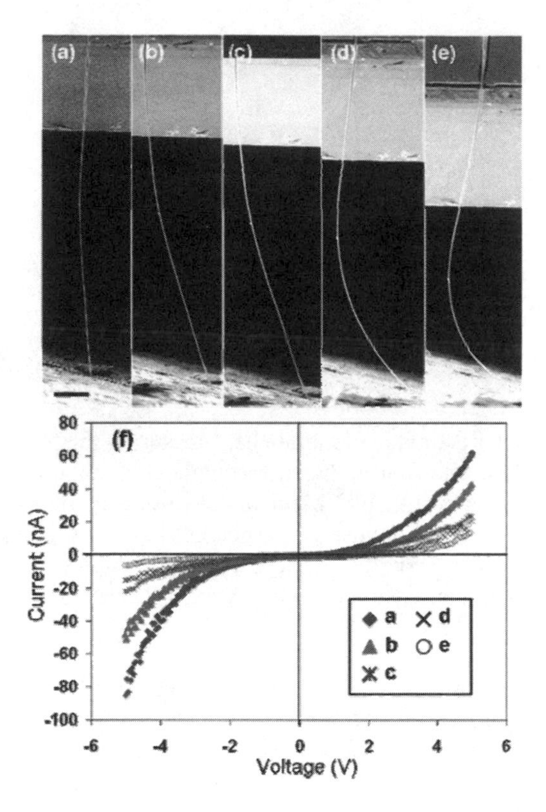

Fig. 6.37 (a-e) SEM images with the same magnification showing the five typical bending cases of the ZnO nanowire; the scale bar represents 10 μm. (f) Corresponding I-V characteristics of the ZnO nanowire for the five different bending cases. This is the I-V curve of the piezo FET. Reprinted with permission from the American Chemical Society publications.[79]

6.5 Nanotechnology Enabled Mechanical Sensors

Micro and Nanoelectromechanical systems (*MEMS* and *NEMS*) hold great promise for a number of scientific and technological sensing applications. Typically, when dimensions of 100 nm or less are used in these systems they are referred to as NEMS. In particular, NEMS oscillators have been proposed for use in ultrasensitive mass detection.[80,81] Such sensors can also be efficiently used for stress and strain measurements. A useful review about NEMS can be found in the paper by Blencowe.[82]

It is always a challenging task to use nano-scale dimensions in mechanical devices, as they deal with small mechanical forces (in the order of micro-Newtons).[83] Commercially available stress sensors are yet to meet this performance level. In addition, the fabrication of free standing and stress-free specimens is generally a difficult task. Micro/nano deposition and etching processes produce pre-stress on these structures. Installation and alignment of nano-scale freestanding parts are also difficult to achieve. Even slightly misaligned specimens may experience unwanted bending moments, large enough to cause failure.

6.5.1 Oscillators based on Nanoparticles

Nanoparticles resonate according to the amount of energy contained in them. Depending on their dimensions, the resonant frequency can be in a range from THz to UV-visible wavelengths. Additional mass on the surface of these nanoparticles or a change in ambient conditions alters the resonant frequency. This resonant frequency is used to detect a measurand.

Energy can be transferred via different means such as electrical, optical, mechanical or chemical excitation. Applying energy to a single nanoparticle is a task which requires a great diligence. One of the best examples is the work reported by Park et al.[84] They described the fabrication of single-molecule transistors based on individual C_{60} molecules connected to gold electrodes. The coupling is due to quantized nano-mechanical oscillations of the C_{60} molecule against the gold surface, with a frequency of about 1.2 THz. Single-C_{60} transistors were prepared by depositing a diluted toluene suspension of C_{60} onto a pair of connected gold electrodes fabricated by electron-beam lithography.

They suggested that the centre-of-mass oscillation of C_{60} within the confinement potential binds it to the gold surface, as shown in **Fig. 6.**38. Theoretical and experimental studies have shown that C_{60} is held on gold by van der Waals interactions, with a C_{60}–gold binding energy of about 1 eV and a distance of about 6.2 Å between the C_{60} centre and the gold surface.[85] Assuming that the C_{60}–gold interaction potential can be described by Lennard–Jones theory, such parameters can be utilized to determine the shape of the potential that describes C_{60}–gold binding.[86] This calculation indicates that the C_{60}–gold binding near the equilibrium position can be approximated by a harmonic potential with an estimated force constant of $k \approx 70$ N m^{-1}, as is shown schematically in Fig. 6.38. This force and the mass M of the C_{60} molecule yield a vibrational frequency of $f = ½\pi(k/M)^{1/2} \approx 1.2$ THz and a vibrational quantum of $hf \approx 5$ meV, where h is the Planck's constant and M represents the mass of C_{60}.

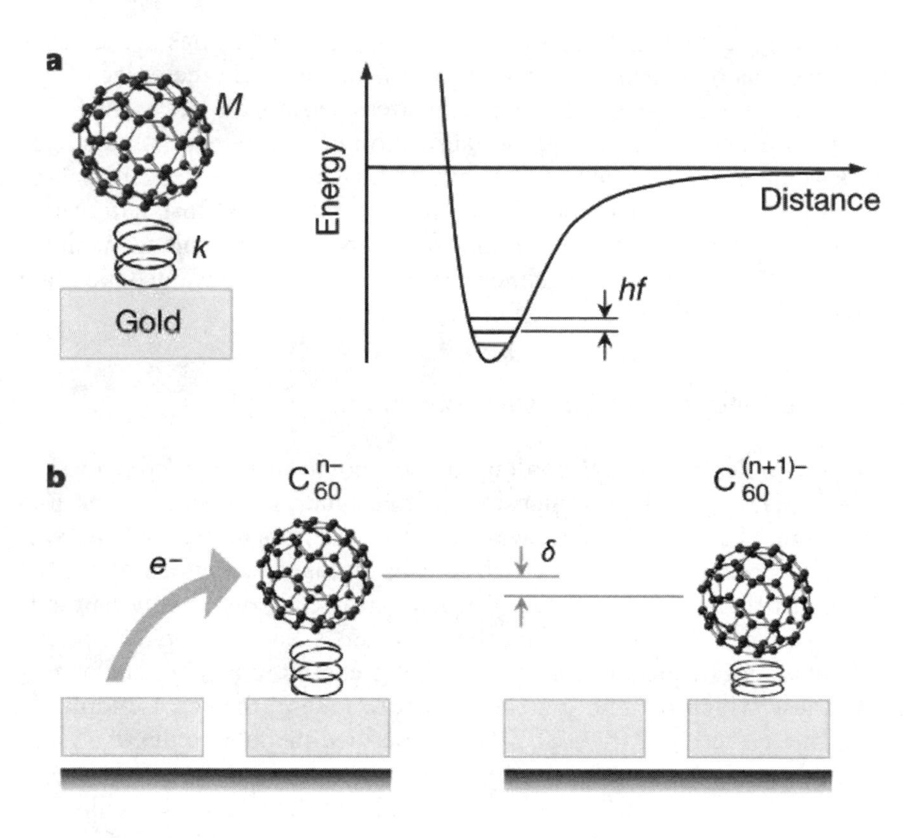

Fig. 6.38 **a**, A C_{60} molecule is bound to the gold surface by van der Waals forces and electrostatic interactions. The interaction potential is shown schematically alongside. **b**, When an electron jumps on to C^{n-}_{60}, the attractive interaction between the additional electron and its image charge on the gold surface pulls the C_{60} ion closer to surface by the distance δ. This electrostatic interaction results in the mechanical motion of C_{60}. Reprinted with permission from the Nature publications.[84]

Such a zero dimensional particle based device can be used as an ultra sensitive mass sensor. The surface of the nanoparticle can be functionalized using chemical means and after interaction with a target analyte the resonant frequency change can be measured.

6.5.2 One-Dimensional Mechanical Sensors

One-dimensional resonators are amongst the most sensitive transducers for mass detection. Decreasing the dimensions of such structures generally increases their operational frequency, which results in higher mass sensitivity. The main limitation of such transducers is that their performance can be deteriorated by the strong dampening effect of the liquid media.

Carbon nanotubes are the stiffest material known, have low density, ultra-small cross-sections and can be synthesized fairly defect-free. Equally important, a nanotube can act as a transistor and thus may be able to sense its own motion.

Sazonova et al[80] reported the electrical actuation and detection of the guitar-string-like oscillation modes of doubly clamped nanotube oscillators. They also showed that the resonance frequency can be widely tuned and that the devices can be utilized to transduce very small forces. **Fig. 6.39** shows a diagram of the measurement geometry and a scanning electron microscope (SEM) image of such a device. The nanotubes (typically single- or few-walled, 1 – 4 nm in diameter and grown by chemical vapor deposition) are suspended over a trench (typically 1.2 – 1.5 μm wide, 500 nm deep) between two metal (Au/Cr) electrodes. A small section of the tube resides on the oxide on both sides of the trench; the adhesion of the nanotube to the oxide provides clamping at the suspension points. A normalized linewidth of $Q^{-1} = \Delta f / f_o = 1/80$, a resonant frequency $f_o = 55$ MHz, and an appropriate phase difference between the actuation voltage and the force on the nanotube was obtained.

Li et al[87] proposed a similar single-walled carbon nanotube-based sensor for measuring strain and pressure on the nanoscale. The resonant frequency shifts are shown to be linearly dependent on the applied axial strain and the applied pressure. The sensitivities of nanotube-based sensors are enhanced by a reduction in tube length and tube diameter for axial strain and pressure sensing, respectively.

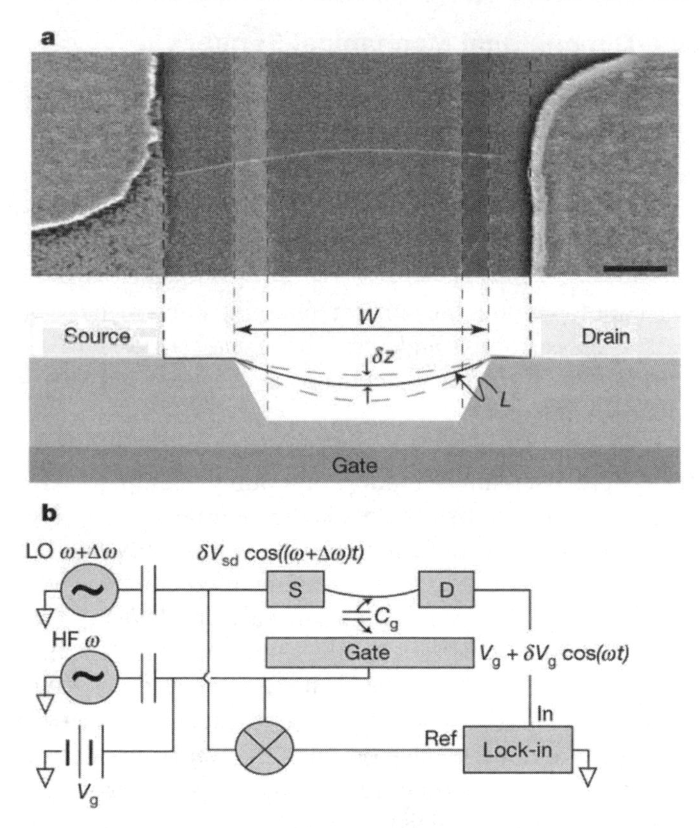

Fig. 6.39 **a**, A SEM image of a suspended device (top) and a schematic of device geometry (bottom). Scale bar, 300 nm. The sides of the trench are marked with dashed lines. A suspended nanotube can be seen bridging the trench. **b**, A diagram of the experimental set-up. A local oscillator (LO) voltage $\delta V_{sd}\omega+\Delta\omega$ (around 7 mV) is applied to the source (S) electrode at a frequency offset from the high frequency gate voltage signal $\delta V_g\omega$ by an intermediate frequency Δω of 10 kHz. The current from the nanotube is detected by a lock-in amplifier through the drain electrode (D), at $\Delta\omega$, with a time constant of 100 ms. Reprinted with permission from the Nature publications.[80]

Ekinci et al[88] described the application of NEMS to ultrasensitive mass detection. In these experiments, a modulated flux of atoms was adsorbed upon the surface of a 32.8 MHz resonator within an ultrahigh-vacuum environment. The mass-induced resonance frequency shifts by these adsorbates were then measured to obtain a mass sensitivity of 2.53×10^{-18} g. In these measurements, this sensitivity is limited by the noise in the transducer with the limitations of the technique being the fundamental phase noise.

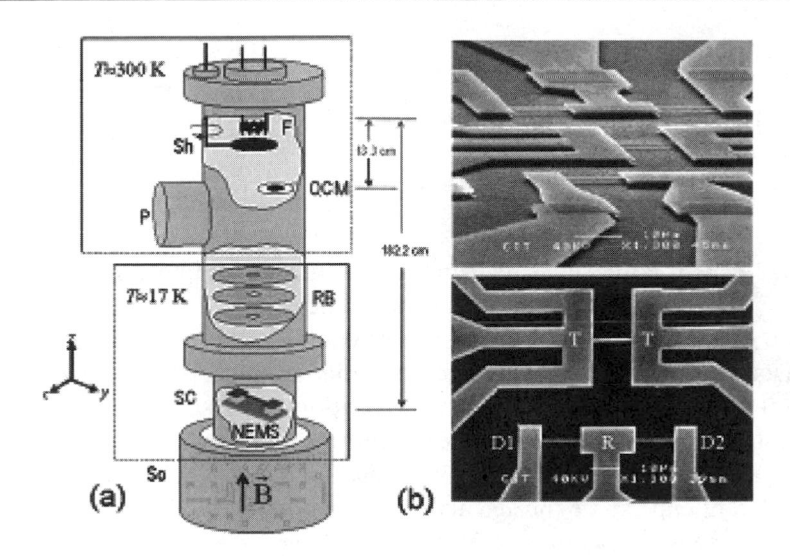

Fig. 6.40 (a) Variable temperature, UHV microwave cryostat for mass sensitivity measurements. The sample chamber (SC) is inserted into the bore of a 6T superconducting solenoid (So) in liquid helium. The radiation baffles (RB) establish a line of sight along the z axis from a room temperature thermal-evaporation source (F) to the bottom of the cryostat. The NEMS resonators are carefully placed in this line-of-sight, some $r_{NEMS} = 182.2$ cm away from the thermal-evaporation source. A calibrated quartz crystal monitor (QCM) at a distance of $R_{qcm} = 13.3$ cm and a room temperature shutter (Sh) are employed to determine and modulate the atom flux, respectively. (b) Scanning electron micrographs of nanomechanical doubly clamped beam sensor elements. The beams are made out of SiC with top surface metallization layers of 80 nm of Al. The beams are configured in an rf bridge with corresponding actuation (D1 and D2) and detection (R) ports as shown. The central suspended structure attached to three contact pads on each side, labeled T, is for monitoring the local temperature. Reprinted with permission from the American Institute of Physics publications.[88]

6.5.3 Bulk Materials and Thin Films Made of Nano-Grains

Study of mechanical characterization of thin films is of great importance due to their extensive use in nano/micro electromechanical systems.[83] Reliability and performance of these systems depend on the thin film materials' response to stresses developed during film deposition, device fabrication processes, and external loading on the devices due to operational and ambient conditions.

Conventional polycrystalline metals and alloys show an increase in yield strength (σ_y) with decreasing grain size (d) according to the well-known *Hall–Petch* (*H–P*) *equation*:[89,90]

$$\sigma_y = \sigma_0 + \frac{k}{\sqrt{d}}. \tag{6.70}$$

where σ_0 is friction stress resisting the motion of gliding dislocation, and k is the Hall–Petch slope, which is associated with a measure of the resistance of the grain boundary to slip transfer. Hardness is a measure of the resistance of a material to plastic deformation under the application of indenting load. The H–P effect in conventional coarse-grained materials is attributed to the grain boundaries acting as efficient obstacles to dislocations nucleated mostly from Frank–Read sources.[89,90] Consequently, a dislocation pileup can be formed against a grain boundary inside a grain.

By decreasing the grain size of metals down to the order of a few tens of nanometers, the H–P slope remains positive but with a smaller value.[91] At ultra-fine grain sizes below ca. 20 nm (**Fig. 6.41**), a reversed softening effect or negative H–P relation is observed for some metals.[89] The interpretation of this inverse H–P effect is still subject to debate. For pore free and dense electroplated Ni samples, ElSherik et al. also observed a deviation from H–P relation. A plateau is found in the hardness versus grain size curve when the grain sizes are less than 20 nm.[91] This raises the issue of whether inverted H–P behavior is inherited from the intrinsic effect of nanograin size or resulted from extrinsic defects introduced into the samples during fabrication.

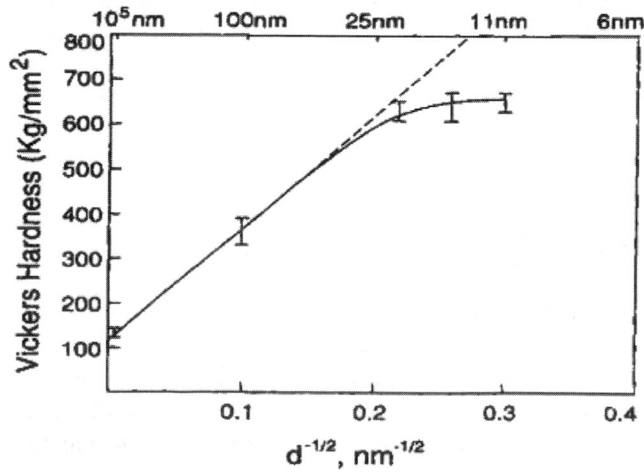

Fig. 6.41. Hall–Petch plot of the hardness of nanocrystalline Ni prepared by electrodeposition. Reprinted with permission from the Elsevier publications.[89,90]

Mass sensors based on thin films were described in Chap. 3. There we observed that cantilevers and TSM are categorized in this family. They can be used for applications such as the evaluation of the viscosity of liquids, the measurement of the mass of biomolecules, and stress measurements. The thinner the thin films become the higher the operational frequency of such devices will be.[92,82]

6.5.4 Piezoresistors

Piezoresistance was described in Chap. 2 where several applications of nanomaterials in the fabrication of piezoresistive sensors were also presented. Sensors based on the piezoresistance can be widely used in the near future as piezeoelectrically generated signals becomes a practical means for the exchange of signals between nanomechanical sensors and external electronic circuits.

A good example of nanosized piezoeresistive elements was the device which was developed by Toriyama et al (**Fig. 6.42**).[93] They fabricated a *p*-type silicon nano-wire piezoresistor with a cross-sectional area of approximately 50 nm × 50 nm for stress sensing applications. They used a combination of thermal diffusion, electron beam direct writing and reactive ion etching in the process. The maximum value of the longitudinal piezoresistance coefficient of the Si nano-wire piezoresistor was found to be 48×10^{-5} 1/MPa at a surface impurity concentration of 5×10^{19} cm^{-3}. They showed that, the longitudinal piezoresistance coefficient of the Si nano-wire piezoresistor increased up to 60% with a decrease in the cross sectional area, while transverse the piezoresistance coefficient decreased with an increase in the aspect ratio of the cross section.

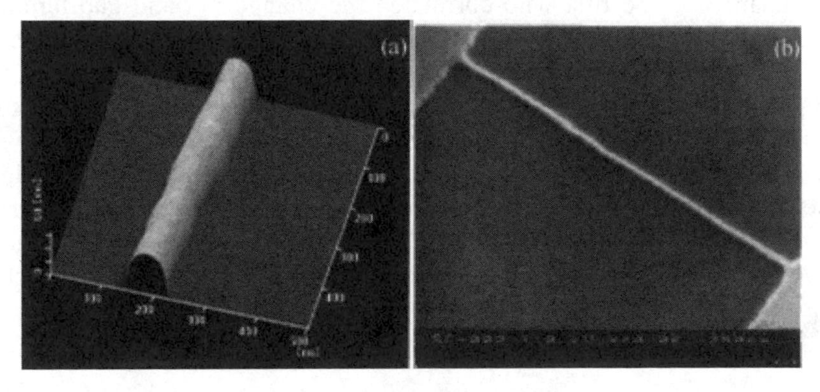

Fig. 6.42 (a) AFM and (b) SEM images of a Si nanowire. Reprinted with permission from the IEEE publications.[93]

6.6 Nanotechnology Enabled Optical Sensors

Optical sensors are amongst the most important types of chemical, biochemical, and physical sensors which utilize the advantages of nanomaterials. As described in Chaps. 2 and 5, optical sensors can be based on observing the shift of wavelengths, reflection and absorption, phase shift and/or attenuation of IR/UV/Vis light wavelengths which interact with the target analytes or travel through a waveguide. In this section, the applications of nanotechnology in the development of such sensors will be presented.

6.6.1 The Optical Properties of Nanostructures

Chemiluminescence (CL) based Sensors

The application of *photoluminescent* (*PL*) properties of nanostructured thin films is the base of many optical nanotechnology enabled sensors. One of the first papers reporting the alteration in the photoluminescence properties of nanostructured thin films was by Canham in 1990.[94] He reported the photoluminescence of Si substrates that were made porous by immersing them in high concentrated in HF solution for differing lengths of time. He showed that the photoluminescence band becomes more intense as it shifts into the visible from infra red. The etched Si surface glows bright red for 6 hours following excitation with 488 nm light of less than 1 mW (**Fig. 6.43**).

Canham was the first who correlated the change in band-gap luminescence to quantum effects. He described the porous Si with bulk materials containing an array of non-interacting cylindrical pores of fixed radius that ran perpendicular to the surface. However, it was later understood that the actual model is far more complex.

Such porous surfaces are widely used for optical sensing applications. One of the first reports was by Sailor et al,[95] describing the PL observed from a porous Si substrate is reversibly quenched upon exposure to organic solvent molecules, indicating that the PL of porous Si is extremely surface sensitive.

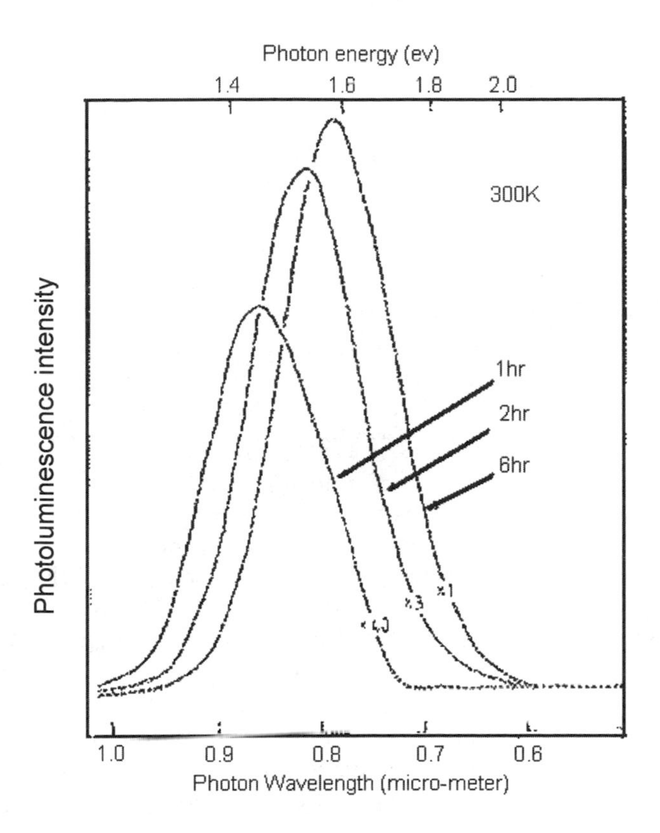

Fig. 6.43 Room-temperature photoluminescence of porous Si substrates with different etching durations. Reprinted with permission from the American Institute of Physics publications.[94]

Fig. 6.44 presents the emission spectra of a luminescent porous Si layer before and after exposure to 160 Torr of tetrahydrofuran (THF) vapor. Immediately after exposure to the solvent, the emission centered at 670 nm decreased in magnitude by a factor of 4 and down shifted to 630 nm. The emission spectrum recovered to the original intensity within seconds of evacuation of the excess solvent vapor.

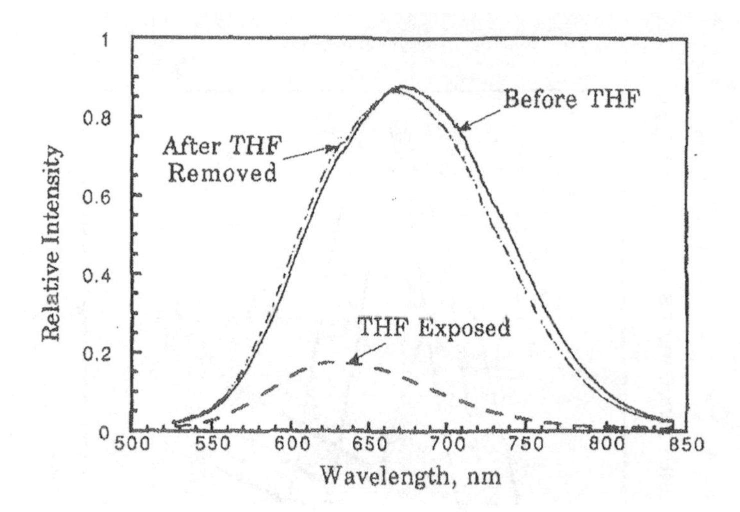

Fig. 6.44 Emission spectra of luminescent porous Si sample before THF exposure (-), after 1 min of THF exposure (- - -), and after removal of THF under dynamic vacuum (--). Excitation source was from a 442 nm He/Cd laser (5 mW/cm^2). Reprinted with permission from the American Chemical Society publications.[95]

A list of chemical sensors that can be developed exploiting the photoluminescence quenching of porous silicon can be found in the review paper by Shi et al.[96] The list includes analytes such a organic solvents, amines, aromatic compounds, metal ions, oxidizing and reducing gas species. *Fabry–Perot interference* based sensors can make use of advantages inherent in nanostructured thin films.[96] The first applications of Fabry–Perot interference were developed by Butler,[97] and were experimentally investigated by Gauglitz and Nahm.[98]

Recent studies have shown that certain electrochemical etches of single-crystal *p*-type Si wafers can produce porous materials that display well-resolved Fabry–Perot fringes in their *reflectometric interference spectrum* (**Fig. 6.45**).[99] A Fabry–Perot fringe pattern is created by multiple reflections of illuminated white light on the air-porous silicon interface and the porous silicon-bulk silicon interface, which is related to the effective optical thickness (product of thickness L and refractive index n) of the film by the equation of $m\lambda = 2nL$ (m is the spectral order and λ is the wavelength of light). This induced shift in the Fabry–Perot fringe pattern can be utilized for sensing. The shift can be caused by the change in the refractive index of the porous medium upon molecular interactions the materials attached within the porous silicon matrix.

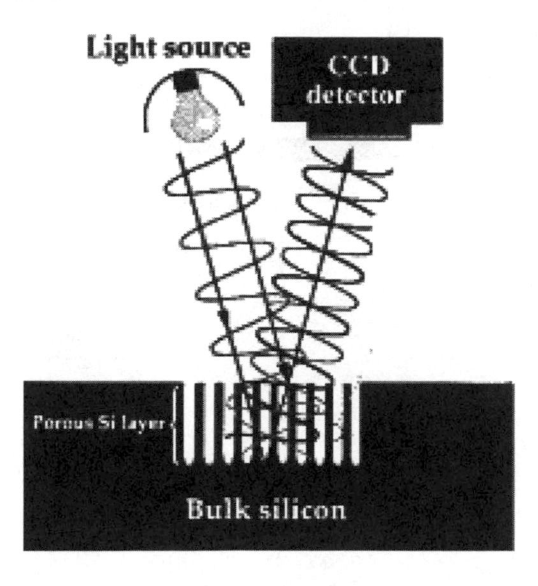

Fig. 6.45 Schematic diagram of a porous Si based optical sensor. The silicon oxide surface of the porous layer can be modified for the recognition of different molecular elements. Reprinted with permission from the Science Magazine publications.[99]

CL phenomenon can also be observed during the catalytic oxidation, *cataluminescence.*[100] In such sensors, material's color alter upon exposure to target molecules occurring via chemical interactions.[96] This effect can be also seen in nanostructured surfaces. McCord et al found that porous silicon treated with nitric acid or persulfate could produce an intense CL.[101]

CL effects can be detected for many nanosized materials including MgO, ZrO_2, TiO_2, Y_2O_3, and $SrCO_3$, when they are exposed to organic vapors. **Fig. 6.46** shows the schematic representation of a CL sensor. The system comprises of:

(1) a nanoparticle-based reactor;

(2) a digital programmable temperature controller; and,

(3) a commercial optical detector.

In such a system, the CL intensity at a certain wavelength can be measured using a photon-counting method through a tunable wavelength optical filter.

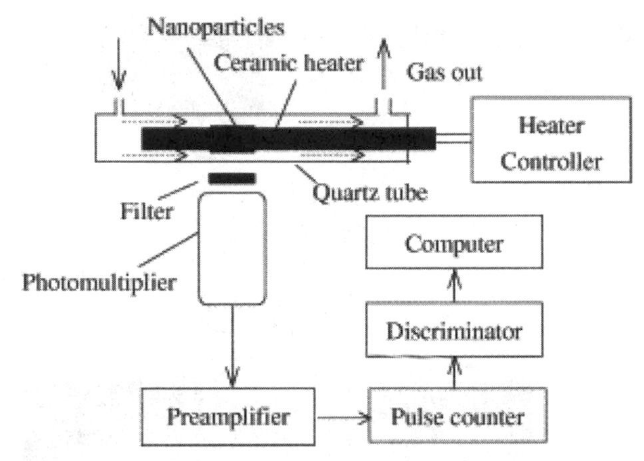

Fig. 6.46 Schematic diagram of a CL detection system. Reprinted with permission from the Elsevier publications.[96]

Photoconductors

A described in Chap. 2 in a photoconductor the increase in the conductivity results from *photo-generation* of electron-hole pairs as well as doping caused by UV light-induced desorption.[102]

Photoconduction can also be seen in 1D structures. If a 1D semiconducting particle, comprising of a metal oxide is irradiated by UV light, it can show a significant increase in its conductivity. This is due to the fact that irradiation affects the bulk of the material. A good example of such a device is the photoconductor developed by Wang et al.[8] This structure is made of SnO_2 nanobelts operating as a diode. Light with a wavelength of 350 nm (E_λ = 3.54 eV) was used, which exceeds the direct band-gap of SnO_2.

6.6.2 The Optical Properties of Nanoparticles

In this section and next section, the plasmon resonance extinction for spherical particles using simplified models will be presented and subsequently some applications will be described.

Metal nanoparticles have been used for centuries to make stained glass. In 1908, Mie used the Maxwell's equations to describe the *extinction* (scattering and absorption) of spherical particles.[103] The spherical solution is important as it is simple to extract and also many nanomaterials are ensembles of such nanoparticles. Despite Mie's simple approach for spherical nanoparticles, real-world problems can be much more complicated as nanoparticles can take a large variety of different morphologies. Also the presence of the supporting substrate, solvent, spacing of particles and aggregations which dictates the electromagnetic coupling coefficient can make such calculations more cumbersome.

6.6.3 Sensors based on Plasmon Resonance in Nanoparticles

When a metallic nanoparticle is irradiated with light, the alternating electric field causes the conduction band electrons to oscillate coherently as shown in **Fig. 6.47**.[104] After irradiation, a restoring force is generated between the electrons and nuclei due to the displacement of the electron cloud. This results in an oscillation of electron clouds relative to the nucleus. The oscillation frequency depends on density of electrons, the shape and distribution of the charge, as well as effective mass of the electron.

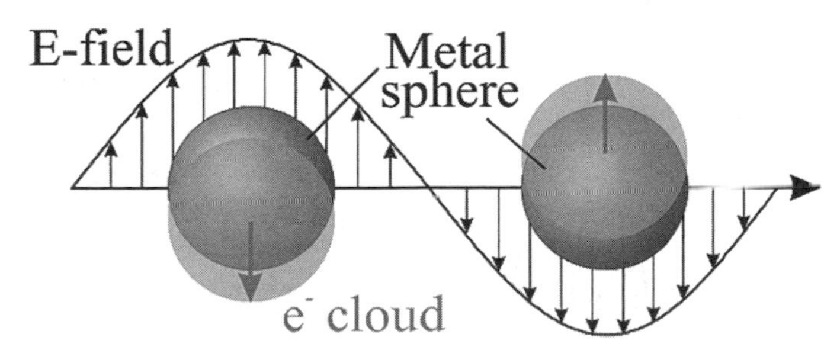

Fig. 6.47 The schematic of plasmon oscillation for a sphere, showing the displacement of the conduction electron charge cloud relative to the nuclei.

The resultant collective oscillation of the electron is referred to as *dipole plasmon resonance* of the particle. Higher mode such as *quardopole* may also occur when the electron cloud moves in parallel to the electric field.

For calculation of the resonant frequency, it is assumed that the size of the nanoparticles are less than the wavelength of the impinging light ($\lambda >> 2R$, for gold $2R < 25$ nm). In this case, the quasi-static electrostatic approximation can be used.

Let's assume the electric field vector, $\overline{E}_0$, is:

$$\overline{E}_0 = E_0\hat{x}, \tag{6.71}$$

where $\hat{x}$ is the unit vector and E_0 is the electric field magnitude in the x direction. The Laplace equation is:

$$\nabla^2 \varphi = 0, \tag{6.72}$$

where φ is the electric potential related to the electric filed via:

$$\overline{E} = -\nabla \varphi. \tag{6.73}$$

The condition in solving the equation is the continuity of the electric displacement which is defined as:

$$\overline{D} = \varepsilon \overline{E}, \tag{6.74}$$

where $\overline{D}$ is the electric field vector.

If we consider the angular momentum of the atomic orbitals to be equal to unity, then using the Laplace and Eq. (6.71) equations and the continuity condition, the electric field outside the sphere is equal to:[104]

$$\overline{E}_{out} = E_0 \hat{x} - \alpha E_0 \left[\frac{\hat{x}}{r^3} - \frac{3x}{r^3}(x\hat{x} + y\hat{y} + z\hat{z}) \right], \tag{6.75}$$

where α is the sphere polarizability, r is the distance from the sphere center and $\hat{x}$, $\hat{y}$ and $\hat{z}$ are the unit vectors. In this equation, the first term is the applied electric field and the second is the induced electric field. For a spherical particle the polarizability is proportional to a^3 with a the radius of the nanoparticles.

In a diluted colloidal solution containing N particles per unit volume, the measured attenuation is given by:[105]

$$a = \log_{10}\left(\frac{I_0}{I_d}\right) = \frac{NC_{ext}d}{2.303}, \tag{6.76}$$

where the path-length is d the and the intensity of the input and output light are I_d and I_o, respectively. In this equation, C_{ext} is the *extinction cross section of a single particle*, which is derived from: [105]

$$C_{ext} = \frac{24\pi^2 a^3 \varepsilon_S{}^{3/2}}{\lambda} \frac{\varepsilon''}{(\varepsilon' + 2\varepsilon_S) + \varepsilon''^2}, \tag{6.77}$$

where ε_S is the dielectric coefficient of surroundings, λ is the wavelength of the impinged wave, and ε' and ε'' are the real and imaginary part of dielectric coefficient of the nanoparticles, respectively.

From Eq. (6.77) it can be derived that resonance occurs when $\varepsilon' \approx -2\varepsilon_S$ and the bandwidth and peak height are roughly determined by ε''.[105] However, the dipole approximation does not show the dependence between size and oscillation frequency. It only describes that size changes the intensity of light.

Experimentally, a strong size dependence of the plasmon bandwidth is observed.[106] As a modification to the Mie theory for small particles, the dielectric function of the metal nanoparticles itself is assumed to become size dependent $\varepsilon' + i\varepsilon'' = f(a, \lambda)$.

The size dependence of the dielectric constant is introduced as the diameter of the particle becomes smaller than the *mean free path* (*MFP*) of the conduction band electrons.

Experimentally, it was first suggested by Kreibig[107] that a $1/a_{bulk}$ dependence of the plasmon bandwidth exists in measurements. His idea was in agreement with experimental results for nanomaterials with sizes down to 2 nm.

A $1/R$ dependence of the plasmon bandwidth is also predicted by a quantum mechanical theory by Persson.[108] When the particle radius is smaller than the mean free path of the bulk metal, conduction electrons are scattered by the surface and an effective radius a_{eff} can be defined as:

$$\frac{1}{a_{eff}} = \frac{1}{a} + \frac{1}{a_{bulk}}, \qquad (6.78)$$

where a_{eff} is calculated from the effect of a_{bulk} and the real a.

For larger nanoparticles (for instance for gold where $2R > 25$ nm) the extinction cross section also depends on higher-order multipole modes within the full Mie equation as a result the extinction spectrum is then dominated by quadrupole and octopole absorption as well as scattering.[109] These higher oscillation modes depend on the particle size. When the size increases the plasmon absorption maximum is shifted to longer wavelengths as the bandwidth increases. The excitation of such higher-order modes can be explained in terms of an inhomogeneous polarization of the nanoparticles by the electromagnetic field as the particle size becomes comparable to the wavelength of the incoming radiation. The broadening of the plasmon band is then usually ascribed to retardation effects.[110] On the other hand, the increased line width (or the faster loss of coherence of the plasmon resonance) could also be qualitatively described as a result of the interactions between the dipole and the quadrupole (and higher-order) oscillatory motions of the electrons, which deteriorates the phase coherence.[110] The size dependence resonance is a useful phenomenon for sensing applications.[111] Some examples are presented below.

Localized surface plasmon resonance (LSPR) bases Sensors

Metal nanoparticles exhibit a strong UV-visible extinction band that is not present in the spectrum of their bulk metal counterparts and is known as the *localized surface plasmon resonance* (*LSPR*).[112,113] LSPR excitation results in wavelength-selective absorption with extremely large molar extinction coefficients (10^{-11} $M^{-1}cm^{-1}$), and enhanced local electromagnetic fields near the surface of the nanoparticle that are responsible for the intense signals observed in all surface-enhanced spectroscopy. As an example, Macfarland et al the localized surface plasmon resonance response of individual Ag nanoparticles to the formation of a monolayer of 1-hexadecanethiol molecules adsorbates.[113] In their work, the adsorption of fewer than 60,000 thiol molecules on single Ag nanoparticles results in a large surface plasmon resonance shift of 40.7 nm.

These nanoparticles could replace existing florescence dyes for imaging. Their resonance is narrow band and they can all be excited with he same source. Some examples were shown in Chap. 2. The applications of these nanoparticles for biosensing will be described in Chap. 7.

6.7 Magnetically Engineered Spintronic Sensors

In *spintronics*, the electron spin carries information and is the base of the operation of devices. In such spintronically enabled devices, standard microelectronics is combined with spin-dependant effects that arise from the interaction between the spin of carriers and the magnetic properties of materials.[114] Major challenges in the field of spintronics include the optimization of electron spin lifetimes, the detection of spin coherence in nanoscale structures, transport of spin-polarized carriers across relevant length scales and heterointerfaces, and the manipulation of both electron and nuclear spins on sufficiently fast time scales.[114]

Magnetoresistance effect was described in Chap. 2. The most widely used type magnetoresistive sensors are the ones which are used for magnetic recording in *hard disk drives* (*HDD*s). In HDDs information is stored in the magnetized regions. Bits are the change of magnetizations in these regions and the read sensors detects their fringing magnetic fields. The read sensor is located just a few nanometers above the recording medium (**Fig. 6.48**).[115] The field is sensed by the change in the resistance of the sensing element.

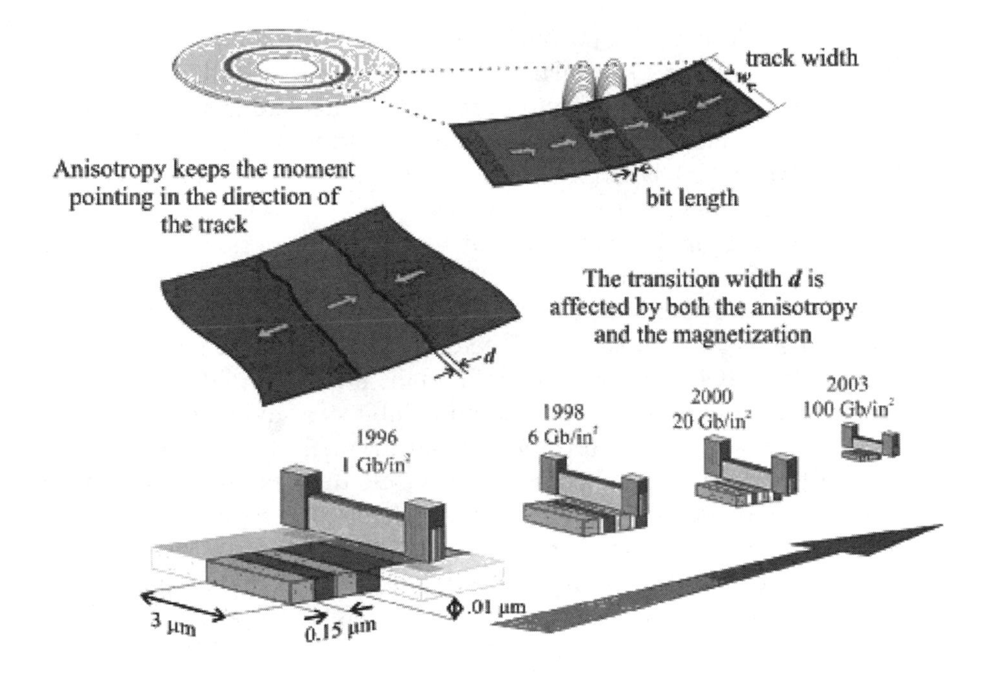

Fig. 6.48 Information is stored by magnetizing regions of a thin magnetic film on the surface of a disk. Bits are detected by sensing the magnetic fringing fields of the transitions between adjacent regions as the disk is rotated beneath a magnetic sensor. As the area of the magnetized region has decreased, the read sensor has had to scale down in size accordingly. Reprinted with permission from the IEEE publications.[115]

There are many technological issues related to these types of sensors. The issues include: reduced flying height; servoying; magnetic shielding; filed strength and speed and noise and data stability.

6.7.1 AMR, Giant and Colossal Magneto-Resistors

In Chap. 2, it was shown that large increases in magneto-resistance are possible using heterostructures. These structures, with largely improved performance, are based on the generation and manipulation of the spin polarized electrons in multilayer structures with magnetic properties. These multilayer structures can act as extremely sensitive magnetic field sensors. Their electrical resistance, in the presence of a magnetic field, demonstrates a much larger change than conventional magnetoresistive materials. These magnetic sensors are generally based on the *giant magnetoresistance*

(*GMR*) effect. These sensors were first commercially utilized by IBM in 1997.[115] They became so popular that in 2006 almost all HDDs use such sensors. GMR is observed in multilayered thin-film materials composed of alternating ferromagnetic and/or nonmagnetic layers.[114] The GMR effect was discovered when the effect of magnetic fields on MBE deposited epitaxial Fe/Cr multilayers was being investigated.[116,117] One of the largest GMR effects appears in multilayers of Co/Cu which is now the basis of many sensors for storage devices.[118] The resistance of the material is lowest when the magnetic moments in ferromagnetic layers are aligned and highest when they are antialigned. The current can either be perpendicular to the interfaces (CPP) or can be parallel to the interfaces (CIP). These materials operate at room temperatures and exhibit substantial changes in resistivity when subjected to relatively small magnetic fields (100 to 1000 Oe). The performance characteristics of GMR devices are improved by several orders of magnitude over earlier technologies, which were generally based on *anisotropic magnetoresistance* (*AMR*).[119]

AMR active components are normally made of ultrathin magnetoresistive ferromagnetic material such as permalloy ($Ni_{81}Fe_{19}$). The plane of this thin film is orthogonal to that of the magnetic media. AMR resistance change is only a few percent. As a consequence of bulk scattering, the real density would be limited to 5 Gb/in^2. In contrast, GMR is more than 100% at room temperature.

The GMR effect is the result of spin-dependant scattering in inhomogeneous magnetic metallic systems. The effect was originally explained by Mott in 1930s.[120] Current in 3*d* transition ferromagnetic metal is carried independently by spin-up (↑) and spin down (↓) electrons. Based on this model, the scattering rate can be different for electrons which use the spin-up channel to those which use the spin-down channels.[121] In a simple model, the current is carried by *d* electrons which generate ferromagnetism. The density of states (DOS) of spin up and down electrons at the Fermi energy are quite distinct. This rationalizes the spin dependant scattering rate. In particular, as shown schematically in **Fig. 6.49**, magnetic multilayers typically have a lower resistance when the magnetic moments of the individual layers are parallel than when antiparallel.[121] The magnetoresistance is defined as (*R* is the resistance):

$$MR = \frac{R_{parallel} - R_{anti-parallel}}{R_{parallel}}. \tag{6.79}$$

Where $R_{parallel}$ and $R_{anti\text{-}parallel}$ are the resistance of the device before and after applying the magnetic field.

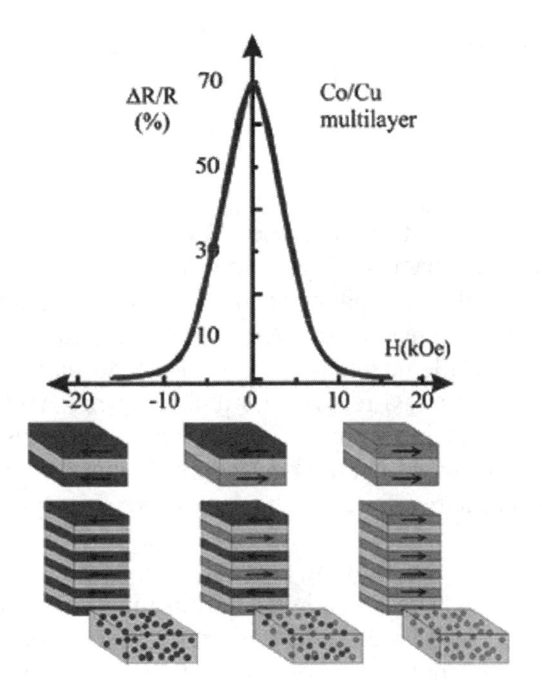

Fig. 6.49 GMR results from interfacial spin-polarized scattering between ferromagnets separated by conducting spacers in a heterogeneous magnetic material, such as a magnetic multilayer or granular alloy. Reprinted with permission from the IEEE publications.[115]

As was mentioned previously, one of the largest GMR effects at room temperature is approximately 100%, which has been found in sputter deposited Co/Cu multilayers.[122] The thinner the magnetic and nonmagnetic layers are the larger the GMR effect will be as the GMR effect is dominated by spin-dependent scattering at magnetic/nonmagnetic interfaces.[123] Thickening these layers largely results in shunting of current away from the interfacial regions.[124]

Manganese oxides with a perovskite structure exhibit a transition between a paramagnetic insulating phase and a ferromagnetic metal phase.[125] *Colossal magnetoresistance* (*CMR*) is associated with this transition which was reported by Jin et al in 1993-1994.[126] In the vicinity of the transition temperature, such materials exhibit a large change in resistance in response to an applied magnetic field. CMR has not been explained by any current physical theories and is currently the focus of ongoing research. CMR materials are able to demonstrate resistance changes by several orders of magnitude.

6.7.2 Spin Valves

A spin valve is a GMR-based device (**Fig. 6.50**). A typical spin valve may contain two ferromagnetic layers (e.g. alloys of nickel, iron, and cobalt) sandwiching a thin nonmagnetic metal (usually copper).[114] In such a structure one of the two magnetic layers is relatively insensitive to moderate magnetic fields while magnetization of the other magnetic layer can be changed by application of a relatively small magnetic field. An antiferromagnetic layer is also in intimate contact with the insensitive magnetic layer. As the magnetizations in the two layers change from parallel to antiparallel alignment, the resistance of the spin valve rises typically from 5 to 10%.

There are other ways of forming a spin-valve. The insensitive magnetic layer can be replaced with a synthetic antiferromagnet which consists of two magnetic layers separated by a very thin (~10 Å) nonmagnetic conductor, such as ruthenium.[118] The magnetizations in the two magnetic layers are strongly antiparallel coupled. As a result, they are also effectively immune to outside magnetic fields. This structure improves both stand-off magnetic fields and the operational temperature of the spin valve. Another method of constructing a spin valve is by using a nano-oxide layer formed at the outside surface of the soft magnetic film. This layer reduces resistance due to surface scattering, thereby increasing the percentage change in magnetoresistance.[127]

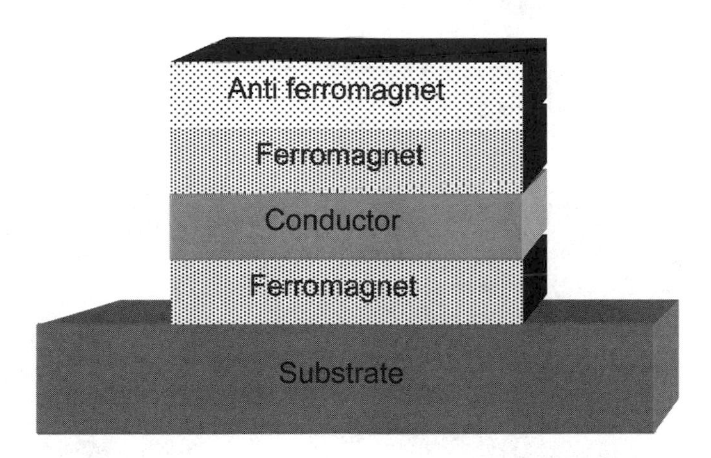

Fig. 6.50 The schematic of a typical spin-dependent transport structures of a spin valve.

6.7.3 Magnetic Tunnel Junctions

The maximum useful magnetoresistance provided by spin-valve read sensors is less than 25%.[128] Sensors with higher efficiencies are needed in order to achieve higher disk drive densities.

Larger values of magnetoresistance can be obtained in *magnetic tunnelling junctions* (*MTJ*).[129] A MTJ is similar to a spin-valve device but the metallic Cu spacer is replaced with a thin insulating barrier. In such a device the sensed current passes perpendicularly through the device (**Fig. 6.51**).[130] *Tunnelling magneto resistance* (*TMR*) can be as large as 50%. Also since the current is perpendicular to the plate the sensor can be attached to the magnetic shield which can also be used as the electrical contact. In contrast, conventional GMR device currents pass parallel to the layers of the device which makes the fabrication of the shielding more difficult. This also increases the size of the sensor.

Because the tunneling current density is usually small, MTJ devices tend to have high resistances. Currently the resistance of MTJ devices is too high to allow them to be used for most commercial applications.

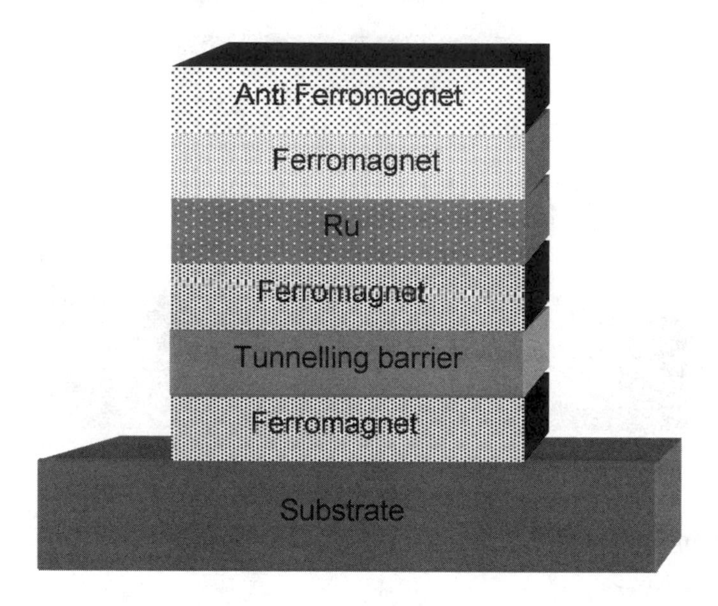

Fig. 6.51 The schematic of a typical spin-dependent transport structures of a MTJ.

6.7.4 Other Nanotechnology Enabled Magnetic Sensors

Similar to many other effects, it is also possible to enhance the Barkhausen effect (see Chap. 2) and use it for sensing applications. Vazquez et al observed bi-stable magnetic behavior which they considered as a giant Barkhausen jump.[131] This effect appears spontaneously in large magnetostriction Fe-based wires and in nonmagnetostrictive Co-based wires under particular thermal treatments. This effect can be also detected in micro-wires as long as 2 mm exhibiting positive magnetostriction in its amorphous state, as well as after transformation into nano or micro-crystalline structures. The effect can be important for improving conventional bistable materials for applications in various sensor devices, such as position sensors, revolution counters, and magnetic codification systems.

Magnetic materials can also be used for gas sensing applications. For instance Punnoose et al[132] demonstrated that a nanoscale magnetic hydrogen sensor can be fabricated based on systematically varying the saturation magnetization and *remanence* (remanence is the left behind magnetization in a medium after removing an external magnetic field) of nanoscale antiferromagnetic haematite (a form of Fe_2O_3 crystal), which changes upon exposure to hydrogen. The saturation magnetization and remanence of the nanoscale haematite sample showed an increase of one to two orders of magnitude in the presence of hydrogen gas at elevated temperatures.

6.8 Summary

In this chapter, different inorganic nanotechnology enabled sensors and transducers were presented. The DOS and NOS were utilized to theoretically describe how alterations in dimensions affect materials properties. This was highlighted with an example regarding the change of conductivity in nanowires. Gas sensors based on nanostructured thin films were presented and it was shown that structural engineering of such materials can be utilized to manipulate the response of such sensors. The effect of phonons in low dimensional structures was described and several examples regarding their applications were presented. Nanotechnology enabled mechanical sensors were discussed with several examples on nanodimensional oscillators and the effect of piezoelectricity in the fabrication of transducers.

It was shown that the performance of optical sensors can be enhanced by utilizing nanotechnology processes. It was theoretically described that the optical properties of nanostructured thin films and nanoparticles are the base of confinement of energy in quantum dimensions which cause resonance and quenching. Following this, several optical sensors based on nanostructured materials were presented. In the final section of the chapter, magnetically engineered spintronic sensors including giant and colossal magneto-resistors, spin valves and magnetic tunnel junctions along with discussions on other magnetic sensors were presented.

The next chapter will focus on organic and bio sensors. The majority of transducers which were presented in this chapter and Chap. 3 along with the purely organic transducers which will be presented in Chap. 7, can be used in organic and bio sensing applications.

References

[1] D. K. Guthrie, T. K. Gaylord, and E. N. Glytsis, Ieee Transactions on Education **39,** 465-470 (1996).

[2] M. A. Omar, *Elemenatary Solid State Physics* (Addison Wesley, Massachusetts, USA, 1993).

[3] C. Kittel, *Introduction to Solid State Physics*, Eight edition ed. (John Wiley & Sons, New York, USA, 2005).

[4] B. J. Vanwees, H. Vanhouten, C. W. J. Beenakker, J. G. Williamson, L. P. Kouwenhoven, D. Vandermarel, and C. T. Foxon, Physical Review Letters **60,** 848-850 (1988).

[5] D. A. Wharam, T. J. Thornton, R. Newbury, M. Pepper, H. Ahmed, J. E. F. Frost, D. G. Hasko, D. C. Peacock, D. A. Ritchie, and G. A. C. Jones, Journal of Physics C-Solid State Physics **21,** L209-L214 (1988).

[6] A. R. Leach, *Molecular modelling principles & applications* (Addison Wesley Publishing Company USA, 1997).

[7] Z. M. Du and N. H. de Leeuw, Surface Science **554,** 193-210 (2004).

[8] Z. L. Wang, Annual Review of Physical Chemistry **55,** 159-196 (2004).

[9] Z. W. Pan, Z. R. Dai, and Z. L. Wang, Science **291,** 1947-1949 (2001).

[10] M. S. Arnold, P. Avouris, Z. W. Pan, and Z. L. Wang, Journal of Physical Chemistry B **107,** 659-663 (2003).

[11] Y. Cui and C. M. Lieber, Science **291,** 851-853 (2001).

[12] P. C. Collins, M. S. Arnold, and P. Avouris, Science **292,** 706-709 (2001).

[13] P. Avouris, Chemical Physics **281,** 429-445 (2002).

[14] J. T. Hu, O. Y. Min, P. D. Yang, and C. M. Lieber, Nature **399,** 48-51 (1999).

[15] H. J. Dai, J. H. Hafner, A. G. Rinzler, D. T. Colbert, and R. E. Smalley, Nature **384,** 147-150 (1996).

[16] M. Brandbyge, J. Schiotz, M. R. Sorensen, P. Stoltze, K. W. Jacobsen, J. K. Norskov, L. Olesen, E. Laegsgaard, I. Stensgaard, and F. Besenbacher, Physical Review B **52,** 8499-8514 (1995).

[17] Q. H. Li, Y. X. Liang, Q. Wan, and T. H. Wang, Applied Physics Letters **85,** 6389-6391 (2004).

[18] E. Comini, G. Faglia, G. Sberveglieri, Z. W. Pan, and Z. L. Wang, Applied Physics Letters **81,** 1869-1871 (2002).

[19] E. Comini, M. Ferroni, V. Guidi, G. Faglia, G. Martinelli, and G. Sberveglieri, Sensors and Actuators B-Chemical **84,** 26-32 (2002).

[20] N. Yamazoe and N. Miura, Sensors and Actuators B-Chemical **20,** 95-102 (1994).

[21] E. Comini, G. Faglia, G. Sberveglieri, D. Calestani, L. Zanotti, and M. Zha, Sensors and Actuators B-Chemical **111,** 2-6 (2005).

[22] C. Baratto, E. Comini, G. Faglia, G. Sberveglieri, M. Zha, and A. Zappettini, Sensors and Actuators B-Chemical **109,** 2-6 (2005).

[23] S. J. Ippolito, A. Ponzoni, K. Kalantar-Zadeh, W. Wlodarski, E. Comini, G. Faglia, and G. Sberveglieri, Sensors and Actuators B-Chemical **117,** 442-450 (2006).

[24] S. J. Ippolito, S. Kandasamy, K. Kalantar-Zadeh, A. Trinchi, and W. Wlodarski, Sensor Letters **1,** 33-36 (2003).

[25] W. Y. Chung and J. W. Lim, Current Applied Physics **3,** 413-416 (2003).

[26] D. S. Lee, Y. T. Kim, J. S. Huh, and D. D. Lee, Thin Solid Films **416,** 271-278 (2002).

[27] D. S. Lee, J. K. Jung, J. W. Lim, J. S. Huh, and D. D. Lee, Sensors and Actuators B-Chemical **77,** 228-236 (2001).

[28] G. Korotcenkov, Sensors and Actuators B-Chemical **107,** 209-232 (2005).

[29] A. Gross, *Theoretical Surface Science - A microscopic Perspective* (Springer, Berlin, 2002).

[30] N. D. Lang and A. R. Williams, Physical Review B **18,** 616-636 (1978).

[31] W. Gopel and K. D. Schierbaum, Sensors and Actuators B-Chemical **26,** 1-12 (1995).

[32] N. Barsan and U. Weimar, Journal of Electroceramics **7,** 143-167 (2001).

[33] N. Barsan, D. Koziej, and U. Weimar, Sensors and Actuators B-Chemical **121,** 18-35 (2007).

[34] T. Sahm, A. Gurlo, N. Barsan, and U. Weimar, Sensors and Actuators B-Chemical **118,** 78-83 (2006).

[35] G. K. Reeves and H. B. Harrison, Electron Device Letters **3,** 111-113 (1982).

[36] U. Hoefer, K. Steiner, and E. Wagner, Sensors and Actuators B-Chemical **26,** 59-63 (1995).

[37] E. Comini, Analytica Chimica Acta **568,** 28-40 (2006).

[38] V. Brinzari, G. Korotcenkov, and V. Golovanov, Thin Solid Films **391,** 167-175 (2001).

[39] G. Korotcenkov, A. Cornet, E. Rossinyol, J. Arbiol, V. Brinzari, and Y. Blinov, Thin Solid Films **471,** 310-319 (2005).

[40] G. Korotcenkov, I. Boris, V. Brinzari, V. Golovanov, Y. Lychkovsky, G. Karkotsky, A. Cornet, E. Rossinyol, J. Rodrigue, and A. Cirera, Sensors and Actuators B-Chemical **103,** 13-22 (2004).

[41] G. Korotcenkov, V. Brinzari, A. Cerneavschi, M. Ivanov, V. Golovanov, A. Cornet, J. Morante, A. Cabot, and J. Arbiol, Thin Solid Films **460,** 315-323 (2004).

[42] G. Korotcenkov, V. Macsanov, V. Tolstoy, V. Brinzari, J. Schwank, and G. Faglia, Sensors and Actuators B-Chemical **96,** 602-609 (2003).

[43] G. Gordillo, L. C. Moreno, W. Delacruz, and P. Teteran, Thin Solid Films **252,** 61-66 (1994).

[44] A. Trinchi, Y. X. Li, W. Wlodarski, S. Kaciulis, L. Pandolfi, S. P. Russo, J. Duplessis, and S. Viticoli, Sensors and Actuators a-Physical **108,** 263-270 (2003).

[45] Y. X. Li, A. Trinchi, W. Wlodarski, K. Galatsis, and K. Kalantar-zadeh, Sensors and Actuators B-Chemical **93,** 431-434 (2003).

[46] K. Galatsis, Y. Li, W. Wlodarski, C. Cantalini, M. Passacantando, and S. Santucci, Journal of Sol-Gel Science and Technology **26,** 1097-1101 (2003).

[47] Y. X. Li, W. Wlodarski, K. Galatsis, S. H. Moslih, J. Cole, S. Russo, and N. Rockelmann, Sensors and Actuators B-Chemical **83,** 160-163 (2002).

[48] K. Galatsis, Y. X. Li, W. Wlodarski, E. Comini, G. Sberveglieri, C. Cantalini, S. Santucci, and M. Passacantando, Sensors and Actuators B-Chemical **83,** 276-280 (2002).

[49] Y. X. Li, K. Galatsis, W. Wlodarski, M. Passacantando, S. Santucci, P. Siciliano, and M. Catalano, Sensors and Actuators B-Chemical **77,** 27-34 (2001).

[50] K. Galatsis, Y. X. Li, W. Wlodarski, and K. Kalantar-zadeh, Sensors and Actuators B-Chemical **77,** 478-483 (2001).

[51] E. Comini, G. Faglia, G. Sberveglieri, Y. X. Li, W. Wlodarski, and M. K. Ghantasala, Sensors and Actuators B-Chemical **64,** 169-174 (2000).

[52] M. Z. Atashbar, H. T. Sun, B. Gong, W. Wlodarski, and R. Lamb, Thin Solid Films **326,** 238-244 (1998).

[53] H. Gleiter, NanoStructured Materials **1** 1-19 (1992).

[54] D. Szczuko, J. Werner, S. Oswald, G. Behr, and K. Wetzig, Applied Surface Science **179,** 301-306 (2001).

[55] D. Szczuko, J. Werner, G. Behr, S. Oswald, and K. Wetzig, Surface and Interface Analysis **31,** 484-491 (2001).

[56] H. Meixner and U. Lampe, Sensors and Actuators B-Chemical **33,** 198-202 (1996).

[57] W. Fliegel, G. Behr, J. Werner, and G. Krabbes, Sensors and Actuators B-Chemical **19,** 474-477 (1994).

[58] D. E. Williams and K. F. E. Pratt, Journal of the Chemical Society-Faraday Transactions **94,** 3493-3500 (1998).

[59] N. Barsan, M. Schweizer-Berberich, and W. Gopel, Fresenius Journal of Analytical Chemistry **365,** 287-304 (1999).

[60] G. Korotcenkov, V. Macsanov, V. Brinzari, V. Tolstoy, J. Schwank, A. Cornet, and J. Morante, Thin Solid Films **467,** 209-214 (2004).

[61] G. Korotcenkov, V. Brinzari, Y. Boris, M. Ivanova, J. Schwank, and J. Morante, Thin Solid Films **436,** 119-126 (2003).

[62] N. Tsud, V. Johanek, I. Stara, K. Veltruska, and V. Matolin, Thin Solid Films **391,** 204-208 (2001).

[63] J. F. McAleer, P. T. Moseley, J. O. W. Norris, D. E. Williams, and B. C. Tofield, Journal of the Chemical Society-Faraday Transactions I **84,** 441-457 (1988).

[64] A. El-Azab, S. Gan, and Y. Liang, Surface Science **506,** 93-104 (2002).

[65] E. A. Symons, *Catalytic Gas Sensors* (Kluwer Academic Publishers, The Netherlands, 1992).

[66] G. Korotcenkov, V. Brinzari, M. Ivanov, A. Cerneavschi, J. Rodriguez, A. Cirera, A. Cornet, and J. Morante, Thin Solid Films **479,** 38-51 (2005).

[67] A. M. Rao, E. Richter, S. Bandow, B. Chase, P. C. Eklund, K. A. Williams, S. Fang, K. R. Subbaswamy, M. Menon, A. Thess, R. E. Smalley, G. Dresselhaus, and M. S. Dresselhaus, Science **275,** 187-191 (1997).

[68] R. A. Jishi, L. Venkataraman, M. S. Dresselhaus, and G. Dresselhaus, Chemical Physics Letters **209,** 77-82 (1993).

[69] K. Barnham and D. Vvedensky, *Low Dimensional Semiconductor Structures Fundamentals and Device Applications* (Cambridge Univesity Press, Cambridge, UK, 2001).

[70] J. Y. Park, S. Rosenblatt, Y. Yaish, V. Sazonova, H. Ustunel, S. Braig, T. A. Arias, P. W. Brouwer, and P. L. McEuen, Nano Letters **4,** 517-520 (2004).

[71] C. L. Kane, E. J. Mele, R. S. Lee, J. E. Fischer, P. Petit, H. Dai, A. Thess, R. E. Smalley, A. R. M. Verschueren, S. J. Tans, and C. Dekker, Europhysics Letters **41,** 683-688 (1998).

[72] Z. Yao, C. L. Kane, and C. Dekker, Physical Review Letters **84,** 2941-2944 (2000).

[73] M. A. Stroscio, M. Dutta, S. Rufo, and J. Y. Yang, Ieee Transactions on Nanotechnology **3,** 32-36 (2004).

[74] K. Baernham and D. Vvedensky, *Low Dimensional Semiconductor Structures Fundamentals and Device Applications* (Cambridge Univesity Press, Cambridge, UK, 2001).

[75] D. McKitterick, A. Shik, A. J. Kent, and M. Henini, Physical Review B **49,** 2585-2594 (1994).
[76] A. G. Kozorezov, T. Miyasato, and J. K. Wigmore, Journal of Physics-Condensed Matter **8,** 1-14 (1996).
[77] M. Cardona and G. Guntherodt, *Light Scattering in Solids V* (Springer-Verlag, Heidelberg, 1989).
[78] P. Poncharal, Z. L. Wang, D. Ugarte, and W. A. de Heer, Science **283,** 1513-1516 (1999).
[79] X. D. Wang, J. Zhou, J. H. Song, J. Liu, N. S. Xu, and Z. L. Wang, Nano Letters **6,** 2768-2772 (2006).
[80] V. Sazonova, Y. Yaish, H. Ustunel, D. Roundy, T. A. Arias, and P. L. McEuen, Nature **431,** 284-287 (2004).
[81] S. Chopra, K. McGuire, N. Gothard, A. M. Rao, and A. Pham, Applied Physics Letters **83,** 2280-2282 (2003).
[82] M. P. Blencowe, Contemporary Physics **46,** 249-264 (2005).
[83] M. A. Haque and M. T. A. Saif, Sensors and Actuators a-Physical **97-8,** 239-245 (2002).
[84] H. Park, J. Park, A. K. L. Lim, E. H. Anderson, A. P. Alivisatos, and P. L. McEuen, Nature **407,** 57-60 (2000).
[85] M. S. Dresselhaus, G. Dresselhaus, and P. C. Eklund, *Science of Fullerenes and Carbon Nanotubes* (Academic, New York, 1996).
[86] R. S. Ruoff and A. P. Hickman, Journal of Physical Chemistry **97,** 2494-2496 (1993).
[87] C. Y. Li and T. W. Chou, Nanotechnology **15,** 1493-1496 (2004).
[88] K. L. Ekinci, X. M. H. Huang, and M. L. Roukes, Applied Physics Letters **84,** 4469-4471 (2004).
[89] A. H. Chokshi, A. Rosen, J. Karch, and H. Gleiter, Scripta Metallurgica **23,** 1679-1683 (1989).
[90] S. C. Tjong and H. Chen, Materials Science & Engineering R-Reports **45,** 1-88 (2004).
[91] A. M. Elsherik, U. Erb, G. Palumbo, and K. T. Aust, Scripta Metallurgica Et Materialia **27,** 1185-1188 (1992).
[92] C. Forster, V. Cimalla, K. Bruckner, M. Hein, J. Pezoldt, and O. Ambacher, Materials Science & Engineering C-Biomimetic and Supramolecular Systems **25,** 804-808 (2005).
[93] T. Toriyama, Y. Tanimoto, and S. Sugiyama, Journal of Microelectromechanical Systems **11,** 605-611 (2002).
[94] L. T. Canham, Applied Physics Letters **57,** 1046-1048 (1990).
[95] J. M. Lauerhaas, G. M. Credo, J. L. Heinrich, and M. J. Sailor, Journal of the American Chemical Society **114,** 1911-1912 (1992).

[96] J. J. Shi, Y. F. Zhu, X. R. Zhang, W. R. G. Baeyens, and A. M. Garcia-Campana, Trac-Trends in Analytical Chemistry **23,** 351-360 (2004).
[97] M. S. Butler and J. A. Piper, Applied Physics Letters **45,** 707-709 (1984).
[98] G. Gauglitz and W. Nahm, Fresenius Journal of Analytical Chemistry **341,** 279-283 (1991).
[99] V. S. Y. Lin, K. Motesharei, K. P. S. Dancil, M. J. Sailor, and M. R. Ghadiri, Science **278,** 840-843 (1997).
[100] M. Breysse, B. Claudel, L. Faure, M. Guenin, R. J. J. Williams, and T. Wolkenstein, Journal of Catalysis **45,** 137-144 (1976).
[101] P. McCord, S. L. Yau, and A. J. Bard, Science **257,** 68-69 (1992).
[102] P. Bonasewicz, W. Hirschwald, and G. Neumann, Journal of the Electrochemical Society **133,** 2270-2278 (1986).
[103] G. Mie, Leipzig, Ann. Phys. **330,** 377-445 (1908).
[104] K. L. Kelly, E. Coronado, L. L. Zhao, and G. C. Schatz, Journal of Physical Chemistry B **107,** 668-677 (2003).
[105] P. Mulvaney, Langmuir **12,** 788-800 (1996).
[106] S. Link and M. A. El-Sayed, Journal of Physical Chemistry B **103,** 4212-4217 (1999).
[107] U. Kreibig, Journal of Physics F-Metal Physics **4,** 999-1014 (1974).
[108] B. N. J. Persson, Surface Science **281,** 153-162 (1993).
[109] J. A. Creighton and D. G. Eadon, Journal of the Chemical Society-Faraday Transactions **87,** 3881-3891 (1991).
[110] U. Kreibig and M. Vollmer, *Optical Properties of Metal Clusters;* (Springer, Berlin, Germany, 1995).
[111] A. P. Alivisatos, Science **271,** 933-937 (1996).
[112] J. J. Mock, D. R. Smith, and S. Schultz, Nano Letters **3,** 485-491 (2003).
[113] A. D. McFarland and R. P. Van Duyne, Nano Letters **3,** 1057-1062 (2003).
[114] S. A. Wolf, D. D. Awschalom, R. A. Buhrman, J. M. Daughton, S. von Molnar, M. L. Roukes, A. Y. Chtchelkanova, and D. M. Treger, Science **294,** 1488-1495 (2001).
[115] S. Parkin, X. Jiang, C. Kaiser, A. Panchula, K. Roche, and M. Samant, Proceedings of the Ieee **91,** 661-680 (2003).
[116] G. Binasch, P. Grunberg, F. Saurenbach, and W. Zinn, Physical Review B **39,** 4828-4830 (1989).
[117] M. N. Baibich, J. M. Broto, A. Fert, F. N. Vandau, F. Petroff, P. Eitenne, G. Creuzet, A. Friederich, and J. Chazelas, Physical Review Letters **61,** 2472-2475 (1988).

[118] S. S. P. Parkin, R. Bhadra, and K. P. Roche, Physical Review Letters **66,** 2152-2155 (1991).

[119] S. S. P. Parkin, Annual Review of Materials Science **25,** 357-388 (1995).

[120] N. F. Mott and H. Jones, *Theory of the Properties of Metals and Alloys* (Oxford University Press, London, UK, 1936).

[121] D. M. Edwards, J. Mathon, and R. B. Muniz, Ieee Transactions on Magnetics **27,** 3548-3552 (1991).

[122] S. S. P. Parkin, Z. G. Li, and D. J. Smith, Applied Physics Letters **58,** 2710-2712 (1991).

[123] S. S. P. Parkin, Physical Review Letters **71,** 1641-1644 (1993).

[124] S. S. P. Parkin, A. Modak, and D. J. Smith, Physical Review B **47,** 9136-9139 (1993).

[125] A. P. Ramirez, R. J. Cava, and J. Krajewski, Nature **386,** 156-159 (1997).

[126] S. Jin, T. H. Tiefel, M. McCormack, R. A. Fastnacht, R. Ramesh, and L. H. Chen, Science **264,** 413-415 (1994).

[127] J. S. Moodera, L. R. Kinder, T. M. Wong, and R. Meservey, Physical Review Letters **74,** 3273-3276 (1995).

[128] W. F. Egelhoff, P. J. Chen, C. J. Powell, M. D. Stiles, R. D. McMichael, C. L. Lin, J. M. Sivertsen, J. H. Judy, K. Takano, A. E. Berkowitz, T. C. Anthony, and J. A. Brug, Journal of Applied Physics **79,** 5277-5281 (1996).

[129] S. S. P. Parkin, K. P. Roche, M. G. Samant, P. M. Rice, R. B. Beyers, R. E. Scheuerlein, E. J. O'Sullivan, S. L. Brown, J. Bucchigano, D. W. Abraham, Y. Lu, M. Rooks, P. L. Trouilloud, R. A. Wanner, and W. J. Gallagher, Journal of Applied Physics **85,** 5828-5833 (1999).

[130] M. Julliere, Physics Letters A **54,** 225-226 (1975).

[131] M. Vazquez and C. GomezPolo, Journal of the Korean Physical Society **31,** 471-476 (1997).

[132] A. Punnoose, K. M. Reddy, A. Thurber, J. Hays, and M. H. Engelhard, Nanotechnology **18** (2007).

Chapter 7: Organic Nanotechnology Enabled Sensors

7.1 Introduction

Nanotechnology, without any doubts, has already shown its impact on the development of *organic sensors*. Such sensors employ organic materials, in particular *biomaterials*, as components of nano devices and nanostructured sensitive layers.

In this book, the definition of an organic sensor is a device that has either, at least, an organic element in its building structure or it utilizes an organic element to sense either a target analyte and/or physical changes. Organic molecules of interest in the fabrication of such sensors consist of a myriad of natural and synthetic materials which include small organic molecules (such as *lipids*, *neurotransmitters* and *carbohydrates*), *monomers* (such as *amino acids*, *nucleotides* and *phosphates*), *synthetic polymers* (such as *Teflon* and *polyaniline*), *biopolymers* (such as *DNA*, *RNA*, *proteins* and *polysaccharides*), and other *synthetic macromolecules* (such as *dendrimers*).

In this chapter, a comprehensive overview of the properties of different surfaces and their interactions of organic molecules will be presented. The properties of the most commonly used surfaces, such as gold, silicon, metal oxides, intrinsically conductive and nonconductive polymers in organic micro/nanotechnology will be briefly explained and the reader will learn how to incorporate them in the development of sensors. Proteins and DNA, as the most important components for organic sensing, will be described and their applications for organic sensing will be illustrated through relevant examples. The applications of organic/organic and organic/ non-organic nanostructured conjugates, assembly of organic/inorganic materials as well as dendritic materials in sensors fabrication will be presented. Eventually mechanical and magnetic effects in nanostructures and manipulating them for the fabrication of sensors will be discussed.

7.2 Surface Interactions

The study of various surfaces (including the surfaces of nanostructures) and understanding how to manipulate those surfaces are the basis of the development of organic sensors. Knowledge about the interactions of target molecules with organic sensors are of great importance as the performance (i.e. the selectivity, durability and stability) of the sensors depends on these interactions.

If the sensor is a *surface-type sensor* (or *affinity sensor*), its active area is prepared by the addition of sensitive and selective layers. In the *bulk-type sensors*, we are dealing with surfaces of nanostructures or nanoparticles within the bulk of the transducer or in the bulk of the target media that are manipulated in order to selectively bind to desired target molecules. In both cases of surface and bulk type sensors, the importance of the surface interactions is obvious. In order to sense biocomponents, such as proteins and DNA, it has to be *immobilized* on the surface of an affinity sensor or on the surface of nanostructures within the bulk of a bulk-type sensor. These surfaces have to become both selective and sensitive to the target component.

Literature on the development of organic thin films and sensitive/selective surfaces is quite extensive and many issues relating to maintaining their activities, optimization and tuning have been studied and addressed.[1-3] Different approaches are possible for the development of such surfaces. The approach taken depends on the transducer type, the nature of organic components, the surface chemistry of the device, the way the device is used, and the nature of the samples.

In this section, the most common processes for the creation of surfaces suitable for interactions with biomolecules and organic materials will be presented. Such interactions can be physical, chemical or a combination of both. The type of interactions can cover a wide range from very strong covalent bonds to weak van der Waals interactions. Furthermore, two major surface fabrication strategies: self-assembly and layer-by-layer techniques will be presented.

7.2.1 Covalent Coupling

When atoms *share* pairs of electrons between them covalent bonds are formed. Many organic derivatives possess different surface functional groups such as amines, thiols, and carboxylic acids. These functional groups can be used to form covalent bonds with the functional groups on the target organic molecules such as proteins **(Fig. 7.)**. A covalent bond can be formed either in a single-step, a two-step (where the surface is first made more reactive), or a multi-step process.

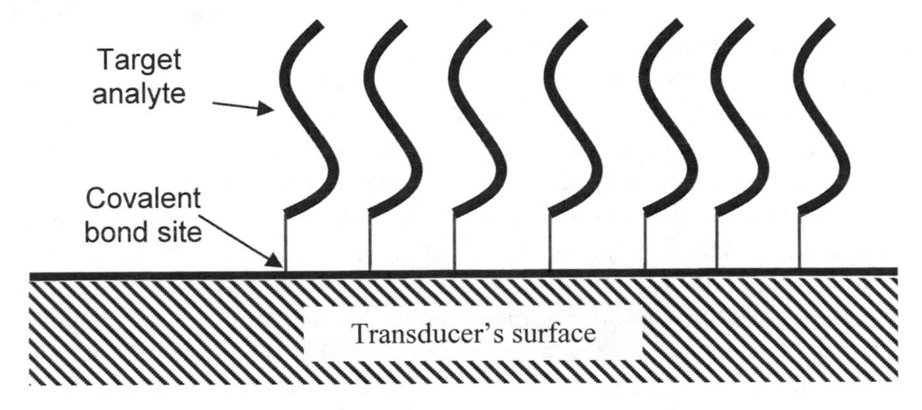

Fig. 7.1 A schematic of covalent coupling to the surface of a transducer.

Cross-linkers can be used to facilitate the formation of a covalent bond. Cross-links are covalent bonds that link one molecule to another (**Fig. 7.2**). They are formed by chemical reactions that are initiated when a form of energy is applied on the surface (such as irradiation of electromagnetic waves, heat, pressure).

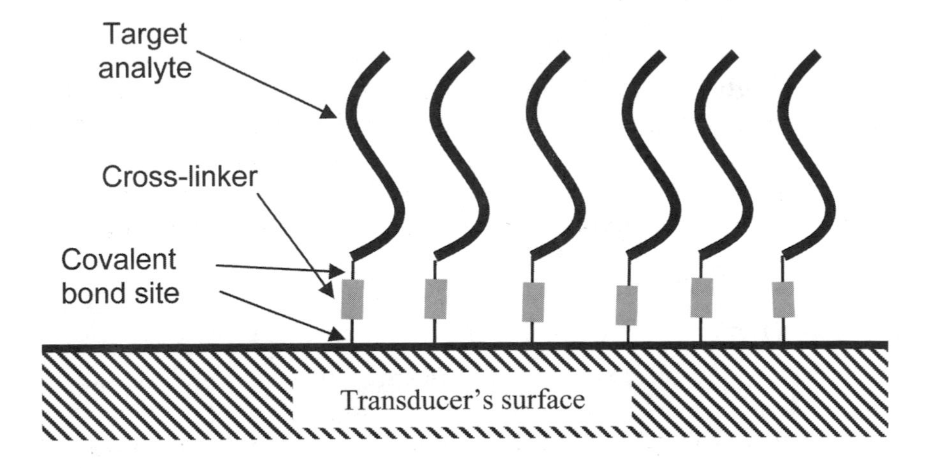

Fig. 7.2 A schematic of cross-linking on the surface of a transducer.

Covalent bonds are strong. The interactions are relatively fast (comparing with other bonds) and very well controlled. However, the processes of their formation are generally complex and require synthetic steps which can be both costly and time consuming.

Cross-linking has similar advantages and disadvantages to covalent bonding. However, it provides flexibility in establishing an interaction by providing relevant functional groups. It is strong but it also has the disadvantage of being able to degrade the material's biocompatibility.

Quite often, the direct coupling of one functional group to another is not energetically favorable; hence functional groups may need to be reacted with an intermediate compound, in order to be made more reactive. This process is called *activation*. For many molecules that are soluble in organic solvents, a wide range of activators and coupling reactions can be employed. Unfortunately, many of these reactions that are widely used in organic synthesis chemistry cannot be utilized for biomolecules such as proteins. This is due to the fact that the coupling processes of such biomolecules need to be carried out in mild conditions in order to avoid deteriorating functionality of the biomolecules which limits our options.

There are a plethora of different types of bonds for covalently coupling functional groups. Some of the most widely utilized covalent coupling strategies will be outlined in this section.

Application of carboxylic acids

Carboxylic acids are organic molecules that contain a carboxyl group, which is a carbonyl group –(C=O)– whose carbon atom is bonded to a hydroxy (–OH) group. It has the formula –C(=O)–OH, usually written as –COOH.[4] In determining the name of a carboxylic acid, the total number of carbon atoms in the longest chain is counted, including the one in the –COOH group. Some examples of such compounds are shown in **Fig. 7.3**.

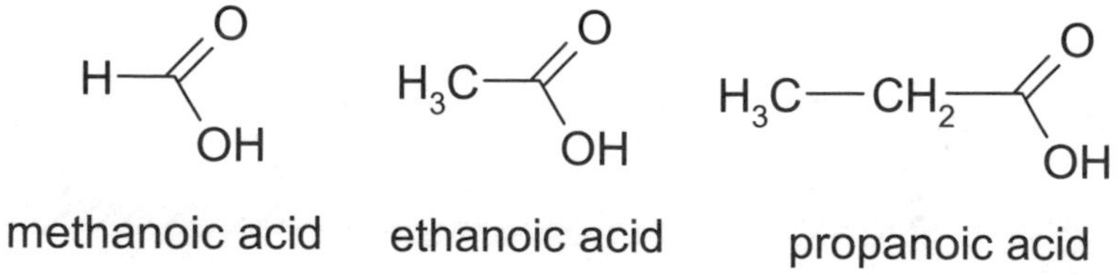

methanoic acid ethanoic acid propanoic acid

Fig. 7.3 Some carboxylic acids.

Carboxylic acids can be covalently coupled with other functional groups to form several types of bonds such as amides, thioesters, acyl esters and aryl esters (see **Fig. 7.4**). In this figure, R′ denotes the rest of the compound (such as a protein) which is supposed to be targeted. R is the compound which is bond to the carboxylic acid (whose other end is attached to the sensor surface), and hence the covalent coupling bond joins both R and R′ which makes the sensing links.

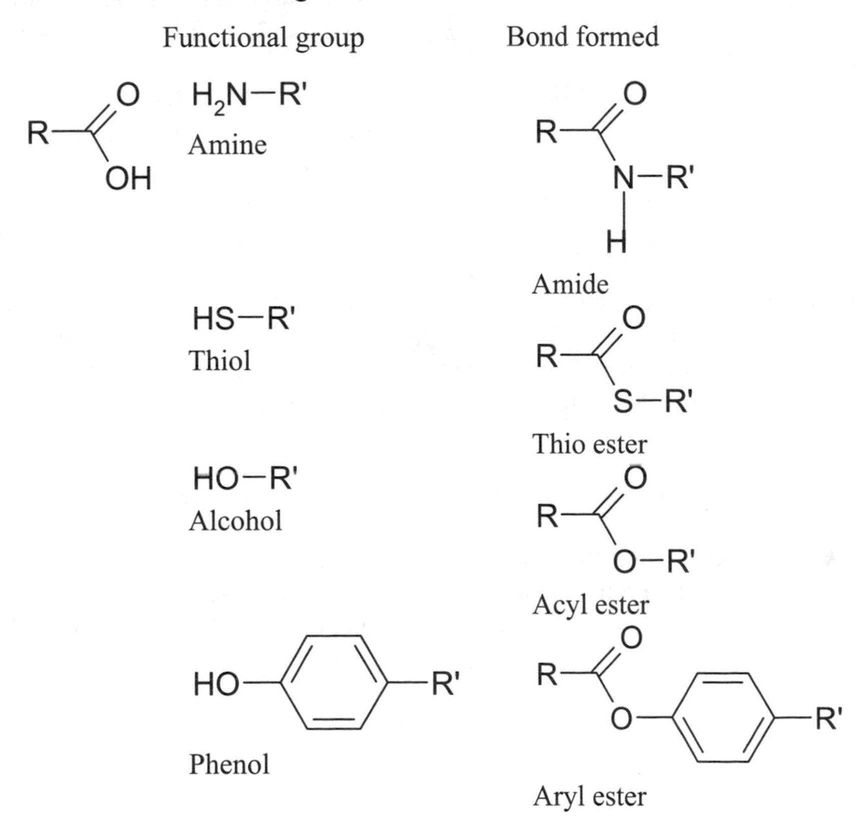

Fig. 7.4 Interactions between carboxylic acids and other functional groups.

Carboxylic acids can be placed on a sensor surface by several different means. Many water soluble molecules contain amine functional groups. When a carboxylic acid covalently bonds with an amine group, it forms an amide bond. The resulting molecule is called an *amide*. In fact, the amide bond is one of the most important bonds in nature, and is the basis of the formation of polypeptides (which will be described later in this chapter and its importance in proteins). Therefore, these bonds are called peptide bonds. The direct reaction between a carboxylic acid and amine functional groups are very slow at room temperature, and hence activation is needed to facilitate the covalent coupling. Some activation strategies for forming amides will be outlined below.

Example 1: Amide bond formation using carbodiimide

Often in synthetic organic chemistry, compounds containing the carbodiimide functional groups are utilized to activate carboxylic acids for the formation of amides or esters. A *carbodiimide* is a functional group which contains –N=C=N–.

Forming such a bond usually involves two steps: in the first step, the carboxylic acid is activated by the carbodiimide reagent to form an O-acylisourea intermediate. This is followed by a nucleophilic displacement of the intermediate by the amine to form the final amide bond.

Several side reactions can be produced in the formation of an amide using a carbodiimide.[5] The carbodiimide may react with the acid to produce: the O-acylisourea, which can is a carboxylic ester and amide and urea. The O-acylisourea can react with carboxylic acid to give a carboxylic anhydride. This carboxylic anhydride may continue the reaction further to produce the stable N-acylurea and the desired amide.

Example 2: Amide bond formation using the displacement of esters

Amide bonds can also be formed by making active ester molecules from carboxylic acids, and then displacing them with amines. An ester (**Fig. 7.5**) in an organic compound in which an organic group (symbolized by R') replaces the hydrogen atom in a carboxylic group.

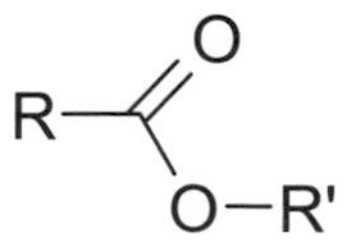

Fig. 7.5 An ester.

If heated, esters can react with primary or secondary amines to produce amides. However, in many cases (such as protein immobilization) we cannot apply heat because the organic components can be damaged. Esters can be formed in NH_2- -carboxyl covalent interaction. The -NH_2 amino coupling method (self-assembled monolayer sample with -NH_2) can be used as shown in **Fig. 7.6**.

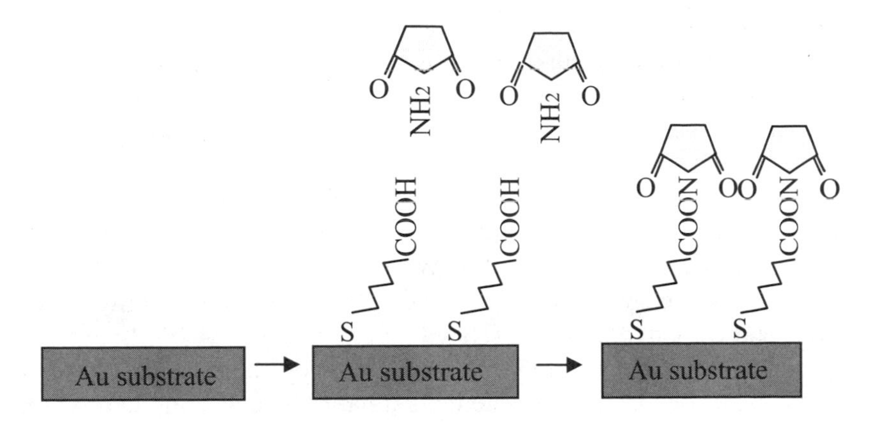

Fig. 7.6 Ester formation in the NH_2- -carboxyl covalent interaction.

Practically however, the carboxylic end group requires activation. The carboxylic acid end group is first made more reactive by converting it into an acid chloride using a chlorination reagent such as thionyl chloride (**Fig. 7.7**). However, acid chlorides are generally not stable, so they are rapidly converted into esters. Such esters are quite commonly referred to as *activated esters*. *Pentafluorophenol* (*PFP*) and *N-hydroxysuccinimide* (*NHS*) are commonly used for forming activated esters as they form excellent displacement groups. Such molecules are well-known for their high stability and their great selectivity for amide formation. The ester forming reaction is carried out in the presence of a base and a typical reaction is shown in **Fig. 7.7**.

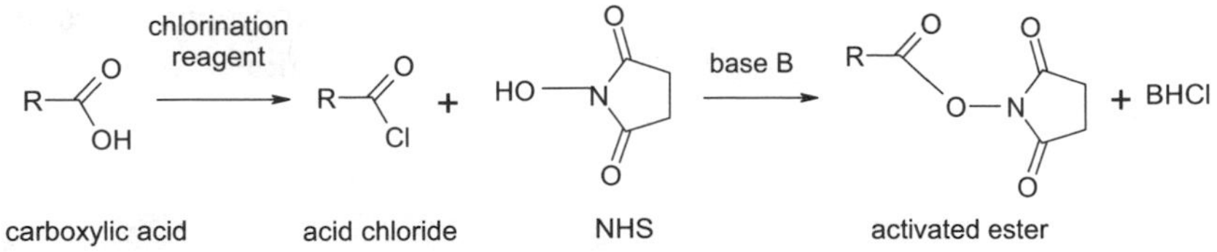

chlorination reagent — base B — + BHCl

carboxylic acid — acid chloride — NHS — activated ester

Fig. 7.7 The formation of an activated ester.

The reaction is continued and the activated ester reacts with an amine to form an amide (**Fig. 7.8**).

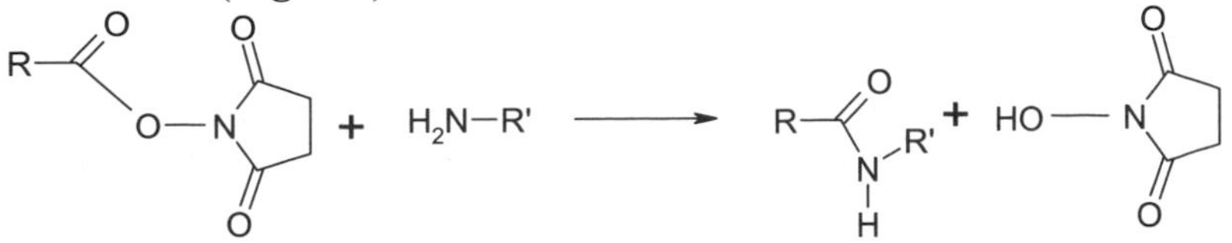

Fig. 7.8 Amide formation by active ester displacement.

Example: Thiol reactions

Thiol groups are very important in the coupling of antibodies as they offer the possibility of forming *Fab′ fragments* with free thiols. Free thiol groups are not very stable. They can be oxidized to disulphides, sulphones and sulphonates. Reagents such as bromine can be used to oxidize the thiol groups and to create an organic disulfide (R-S-S-R). More powerful reagents such as sodium hypochlorite can be employed to transform them into *sulfonic acids* (R-S(=O)$_2$-OH). When compared to carboxylic acids, sulfonic acids are generally much stronger and tend to tightly bind to proteins and carbohydrates.

Thiol groups play an important role in biological systems. When the thiol groups of two cysteine (cysteine is a naturally occurring hydrophobic amino acid which contains a thiol group and is found in most proteins) residues (as in monomers or constituent units) are brought near each other in the course of protein folding, an oxidation reaction can create a cystine unit with a disulfide bond (-S-S-).

Disulfide bonds can contribute to a protein's tertiary structure if the cysteines are part of the same peptide chain, or contribute to the quaternary structure of multi-unit proteins by forming fairly strong covalent bonds between different peptide chains. The heavy and light chains of antibodies are held together by disulfide bridges.

Two thiol compounds can be covalently coupled either through a disulfide formation or via thio-ether formation. For the disulphide formation, a two step process is used. First the surface thiol group reacts with a symmetrical disulphide such as 2,2′-dithiopyridine (**Fig. 7.9**). Next, the protein or ligand, which contains a thiol group and that we wish to immobilize on the surface of sensor, is added to the environment. This then interacts with the surface-bound asymmetric disulphides to make the ligand or protein attachment (**Fig. 7.10**).

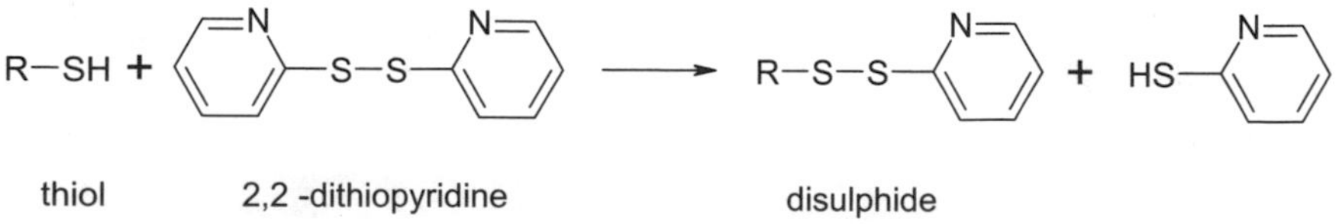

R—SH + [2,2′-dithiopyridine] ⟶ R—S—S—[pyridyl] + HS—[pyridyl]

thiol 2,2 -dithiopyridine disulphide

Fig. 7.9 Reaction of a thiol group with a symmetrical disulphide to form a new disulfide.

R—S—S—[pyridyl] + HS—R′ ⟶ R—S—S—R′ + HS—[pyridyl]

Fig. 7.10 Thiol-disulphide exchange between a surface-bound (R-) disulphide group and the thiol group of the ligand or protein (-R′).

There are numerous other examples of surface-immobilization by covalently coupling that use other functional groups such as aldehydes, hydroxides, sulfonyl chlorides, and alcohols.[1-3]

Photo-crosslinkers

There are also chemical groups such as aryl azides, aryl diazirines and benzophenons that can be functionalized and to act as intermediators when light impinges upon them. Such groups can function as excellent *photo-crosslinkers*.

Arryl azides usually absorb photons in the UV range between 330-370 nm.[6] After receiving photons, it transforms into nitrene which is highly reactive. This reacts instantly with any chemical groups in its vicinity including the solvent; hence the efficiency of this crosslinker is low. Aryl diazirines can generate highly reactive carbine species upon the absorption of a phonon in UV-range frequencies. Carbenes can react with acidic species such as R-NH_2 and R-OH which are abundant in proteins.[7]

The first application of benzophenones in biological systems was reported in 1974.[8] Benzophenone can react with UV light of the wavelength of 360 nm to produce a biradical at the ketyl centre.[9] Such radicals are abundant in proteins but generally absent in solvents. As a result, this material is a highly efficient as a protein-selective cross-linker.

7.2.3 Adsorption

Molecules may be adsorbed onto different surfaces using *hydrophobic*, *ionic* and/or *Van der Waals* interactions (**Fig. 7.11**). For instance, proteins can be directly adsorbed onto carbon or gold surfaces. Many other organic molecules also show this capability. Adsorption is simple to implement and inexpensive for many sensing applications. However, it is relatively unstable, it may cause the protein to denature (which will be explained later in this chapter) on hydrophobic surfaces, and the rate of formation depends on the analytes used.

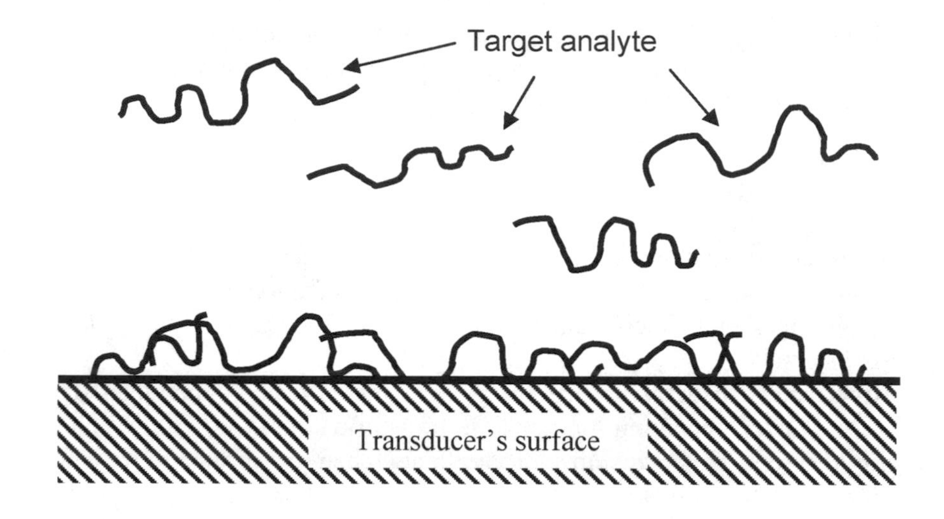

Fig. 7.11 The Schematic of surface adsorption of biomolecules.

7.2.4 Physical Entrapment

In this method, a semi-permeable membrane is generally used. Small molecules can diffuse freely in and out of the bulk of the membrane, but larger molecules are held inside (**Fig. 7.12**). Physical entrapment can be used in mass production and it requires mild conditions for the interactions which can be biocompatible. However, it is difficult to reproduce and the diffusion barrier slows the response.

Large biomolecules such as proteins can be entrapped in the bulk of a polymer hydrogel. Some polymers such as poly (vinyl alcohol) can be dissolved in solvent at elevated temperatures and can gel at low temperatures due to hydrogen bonding.[10] For larger bioparticles such as organelles and whole cells, materials such as agarose are used.

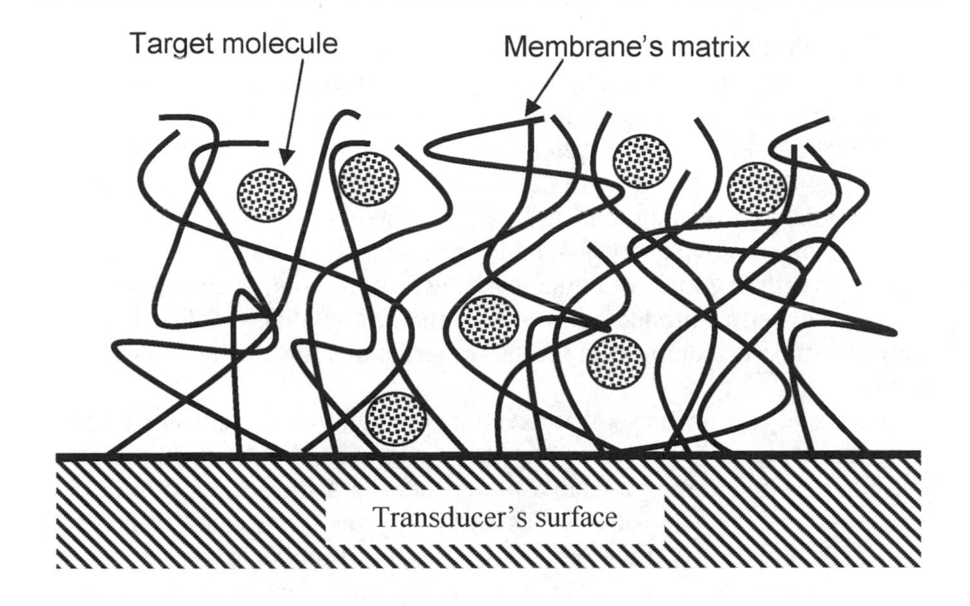

Fig. 7.12 The schematic of physical entrapment.

7.2.5 Chemical Entrapment

Entrapment can also occur with the help of chemical means. This is when the inner sites of the porous membrane make strong bonds with trapped molecules.

In order to make stronger trapping conditions it is possible to produce covalent links between the membrane internal sites and the biomolecules. A common entrapment procedure is via the use of silica sol-gel glasses to create a porous network under mild conditions. An alkoxide precursor can be hydrolyzed/condensed by an acid or a base to form these networks.[11]

7.2.6 Self-Assembly

Self assembly is a process in which atoms, molecules, nano-particles and other building blocks attach to functional systems in an ordered manner. Self-assembly is generally self-driven by the energies of such systems.[12] In nature, many biological systems use this process to form bio-structures. By imitating the strategies that occur in nature, we can use self-assembly to modify the surfaces of sensors, and to create novel molecules and nano-structures as well as supramolecules.

Although the concepts of self-assembly are applicable to any material, currently the most promising avenues for self-assembly are those concerned with organic components. Such self-assembled materials are used in controlling the growth of nanomaterials, fabricating electrical insulators, making sensitive and selective layers, developing superlattices of organic compounds with accurate thicknesses, etc. There are also several other advantages of employing self-assembly in nanotechnology:[13] it can be utilized to directly incorporate biological structures as components in the final systems as well as producing structures that are relatively defect-free and flawless since it requires that the target structures are thermodynamically stable.

Despite all the advances that have been made, the mechanisms responsible for self-assembly have not yet been fully understood. In addition, it is still not possible to mimic many of the processes known to happen in biological systems. As a result, self-assembling processes cannot, in general, be designed and carried out on demand. Many of the ideas that are essential to the development of this area are simply not yet under control such as the molecular conformation, the entropy and enthalpy relationships, and the nature of the non-covalent forces that connect the particles.

Self-assembled monolayers (*SAMs*) are surfaces that consist of a single (mono) layer of molecules on a surface. SAMs are becoming increasingly useful in different technologies, especially in developing sensors and fabricating accurate micro/nano devices. This is due to the fact that, they allow the film thickness and the composition to be precisely controlled at the scale of ~ 0.1 nm.

Amphiphilic molecules, which are molecules that contain both hydrophilic and hydrophobic groups, are widely used in self-assembly processes. Due to their hydrophobic-hydrophilic nature, amphiphilic molecules find a plethora of applications in our daily lives such as paint dispersants, cosmetic ingredients, detergents, soaps. SAMs can use these molecules to form surfaces suitable for molecular immobilization on transducers. Such molecules should consist of two different head and tail functional groups where one end sticks to the sensor's surface and the other end interacts with the analyte biomolecules. Hence, they make a strong link between the biomolecules and sensor.

For the formation of SAMs, the substrate is generally immersed into a dilute solution of the self-assembling molecules and a monolayer film gradually forms (**Fig. 7.13**). The monolayer formation rate can range from just a few minutes to several hours.

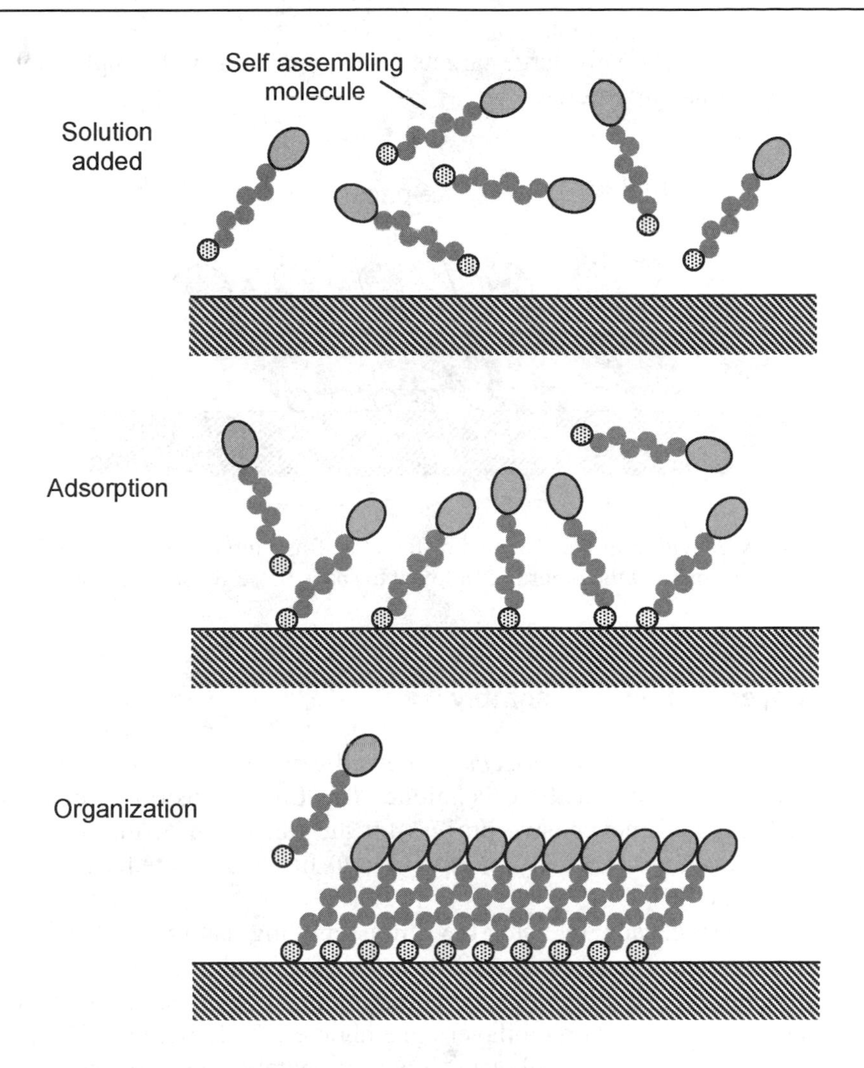

Fig. 7.13 Formation of SAMs.

SAMs cause compressive surface stress change on surfaces which can be monitored with acoustic sensors and AFM measurements. SAMs can be fluorescently tagged to be used with optical sensors.[14]

By mixing self-assembling molecules with different end groups in the preparation solution; we can produce mixed SAMs which show interesting sensing properties: one end group is sensitive to one target analyte and the other end group is sensitive to a totally different target analyte (**Fig. 7.14**). This produces sensitive layers which can be used for detecting different targets. SAMs with no active end groups or different chain lengths can be used as spacers. These inert spacers are sometimes necessary when we intend to sense relatively large targets (in comparison with single molecules) such as nano-particles.

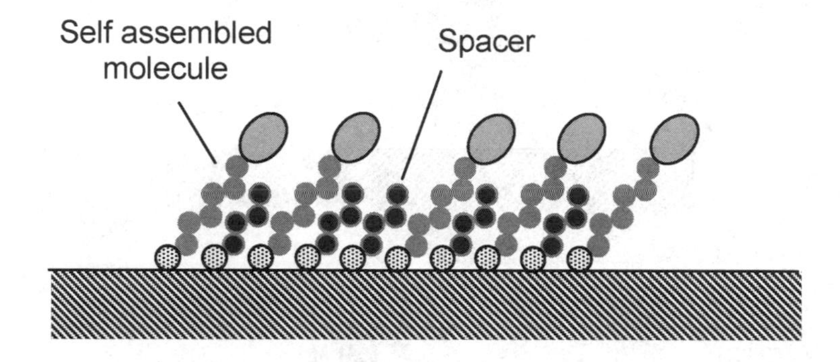

Fig. 7.14 SAMs with a mixture of two different self-assembled molecules. One is utilized for sensing and the other is employed to space those sensing molecules.

7.2.7 Layer-by-Layer Assembly

The *layer-by-layer* (*LbL*) *assembly* method can be regarded as a versatile bottom-up nanofabrication technique. The LbL deposition technique offers an easy and inexpensive process for the formation of multilayered materials and allows a variety of materials to be incorporated within the film structures.[15]

Over the past decade, by using the self-assembling nature of artificially-designed molecules, many types of nanoscale molecular assemblies have been developed.[16] This includes molecular recognition-directed molecular assemblies,[17,18] surfactant bilayer membranes,[19] Langmuir–Blodgett films,[19] self-assembled monolayers,[20] and alternately deposited polyelectrolyte multilayers.[21] LbL assembly is an approach based on the alternating adsorption of different materials containing complementary charged or functional groups to form ultratin layers. This method generally uses oppositely charged species and the level of control over the film composition and structure is down to several nanometers.[22]

The classical LbL approach developed by Decher is an interesting example.[21] They used the electrostatic attraction between oppositely charged molecules as a driving force for the multilayer buildup. The process is depicted in **Fig. 7.15** for the case of polyanion-polycation deposition on a positively charged surface. Strong electrostatic attraction occurs between a charged surface and an oppositely charged molecule in solution. In principle, the adsorption of molecules carrying more than one type of charged elements allows for charge reversal on the surface, which has two important consequences: (i) the repulsion of liked-charged molecules and thus self-regulation of the adsorption occurs and this restricts the deposition to a single layer, and (ii) the ability of an oppositely-charged molecule to be adsorbed in a second step on top of the first one. Cyclic repetition of both adsorption steps leads to the formation of multilayer structures.

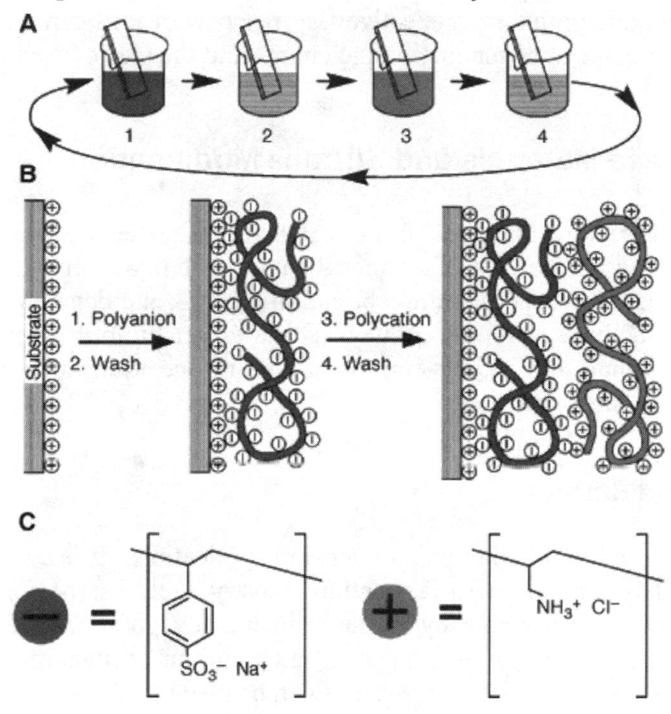

Fig. 7.15 (**A**) A schematic diagram of the film deposition process using slides and beakers. Steps 1 and 3 represent the adsorption of a polyanion and a polycation, respectively, and steps 2 and 4 are washing steps. These four steps are the basic buildup sequence for the simplest film architecture, $(A/B)_n$. The construction of more complex film architectures only requires additional beakers and a different deposition sequence. (**B**) A simplified molecular picture of the first two adsorption steps, depicting the film deposition starting with a positively charged substrate. Counterions are omitted for clarity. The polyion conformation and the layer interpenetration are an idealization of the surface charge reversal with each adsorption step. (**C**) The chemical structures of two typical polyions, firstly the sodium salt of poly(styrene sulfonate) and secondly poly(allylamine hydrochloride). Reprinted with permission from the Science Magazine publications.[21]

Using this method semi permeable membranes can be made for the separation of gases in sensing applications. For instance, it has been reported that in order to block O_2 and make the layer permeable for H_2 poly(allyl amine)/sulfonated polystyrene (PAH/SPS) polyelectrolyte multilayers can be used.[23] Selective permeation in multilayer thin films can also be employed for the separation of small biomolecules. Liu and Bruening reported the separation of glucose and sucrose using a multilayer film in this way.[24]

Another interesting application of these multilayers is in the area of sensing organic pollutants. Selective ion transport of ions can be observed in such multilayers, depending on the charge and the size of the ions.[25]

7.3 Surface Materials and Surface Modification

There are many materials from which the transducer surface can be made of. The choice of surface modification technique depends not only on these materials, but also the fabrication method, the degree of sensitivity and selectivity required, as well as the target biomolecules. Typical materials which are widely employed for surface modification will be presented in this section.

7.3.1 Gold Surfaces

Gold is used in a wide range of sensing applications. It is a noble metal and is widely used in micro-fabrication industry. Gold can be deposited on the active area of a sensor by methods such as evaporation or sputtering. To have a good adhesion to the surface a chromium or titanium intermediate layer is needed. Gold nanoparticles can be placed on a sensor surface to form an array or they can be embedded into thin films (such as polymeric thin films). Freshly deposited gold is quite hydrophilic; however, organic molecules rapidly adsorb onto it and hence it becomes hydrophobic.

The deposition of gold thin films and the utilization of gold surfaces is well studied. In addition, the fabrication of gold nanoparticles is also a well established science. Gold is a suitable substrate for forming SAMs. The phenomenon of self-assembly on gold was first introduced by Nuzzo and Allara in the early 1980s.[26] A common example is the formation of *thiol SAMs* on gold. A thiol is a compound that has a functional group composed of a sulfur atom and a hydrogen atom (-SH). Alkane-thiols are among the most used thiols in sensing applications. An alkane-thiol is an alkane with a thiol functional group ($HS\text{-}C_nH_{2n+1}$). An alkane is an acyclic saturated hydrocarbon that has the general formula of C_nH_{2n+2}.

Sulfur has particular affinity for gold with a binding energy of approximately 20-35 kcal/mol. The thiol head group will attach to the gold surface. The alkane tails stand extended from the surface at an angle of ~30° from the perpendicular of the substrate, and are packed together due to van der Waals forces. In general, thiols (of concentrations in the order of a few mM) are adsorbed onto the substrate from alcohol solutions. The transducer with a gold layer is simply immersed in the solution and washed subsequently. Thiols can deprotonate upon adsorption onto the gold surface:

$$\text{R-SH+Au} \rightarrow \text{R-S-Au} + e^- + H^+ \tag{7.1}$$

This creates a strong thiolate bond with the surface. Thiols assemble in the form of dense monolayers with a 2D order. With time, the SAMs undergo reorganization, which produces a more perfect and well orientated layer (**Fig. 7.13**). In this case, the self-assembly molecules are alkanethiols and the substrate is gold. Other molecules such as thiols (R-SH), sulphides (R-S-R) and disulphides (R-S-S-R) also able to self-assemble well on gold structures.

Other metals such as silver (Ag), platinum (Pt) and copper (Cu) can also be used as substrates in the formation of SAMs. However, gold is the most popular choice, as Ag and Cu oxidize rapidly and Pt is not the most common material in the microfabrication industry. Because of widespread interest in SAM technology, many different types of thiols with different chain lengths and end group functionalities are commercially available.

7.3.2 Silicon, Silicon Dioxide and Metal Oxides Surfaces

Materials such as *Silicon, silicon dioxide* (*SiO_2*), and many metal oxides are the backbone of the semiconductor industry and sensor fabrication. They are well studied materials and a result their properties are known and there are standard tools available for the fabrication of devices which incorporate these materials.

Certain forms of silicon dioxide, such as silica, quartz, and glass are transparent. Since many sensors rely on light absorbance or fluorescence measurements, it is important that the substrate is transparent so as not to interfere with the measurements being taken. Most metal oxides are transparent to visible light. Silicon, which is currently the material of choice in the semiconductor industry, is not transparent to visible light; however it is transparent to UV light. Silicon is easily oxidized in air to form a thin native oxide layer on the surface.

It is possible to generate oxygen-bridged metal atoms and hydroxyl (-OH) groups on the surface of SiO_2 and metal oxides (**Fig. 7.16**). The presence of the hydroxyl groups makes the surface quite hydrophilic.

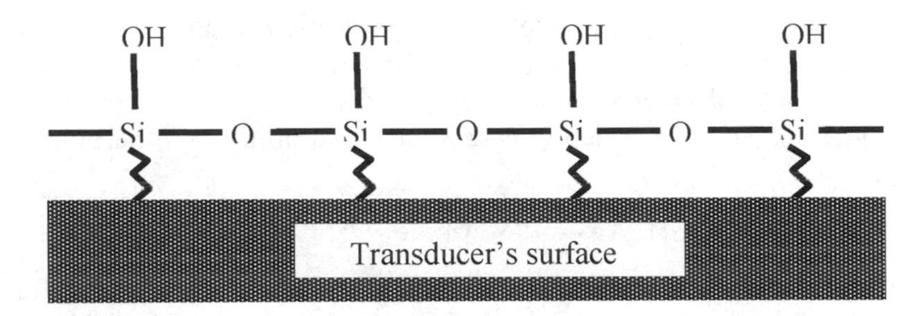

Fig. 7.16 Oxygen-bridged Si atoms terminated with hydroxyl groups on the surface of SiO_2.

This group also provides the opportunity to modify the surface with compounds such as *silanols* which can be used as intermediate links between the surface of the sensor and the target analytes. For example, in the presence of water, silane-coupling agents (silicon-hydrocarbyl derivatives such as amoninopropyltrimethoxsilane) produce highly reactive silanols (**Fig. 7.17**). Subsequently, these silanols begin to condense, forming oligomeric structures. At the same time, weak hydrogen bonds also form on the substrate's surface. Further condensation and dehydration between the coupling agent and the surface results in multiple strong, stable, covalent bonds being formed on the surface. Silane-coupling agents are commercially available with many different terminal functional groups such as the amine ($-NH_2$), thiol (-SH) and chloride (-Cl) groups. These terminal groups can be utilized to couple organic molecules and organic nanoparticles to the surface. These attributes make them very attractive for the modification of sensor surfaces which are compatible with the silicon industry.

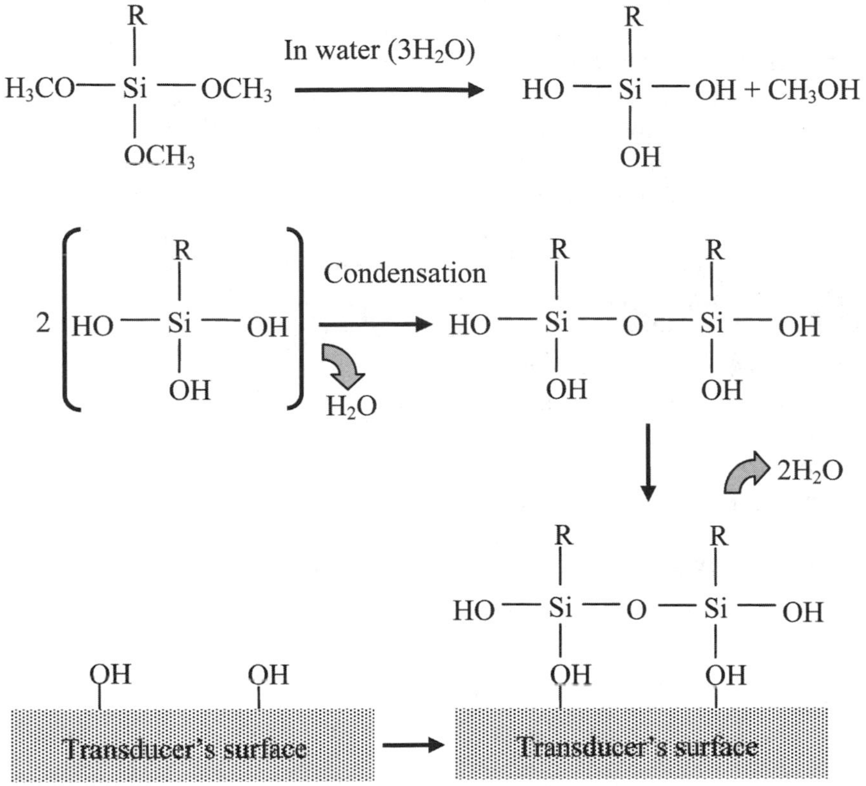

$$R\text{-}Si(OCH_3)_3 \xrightarrow{\text{In water } (3H_2O)} R\text{-}Si(OH)_3 + CH_3OH$$

$$2\,[R\text{-}Si(OH)_3] \xrightarrow{\text{Condensation}} HO\text{-}Si(R)(OH)\text{-}O\text{-}Si(R)(OH)\text{-}OH + H_2O$$

$$\downarrow \; 2H_2O$$

Transducer's surface → Transducer's surface

Fig. 7.17 Application of silanols to produce a modified surface.

7.3.3 Carbon Surfaces

Carbon is widely used as an electrode for electrochemical sensors. Additionally, it is a very favorable material with the emergence of carbon nano-tubes. Carbon surfaces are very hydrophobic and can adsorb many organic species. This increases the non-specific binding, which in turn reduces both the selectivity and the sensitivity of the sensing system. Unfortunately, the surface of a carbon species is also difficult to modify, as oxidation of the surface is not an easy task. Generally, an intermediate layer is used to modify the surface.

7.3.4 Conductive and Non-Conductive Polymeric Surfaces

Polymers are large organic molecules that consist of repeating units, called monomers, which are covalently connected together. They are attractive materials for the development of sensors as their chemical and physical properties can be tailored over a wide range.[27,28] Some of their advantages include the following: low fabrication cost, an ability to take different forms including nanosized formation, biocompatibility and their ability for chemical and biosensing at different temperatures.

Both *intrinsically-conducting* and *non-conducting polymers* (*ICPs* and *NCPs*) can be used for the development of sensors. They can be employed in the sensor structure, they can directly participate as the sensitive layer, and they can also be used as the media for immobilizing biomaterials on a surface.

There are a multitude of possible polymeric materials which can be used in sensing applications. As fabrication technologies such as nanoimprinting are becoming more and more mature, and due to the emergence of new thermal and UV-curable polymers, the attraction of working with polymers is becoming increasingly favorable in nanotechnology.

Non Conducting Polymers (NCPs)

NCPs are becoming increasingly attractive in the fabrication of transducers and sensors. They are inexpensive, simple to prepare and it is relatively easy to manipulate their structure. Screen printing, molding and stamping are generally employed to carry out such tasks. With a plethora of photo and thermally curable NCPs available, there are many possible sensor designs that can be formed. NCPs can also be employed as selective layers in bio and ion selective sensing. In addition, NCPs have many applications in biosensing. They can be used for the entrapment of biomolecules and as a membrane for the formation of bio-selective layers in order to immobilize biomolecules such as proteins and DNA. NCPs are also very attractive in the fabrication of ion-selective membranes. This is achieved by placing *ionophores* within polymeric membranes. These ionophores can be organic salts, macrocyclic compounds such as crown ethers, calixerenes, and antibiotics. The most commonly used polymeric matrix for making ion selective membranes is *polyvinyl chloride* (*PVC*). The PVC is mixed with a plasticizer in order to soften it and then the, ionophores are added to provide selectivity towards the target ion.[29] An example of such ion-selective membranes can be found in the report by Shamsipur et al who fabricated a PVC membrane electrochemical sensor for monitoring cerium(III) ions.[30] The sensor was based on a 1,3,5-trithiane membrane carrier and had detection limit of 3.0×10^{-5} M.

With polymers a wide range of surface functionalities can be obtained. Such surfaces can be directly used as sensitive components of transducers or can be manipulated using a variety of surface modification techniques that have been previously discussed.

Surfaces of polymers can be extremely diverse, ranging from extremely hydrophobic to extremely hydrophilic. For instance, *polysaccharides* such as *cellulose* and *agarose* are very hydrophilic due to the predominance of hydroxyl (-OH) groups. Biomaterials such as proteins do not adsorb directly onto these surface, which means that only a small amount of non-specific binding occurs. As a result, these biomaterials need to be immobilized by covalent bonding. However, hydroxyl groups are difficult to be employed for direct covalent bonding. Therefore, we need to make the surface more reactive using activating reagents, which prepares the surface for direct coupling. Many methods can be found in literature for the manipulation of these polymeric surfaces.[31,32]

Polymers, such as *carboxymethyl dextran* (a cosmetic ingredient in hair care and skin care products), can efficiently increase the concentration of ion exchange groups near the surface. This approach has been highly successful in using such surfaces for sensing applications, as introduced by BIAcore in their *surface plasmon resonance* (*SPR*) sensing system.[33]

Molecular imprinting

Molecular imprinting (*MIP*) is defined as a technique of creating template-shaped moulds/recesses in polymer matrices with the memory of the template molecules. In this method, selective recognition sites are formed in synthetic polymers within templates (atom, ion, molecule, complex or a molecular, ionic or macromolecular assembly).[34,35]

The key step of this technique is the polymerisation of monomers in the presence of a templating ligand (**Fig. 7.18**).[36] Subsequent removal of these templating ligands leaves behind *memory sites*, or imprints, in the polymeric network. It is believed that the functional monomers become spatially fixed in the polymer via their interaction with the imprint species during the polymerisation reaction. The result is the formation of imprints in the polymer, which are complementary, both sterically and chemically, to the templating ligand.

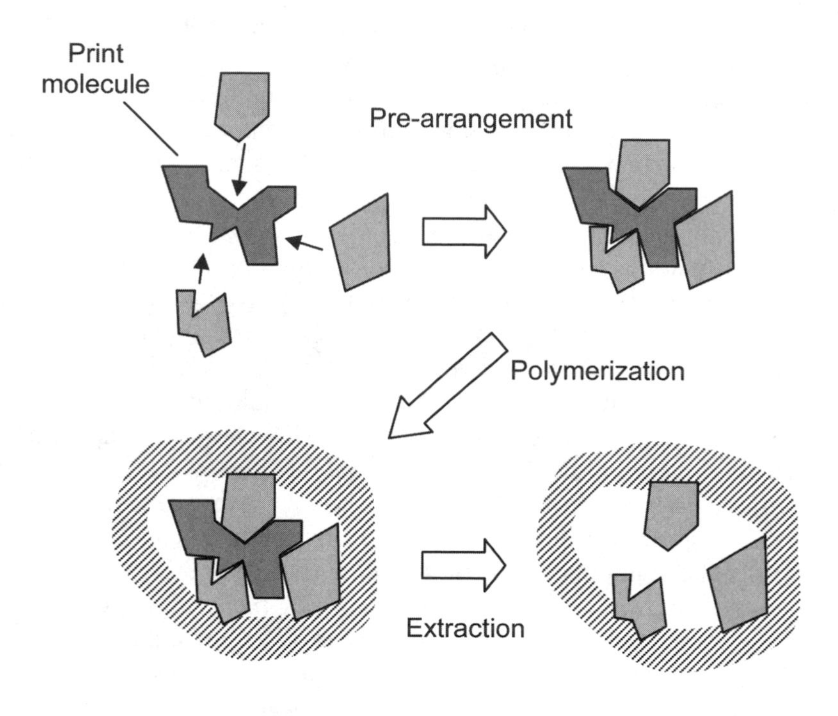

Fig. 7.18 Schematic depiction of the preparation of molecularly-imprinted polymers.

These imprints enable the polymer to rebind selectively to the molecules from a mixture of similar compounds. MIP is a powerful tool for analytical chemistry and sensing applications.[37] It has been applied as highly selective sensing components in bulk acoustic wave (BAW) and surface acoustic wave (SAW) devices[38-41] as well as field effect transistor transducer.[42] They are also used in conductometry and fluorometry sensing processes. [43,44]

Intrinsically-Conductive Polymers (ICPs)

ICPs (or *synthetic metals*) can show electrical, magnetic and optical properties similar to those of metals and semiconductors.[45] ICPs, such as *polyaniline, polythiophene, polypyrrole,* and *polyacetylene*, are different from *conductive polymers* (*CPs*) in that CPs are generally referred to as physical mixtures of non-conductive polymers and conductive materials.

ICPs continue to find market niches as they emerge from the research laboratory. Several of these aspiring applications can be found in MacDiarmid's lecture (one of the Nobel Prize winners in 2000 for his contributions in the field of chemistry) on conductive polymers.[45] ICPs play a significant role in the development of sensors as their electrical, mechanical and optical properties change when they are exposed to different environments and target analytes.

For a polymer to be conductive it has to alternate single and double bonds along the backbone of the polymer, which is called *conjugation*, and the resultant polymer is described as *conjugated* (**Fig. 7.19**). Generally, the π bonds give the conjugated polymer the properties of a semiconductor.[46]

Fig. 7.19 The conjugated structure of polyacetylene: (a) cis and (b) trans forms.

Fig. 7.20 shows the degree of conductivity of different forms of polyaniline in comparison with conventional materials.

Material	Conductivity (S/m)
Ag	10^{6}
In	10^{3}
Ge	1
Si	10^{-6}
Glass	10^{-9}
Diamond	10^{-12}
Quartz	10^{-15}

Doped-polyaniline 10^{5} (S/m)

Polyaniline 10^{-10} (S/m)

Fig. 7.20 Conductivity of polyaniline.

When dealing with ICPs for sensing applications, doping is an important process as the user can tailor the conductivity of polymer to suit their needs. The conductivity of ICPs can be manipulated by doping it with certain certain atoms or molecules.[46,47] There are also other methods for doping ICPs. The most common doping processes are as follows:

Redox doping (*ion doping*): all conductive polymers can undergo either p- and/or n- redox doping by chemical and/or electrochemical processes. These doping processes change the number of electrons associated with the polymer backbone; hence the conductivity of the polymer is also changed.

Photo and charge-injection doping: when ICPs are exposed to radiation whose energy is greater than its band-gap energy of the polymer, electrons hop across the gap and the polymer undergoes photo-doping. In addition, charge-injection doping can also be conducted using a *metal/insulator/ semiconductor* (*MIS*) configuration involving a metal and a conductive polymer separated by a thin layer of an insulator with a high dielectric dielectric strength.

Non-redox doping: in this process, the number of electrons present in a polymer does not change but instead the energy levels of these electrons are rearranged during doping. An example of this occurs when an ICP such as emeraldine base form of polyaniline becomes highly conductive by immersing the polymer in an acid. Emeraldine base interact with aqueous protonic acids to produce the protonated emeraldine which is highly conductive.[48,49] This process can increase the conductivity of the polymer by several orders of magnitude (as demonstrated in **Fig. 7.20**).

Example: polyaniline

Amongst intrinsically-conductive polymers, polyaniline is one of the most attractive ones.[50] Polyaniline is easy to synthesis and process and is environmentally stable. In order to create more efficient sensing properties for polyaniline, it is possible to synthesis and use its different nanostructured forms. Such a nanostructured polyaniline can be employed as a sensitive film, where the smaller dimensions allow the target analyte to diffuse faster and more easily into the sensitive layer and interacts with the functional sites.

Polyaniline is composed of reduced benzenoid and oxidized quinoid units.[50] It contains amine (-NH-) and imine (=N-) functional groups in equal proportions. Polyaniline has the ability to exist in a wide range of oxidation states. If we assume that the average oxidation state is given by 1-y as shown in **Fig. 7.21 – a** then: when y = 1, polyaniline is in the *leucoemeraldine* (*LM*) base state **(Fig. 7.21 – b)**. It is completely oxidized when y = 0 as **Fig. 7.21 – c** and is called *pernigraniline* (*PNA*) base. The form that can be doped into a highly conductive state is called *emeraldine base, EB* **(Fig. – d)** with y = 0.5. In this state, the structure consists of an alternating sequence of two benzenoid and one quinoid units. The emeraldine base form can undergo a non-redox doping with an acid in order to obtain the conductive *emeraldine salt* state of polyaniline as in **Fig. 7.21 – d**. The resulting structure is that of a *polaron* lattice, which is delocalized over the polymer. A polaron can be considered as a type of electronic defect that occurs within the π orbitals of the polymer backbone. This electronic dcfcct is responsible for the high conductivity of polyaniline as it makes polyaniline asymmetric. For the emeraldine state the polarons overlap to form a mid-gap band. The electrons are thermally promoted at ambient temperatures to the unfilled bands, which permit conduction across the polymer.[51] Polyaniline then behaves like a conducting material in which metallic islands are dispersed throughout a non-conductive media. In such a material conduction of electrons may occur through the tunneling effect between these islands.

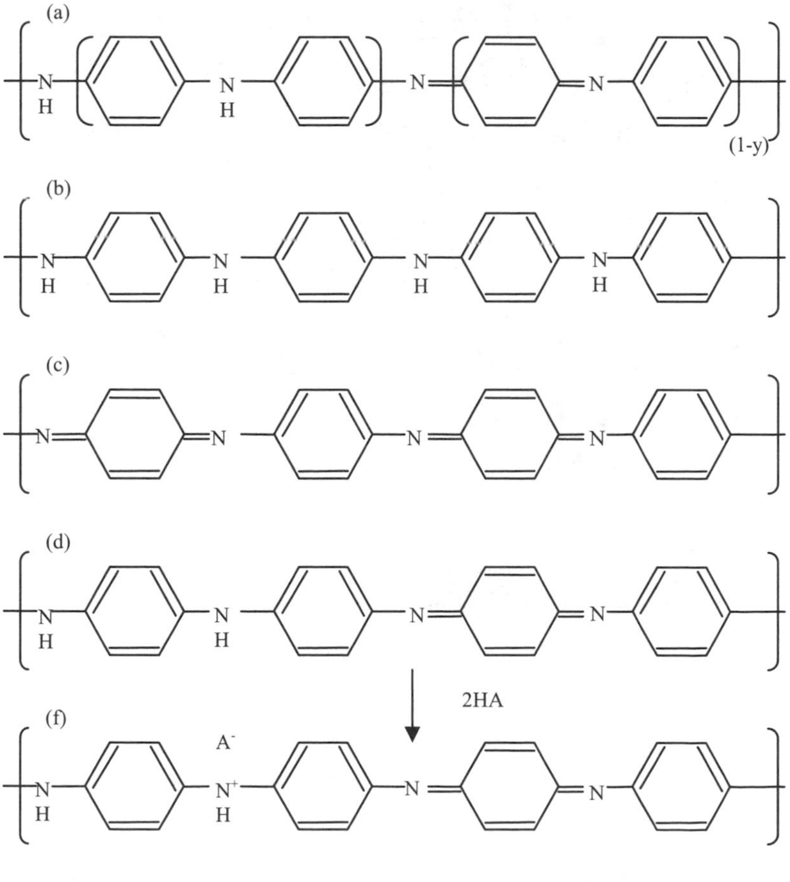

Fig. 7.21 Certain oxidation states of polyaniline.

There are two major methods for the polymerization of polyaniline: electropolymerization and chemical polymerization. There are also other methods such as electro-spinning,[52,53] Langmuir-Blodgett,[54] vacuum deposition,[55] nanofiber seeding,[55] and sol-gel processes.[56]

Electropolymerization of polyaniline is generally carried out in a highly acidic solution using aniline as its sole monomer.[57,58] Aniline monomer is soluble in water.

It is possible to synthesize arrays of uniform and well-oriented polyaniline nanofibers using electropolymerization. In this procedure, the deposition can be performed by either using supporting porous templates to support and confine the polymer[59] or by controlled nucleation and growth using a step-wise deposition process.[60] An SEM image of electrodeposited nanofibers is shown in **Fig. 7.22.**

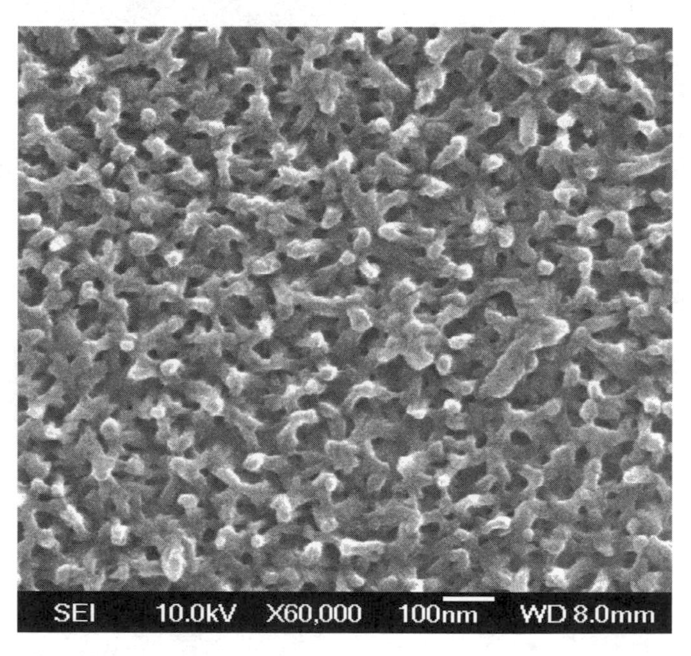

Fig. 7.22 Polyaniline nanofibers grown on a gold surface.

The chemical polymerization process generally occurs when an oxidizing agent, such as $(NH_4)S_2O_8$[61] or KIO_3[62], is fed into the acidic solution which contain the monomer. In order to reduce the size of polyaniline to nano-dimensions it is possible to use interfacial polymerization at an aqueous/organic interface.[63,64] Several other chemical polymerization methods exist for developing polyaniline nanostructures which make use of templates, such as zeolites,[65] and nanoporous membranes.[66] A TEM image of such nanofibres is shown in **Fig. 7.23.**

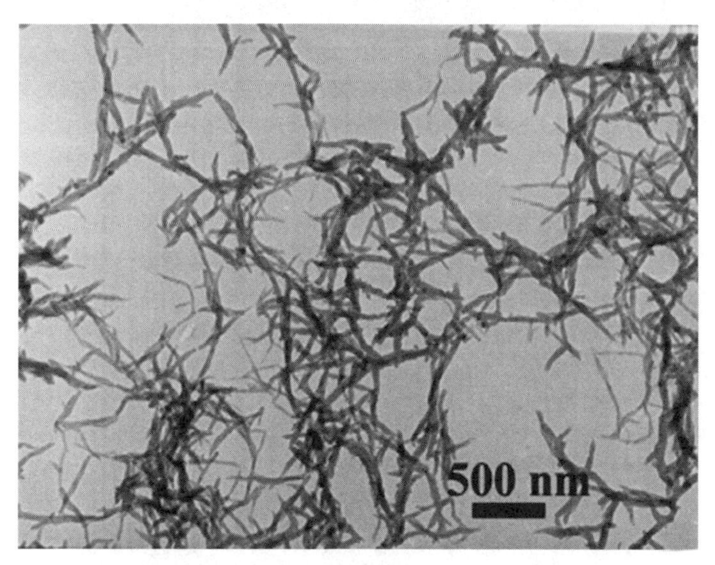

Fig. 7.23 Polyaniline nanofibres grown by a chemical polymerization technique.

Angelopoulos et al have compared the IR spectra of emeraldine and leucoemeraldine base polyaniline films (**Fig. 7.24**).[67,68] They concluded that the major spectroscopic signatures for polyaniline are generally located between 600 to 1700 cm^{-1} and 3000 to 3500 cm^{-1}.

In order to investigate the directionality of polyaniline nanofibres IR spectroscopy can also be employed. Liu et al[60] showed that for nanofibres which are aligned to the surface, there is a strong p-polarized peak at 3350 cm^{-1} which cannot be observed in the bulk polyaniline spectrum. This peak implies that, in the films that were deposited, the N-H bonds are nearly perpendicular to the surface.

ICPs posses the ability to change color when their conductivity changes during a redox process (after interacting with a target analyte) and this property can be utilized in the fabrication of optical sensors. As can be seen **Fig. 7.25**, the UV-visible absorption spectrum of the emeraldine base polyaniline contains two peaks at approximately 330 nm (3.75 eV) and 650 nm (1.95 eV). The absorption spectra for the pernigraniline base state show peaks at approximately 280 nm (4.3 eV), 330 nm (3.8 eV) and 530 nm (2.3 eV). For the leucoemeraldine base state only one peak at 340 nm (3.6 eV) is observed.[67] These values change dramatically after the polyaniline is doped or when it is in nanofibre form.

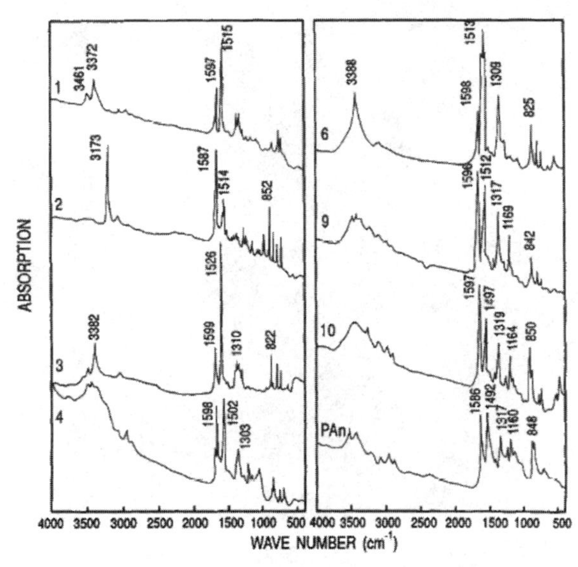

Fig. 7.24 Spectroscopic signatures for different forms of polyaniline. Reprinted with permission from the Elsevier publications.[67,68]

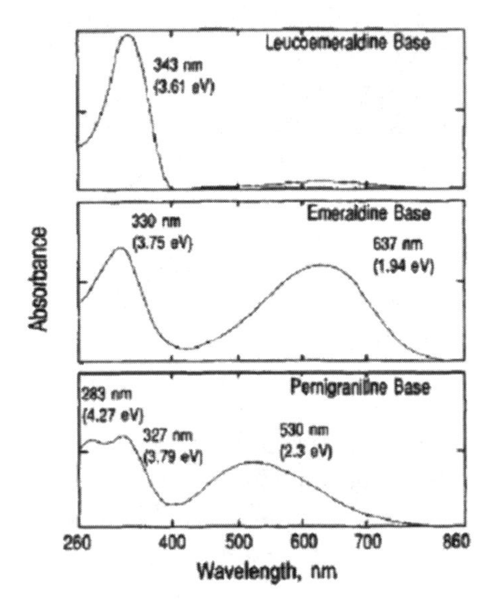

Fig. 7.25 UV-visible absoption spectra for the LM, EM and PNA bases of polyaniline. Reprinted with permission from the Elsevier publications.[67,68]

The electrical, mechanical and optical properties as well as surface affinity towards target analyte of conductive polymers can be affected when they are exposed to different chemicals. For example, redox-active chemicals change the inherent oxidation of ICPs. This unique property can be used for the fabrication of gas and liquid phase sensitive layers.

Virji et al [69] used the template-free, interfacial method to synthesize polyaniline nanofibers using chemical oxidative polymerisation of aniline. They developed nanofiber based conductometric sensors and then compared them to conventional polyaniline sensors. They showed that the polyaniline nanofibers performed better than the conventional thin films as the response time was shorter and the sensor showed a superior sensitivity which was a few orders of magnitude larger than their conventional thin film counterparts as shown in **Fig. 7.26.** Interestingly, the sensor's response was independent of the polyaniline film thickness when it was made of nanofibres.

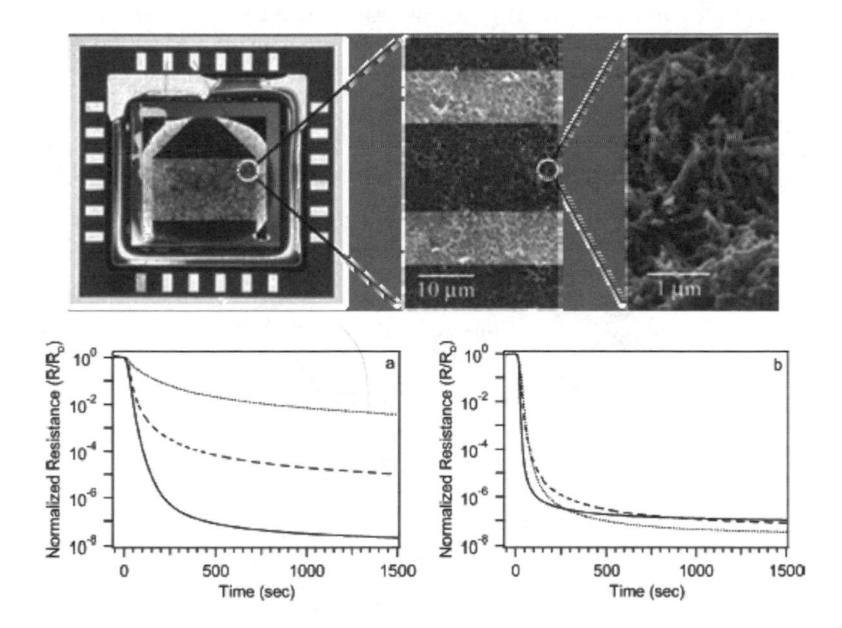

Fig. 7.26 (Top) A sensor with the polyaniline nanofibres and (bottom) the response of polyaniline films to 100 ppm of gaseous HCl. Film thicknesses are as noted: (a) conventional polyaniline, 0.3 μm (black line), 0.7 μm (dashed line), 1.0 μm (gray line); (b) nanofiber polyaniline, 0.2 μm (black line), 0.4 μm (dashed line), 2.0 μm (grey line). Reprinted with permission from the American Chemical Society publications.[69]

ICPs also have numerous applications in biosensing. These polymers are attractive materials to use in biosensors due to the considerable flexibility that exists in their chemical structures and their redox characteristics. Such characteristics of conductive polymers are very useful for the development of enzyme-based biosensors (enzyme-based sensors will be discussed in the coming sections) where rapid electron transfer at the electrode surface is required. In fact, ICPs such as polyaniline, polypyrrole, polyindole and polythiophene have already been widely used in enzyme-based biosensors.[70-72]

The mild conditions used for the polymerization of such conductive polymers are ideal for the incorporation of enzymes, antibodies or even whole living cells. Polymers such as polypyrrole,[73] polyaniline[74] and polyphenylenediamine[75] can be used for the physical entrapment of enzymes during the electrodeposition of the polymers.

ICPs in nano-form also enhance the biosensing properties of existing transducers. For instance, Kim et al[76] reported producing a conductometric immunosensor with polyaniline-bound gold colloids (**Fig. 7.27**). They introduced polyaniline as a conductive agent on the gold surface after immobilizing an antibody specific to human albumin on transducer. This polyaniline-bound gold system can amplify the conductometric signal several times compared with the plain gold system.

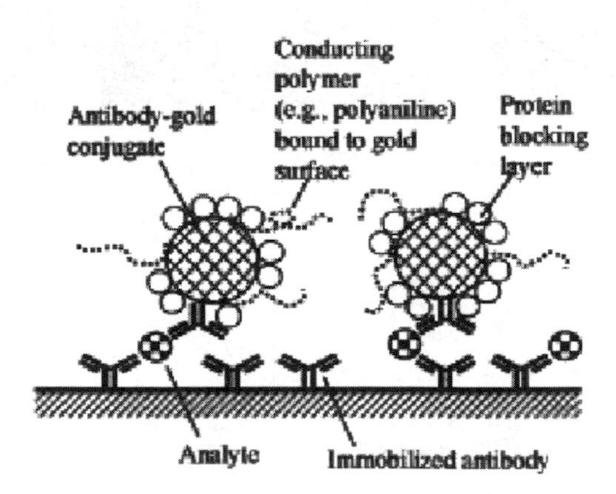

Fig. 7.27 The concept of enhanced electron transfer with a conducting polymer adsorbed onto the surface of gold colloids for biosensing applications. Reprinted with permission from the Elsevier publications.[76]

7.3.5 Examples of Surface Modifications in Biosensors

As was mentioned earlier in this chapter, in order to develop a sensitive and highly stable biosensor, a suitable immobilization method must be used.[77] The surface modification protocols that were presented in the previous sections can be utilized to functionalize the surface of nanoparticles or the surface of the transducers.[78]

Commercial bioaffinity sensors which are based on surface plasmon resonance (SPR) effect have provided powerful tools both for research applications and for the pharmaceutical industry.[79,80] As was described in Chap. 3, SPR systems can be used to sense evanescent plasmon waves perturbations which are brought about by changes in the refractive index of the surface. Such perturbations are caused by the interaction of biocomponents in the solution with the surface of the sensor.

The first commercial SPR system for biosensing was presented by BiaCore Pty Ltd. Recent advances in optical affinity biosensors include the Texas Instrument Spreeta SPR (**Fig. 7.28**) which is small in size and is much less expensive than its counterpart, BioCore, and also the development of integrated Mach-Zehnder interferometers on silicon substrates. Both BiaCore and Texas Instruments now commercially provide a wide range of functional surfaces for the immobilization of biochemical components which also includes nanoparticles. These surfaces range from pure gold for thiolation processes to a matrix of polymers for physical entrapment processes.

The available surface chemistry modifications can provide high functionality in conjunction with minimal non-specific binding to the surface of the SPR sensor. There are plenty of examples that can be found in scientific literatures for SPR systems. The gold surface may be functionalized by the electrostatic immobilization of biochemical components such as liposomes. A streptavidin monolayer can be formed on a gold surface using biotin and the monolayer can then be functionalized with biotinylated biomolecules.[81] A common approach is to form SAMs of thiol-containing molecules with functional tail groups on a gold surface.[82] Target biochemical components such as proteins may be bound to these tail groups after further activation of the surface has been completed. SAMs can also be used to allow the covalent binding of materials such as dextran molecules to the surface. After treating the dextran with iodoacetic acid, the resulting carboxylic groups are used to immobilize bio chemical components. As was described previously, a common method is to activate these carboxylic groups using NHS and ethyl-methyl-diaminopropyl-carbodiimide (EDC). This makes the surface prone to form covalent bonds with a functional group such as $-NH_2$. If the dextran molecules contain streptavidin then they can bind to biotinylated molecules.[83] A review of SPR sensing methods for controlled coupling to carboxymethyldextran surfaces can be found in the paper by Löfas et al.[84]

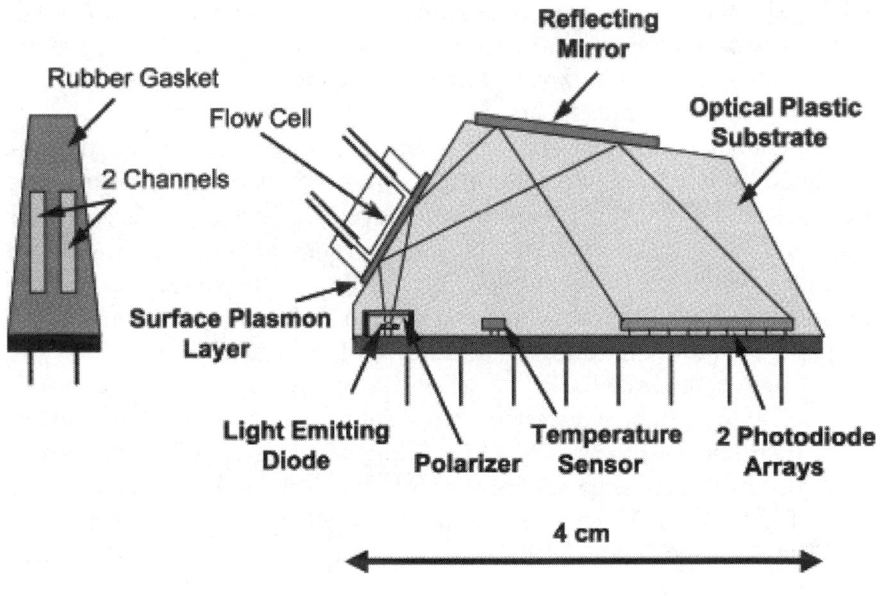

Fig. 7.28 A schematic diagram of the two-channel integrated miniature SPR sensor with a detachable flow cell. On the left, the frontal view shows the two channels which are laser-cut in the silicone rubber gasket. On the right is the side view of the sensor with the flow cell attached, showing the functional components of the sensor. Reprinted with permission from the Elsevier publications.[85]

The incorporation of nanoparticles of predetermined dimensions which are bound to organic molecules in immunoassays can be used for increasing the sensitivity of conventional chromatic assays. Commercial mass sensors based on quartz crystal microbalance (QCM) or surface acoustic wave (SAW) devices, which are used for the sensing affinity between different organic molecules, are becoming more available.

QCM electrodes are generally made of gold. Such a gold surface is excellent for the formation of SAMs, electrostatic bonds, etc. For instance, Carusu et al[86] used QCMs with gold electrodes for DNA sensing. They employed both covalent bonds, which were established using carbodiimide hydrochloride and N-hydroxysuccinimide (NHS), and electrostatic bonds using poly(allylamine hydrochloride) (PAH). Immobilisation by silanising the surface, for instance with γ-aminopropyltriethoxysilane activated with glutaraldehyde, is also frequently employed on QCMs.[87,88]

Electrochemical sensors can also take advantage of surface Fictionalization. Recently, the electrochemical properties of nanomaterials such as *carbon nanotubes* (*CNTs*) have been investigated with promising results.[89] Nanomaterials such as CNTs showed a high elcctrocatalytic effect and a fast electron-transfer rate.

Lin et al developed glucose biosensors based on CNT nanoelectrode (NEs).[90] They covalently immobilized glucose oxidase (whose function will be explained in the coming sections) on CNT NEs using carbodiimide chemistry by forming amide linkages between their amine residues and the carboxyl groups on the CNT tips (**Fig. 7.28**). They showed that the catalytic reduction of hydrogen peroxide (which was liberated from the enzymatic reaction of glucose oxidase with glucose and oxygen on CNT NEs) can be employed for glucose sensing. Their biosensor performed a selective electrochemical sensing of glucose in the presence of interferents such as acetaminophen, and uric and ascorbic acids. Such usage of nanomaterials eliminates the need for selective membrane barriers or artificial electron mediators. As a result, they greatly simplify the sensor fabrication process.

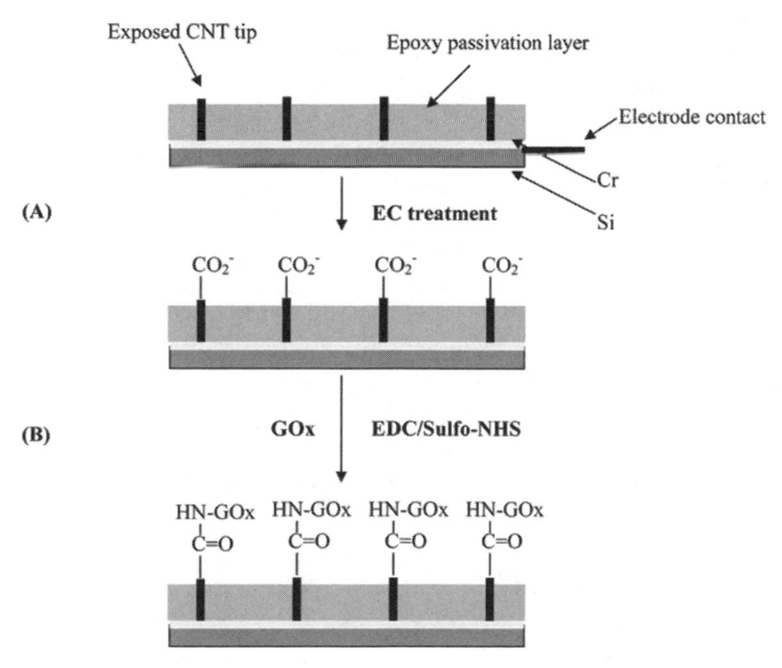

Fig. 7.29 The fabrication of a glucose biosensor based on CNT nanoelectrodes: (A) The electrochemical treatment of the CNT NEs for functionalization, and (B) the coupling of the glucose oxidase enzyme (GOx) to the functionalized CNT NEs. Reprinted with permission from the American Chemical Society publications.[90]

7.4 Proteins in Nanotechnology Enabled Sensors

As nanosystems, proteins are many tasks that proteins can perform for us.[91] Their use in sensitive materials, pores with triggers, switches, self-assembling arrays, and motor which are harnessed for specific tasks are just a few niche applications in nanotechnology for which proteins can be employed. For instance, the ion channelsion the cell membrane can keep the ion balance across our cells by letting single ions pass through. Proteins which operate as nano-motors can use the energy it gains from small molecules to throw themselves forward in order to walk on a surface within a cell. It is also possible to make artificial proteins of a desired sequence which can function as drugs. A vaccine protein, for example, can bind to a viral surface to inactivate/activate it. The electrical, mechanical and chemical activities within cells are conducted by proteins. They are also building blocks from which cells are assembled and they provide cells with their shape and structure. Proteins can perform a large number of functions as they form huge number of different structures.

Despite the above mentioned advantages, there are still many unresolved problems for controlling, keeping the proteins active and external communications with them hinder the fabrication of complex nanobioelectronic circuits, and nanobiomechanical gadgets using these nano tools.

7.4.1 The Structure of Proteins

The total number of protein types is different in various organisms. For instance, yeast has approximately 6,000 proteins but a human being has approximately 32,000 different types of proteins.[92]

Amino acids are subunits of proteins. A simple amino acid, alanine is shown **Fig. 7.30.** Alanine consists a chiral carbon which is bound o an H_2N– amino group, a –COOH carboxyl group and $-CH_3$ methyl the side chain

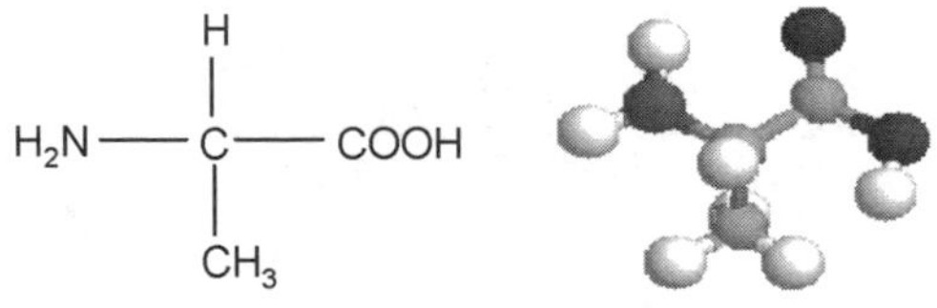

H
H_2N — C — COOH
CH_3

Fig. 7.30 The structure of alanine.

All amino acids possess a carboxylic acid group and an amino group, both linked to the same carbon which is called the α-carbon. The chemical variety in amino acids comes from the side chain which is also attached to the α-carbon.

The covalent bond between two adjacent amino acids is called a *peptide bond* **(Fig. 7.31)**. The chain of amino acid is called a *polypeptide*. The combination of amine in a carboxyl groups is given the polypeptide directionality.

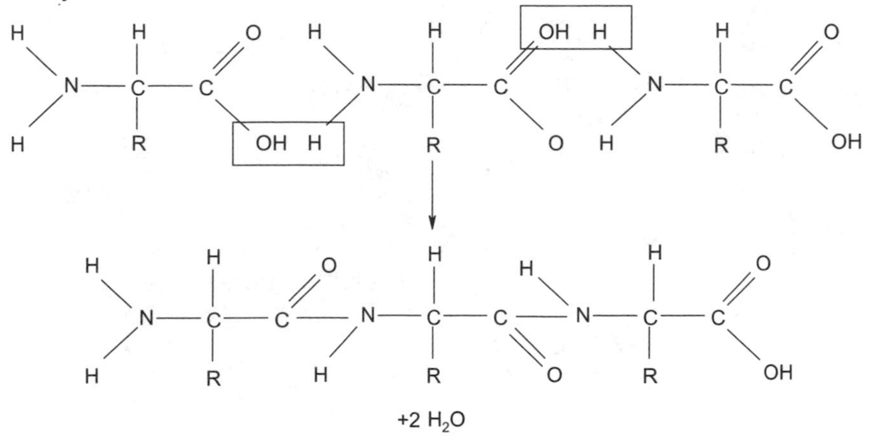

Fig. 7.31 The joining three amino acids through peptide bonds which forms a polypeptide.

There are twenty different types of amino acids which are commonly found in proteins, each with different side chain attached to the α-carbon. One of the mysteries of evolution is that only twenty amino acids occur over and over again in all proteins whether they be from humans or from bacteria.[92] These twenty standard amino acids provide a remarkable amount of chemical versatility to proteins. Five of these amino acids can ionize in solutions the other are unchanged. Some amino acids are polar and hydrophilic and some are nonpolar and hydrophobic.

A protein is made from long chains of amino acids. As such, proteins are polypeptides. Each type of protein has a unique sequence of amino acids. Long polypeptides are very flexible. Many of the covalent bonds that link the atoms in an extended chain of amino acids allow the free rotation of the atoms they join. As a result, proteins can fold in an enormous number of ways. Once folded, each chain is restrained by noncovalent bonds. These noncovalent bonds, which maintain the protein shape, include hydrogen bonds, ionic bonds, and van der Waals attractions **(Fig. 7.32)**.

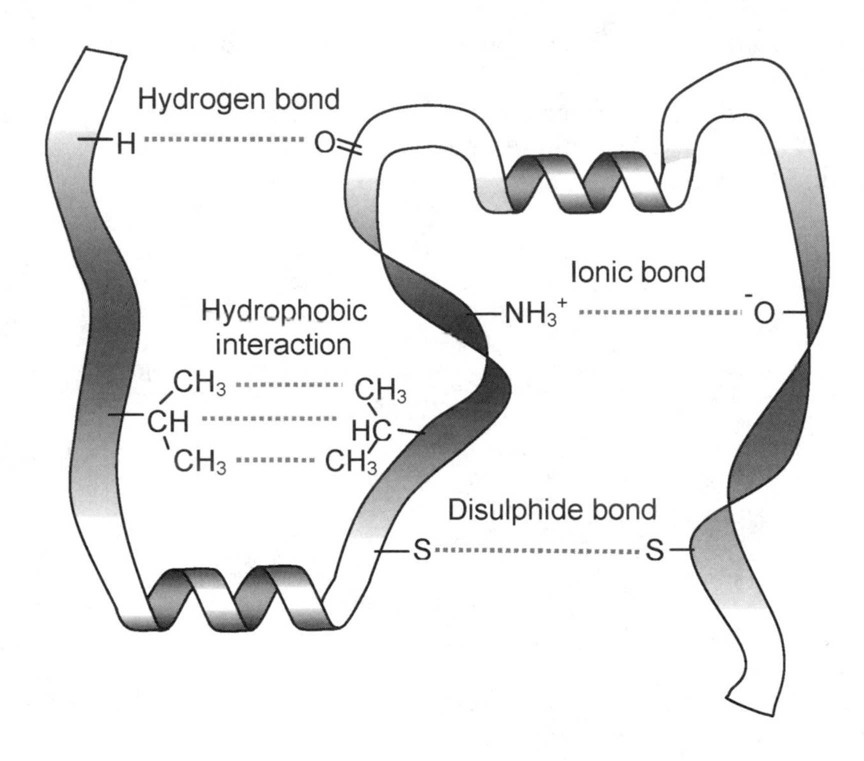

Fig. 7.32 The noncovalent bonds within the structure a protein.

Proteins fold into a conformation of lowest energy in order to minimize their free-energy. These conformations are the most stable states at which the proteins exist. Nevertheless, this stable condition can change when a protein interacts with the other molecules in a cell. When a protein folds improperly, it can form aggregates which can damage cells and even whole tissues. *Alzheimer's* and *Mad Cow* diseases are caused by the aggregation of proteins.[92] In living organisms, generally the protein folding process is assisted by special proteins called *molecular chaperons*. Proteins can also be unfolded or denatured by certain solvents that can disrupt the noncovalent interactions **(Fig. 7.33)**.

Proteins can be directly adsorbed onto the surface of a sensor; however, in this process they may become denatured. As a result, the protein may lose its functionality, which can be a major problem in sensing applications. For example, an antibody (which will be explained later) must retain its shape in order to form a selective layer.

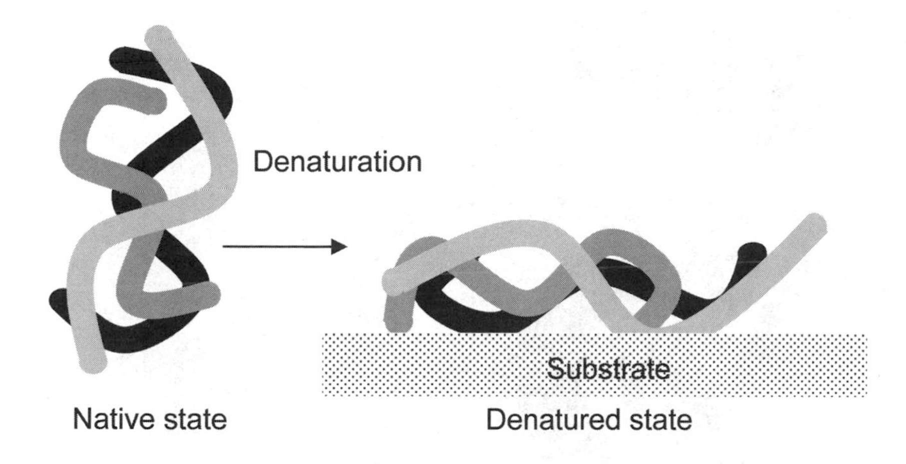

Fig. 7.33 An example of proteins in native and denatured states.

Proteins are the most diverse macromolecules in a cell. Their size can range from approximately 30 amino acids to 10,000, they can be globular or fibrous, and they can form filaments, sheets, rings or spheres.[92]

Revealing the *amino acid sequence* is an important part of analyzing a protein and its functions. The first protein to be sequenced was beef insulin by Fred Sanger the 1958 Nobel Prize winner from the University of Cambridge. The analysis of a protein begins with determining its amino acid sequence. The cells are broken open and the proteins are separated and purified so as to conduct a direct analysis on the chemical components. Alternatively, the genes that encode proteins can be sequenced instead. Once the order of the nucleotides in the DNA that encode the protein is known, the information can be converted to the corresponding amino acid sequence. Generally a combination of these direct and indirect methods is used for a comprehensive analysis of a protein.

The α-*helix* and the β-*sheet* are common folding patterns for proteins **(Fig. 7.34)**. These two patterns are found in the structure of most of proteins. The α-helix was first found in a protein called α-*keratin* which is abundant in skin. The β-sheet is the second folded structure which was first found in protein *fibroin*, which is the major constituent of silk.

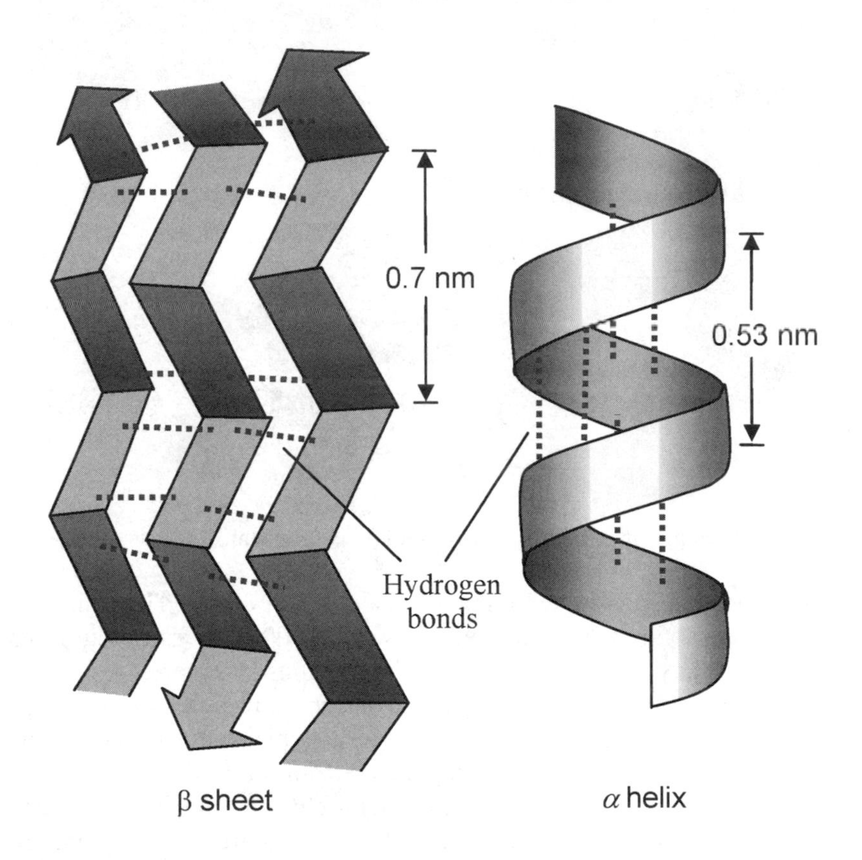

Fig. 7.34 Schematic of α-*helix* and the β-*sheet* folding patterns.

A helix resembles a spiral staircase. For α-helix a single polypeptide chain turns around itself to form a structurally rigid cylinder. A hydrogen bond is made between every fourth peptide bond linking the C = O group of one peptide to the N–H group of another. The sides of the amino acids which contact the hydrophobic hydrocarbon tails of the phospholipid molecules are hydrophobic. However, the hydrophilic parts of the polypeptide backbone forms hydrogen bonds in the interior of the helix.

β-sheets are made when hydrogen bonds are formed between segments of polypeptide chains which are lying side-by-side. They form rigid structures at the core of many proteins and perform functions such as giving silk fibres their tensile strength and keeping insects from freezing in the cold.

There are many higher levels of organizations for proteins rather than just α-helix and β-sheets. The order in which amino acids are linked give the primary structure of the protein. The next level up is the *secondary* which is concerned with the shape of the protein in localized areas and takes into account the α-helix and β-sheets that forms within certain segments of the polypeptide chain. The full three-dimensional protein structure is the *tertiary structure* and it consists of all α-helix and β-sheets, coils, loops and folds in the protein. The *quaternary structure* of a protein is the overall structure in the case that the protein comprises more polypeptide chain. An example of this is hemoglobin which has four polypeptide chains in its quaternary structure.

7.4.2 The Analysis of Proteins

When analyzing a protein, the fundamental process is that cell should be broken open, all of the soluble proteins should be separated by centrifugation, and then passed through a column matrix that contains the pure target protein. Isolating these proteins by electrophoresis through a polymer gel can be the next step, which separates polypeptides on the basis of their size.[92] Once the protein has been identified, we are ready to determine its amino acid sequence. In order to do this, the protein is broken up into smaller pieces by using selective enzyme called *protease*. For instance, the *enzyme trypsin* cleaves polypeptide chains on the carboxyl side of any lysine or arginine units.

Mass spectrometry (Chap. 5) can be utilized to determine the exact mass of each peptide fragment. This method allows identifying the first finger print. For structural analysis generally a large amount of the protein is needed. Using the *recombinant DNA* method the proteins can be inserted into cells of bacteria to get the cell to produce large amounts of the protein.

To determine the proteins' structure using X-ray crystallography, they have to be stacked together to form crystals which are large, highly ordered arrays. The patterns are generally very complex and computer programs are required to analyse these patterns. If the protein is small the structure can also be determined by using nuclear magnetic resonance (NMR) spectroscopy.

7.4.3 The Role of Proteins in Nanotechnology

Proteins represent an exciting area in nano-biotechnology as they have ideal properties for engineering tasks such as processing sophisticated architectures at nanoscale dimensions, have rich chemistry and provide us versatile functionalities. By employing the knowledge of biochemical engineering, it is possible to harness the protein's power so as to create new components for materials and devices. It is also possible to synthesis basic artificial proteins of a desired sequence which can function as drugs. For instance, a vaccine protein can bind to a viral surface and inactivate it. However, sophisticated protein-based nanodevices have not yet been synthesized. There are still many unknowns which have to be investigated before these nanodevices reach their potential.

Nanotechnology can make use of proteins as they can perform the following tasks:

- Generate movements in cells and tissucs via moto proteins: M*yosin* and *Kinesin* are such proteins. Myosin, which is found in skeletal muscle cells, propels organelles throughout the cytoplasm. It uses *ATP hydrolysis* to generate the force that it needs to be able to walk along the filament. Kinesin attaches to microtubules, and moves along the tubule in order to transport cellular cargo hloders, such as vesicles.
- Transport materials, such as small molecules and ions: in the bloodstream *Serum Albumin* carries lipids, *Hemoglobin* carries oxygen and *Trasferrin* carries iron.
- Receive signals: some proteins can detect signals and send them to the cell's response machinery. *Rhodopsin* in the retina, for example, detects light.
- Store small molecules or ions: iron is stored in the liver within the small protein *Ferritin.*
- Promote intermolecular chemical interactions: enzymes generally catalyze the breakage and formation of covalent bonds, as in *DNA polymerase* which is used in copying DNA.
- Function as selective valves: proteins embedded in the plasma membrane form channels and pumps that control the passage of nutrients and other small molecules into and out of cell.
- Carry messages from one cell to another: *Insulin* is a small protein that controls the glucose levels in blood.
- Serve as nanomolecular machines: *Topoisomerase* can be used for untangling DNA molecules.
- Act as antibodies.
- Perform as toxins.
- Act as antifreeze molecules: for instance the proteins arctic fish which help them to avoid freezing in colder waters.
- Form elastic fibres.
- Generate light through luminescent reactions.
- Form glues: The glue that forms in mussels is a protein which is produced in order for a mussel to attach to other objects.
- And many more.

7.4.4 Using Proteins as Nanodevices

In this section we will present a selection of protein-based components for the development of nanodevices as well as nanomaterials which can be useful in sensor technology.

The conformations of proteins give them their uniquc functions based on their chemical properties. Proteins units can be engineered for conducting specific chemical reactions. The chemical and physical properties of a protein give it the ability to perform extraordinary dynamic processes.

The biological properties of a protein depend on how it interacts physically and chemically with other molecules. For instance, *antibodies* attach to certain viruses or bacteria, *enzymes* can interact with substrates and molecules such as ATP (which will be explained later in this chapter) to catalyze reactions in proteins. *Actin* molecules attach to each other to form an actin filament which are extremely abundant in eukaryotic cells where it forms one of the major filaments of the cytoskeleton which gives shape to a cell.[92]

The binding of proteins to other chemicals is not always strong; in many cases it is actually weak. However, the binding always shows specificity, which means that each protein can only bind to one or at most a few molecules that it encounters. Such a nature can be efficiently used for making sentitive surfaces. The substance which is bound by a protein is called a *ligand* for that protein. This ligand can be an ion, a small molecule or a macromolecule. *Ligand-protein binding* is of great importance in sensor technology.

Noncovalent bonds such as hydrogen bonds, ionic bonds, and the Van der Waals attractions as well as hydrophobic interactions are responsible for the ability of selective bind of a protein to a ligand. The effect of each bond can be weak but the simultaneous formation of many weak bonds between the protein and the ligand form the selective binding. Even the matching of the surface contour of the protein and ligand can be the reason for selective binding (**Fig. 7.35**), as molecules of the wrong shape cannot approach the active sites closely enough to bind, if indeed they could. The region of a protein that associates with a ligand is called a *binding site*. These sites usually consist of a cavity in the protein surface which is formed by the folding arrangement of the polypeptide chains.

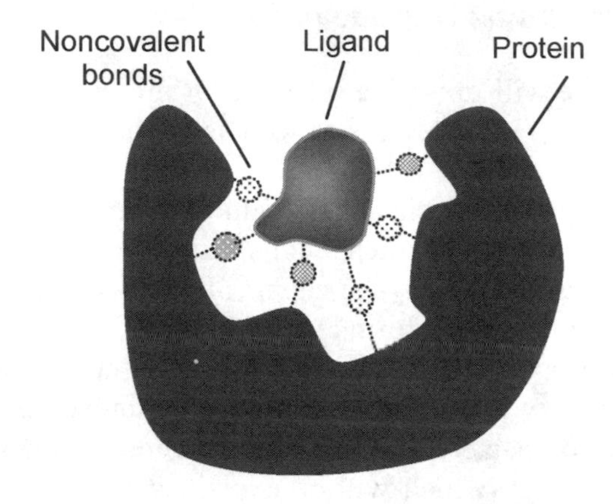

Fig. 7.35 The protein-ligand interaction is the result of a number of separate of noncovalent bonds.

7.4.5 Antibodies in Sensing Applications

Antibodies (which belong to the family of *immunoglobulins*) are proteins which are produced by the immune system in response to *foreign materials*. The most common of these foreign molecules are invading *micro-organisms*. Antibodies either render them inactive or prepare them for destruction. The proteins of the antibody family are highly developed proteins with the capacity to bind to particular ligands. The antibody family consists of five subdivisions: *IgG, IgM, IgA, IgE* and *IgD*. An antibody recognizes the target, which is called an *antigen*, with a high specificity. Antibodies are Y-shaped molecules (as seen **Fig. 7.36**).[92] As can be seen, antibodies have two identical binding sites. These binding sites are complementary to a small portion of the antigen. The antibody binding sites are formed from several loops of polypeptide chains, which are connected to the end of the *protein domains*. These protein domains are comprised of four polypeptide chains, two of which are identical heavy chains and the other two are identical light chains. These chains are all held together by disulfide bounds.[92]

The amino acids in the binding sites can be changed by *mutation* without altering the domain structure of the antibody. A large number of antibody binding sites can be formed by changing the length and the amino acid sequence of the loops.

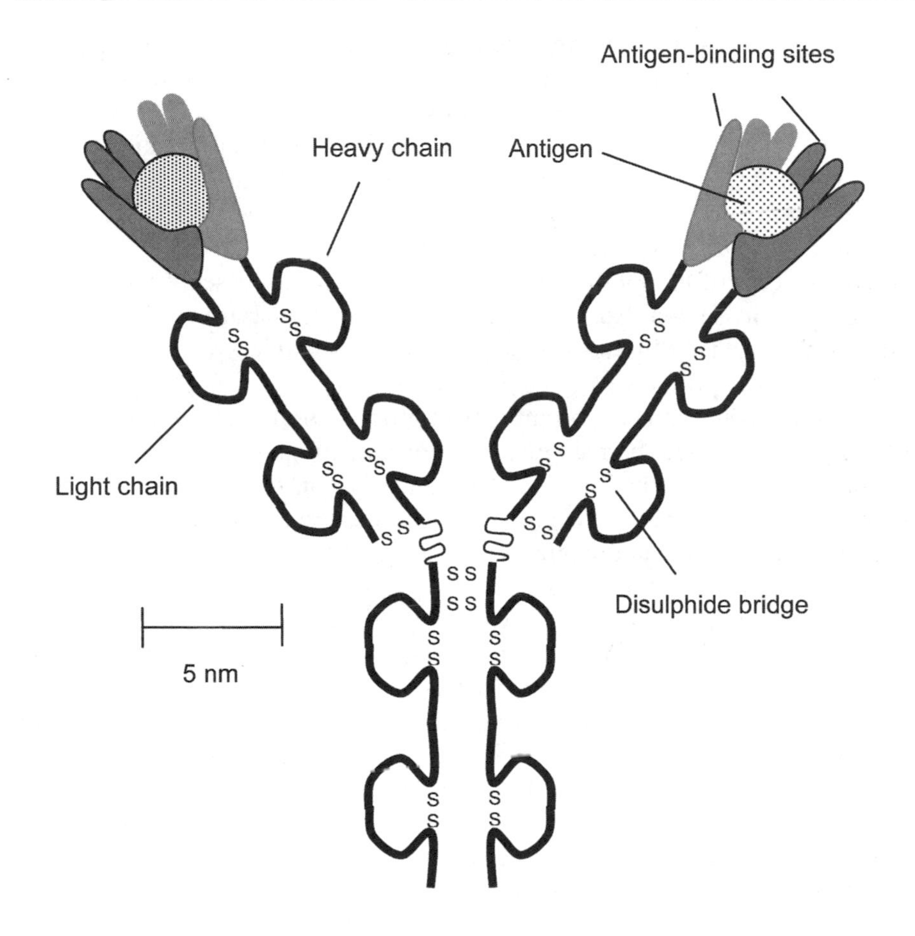

Fig. 7.36 The general structure of an antibody.

The reactivity of an antigen is restricted to specific parts of the molecules which are called *epitopes*. These epitopes specifically bind to the antibody binding sites, which are also known as *paratopes*.

The knowledge of the structural formation of antibody-antigen interactions is mainly based on X-ray diffraction (XRD) investigations.[93] The area of the antigen epitopes surface comprising between fifteen to twenty two amino acids interacts with the same number of amino acids on the antibody paratopes. Many intermolecular hydrogen bonds, ionic bonds, and salt bridges participate in these interactions and apparently water molecules also participate in these interactions.

When suspended in water the protein molecules initially repel each other due to their hydrophilic nature. This force must be overcome before the epitopes and paratopes are brought together. The major reasons for the attractions are probably electrostatic interactions and still several available hydrophobic sites on antibodies and antigens.

For developing biosensors which are sensitive to specific antibodies or antigens their corresponding antibodies or antigens should be either immobilized on the active area of the affinity-type sensors or embedded within the bulk of the bulk-type sensors.

The concept of immobilization has already been discussed earlier in this chapter. It was shown that different types of functional groups can be created on the surface of transducers which interact with corresponding functional groups in proteins.

For successful antibody or antigen immobilizations the functional sites of the proteins should be recognized. Consequently, a procedure should be developed to realize the binding of the proteins on the surface that keeps them active. In the immobilization of antibodies the epitopes and paratopes should remain active and available while the other site (the stem) should perform the immobilization task.

Protein immobilization can be carried out using hydrogen bonds, covalent bonds, via Van der Waals forces, ionic bonds, etc. If the protein adsorption occur on a hydrophobic surface, it is often followed by a small unfolding of a protein structure (e.g. from a quaternary which causes the amount of hydrophobic polypeptide chain to increase). In many cases, this phenomenon is undesired, as it leads to the denaturing and deactivation of the protein. Fortunately, many antibodies are quite resistant to being unfolded during the adsorption process, as they have a very rigid tertiary structure. The hydrophobic adsorption method is very popular for well-developed antibody-based immunoassays such as *radioimmunoassay* (*RIA*) and *enzyme-linked immunosorbent assay* (*ELISA*). Another problem with a hydrophobic surface is that any protein in the solution will tend to bind to the surface, which is also referred to as *non-specific binding* (*NSB*). For a sensor this is problematic as it causes a high background signal. Proteins can bind to a charged surface due to ion-pair interactions, as some proteins have ionized surface groups which can interact with the charges on the surface. This interaction occurs on many different types of surfaces, even those that are weakly charged.

As described previously, semi-permeable polymeric and non-polymeric membranes can be employed for trapping proteins such as enzymes and antibodies. Generally the proteins which are captured this way are small proteins, having molecular weights less that 10 kDa. Several well studied membranes include polycarbonate and nylon. Proteins may also become physically entrapped within the volume of a hydrogel. A hydrogel can be made of a polymer which is dissolvable in warm water and gels when cooled. Gelling is due to hydrogen bonding. The most widely used polymer is *agarose*, which gels at approximately 40°C. It is high porous which makes it more suited to micro-organisms and organelles as opposed to single proteins.

An interesting example of entrapment involves the use of polyvinyl alcohol. It is a high molecular weight polymer that can completely dissolve in boiling water. It remains dissolved at room temperature, however, it is insoluble in cold water. A sensitive layer can be made from such a material. Proteins are mixed with the polymer solution and then deposited onto the transducer surface. After drying, a film is left on the surface which contains these proteins, which can be used as a sensing layer for the entrapped enzymes and antibodies.

There are also many biological assays involving antibody-antigen interactions which are used in sensing procedures. The most common of these assays are *direct assay* and *competitive assays*. In direct assays, the antibodies or other receptor molecules are immobilized onto the sensor's surface. The target molecules can then selectively bind to these molecules (**Fig. 7.37**). Direct assays find applications in affinity sensors such as piezoelectric and optical waveguide-based sensors where the evanescent waves penetrate into the added layer and in doing so produce a response. No optical tagging is needed in these sensors as they are able to sense the mass (in fact the thickness perturbations) of the added layer.

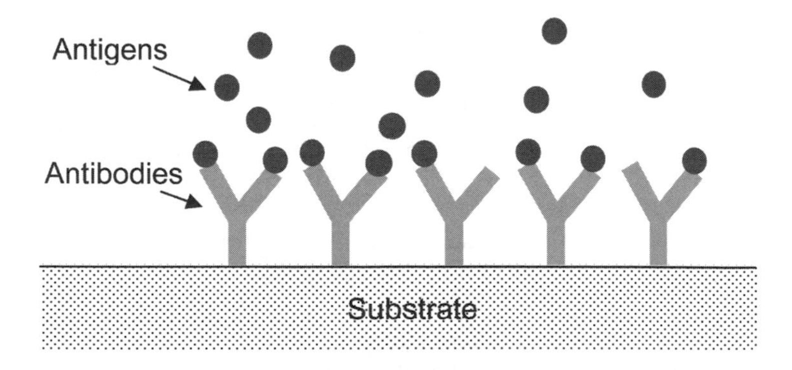

Fig. 7.37 A simplified schematic of a direct assay.

In a competitive assay only a small number of antibody sites are available for antigens to bind to. Firstly the antibody is immobilized on the substrate's surface. Then a known quantity of labeled analytes, which are similar to the target molecules, is added and they compete with the unlabelled molecules for the available antibody bonding sites. The result can be optically investigated and interpreted using the relationship that the lower the optical intensity means the more of target analytes are available. It is obvious that such an assay is more suitable when the concentration of the target analyte is small (**Fig. 7.38**).

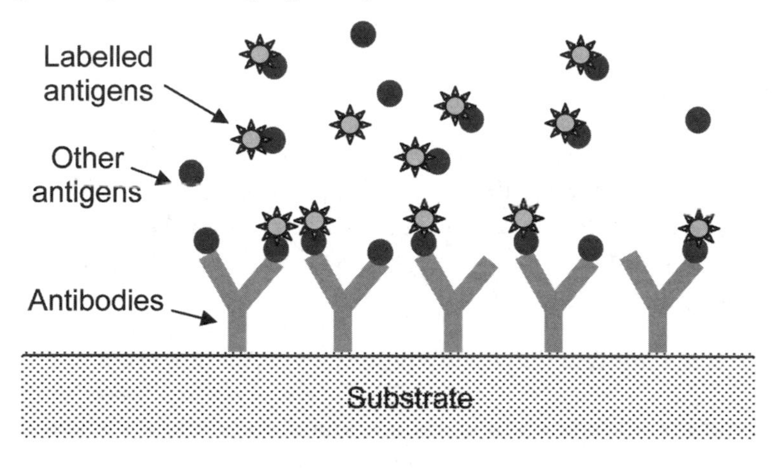

Fig. 7.38 A simplified schematic of a competitive assay.

In using antibodies and antigens as selective layers the cross-sensitivity for the antibody-antigen reaction becomes an important issue. This gives indication about how selective the sensor can become. Berzofsky et al described that the antibody specifity concept is complementary to the antigen cross-sensitivity concept. [94] In practical sensing applications, the specifity is very important as it rates the sensor's level of discrimination for recognizing a particular analyte amongst other species. Antibodies with greater specifity also have a higher affinity to their complimentary antigens.

Electrochemical-based affinity biosensors offer some of the most sensitive immunoassays that are currently available. By exploiting enzyme amplification technique in low-volume capillaries, they can measure as little as 10^{-21} moles of an analyte.[95,96] A very recent award-winning innovation comes from the laboratories of Schumann (Ruhr-Universität Bochum, Germany) and Csoregi (University of Lund, Sweden), which have developed an affinity biosensor in which the diffusion coefficient of an electroactive label alters upon the recognition of its complementary binding partner in an antibody-antigen intereaction.[97] The increase in the molecular weight and the associated decrease in the diffusion coefficient of the label are monitored by determining the diffusion-limited current by cyclic voltammetry. The signal is then obtained by redox-recycling at a microelectrode.

Another which has drawn a lot of attention is the use of antibodies on mass sensitive devices (such as QCMs) where they offer their inherent bioselectivity. The first piezoelectric immunosensor was developed by Shons et al.[98] Since then, a plethora of methods for the immobilisation of antibodies on the crystal surface have been investigated.[99,100] The use of QCMs to monitor immunoliposome-antigen interactions,[101] and the detection of antibodies such as HIV[102,103] are just a few examples. Protein A is one of the most widely used pre-coatings which aids the immobilisation of the antibodies and has also been employed in the detection of pesticides,[104] bacteria[105] and drugs.[106,107]

In addition to piezoelectric devices, antibody based immunosensors can also be based on electrochemical[108] and amperometric[109] transducers. In addition, they can be fabricated using devices such as optical transducers, for pesticides,[110] bacteria,[111] and medical sensing applications.[111-113]

SPR (Chap. 3) biosensors, which are optical sensors, have been widely implemented for sensing small analytes such as pesticides (simazine and atrazine), medical components (morphine, methamphetamine, and theophyline), and food components (fumonisin B1, sulfamethazine, and sulfadiazine).[114] As one of the early examples, Minunni and Mascini used an SPR sensor and binding inhibition assay to detect atrazine using monoclonal antibodies against atrazine. They used a sample containing a mixture of atrazine and an antibody and exposed the sample to an atrazine derivative-coated SPR biosensor. In these experiments, they obtained detection limit of several pg mL^{-1} of atrazine.[115] **Fig. 7.39** shows typical responses for binding inhibition assay detection of atrazine using a wavelength-modulated SPR sensor.

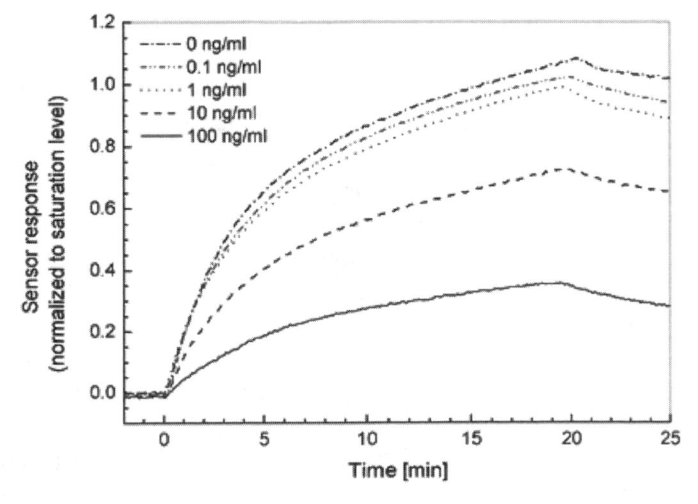

Fig. 7.39 Detection of atrazine using an inhibition assay.[115]

SPR systems are also frequently used in immunosensing applications which uses the antibody-antigen interactions. For such immunosensing experiments, the minimum detection limits in the order of several ng mL^{-1} can be obtained. A typical kinetics of an assay using a SPR system is shown in **Fig. 7.40**. It is a sandwich assay for the detection of staphylococcal enterotoxin B which is a potent gastrointestinal toxin.[116,115]

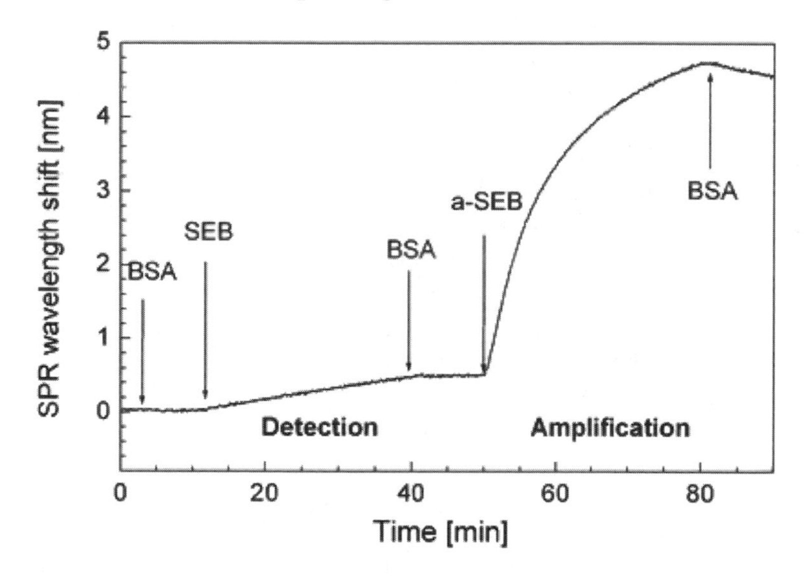

Fig. 7.40 Detection of Staphylococcal enterotoxin B using a sandwich assay. [115]

7.4.6 Antibody Nanoparticle Conjugates

When arrays are incorporated in the structure of nanotechnology enabled sensors, fabrication methods are needed that offer the capability of working with both inorganic and organic materials.

Atomic force microscopes are amongst the most common tools which provide such a capability via the dip-pen nanolithography procedures (Chap. 5). Mirkin et al demonstrated that the dip-pen nanolithography can be efficiently implemented for fabricating and functionalizing gold nanostructures in conjugation with proteins.[117] They created an array of protein-nanoparticles conjugates (rabbit IgG and gold) on a semiconducting substrate. Subsequently, they studied the bioactivity of the array by monitoring its reaction with fluorophore-labelled anti-rabbit IgG.

They thermally evaporating 8–10 nm of Au on a Ti-coated (1 nm) oxidized silicon (500 nm of oxide) substrate. These Au substrates were then patterned with 16-mercaptohexadecanoic acid (MHA) by using the dip-pen nanolithography as shown in **Fig. 7.41**. Afterwards, they functionalized the surface using the process shown in **Fig. 7.42**.

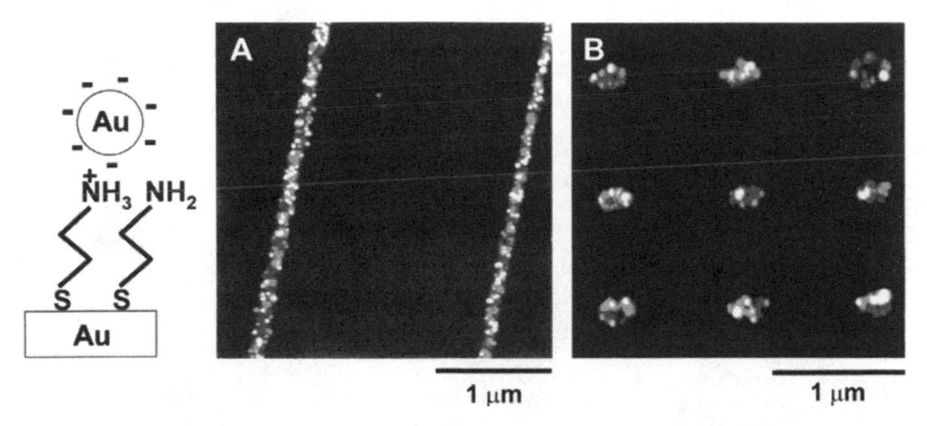

Fig. 7.41 AFM images of individual Au nanoparticles adsorbed on NH_2-SAM-modified nanopatterns of lines (A) and dots (B). Reprinted with permission from the Institute of Physics Journals publications.[117]

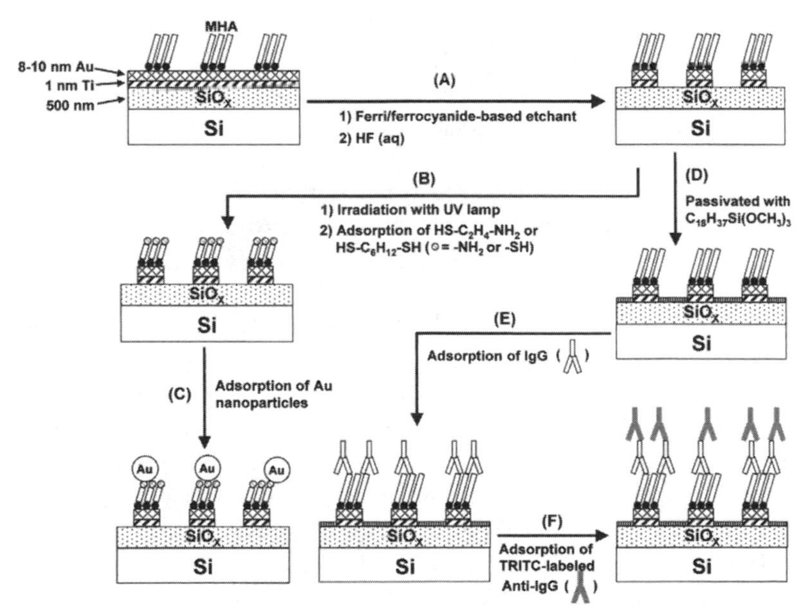

Fig. 7.42 The procedure for preparing biofunctionalized nanostructures using dip-pen nanolithography. Reprinted with permission from the Institute of Physics Journals publications.[117]

7.4.7 Enzymes in Sensing Applications

There are many other proteins which are different from antibodies for which their attachment to a ligand is only the first step of their fnction. This is the case for the large class of proteins which are called *enzymes*. Enzymes bind to one or more ligands, which are called *substrates*, and convert them into *chemically modified products* (**Fig. 7.43**).

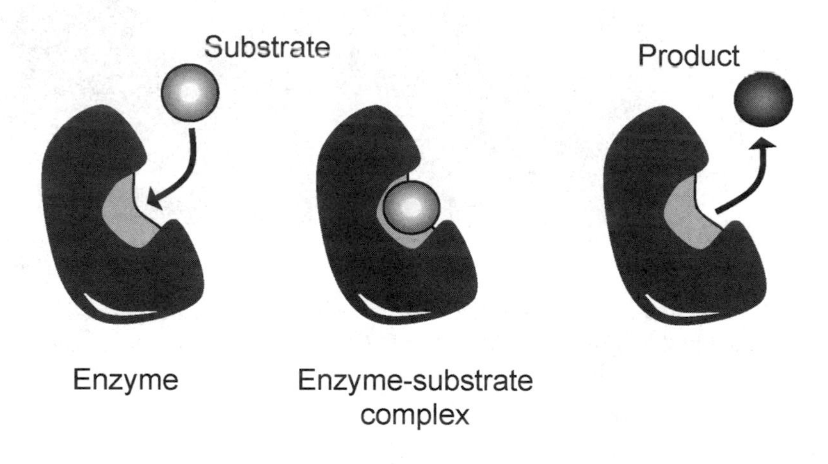

Fig. 7.43 A schematic of an Enzyme-substrate interaction.

Enzymes speed up chemical reactions without being used up themselves. As a result, they are categorized as *catalysts*, which are materials that accelerate a chemical reaction without themselves undergoing a net change.

Each type of enzyme is highly specific, catalysing only a single type of reaction. For instance, the blood clotting enzyme *thrombin* cleaves a particular type of blood protein in a specific place and nowhere else. Enzymes often work in teams, where the product of one enzyme-substrate interaction can become the substrate of another enzyme-substrate reaction.

There are many biological assays which are used in sensing the presence of an antibody or antigen with the help on enzyme. Assays such as *enzyme-linked immunosorbent assays* (*ELISA*))are commonly used in optical sensing. ELISA generally uses two antibodies: one antibody is specific to the antigen, and the other antibody reacts with antigen-antibody complexes, and is coupled to an enzyme. This second antibody, which is generally labeled with fluorescent or chromatic molecules, is used for producing a detectable signal. A sandwich ELISA procedure is shown in **Fig. 7.44.**

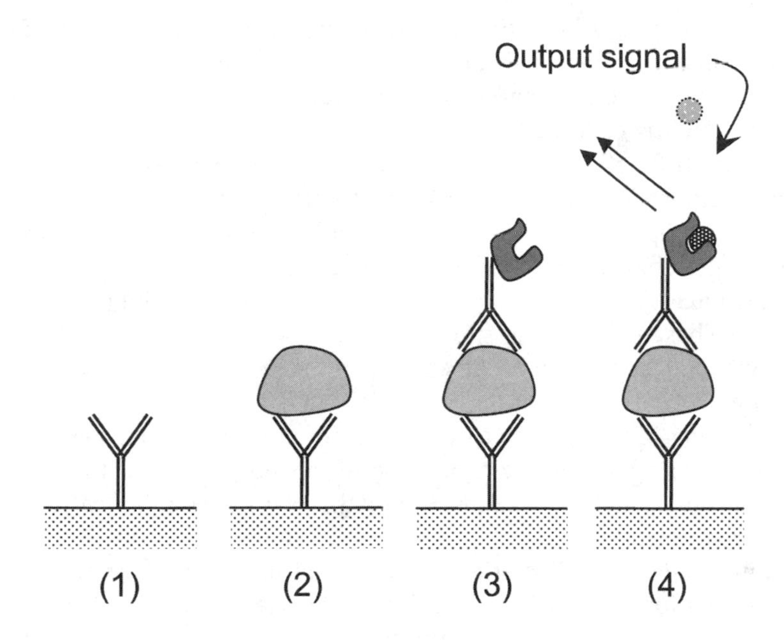

Fig. 7.44 A schematic of an ELISA. (1) The first antibody is immobilized on the substrate; (2) thc antigen is added and interacts with the antibody, (3) the detecting antibody (second antibody) is added which binds to the other side of antigen. The secondary antibody is enzyme-linked (4) the substrate is added. It is the substrate-enzyme interaction which produces a detectable signal.

Inside a cell, most proteins do not work continuously; instead their activities are regulated. As a result, the proteins in a cell are usually in a state of equilibrium. The activities of cellular proteins are controlled in an integrated manner. For instance, the catalytic activities of enzymes are regulated by other molecules. The system can be very elaborate. Many enzymes may operate at the same time in a cell. Many of them may compete for the same substrate at the same time. Their factions should be coordinated.

The regulation of enzymes can occur at different levels. At one level, the cell controls how many molecules of each enzyme are made. At the second level, the cell can control the enzymatic activities by confining them to sub-cellular compartments. The third level, which is the most rapid control system, is where the cell adjusts the reaction rate of each enzyme.

The third level of controlling enzymes finds applications in nanotechnology as it can be programmed by the user outside a cell. The control functions as follow: molecules other than substrates, when added to the environment, may become attached to the binding sites and thus hinder the enzymatic process. Alternatively, when a molecule other than the substrate binds to an enzyme, at a special regulatory site outside the active site, it can alter the conversion rate of the substrate. This is a negative feedback process which is called a *feedback inhibition*, and its role is to prevent an enzyme from functioning. Alternatively, enzymes can also be subjected to positive regulations where the activity is stimulated by a *regulatory molecule*.[92] The shapes of the regulatory molecules are totally different from the shape of the enzyme molds and are called *allostery* (means other-solid in Greek language). Enzymes have at least two different binding sites on their surfaces. These two binding sites can communicate with each other, when binding occurs at one site of the enzyme it causes a change in the protein's structure.

The *competitive inhibitor* can be directly placed into the main binding site (to fill and block the enzyme) or it can be placed in the secondary bonding site to alter the dimensions of the main ones (also called a *non-competitive inhibitor*).

Many proteins are allosteric. It means that they can adopt two or more different conformations. By changing from one conformation to another they are able to regulate their activities. The chemistry is simple as each molecule that binds the enzyme brings about a slight change in the overall forces within the molecule hence which produces a new conformation.

The other method of regulating the activities of proteins involves the attachment of phosphate groups, which covalently bonds to one of the amino acids in the protein. The phosphate group carries two negative charges and as a result, its attachment can cause a major *conformational change* in the protein. For instance, it can attract a cluster of positively charged amino acid side chains thus greatly altering the protein's activity. This process is reversible and is called protein *phosphorylation.* The addition or removal of a phosphate group to or from a specific protein often occurs in response to signals that specify changes in a cell's state. Many of these signals are generated by hormones and neurotransmitters and carried out from within the cell's membrane by a cascade of phosphorylation events.

Protein phosphorylation is conducted by a catalyst protein called *kinase.* It transfers the terminal phosphate group of an *adenosine triphosphate* (*ATP*) molecule to the hydroxyl group of an amino acid. The removal of the phosphate group, which is the reverse reaction, is catalysed by the protein *phosphatase.*

ATP is a *nucleotide*, which is a molecule that consists of a nitrogen-containing ring that is linked to a five-carbon sugar: either *ribose* or *deoxyribose.* Nucleotides can act as short-term carriers of chemical energy. ATP participates in the transfer of energy in molecular interactions. ATP is formed in the process of the generation of energy using the oxidative breakdown of food materials. The three phosphates in ATP are linked in series and the rupture of these bonds releases a large amount of energy. When the terminal phosphate group is split off by hydrolysis, *adenosine diphosphate* (*ADP*) is produced plus the released energy (**Fig. 7.45**). For many proteins, when a phosphate group is added or removed to their side in a continuous manner, it can switch rapidly from one state to another (like motor proteins).

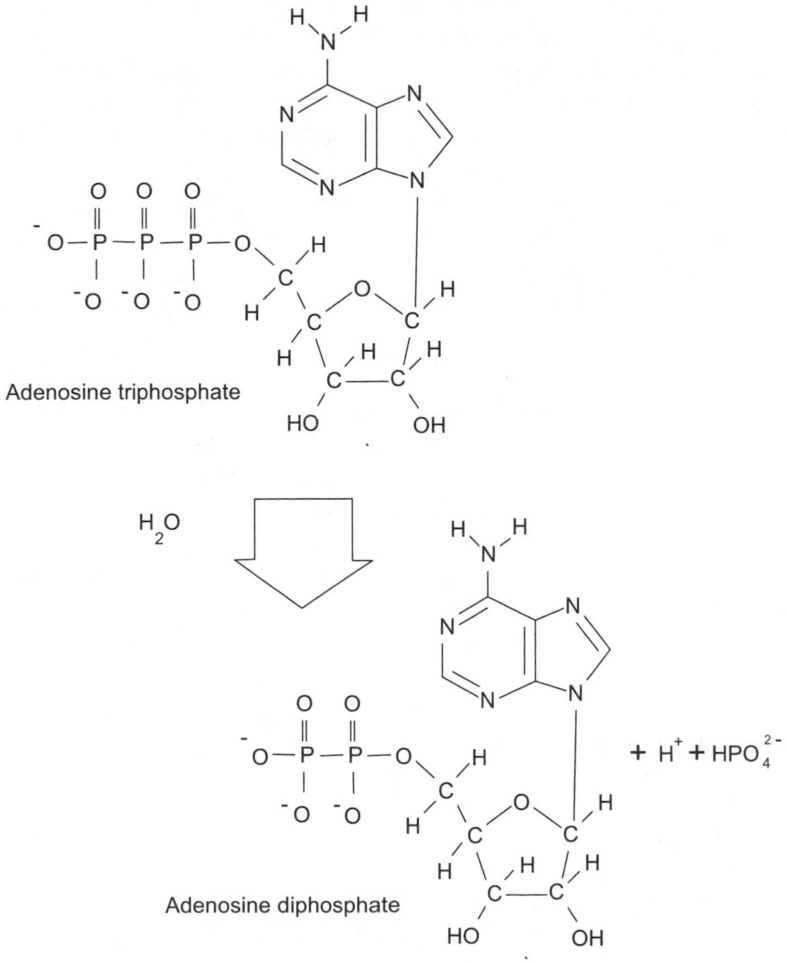

Adenosine triphosphate

H_2O

Adenosine diphosphate

$+ H^+ + HPO_4^{2-}$

Fig. 7.45 The conversion of ATP to ADP, which yields approximately 7.3 kcal/mol of ATP and the source of energy for many intercellular activities.

A protein can also inactivate itself by the hydrolysis of its internal *guanosine triphosphate* (*GTP*) to *guanosine diphosphate* (*GDP*) without the aid of a catalyst. The GTP-binding protein is active and by releasing a phosphate and turning into GDP it changes into an inactive state.

There are a large number of related GTP-binding proteins that function as molecular switches in a cell. The dissociation of GDP and its replacement by GTP can turn a switch on. The GTP-binding proteins often bind to other proteins in order to control their enzymatic activities and they play an important role in the intercellular signaling pathways.

A large conformational change can be produced in response to nucleotide hydrolysis. *Elongation factor Tu* (*EF-Tu*), a GTP-binding protein, elongates a polypeptide chain during the cellular synthesis processes. The hydrolysis of bound-GTP in EF-Tu causes only a small change in the position of the amino acid at the nucleotide binding site of approximately 0.1 nm. The release of the phosphate group changes the intramolecular bond similarly to a latch and creates a major change in the shape, which in turn releases the *tRNA* molecule to allow the synthesis process to be conducted.

One of the most important enzyme types in sensing applications are the redox enzymes groups. In redox reactions inside the body of a living organism a substrate becomes oxidized or reduced. This action is uses energy and support life within the body of the living organism. Redox enzymes accelerate the redox reactions to render them biologically useful. Redox enzymes also control the highly reactive intermediates. As a result, in such enzymes generally only a single substrate is employed and a single specified product is produced.

Redox enzymes are used in many sensing applications such as the glucose sensors. However, these redox enzymes usually lack direct communications with electrodes. As a result, they have to be used along with materials which mediate the diffusion of electrons into electrodes.

Example: Blood Glucose Sensors

One of the most popular electrochemical sensors in the market is the blood glucose monitor.[118] Patients with diabetes require the monitoring the glucose level in their blood. This monitoring has to be conducted several times a day to allow them to control their disease through insulin injections.

Early types of home glucose sensors consisted of a three layer working electrode (**Fig. 7.46**) and an Ag/AgCl reference electrode (Chap. 3).

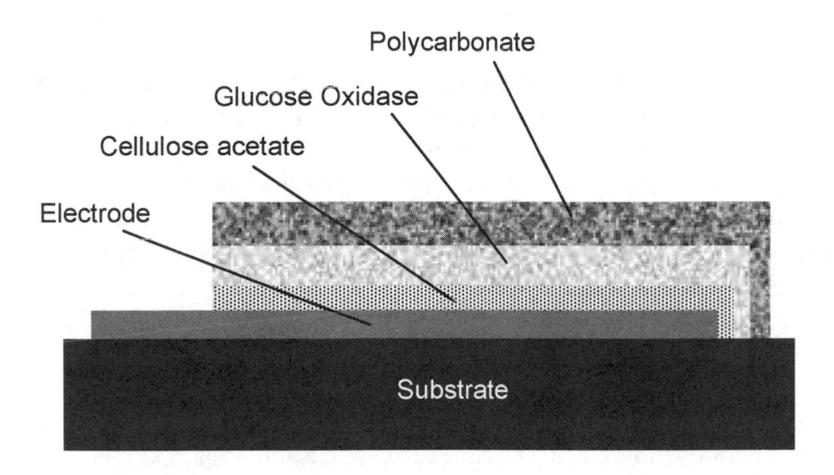

Fig. 7.46 A schematic of a working electrode for the early type of glucose electrochemical sensors.

In the working electrode, the outer layer is polycarbonate. The surface of polycarbonate is hydrophilic and permeable to glucose but not to the other constituents of blood (such as proteins). In the sensing process, a drop of the patient's blood is placed on this surface. The middle layer of the working electrode consists of an enzyme, *glucose oxidase,* which catalyses the reaction of glucose with O_2 forming gluconolactone (a gluconic acid) and hydrogen peroxide (H_2O_2). The bottom layer is made of cellulose acetate which is permeable to H_2O_2. The hydrogen peroxide is oxidized at the working electrode which is held at a positive voltage (approximately +0.6 V) versus the Ag/AgCl electrode. This produces a current according to the redox half-equation:

$$H_2O_2 \rightarrow O_2 + 2H^+ + 2e^- \quad (7.2)$$

The resulting current is proportional to the glucose concentration of the sample, and so the blood sugar levels of diabetics can be monitored.

7.4.8 Enzyme Nanopraticle Hybrid based Sensors

As described in the previous section, redox enzymes can be efficiently utilized in the fabrication of sensors. However, such enzymes generally lack the capability of proper alignment on electrodes, have difficulties in directly communicating with electrodes, and the electron transfer rate via conventional *mediators* can be low which all translate to poor sensing performance.

It is possible to increase the efficiency of such sensors with nanomaterials. The application of Au nanoparticles co-deposited with redox enzymes is one solution.[119] In such systems, the metal nanoparticles operate as nanoelectrodes that communicate with the enzyme redox sites. Such combinations increase the efficiency of electrical communications with electrodes. This metal supports also makes the enzyme electrode less sensitive to contaminants such as ascorbic acid.[120]

It is also possible to use semiconducting nanoparticles to induce an enhancement in response. In this case, activation can occur photoelectrochemically. A good example is the work of Curri et al who showed that the direct electron transfer from the enzyme active center to the CdS photogenerated holes can be achieved.[121] In such systems, electrons and holes, generated respectively in the conduction and valence bands of the semiconductor, are used by the enzyme for the reduction and oxidation of the substrate (**Fig. 7.47**). Transfer of charge carriers from the semiconductor surface to the active center of the enzyme is performed by mediators enables reversible redox processes. Curri et al biological-inorganic hybrid could perform the catalytic oxidation of formaldehyde using formaldehyde dehydrogenase (FDH) enzyme and CdS nanocrystals. The semiconductor particles were also used as charge carrier mediators instead of the nucleotidic co-factors, which played the same role in the biological apparatus. The selectivity and peculiar properties of inorganic nanocrystalline moiety also allow overcoming the problems due to the poor electrochemical performances of the NAD+/NADH couple, which exhibits kinetically unfavored processes of oxidation and reduction. The direct usage of thee NAD+/NADH-dependent redox enzymes create practical problems due to their poor stability and electrical contacting with electrodes.

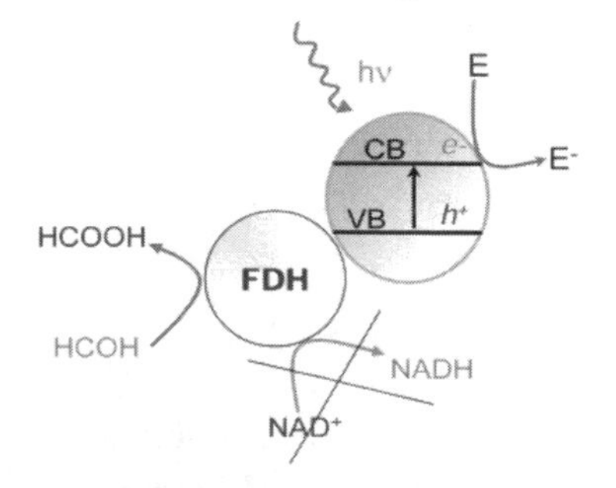

Fig. 7.47 The Schematic illustration of photoinduced oxidation of formaldehyde to formic acid using FDH coupled to CdS as charge carrier mediator

instead of NAD^+ co-factor. Reprinted with permission from the Elsevier publications.[121]

7.4.9 Motor Proteins in Sensing Applications

Motor proteins are interesting tools in nanotechnology. Apart from the enzyme regulation and the cell signaling, they also play an important role in moving molecules in cells. They generate the force which is responsible for movements in cells and muscle contractions. Examples are activities such as the movement of organelles along the molecular tracks in a cell, enzymes along a DNA strand, and moving chromosomes during *mitosis* (the division of molecules in eukaryotic cells). Plenty of applications are emerging for such proteins in the field of sensing. In an interesting work, Oliver et al have shown that the electromotility of the outer cells of the mammalian cochlea can be mediated by a voltage-sensitive motor molecule for cochlear amplification.[122] These proteins can also be associated with microtubules, contributing to the sensory mechanism that activates the checkpoints for bio-events that are completed in the correct order.[123]

The motor proteins' movements use energy from other sources. Without a foreign energy provider the protein only moves randomly. In order to make a movement in one direction, one of the movement steps should be irreversible. Many motor proteins use the hydrolysis of an ATP molecule which is bound to the protein. When the ATP is hydrolyzed, it releases a phosphate and ADP. Many motor proteins generate movement in this way. Myosin, which is a muscle motor protein, can run along the actin filaments to generate muscle contraction.

Kinesin (**Fig. 7.48**) is involved in the chromosomes movements during mitosis. Some of these proteins which are involved in the replication of DNA, can move themselves along DNA strands at a rate of 1000 nucleotides per second which is a phenomenal rate.

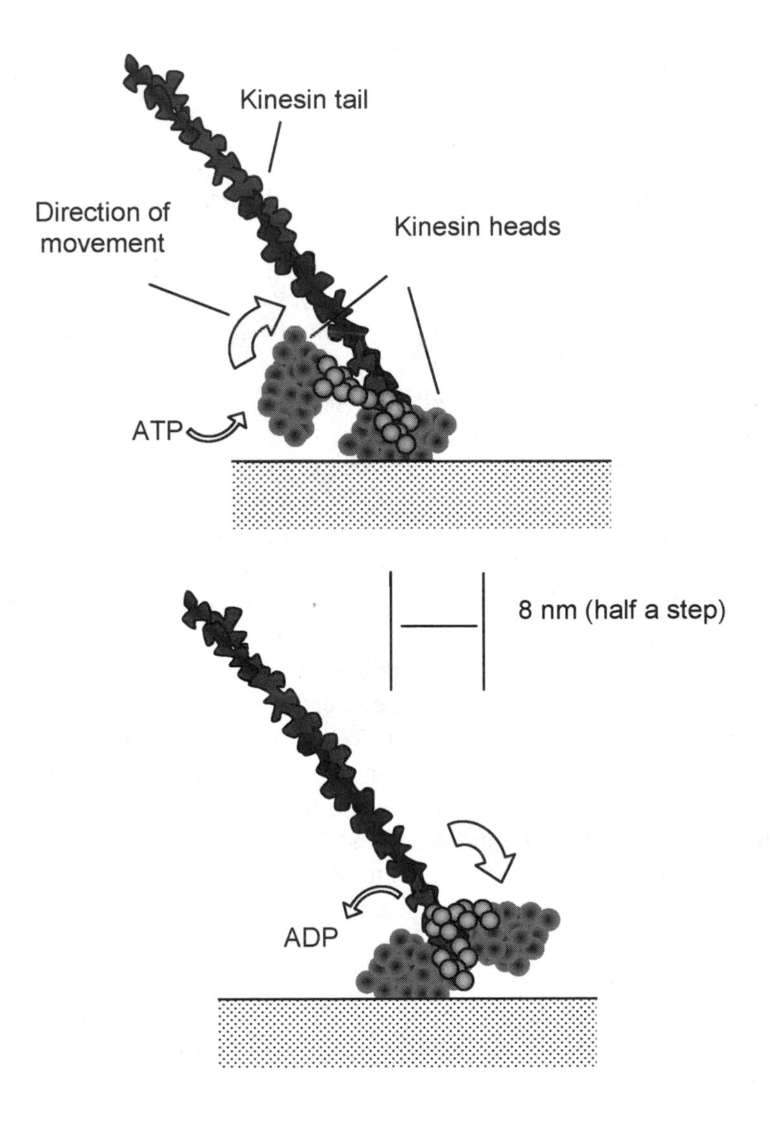

Fig. 7.48 How Kinesin functions as a motor protein.

7.4.10 Transmembrane Sensors

The membrane of a cell contains a number of integral membrane proteins. These proteins can be integrated within, placed on the outside of, or throughout the membrane (**Fig. 7.49**).

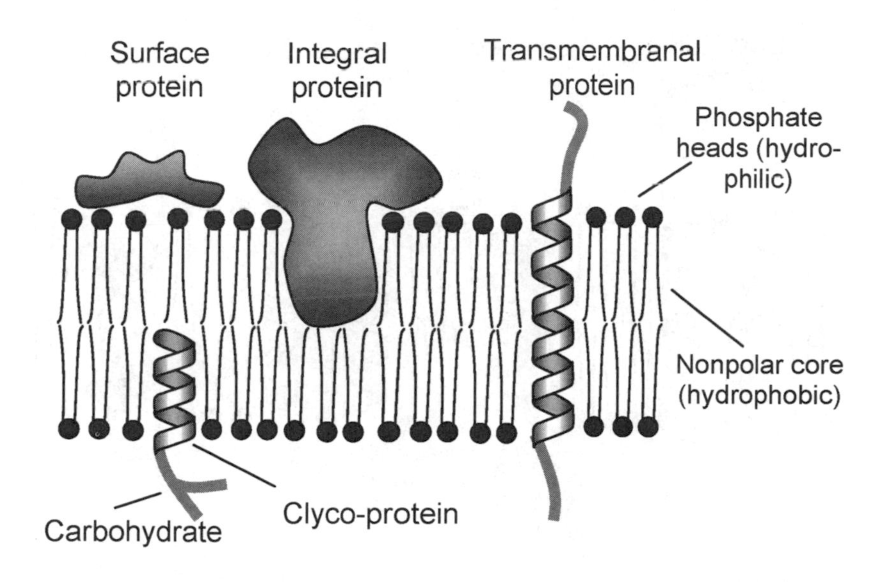

Fig. 7.49 **A** cross section of a membrane showing the phospholipid bilayer.

A group of the *transmembrane proteins* consist of polypeptide chains which cross the bilayer and usually have an α-helix form. In many trans-membrane proteins the polypeptide chain crosses the membrane only once. This type of protein can be a receptor for the *extracellular signals*. Other trans-membrane proteins form pores that allow water-soluble molecules to cross the membrane. Such pores cannot be formed by proteins with a single, uniformly hydrophobic transmembrane α-helix. Instead, these proteins generally have α-helices that cross the bilayer a few times. Some of these α-helices have both hydrophobic and hydrophilic side chains. The hydrophobic chains are close to the bilayer lipids and hydrophilic chains are located inside the cell. These pores function as selective transporters of small water-soluble molecules across the membrane. The subunits made of such helices are denoted as β, γ, and so on.

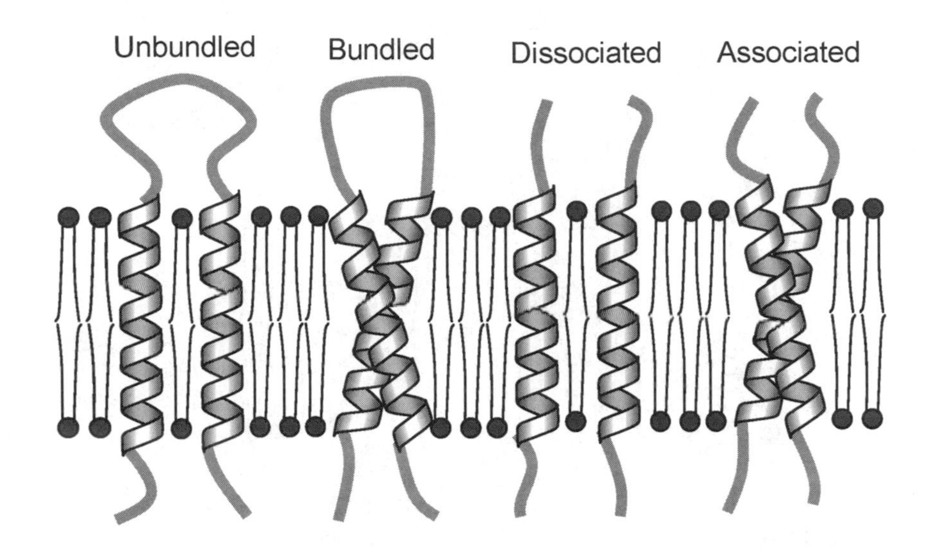

Fig. 7.50 Helix formation and bundling/association within membranes.

There is another type of proteins which acts as *transmembrane transporters*. These proteins allow nutrients, metabolites and ion to travel across the *lipid bilayers* (**Fig. 7.51**). These proteins are necessary as the lipid bilayers, which are the building blocks of the membrane structures, are highly impermeable to all ions and charged molecules and many nutrients and wastes such as sugars, amino acids, nucleotides and many cell metabolites.

Each transport protein provides a private passageway across the membrane for a special class of molecules. These membrane proteins extend through the bilayer, with part of their mass on either side. They have both hydrophobic and hydrophilic regions. The hydrophobic region is in the interior of the bilayer which binds to the hydrophobic tails of the lipid bilayer. The hydrophilic regions of these proteins are exposed to the aqueous environment on either side of the membrane. It is possible to remove these proteins from the membrane by disrupting the lipid bilayer (i.e with a detergent).

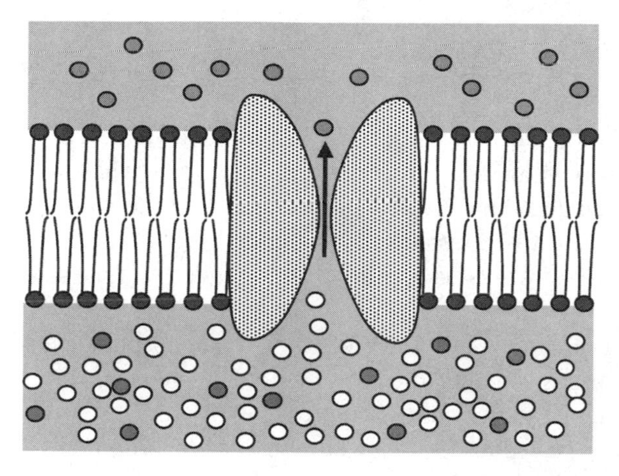

Fig. 7.51 A transmembrane transporter allowing the selective passage of particles.

The transportation of molecules can be either *passive* or *active*. Protein is passive when it automatically, and without any control, allows the passage of molecules, when the concentration of the target molecule is higher on one side of than the protein gate than the other side. However, if the transportation is against the concentration gradient then energy must be provided to the system to move these molecules to higher concentration site which is termed as *active transport*.

One example of *passive transportation* is the transportation of glucose in the plasma membrane of mammalian liver cells. This transport protein can adopt two different conformations: in one conformation glucose binds to the exterior of the cell and in the other it is exposed to the interior of the cell. As glucose is an uncharged particle, the mass concentration has to be measured then send signals must be sent to the valve proteins.

Another example of passive transportation is the movement of ions across membranes. Many cell membranes have an embedded a voltage across them, which is referred to as the *membrane potential*. This difference in potential exerts a electrical force on any molecules that carry charges. The net force is called the *electrochemical gradient* of the membrane. Depending on the direction of this gradient it may cause the movement of ions from one side to the other or vice versa.

Passive transport cannot satisfy all the needs of a cell, as in many occasions the direction of transport should be against its natural way. The major ways that *active transports* can be conducted are: (1) *coupled transport,* (2) *ATP-driven pumps* and (3) *light-driven pumps*.

Special types of ion channels can be the interesting examples of such active transportation gates. These ion channels concern the exclusive transport of ions such as Na^+, K^+, Cl^- and Ca^{2+}. In order to do this, they have to show incredible selectivity. Narrow channels do not allow large ions. Channels with negatively charged interiors deter negative ions. However, channels should be more selective letting only one type of ions to pass through and only in one direction. For example, when ions pass through these channels they make transient contact with the walls. This contact allows the protein to recognize them and only allow the right ions to pass. This contact also works as a counter that generates a passage rate count for a cell.

These ion channels are not continuously open but instead are gated. A specific stimulus triggers them to switch between an opened and closed state by a change in their conformation.

Membrane transporters are useful in the development of accurate and selective sensors. They are extremely selective and can respond to even a single molecule or ion. The ion channels described above are also called *nanopores*. Many examples of their applications as selective membranes can be found in the literature. There are many practical ways of using transporters. In the case of ion channels, it is possible to measure the current which is generated by the passage of single ions through these channels.[124,125] Single pores are used in stochastic sensing where individual interactions between the pores and analyte molecules are detected.[126]

One of the procedures for conducting such measurements is *patch-clamp* recording. In a patch-clamp recording, a glass tube with an open tip, of a diameter of only a few micrometers is used (**Fig. 7.52**). The glass electrode is filled with an aqueous conductive solution. This tip is pressed against the wall of a cell membrane. With gentle suction a small part of this membrane can be removed. Neher and Sakmann developed the patch clamp in the late 1970s and further perfected in the early 1980s.[127,128] They received the Nobel Prize in Physiology or Medicine in 1991 for this work. A metal wire is inserted at the other end of the tube. Finally, this *microtube electrode* can be used in a solution of ions along with other metal electrodes with accurate circuits connected to measure the generated currents **(Fig. 7.52)**.

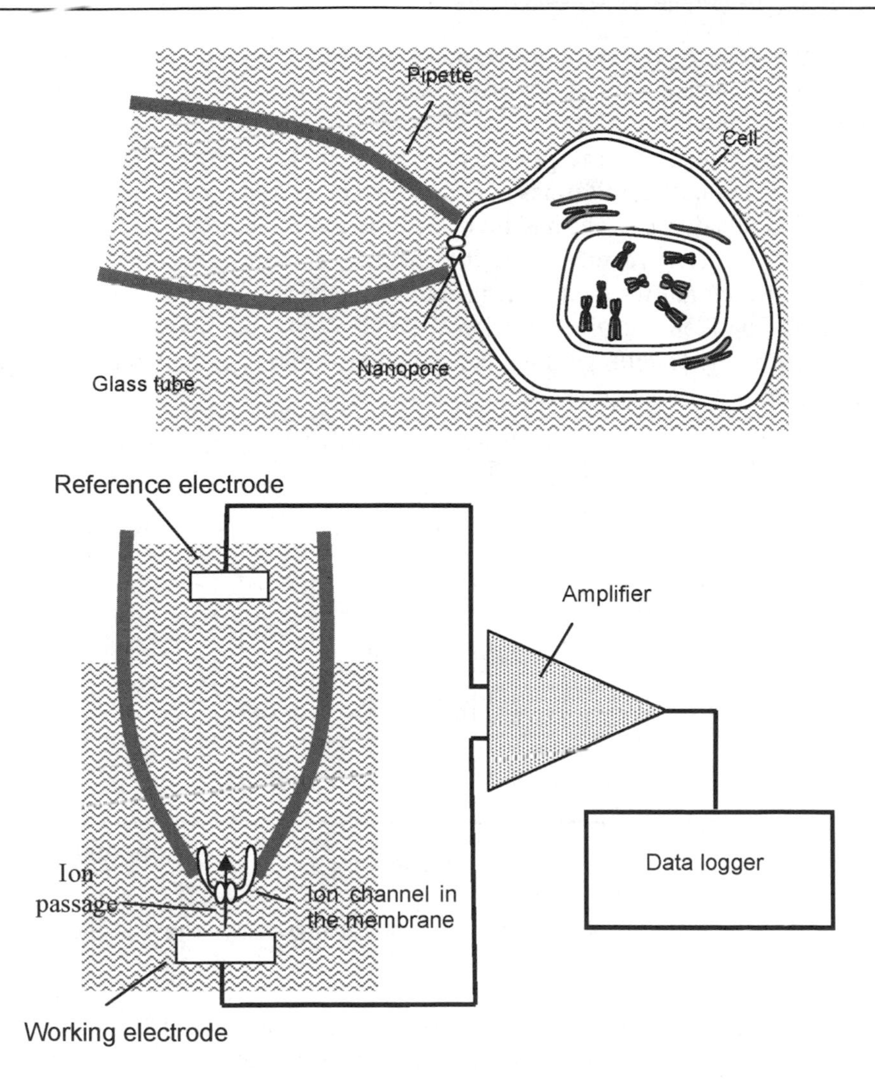

Fig. 7.52 The patch and clamp procedure (above) and the set up as an ion sensor (below).

By varying the concentration of the ions, it is possible to see a voltage generated. The current measured can be in the range of pico-amps which almost represents for the passage of single ions. Even if the conditions of measurements are kept constant, a random square wave signal is always recorded. This shows that the ion channels snap between their closed and opened states randomly (**Fig. 7.53**). The channel activity can be modulated by several classes of agents. One class comprises channel blockers, which can be natural in origin and endogenous to the tissue in which the channel is located (e.g. cytoplasmic polyamines and Mg^{2+}, which act on many eukaryotic channels), natural but exogenous (e.g. plant and animal toxins), or synthetic (e.g. various therapeutic agents).[126] Channels can exist in opened and closed states, and the presence of a channel blocker reduces (partial block) or eliminates (full block) the conductance of an open state.

This modulated current has high information content. In particular, the modulated current reveals both the identity and the concentration of a blocker, which allows a channel protein to be used as a stochastic sensing element. The identity of the blocker depends on the nature of the blocking events, e.g. the dwelling time, the extent of blockage, and their voltage dependences.

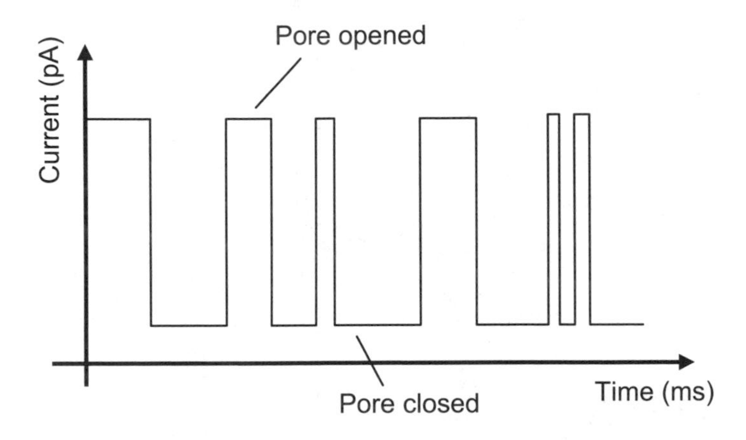

Fig. 7.53 The opening and closing of the pore changes the current in the patch-and-clamp set-up.

Another elegant approach for the development of biosensors using selective membrane proteins was proposed by Cornel et al.[129] In their work the detection of analytes possessing multiple recognition sites was performed using the structure shown schematically in **Fig. 7.54**. This structure was assembled using a combination of sulphur–gold chemistry and physisorption as described in **Fig. 7.54**. The membrane consists of lipids and channels, some of which are immobilized on the gold surface and some of which diffuse laterally within the plane of the membrane. The antibodies on the mobile channels scan an area of the order of 1 μm^2 in less than 5 minutes. As a result, with a low density of channels and a high density of immobilized antibodies, each channel can access up to 10^3 times more captured antibodies than if the gating mechanism were triggered by a directing binding of the analyte to the channels. The speed and sensitivity of the biosensor response can be adjusted in direct proportion to the number of binding sites accessible to each mobile channel. This allows for the quantitative detection of an analyte from sub-picomolar to micromolar concentrations in less than 10 minutes.

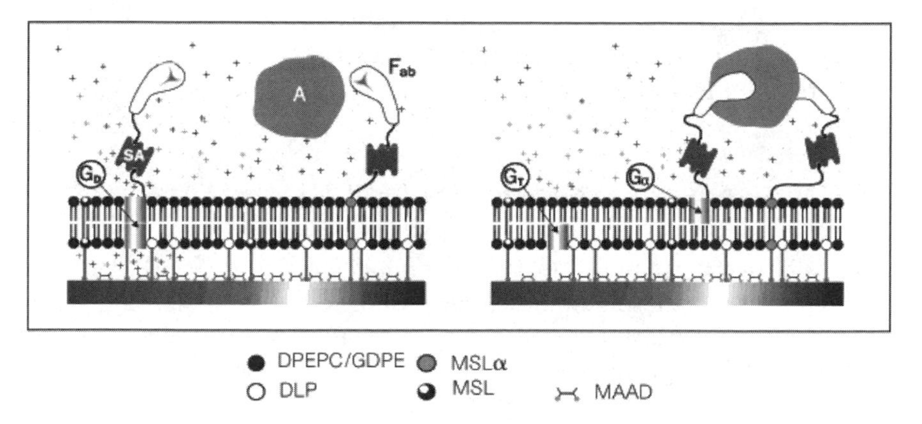

Fig. 7.54 A schematic of two-site sandwich assay. Immobilized ion channels (G_T), synthetic archaebacterial membrane-spanning lipids (MSL) and half-membrane-spanning tethered lipids (DLP) are attached to a gold surface via polar linkers and sulphur–gold bonds. Polar spacer molecules (MAAD) are directly attached to the gold surface using the same chemistry techniques. Mobile half-membrane-spanning lipids (DPEPC/GDPE) and mobile ion channels (G_α) complete the membrane. The mobile ion channels are biotinylated and coupled to biotinylated antibody fragments (Fab') using streptavidin (SA) intermediates. Some of the membrane spanning lipids (MSL_α) also possess biotin-tethered Fab'. In the absence of the analyte (A), the mobile ion channels diffuse within the outer monolayer of the tethered membrane, intermittently forming conducting dimers (G_D). The addition of the targeted analyte crosslinks the Fabs onto the MSL_α and G_α, and forms complexes that tether the G_α distant from their immobilized inner-layer partners. This prevents the formation of channel dimers and lowers the electrical conductivity of the membrane. Reprinted with permission from the Nature publications.[129]

It is also possible to engineer nanopores which are similar to cell membrane channels. However, engineering these pores is still technologically difficult. The structure and the locations of the polypeptides must be accurate in order to obtain a nanopore with the correct function. Individual synthetic helices were devised by Akerfeldt et al,[130] whilst template-assembled synthetic proteins were introduced by Grove et al[131] and Pawlak et al.[132] Literature can also be found on single chain bundles,[133] but the structures are not yet stable. Recently, progress has been made on barrel-like structures formed from helices,[134] and we can expect more work in this area in the near future. As molecular modeling would suggest, the more open barrel structure, which consists of β-barrels, is more amenable to amino acid substitutions. For example, the entire transmembranal domain of α-hemolysin can be replaced by a barrel made from a reversed amino acid sequence.[133] It is unlikely that such a manipulation would work with a helix bundle in which amino acid side-chain interactions are important for structural stability of the complex.

7.5 Nano-sensors based on Nucleotides and DNA

The discovery of the fact that *deoxyribonucleic acids* (*DNA*), the building blocks of chromosomes, are the genetic materials in a cell (**Fig. 7.55**), was a fundamental leap forward for the biological sciences. The discoveries around DNA structure and its functionality are recognized as one of the greatest achievements of our recent scientific history.

DNA structures have numerous applications in the development of sensors and are believed to have applications as the building blocks of future nano-devices. This section will present an introduction of these fascinating structures and will present some of their applications in sensor field.

DNA fragments are used for decoding gene expressions with microarray sensors. They can be used as selective layers for conductometric sensors. It changes its properties when exposed to an external source of energy. DNA has an excellent binary structure with embedded data for the selective detection of protein structures. DNA fragments are ready-made tools in biology for the construction of bio-nanomaterials. There are many more to add to this list.

The ability of cells to *store*, *retrieve* and *translate* the genetic information is what our lives depend upon. The combination of these activities maintains a *living organism* and differentiates it from other materials. At cell division, this hereditary information is passed on from a cell to its *daughter cell*. The information is stored in the *genes* within a cell, which are information-containing elements for the synthesis of particular proteins. As such they determine the characteristics of a species. The number of genes is approximately 30,000 for humans. The information in genes is copied and transmitted from cells to daughter cells again and again during the life of a multicellular organism which is carried out with a supreme accuracy. This process keeps the genetic *code* fundamentally unchanged during the life an organism.

Genetic science emerged in the middle of the twentieth century. In 1940, it was found that in simple fungi the genetic information consists chiefly of instructions for producing proteins. At the same time, it was found that DNA is the carrier of genetic information; however, the great breakthrough happened in 1953 when Watson and Crick (Nobel Prize in Physiology or Medicine in 1962 for this work) determined the structure of DNA. Soon scientists realized how DNA is replicated and how it is used for encoding the instructions for the creation of proteins.

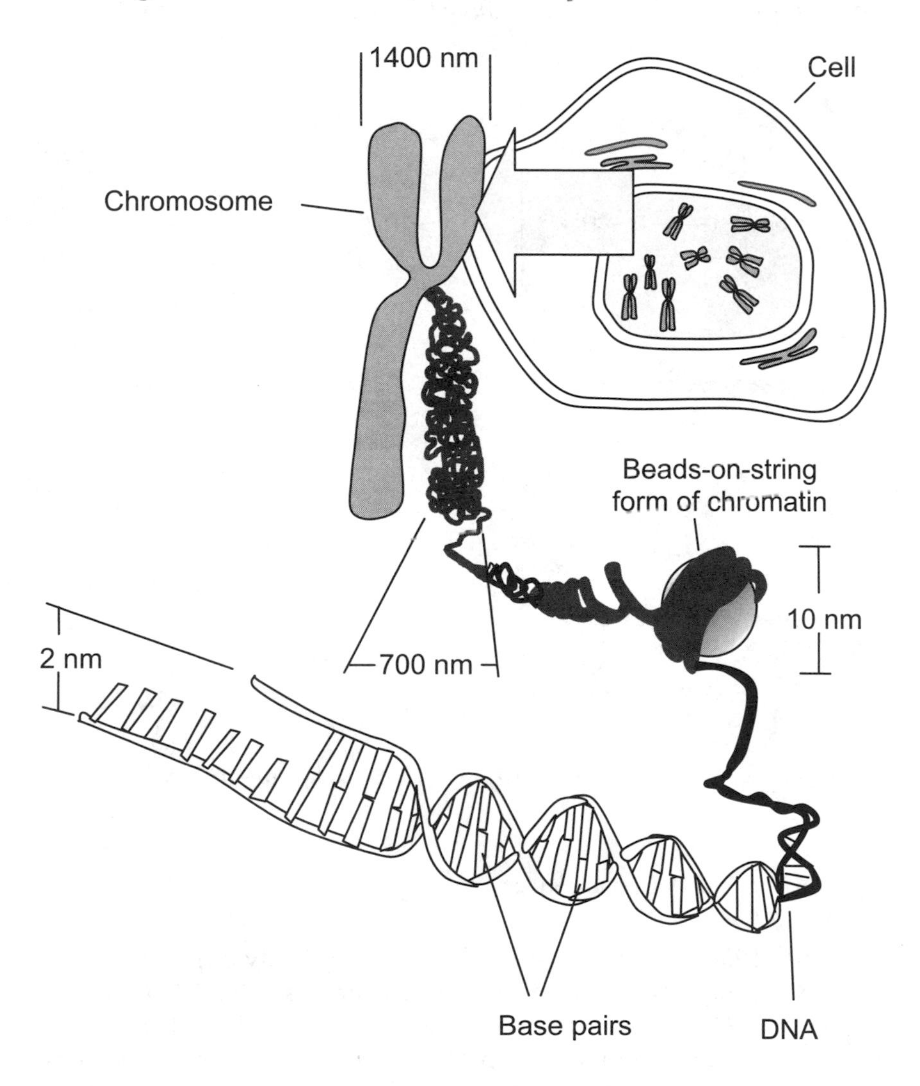

Fig. 7.55 A schematic showing the relationship between chromosome and DNA.

7.5.1 The Structure of DNA

The main building blocks for DNA are *nucleotides* which are composed of: (1) a *phosphate* (2) a *sugar* and (3) a *base*.[92] The nucleotides are covalently linked together into a *polynucleotide chain*. The *sugar-phosphate* chain is the backbone from which the bases are extended. The four bases are *adenine* (*A*), *cytosine* (*C*), *guanine* (*G*) and *thymine* (*T*) (**Fig. 7.56**).

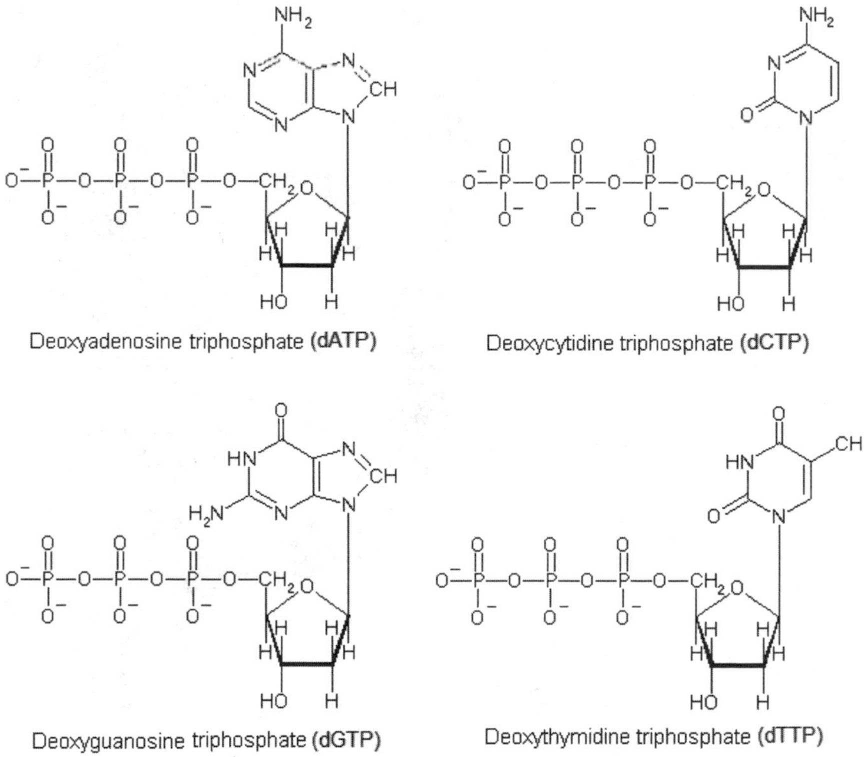

Deoxyadenosine triphosphate **(dATP)**

Deoxycytidine triphosphate **(dCTP)**

Deoxyguanosine triphosphate **(dGTP)**

Deoxythymidine triphosphate **(dTTP)**

Fig. 7.56 The four DNA nucleotides.

In early 1950, the examination of DNA using X-ray diffraction revealed that it was composed a helix of two twisted strands (**Fig. 7.57**). The model proposed by Crick and Watson described the possibility of using such a helix for encoding and replication of proteins. A DNA molecule consists of two long chains known as *DNA strands*. These strands are composed of four types of *nucleotide subunits* and they are held together by *hydrogen bonds*.

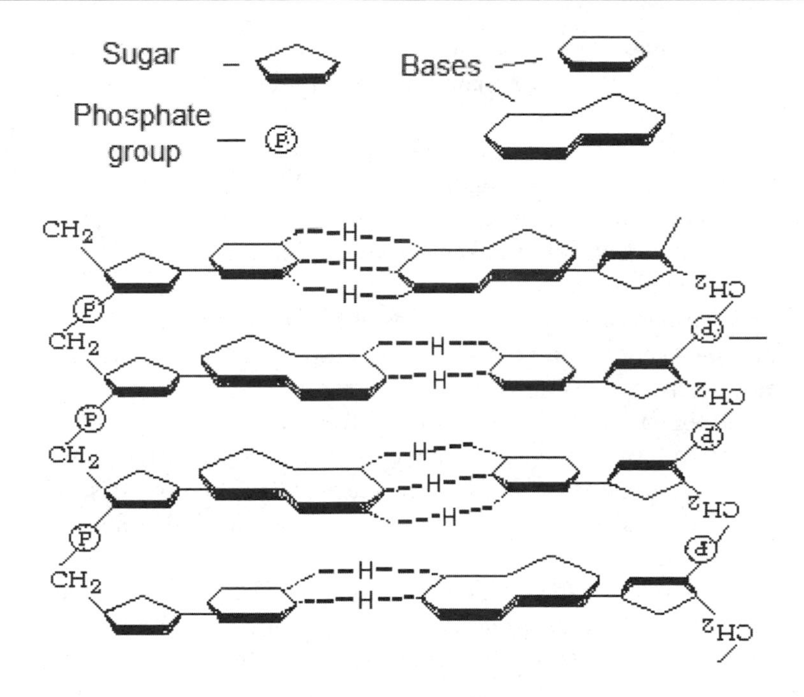

Fig. 7.57 The structure of DNA strands.

Nucleotide subunits are linked together at a directional way (**Fig. 7.57**) and as a result, the DNA strands have polarity. At the ends of chains are the *3′hydroxyl* and the *5′phosphate* groups, which are referred as the *3′ end* and the *5′ end.*

The bases are paired by hydrogen bonds: A selectively pairs with T and C selectively pairs with G. The two sugar-phosphate backbones twist around one another to form a double helix containing 10 bases per helical turn. Each strand of the DNA molecule contains a sequence of nucleotides which is exactly *complementary* to the nucleotide sequence of its partner strand.

The formation of nucleotides in a DNA strand can be decoded for the proteins' synthesis in a living organism. There is a correspondence between the binary format of the nucleotides and the 20 amino acids which are available to synthesize proteins. This correspondence is described as the *gene expression*. It is used when a cell converts a nucleotide sequence from a gene into the amino acid sequence of a protein.

A *genome* is the complete set of information in the DNA of an organism. Each human cell contains 2 meters of DNA strands which are tucked into a 5 to 8 μm diameter cell nucleus **(Fig. 7.55)**. Cells, by generating a series of coils and loops, pack these strands into *chromosomes*. The *human genome* consists of approximately 3.2×10^9 nucleotides distributed over 24 chromosomes.

A *Chromatin* is a complex of DNA and proteins. Except in the germ cells (sperm and eggs) and highly specialized eukaryotic cells (such as red blood cells) all other human cells contain two copies of each chromosome. One is inherited from the mother and one is inherited from the father and they are called *chromosomes*.

Chromosomes carry genes as codes embedded into their different sites. In addition to genes, a large excess of interspersed DNA is available which does not seem to carry any useful information. These excessive portions may have appeared in a long-term evolutionary process of different species. In general, the more complex the organism is the larger is its genome. The human genome is about 200 times larger than that of yeast. However, this is not always true; amoeba's genome is about 200 times larger than that of a human!

DNA replication is the duplication process of DNA which occurs before a cell divides into two daughter cells which are genetically identical.[92] The replication process needs elaborate consequence of functions by the organelles and proteins within a cell. In addition, specialized enzymes from a cell are needed for carrying out repairs when the components of a cell are damaged during the replication process. Despite all the pre-emptive measures by a cell, permanent damage may still occur during the replication. Such changes from the norm are referred to as *mutations*. The mutations are often detrimental, having the ability to cause genetic diseases such as many types of cancers. Nevertheless, mutations can be advantageous as well. For instance, bacteria can make their next generation resistant to antibiotics.

Base-pairing is fundamental to the DNA replication process. For the synthesis of a new complementary DNA strand, the initial strand functions as a template. In order to produce these single strands, the double helix must unwind. The hydrogen bonds make the DNA double helix a stable structure, so sufficient energy is needed to separate these bonds and can be applied using exposure to mechanical energy, thermal energy, and radiation by electromagnetic waves.

The process of DNA replication is initiated by *initiator proteins* in a cell. They first bind to the DNA and then separate the two strands apart. Firstly, these proteins separate a short length of the strand at room temperature at a segment which is called the *replication origin*. After starting the process, the initiator protein attracts other proteins, whose duty is to continue the replication process.

The replication is a bidirectional process which has a rate of replication of approximately 100 nucleotide pairs per second. DNA polymerase, which is the main protein in synthesizing a DNA, catalyses the addition of nucleotides to the 3′ end. It forms a *phosphodiester bond* between this end and the 5′- phosphate group of the incoming nucleotide. The energy-rich nucleotide triphosphate provides the energy for the polymerization, which links the nucleotide monomers to the chain and releases pyrophosphate (PP_i). Pyrophosphate is further hydrolyzed to inorganic phosphate (P_i) which makes the polymerization reaction irreversible.

The replication generates a closed loop which is known as *Okazaki fragment*. The DNA polymerase is self-correcting which can correct its mistakes. Before the enzyme adds a nucleotide it checks whether the previous nucleotide was base-paired correctly which significantly improves the accuracy of the replication process.

7.5.2 The Structure of RNA

For making a new strand of DNA, *ribonucleic acid* (*RNA*) of a length of 10 nucleotides long is used as the *primer*. *Primase* is the enzyme which uses RNA primer to synthesize a strand of DNA. A strand of RNA is similar to a strand of DNA except that it is made of *ribonucleotide* subunits. In RNA the sugar is ribose and not deoxyribose as in DNA. RNA also has another difference as well: the base thymine (T) is replaced by the base *uracil* (*U*).

Three more enzymes are needed to produce a continuous new DNA strand from pieces of RNA and DNA. They are *nuclease* which removes the RNA primer, *repair polymerase* which replaces it with DNA and *DNA ligase* which joins the DNA fragments together. For DNA ligase to be able to function, ATP or *NADH* is required to provide the necessary energy *Nicotinamide adenine dinucleotide* (*NAD*) and *nicotinamide adenine dinucleotide phosphate* (*NADP*) are two important *co-enzymes* found in cells. NADH is the reduced form of NAD, and NAD^+ is the oxidized form of NAD. Eventually in the replication process, *telomerase* puts in the ends of the eukaryotic chromosomes.

Cells also have an extra protein to reduce the occurrence of errors in the DNA replication which is called *DNA mismatch repair*. DNA is continually undergoing thermal collisions with other molecules, which results in changes and thus damage to the DNA. In addition, amino groups may be lost from cytosine. UV light is also another source of damage as it may produce a *thymine dimer* by covalently linking two adjacent pyrimidine bases.

7.5.3 DNA Decoders and Microarrays

The DNA sequence in chromosomes does change with time and can be rearranged.[92] This process may allow organisms to evolve in response to a change in their environment. This arrangement is called genetic recombination.

These days, new methods for analysing and manipulating DNA, RNA and proteins are fuelling an information explosion. By knowing the sequence of the nucleotides in biostructure, we have obtained the genetic blueprint of many organisms.

In the 1970s it became possible to isolate pieces of DNA in chromosomes. It is now possible to generate new DNA molecules and introduce them back into living organisms. This process is referred to with different expressions such as *recombinant DNA*, *gene splicing* and *genetic engineering*. Using the process, we can create chromosomes with combinations of genes which are not present in nature.

In sensing applications, the recombinant DNA techniques can be utilized for revealing the relationships between phenomenons in cells, detecting the mutations in DNA which are responsible for inherited diseases, and identifying possible suspects.

Recombinant DNA techniques are used for unlocking the genome by breaking large pieces of DNA into smaller pieces. Codes in the genome are embedded within these fragments and so by studying these fragments we can gain an understanding of which proteins the DNA code for.

A class of bacterial enzymes known as *restriction nucleases* is used to cut DNA at particular sites. These sites are identified by a short sequence of nucleotide pairs. Restriction nucleases are now commonly used in *DNA technology*, with hundreds of them available in the market. Each one is able to cut DNA at a particular site giving researchers powerful tools to investigated DNA-related areas.

Gel electrophoresis was one of the first procedures that was utilized to separate cleaved DNA pieces. The most common gels are made of *agarose* and *polyacrylamide*. After gene splicing, the mixture of DNA fragments are loaded at one side of the gel slab and a voltage is applied. The fragments of DNA travel with velocities which are proportional to their mass and consequently to their sizes. As a result, the fragments become separated according to their sizes. This produces a pattern which can be decoded and used as the expression of a gene.

The two strands of DNA are held together by hydrogen bonds that can be broken at temperatures higher than 90°C or at extreme pHs. If the process is reversed the complementary strands will re-form double helices in a process called *hybridization*.

The hybridization of DNA is used for diagnostic applications. For this propose, *DNA probes* are used. A DNA probe is a single stranded DNA molecule, typically 10-1000 nucleotides long. Those probes are used in detecting nucleic acid molecules, which contain complementary sequences.

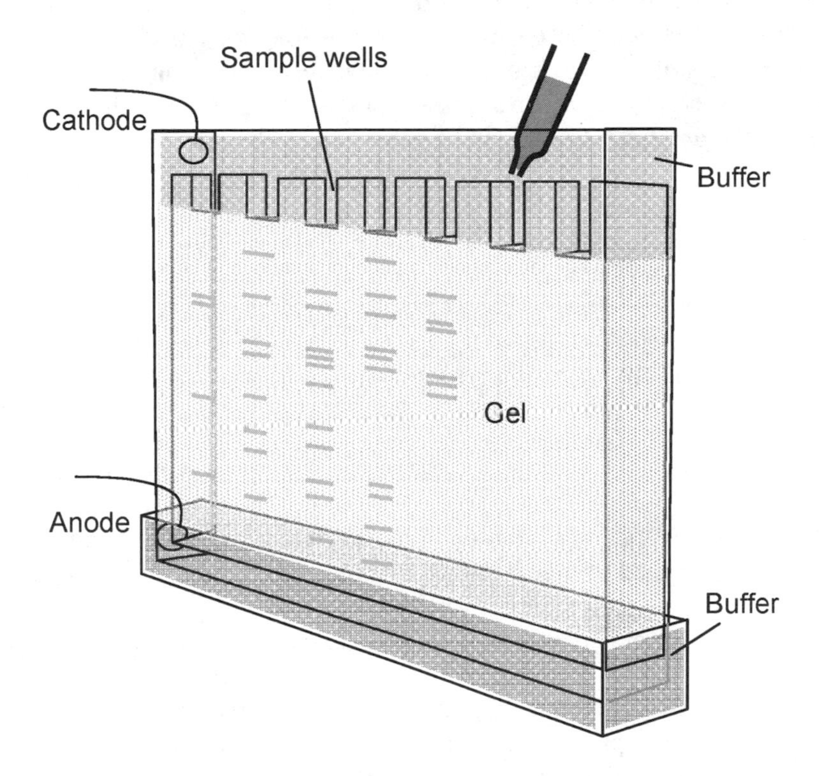

Fig. 7.58 A schematic of gel electrophoresis. In order to visualize the DNA bands the gel is soaked in a dye which binds to DNA. This particular dye can fluorescence under UV light.

One of the most important applications of DNA probes in diagnostics is for identifying the carriers of genetic diseases. For instance, for *sickle-cell anemia* the exact nucleotide change in a mutant has been determined. If the sequence G-A-G is changed to G-T-G at certain positions of the DNA strand which codes for the β-globin chain of haemoglobin then the person is diagnosed with the disease.

A common laboratory procedure utilized for hybridization in conjunction with electrophoresis for the decoding of DNA fragments is called *Southern blotting* (**Fig. 7.59**). In Southern blotting the unlabelled fragments of DNA, which are separated using electrophoresis, are transformed onto a nitrocellulose paper and then probed with a known gene or fragment for decoding.

DNA microarrays, developed in the last two decades, have revolutionized the way we analyse genes, by allowing the DNA fragments and the RNA products of thousands of genes to be investigated simultaneously.

By observing the microarray measurements, the cellular physiology in terms of the *gene expression* can be patterned.[92] The mechanisms that control gene expression act as both an on/off switch to control which genes are expressed in a cell as well as a *volume control* which increases or decreases the level of expression of particular genes as is necessary.

DNA microarrays are substrates with a large number of DNA fragments on them in contained areas. Each of these fragments contains a nucleotide sequence that serves as a probe for a specific gene. Some types of microarrays carry DNA fragments corresponding to entire genes that are spotted onto the surface by robots. Others contain short oligonucleotides which are synthesized on the surface. An *oligonucleotide*, or *oligo* as it is commonly called, is a short fragment of single-stranded DNA that is typically 5 to 50 nucleotides long.[135]

In oligonucleotide microarrays (or single-channel microarrays), the probes are designed to match known or predicted parts of the sequence of *RNAs*. There are commercially available designs that cover complete genomes from companies such as GE Healthcare, Affymetrix, or Agilent. These microarrays give estimations of the gene expression (**Fig. 7.60**).

The wafer (which is generally quartz or glass) is patterned with a photoresist and the appropriate mask is put in place. The mask is designed with 1-25 μm^2 windows that allow light to pass through areas where a specific nucleotide is needed.

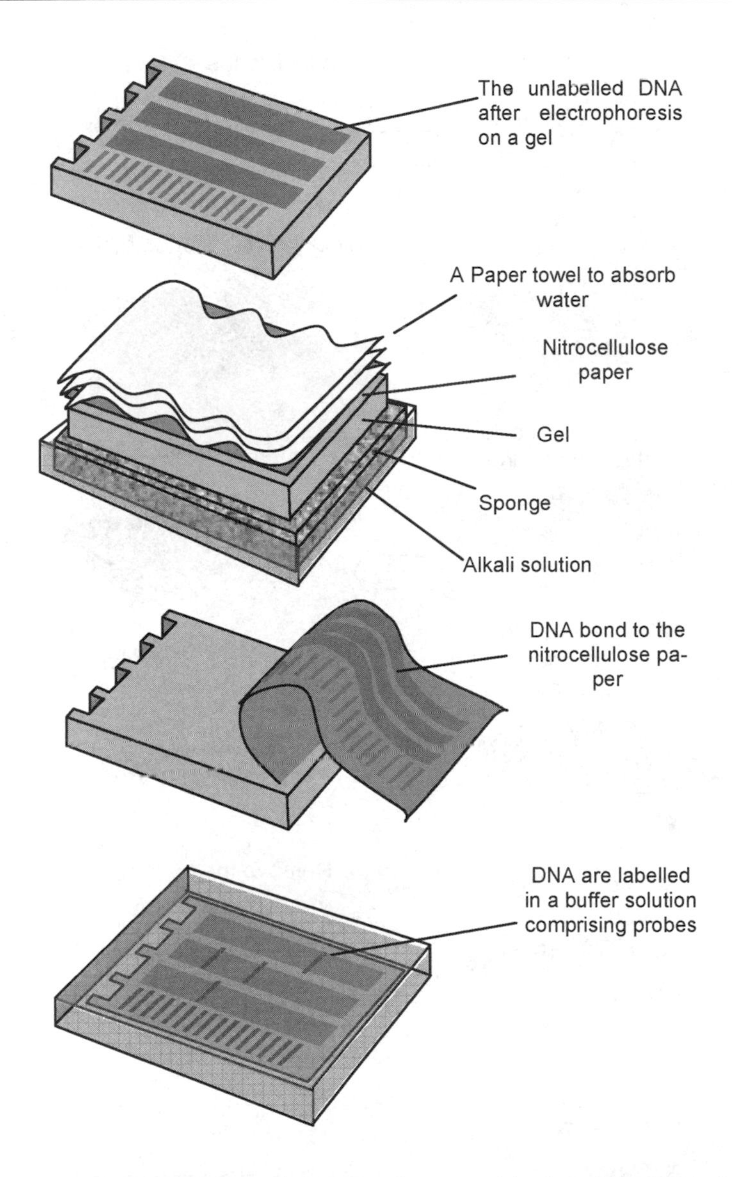

Fig. 7.59 DNA fragment detection using Southern blotting: The DNA fragments are separated by electrophoresis. A sheet (commonly nitrocellulose and/or nylon) is placed on the gel, and blotting process transfers the DNA fragments. An alkali solution is sucked through the gel and the sheet by a stack of paper towels. The sheet containing the bound single-stranded DNA fragments is removed and placed in a buffer containing the labeled DNA probe. After hybridization, the DNA that has hybridized to the labeled probe shows the code.

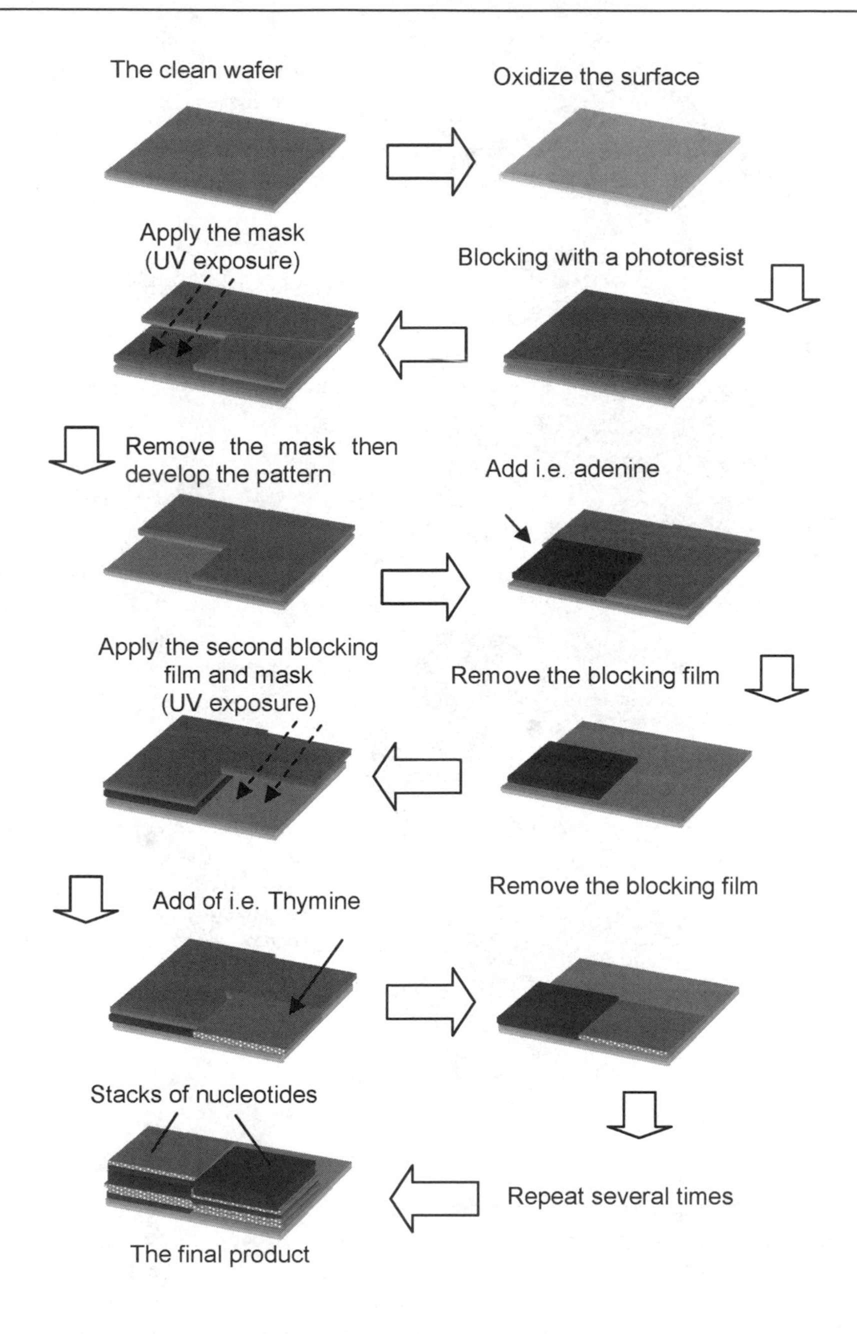

Fig. 7.60 The fabrication of an Affymetrix Genechip through the use of photolithography and combinatorial chemistry-specific probes.

To use a DNA microarray to simultaneously monitor the expression of a gene in a cell, *messenger* (*mRNA*) which is the RNA that encodes and carries information from DNA during transcription to the sites of protein synthesis, is extracted from the cells and copied into a *complementary DNA* form. As working with mRNA is difficult (mRNA is unstable and is easily degraded by *RNases* which can be found even on the skin), an enzyme called *reverse transcriptase*, which produces a DNA copy (*complementary DNA* or *cDNA*) of each mRNA strand, is used instead. The cDNA is also easier to manipulate than the original RNA. The cDNA can be labeled with a fluorescent probe.

The microarray is incubated with the labeled cDNA sample, and hybridization occurs. The array is then washed to remove the unbound molecules, and a scanning laser microscope can used to find the fluorescent spots. A typical gene microarray-based experiment is shown in **Fig. 7.61**. The array is scanned, for example by confocal microscopy, and the scanned image is then interpreted. The final image is of spots of differing intensity of the label (shown by the brightness of the fluorescent labels) that are related to the degree of hybridisation between the probe and the target DNA.[136]

The array positions are then compared to particular genes whose expressions are known. The cDNAs which are the reverse of the transcribed mRNAs, can be found in sets of libraries. A *cDNA library* refers to a nearly complete set of all the mRNAs contained within a cell or organism. Such a library has several uses: it is important for analyses in *bioinformatics*; in addition, the cDNA sequence gives the genetic relationship between organisms through the similarity of their cDNA. By using all of this data novel DNA molecules can be engineered.

The example shown in **Fig. 7.61** is a comparative gene expression experiment. Two sets of cells, for example diseased and healthy cells, provide the initial sample. The aim of the experiment is to investigate the gene expression in the cell populations by monitoring the mRNAs in the cell nucleus. In the first stage of the experiment, mRNAs are extracted from the cells, and reverse-transcribed into the more stable cDNA. The cDNAs from each cell population are then labeled, typically with different colored fluorescent dyes. The cDNAs are then hybridized to a DNA microarray and the hybridized array is washed and scanned. The positions and intensities of the spots seen provide information regarding the genes which are expressed by the cell.

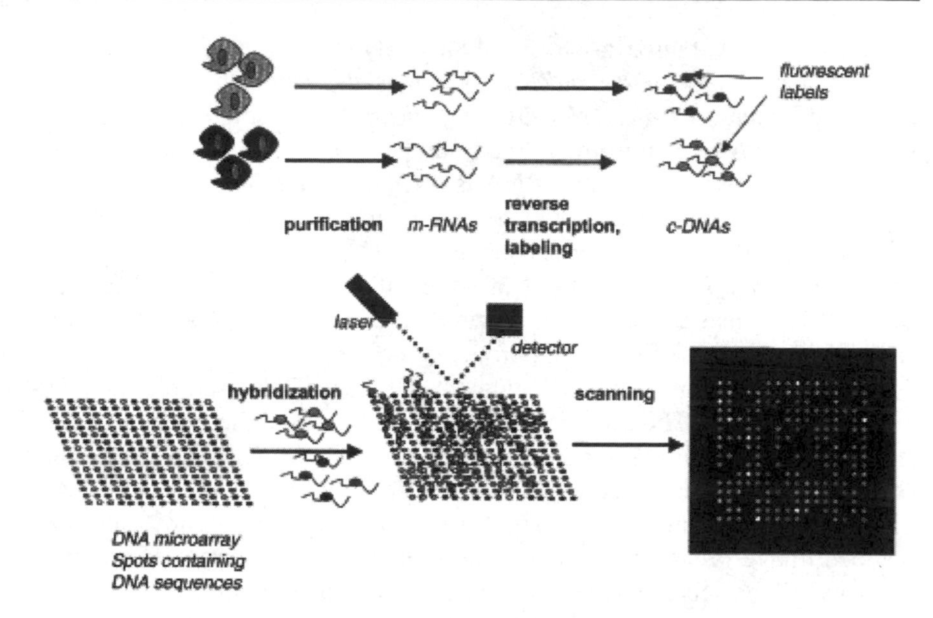

Fig. 7.61 A typical gene microarray experiment. In this example, a comparative gene expression experiment is shown. Reprinted with permission from the Elsevier publications.[136]

Generally, for DNA tests it is necessary to amplify the number of DNA strands and increase their number to a sufficient quantity. The *polymerase chain reaction* (*PCR*) is a powerful form of DNA amplification. The PCR process usually consists of a series of twenty to thirty five cycles, each of which consists of three steps (**Fig. 7.62**):

(a) The double-stranded DNA has to be heated to 94-96°C in order to separate the strands. This step is called *denaturing*; it breaks apart the hydrogen bonds that connect the two DNA strands.

(b) After separating the DNA strands, the temperature is lowered so that the *artificial primers* (often not more than fifty and usually only 18 to 25 base pairs long nucleotides that are complementary to the beginning and the end of the DNA fragment to be amplified) can attach themselves to the single DNA strands. This step is called *annealing*. The temperature of this stage depends on the primers being used (45-60°C).

(c-d) Finally, the DNA Polymerase has to copy the DNA strands. It starts at the annealed primer and works its way along the DNA strand in a step called *elongation* (c). The elongation temperature depends on the DNA Polymerase, which produces the double strand from the single strand DNA.

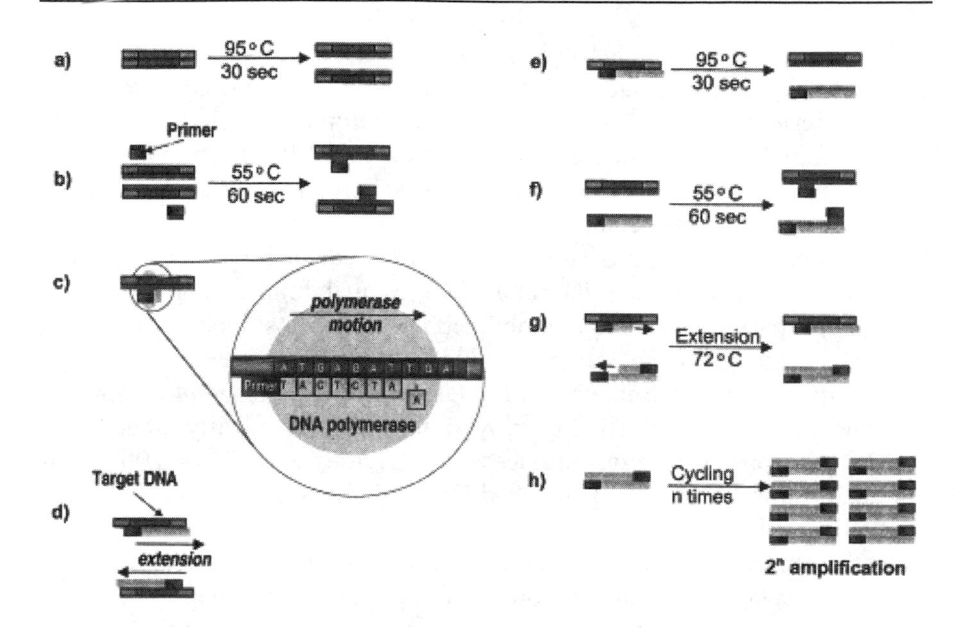

Fig. 7.62 A schematic diagram of the PCR cycle: (a) denaturing (b) annealing, (c-d) Elongation. (e-f-g) Repeat. Reprinted with permission from the Elsevier publications.[136]

The rapid evolution of nucleic acid based assays in the form of DNA chips is one of the latest developments in biosensor research. The concept of a million hybridization assays performed simultaneously on a one-square centimeter chip has much in common with the goal of high-density sensor arrays.

7.5.4 DNA-based Sensors

A large number of DNA-based biosensors have been described in the literature with electrochemical,[137-140] optical,[114,141-143] and piezoelectric[144] transducing elements. Such biosensors generally take advantage of the DNA strands' lock and key properties to achieve their selectivity. In such sensors the sensitive/selective layers can also comprised of DNA single strands which interact with functionalized antibodies and/or enzymes.

Basically most of the recent biochips, that were described in the previous section, are optical sensors as the DNA are labeled with optical tags which are detected using the irradiation and measurements of the absorbing light.

DNA can also help to amplify or quench the optical signals when it is attached to a porous surface. Such a phenomenon can be used in reflectometric interference based sensors. A classic example of such a sensor was described in Chap. 6, when a nanostructured thin film, such as a porous silicon surface, displays well-resolved Fabry-Perot fringes in its reflectometric interference spectrum.

The same structure can also be efficiently used for biosensing applications. Lin et al showed that the binding of an analyte to its corresponding recognition partner which is immobilized on a porous silicon substrate results in a change in the refractive index of the layer medium.[142] This change in the refractive index can be detected as a corresponding shift in the interference pattern. They employed an electrochemically etched silicon layer to produce porous surfaces with cavities as wide as 200 nm in diameter. Subsequently, they attached DNA oligonucleotides onto this porous silicon film.

In the presence of cDNA sequences, pronounced wavelength shifts in the interference pattern of the porous silicon films were observed. However, in the presence of non-cDNA sequences, but under similar conditions, no significant shift in the wavelength of the interference fringe pattern was seen. The lowest DNA concentration measured with this sensor was 9 fg/mm^2 (**Fig. 7.63**) which is far smaller than those of the sensing platforms described in Chap. 3.

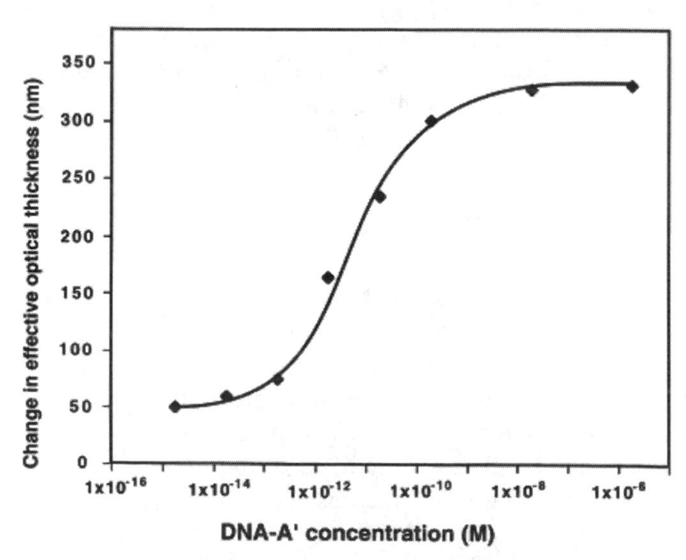

Fig. 7.63 Change in the effective optical thickness vs. DNA concentrations for a porous silicon reflectometric interference spectrum biosensor. Reprinted with permission from the Science Magazine publications.[142]

DNA can also be used in electrochemical sensors, which can be employed for the detection of specific viral or genetic anomalies. The main advantage of these type of sensors is that the set up is generally low cost and easy to implement.

The electrochemical DNA-based biosensor developed by Wang et al is a good example.[139] It is an electrochemical sensor for the detection of short DNA sequences related to the human immunodeficiency virus type 1 (HIV-1). The sensor relies on the immobilization and hybridization of the 21- or 42-mer single-stranded oligonucleotides from the HIV-1 U5 long terminal repeat sequence at either carbon paste or strip electrodes.

Fig. 7.64 demonstrates the potential of the sensor for quantifying nanomolar concentrations of the HIV-1 target sequence. Measurements of 2×10^{-8} M concentrations of the 42- and 21-mer targets (traces A and B) yield defined increases in the indicator peak (following a 30 min hybridization) compared to the response for the blank solution (the dashed line). They also used poly(ethylene glycol) (PEG) to accelerate the hybridization rate. The best results were obtained with a 1% w/w PEG (MW = 10,000) solution. The improved detectability accrued from the use of PEG is illustrated in **Fig. 7.64-C** for the sensing of the 1×10^{-8} M concentration of the 21-mer HIV-1 DNA target (following a 30 min hybridization period). A detection limit of 4×10^{-9} M can be expected from the signal-to-noise characteristics (S/N = 3) of this data. A total quantity of 1.35 ng (200 fmol) can thus be detected and even lower detection limits are expected for longer hybridization times.

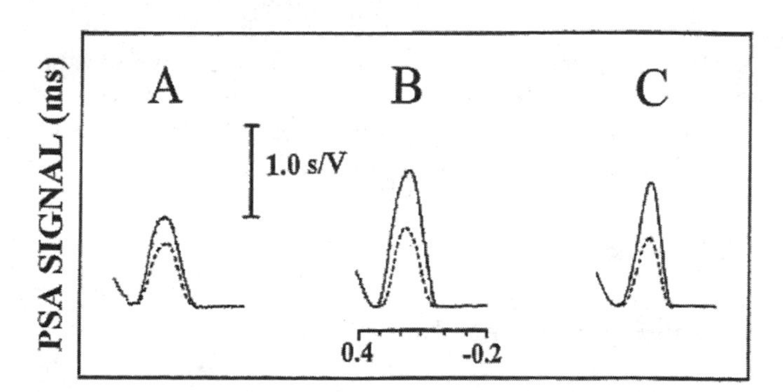

Fig. 7.64 Potentiograms for the targets: 2×10^{-8} M 42-mer HIV DNA: (A) 2×10^{-8} M (B), and 1×10^{-8} M (C) 21-mer HIV DNA in the absence (A and B) and presence (C) of 1% PEG (MW = 10,000), using corresponding complementary strands as the probes. Dashed peaks denote the chronopotentiometric responses for the corresponding blank solutions. Reprinted with permission from the American Chemical Society publications.[139]

Piezoelectric transducers can also be utilized along with the lock-and-key capability of DNA strands to develop highly specific sensors. As described in Chap. 3, Caruso et al were the one of the first groups who used this ability with quartz crystal transducers.[86] They used both single-layer and multilayer DNA. They monitored the DNA immobilization and the hybridization of the immobilized DNAs in situ from the QCM) frequency changes (**Fig. 7.65**). Equal frequency changes were observed for the DNA immobilization and hybridization steps for the single-layer films, which was an indication that the DNA probe-to-hybridized DNA target was in a ratio of 1:1. The multilayered DNA films also exhibited DNA hybridization, with a greater quantity of DNA hybridized compared with the single-layer films.

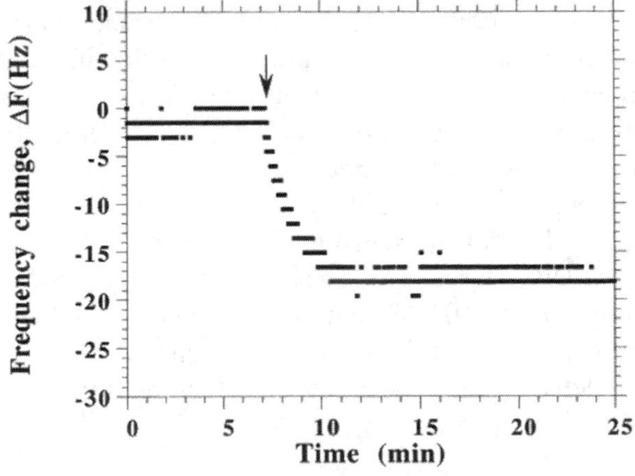

Fig. 7.65 The QCM frequency change versus time for the immobilization of biotinylated-DNA from HEPES buffer onto an avidin-modified QCM electrode. The arrow indicates the time at which the biotinylated-DNA was injected into the HEPES solution. Reprinted with permission from the American Chemical Society publications.[86]

7.5.5 DNA-Protein Conjugate-based Sensors

DNA is a ready-made engineered material that can be employed for sensing applications and as a building block for making nanosized transducing platforms. Until researchers learn to intelligently control the bottom-up synthesis of complex nanostructures, they have little choice but to imitate biology, and use bio-components such as DNA and proteins which can be put together on demand. The pool of knowledge about such molecules allows researchers to execute their needs with control, precision and guidance.[145] Apart from smart structures such as DNA and proteins, most of the other nanosized structures, especially the inorganic ones, lack any guidance mechanism. This issue makes it extremely difficult to have bottom-up control over the dimensions and shape of these structures.

DNA strands, as intelligent and highly coded materials, can also give directionality to materials which are conjugated with them. *DNA-protein conjugates* are versatile molecular tools which can be utilized in applications such as the self-assembly of high-affinity reagents for immunological detection assays, the fabrication of highly functionalized laterally microstructured biochips and the synthesis of supramolecular devices with nanostructured subunits.[146]

DNA-protein conjugate syntheses and their applications are exemplified in the research work conducted by Niemeyer et al as shown in **Fig. 7.66**.[146] They reported the self-assembly of streptavidin (**Fig. 7.66 – 1**) and 5',5'-bisbiotinylated double-stranded DNA fragments (**Fig. 7.66 – 2**).[147] They cross-linked bivalent double-stranded DNA (dsDNA) molecules to streptavidin, generating three-dimensionally linked networks. Their studies with nondenaturing gel electrophoresis and scanning force microscopy indicated that oligomeric DNA-streptavidin conjugates were formed (**Fig. 7.66 – 3**), which bridge adjacent DNA fragments. As a consequence of the streptavidin's low valency, the oligomeric conjugates have a large residual biotin-binding capacity that can be utilized for further functionalization of these complexes. Therefore, the resulting conjugates can be applied as reagents in immunoassays. One of the main advantages of such a technique is that due to the possibility of DNA amplification using the PCR, the DNA-protein conjugates can be detected at levels far below those available for the detection of proteins by methods such as conventional *antibody-based enzyme-linked immunosorbent assay* (*ELISA*).[146]

The combination of the ELISA with the PCR amplification (also called as *immuno-PCR* (*IPCR*)) was originally suggested by Sano et al.[148] This method can generally provide up to a 1000-fold enhancement in the detection limit when compared with an analogous ELISA expriment.[146]

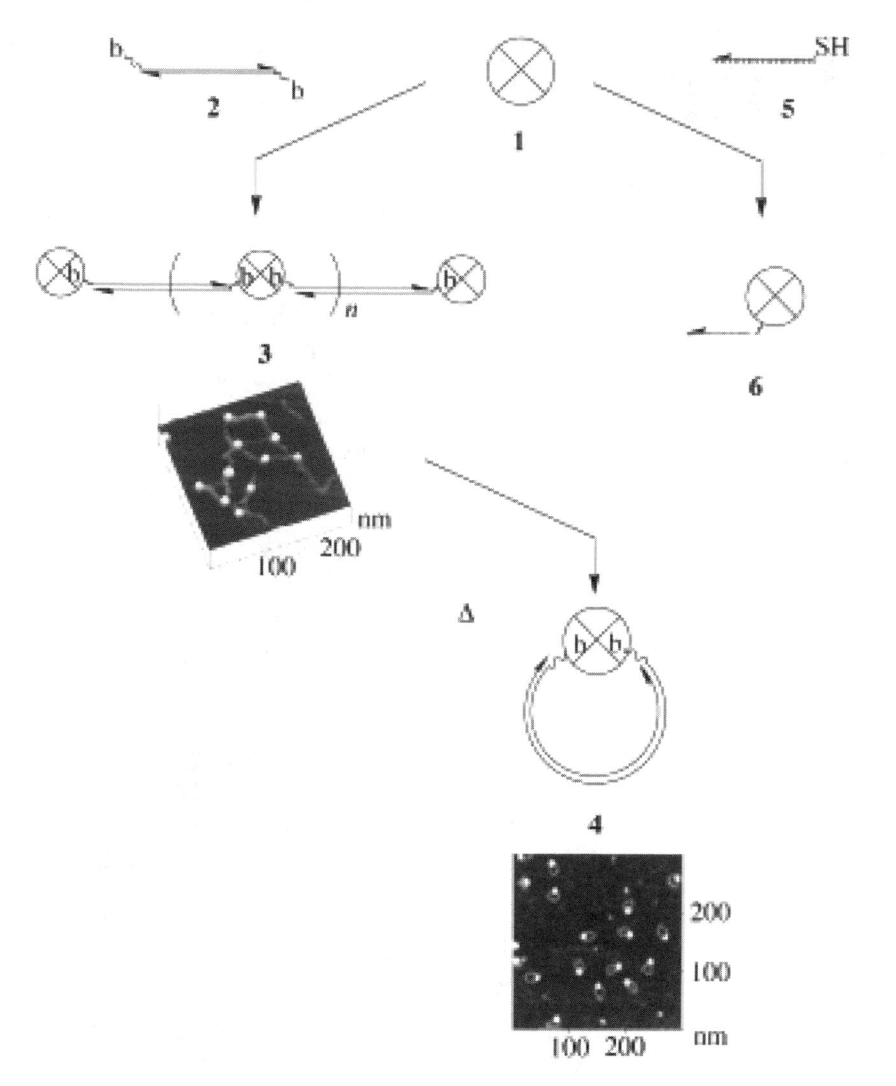

Fig. 7.66 The synthesis of DNA-streptavidin conjugates. Reprinted with permission from the Wiley InterScience publications.[146]

The DNA-streptavidin conjugates can also be used as a molecular templates for the development of DNA-based nanomaterials.[146] For instance, the biotin-binding capabilities can be employed for the attachment and positioning of biotinylated macromolecules, such as enzymes and antibodies[149] as well as low molecular weight peptides and fluorophores,[150] which both have applications in sensing.

As the DNA-streptavidin conjugates are formed by statistical self-assembly, distinct supramolecular species can be isolated and further functionalized by controlling the self-assembly process. As an example of the latter, a network of DNA-streptavidin conjugates can be effectively transformed into supramolecular DNA-streptavidin nanocircles as shown in (**Fig. 7.66** – **4**) by thermal treatment (**Fig. 7.66**).[151] Since the endogeneous protein molecule within the DNA-streptavidin nanocircles allows for the convenient attachment of other functional molecules, potential applications of such nanostructures obviously include their use as molecular tools in novel immunological assays.[152]

In addition, functional analogues of the conjugates described above can be obtained from the use of streptavidin covalently functionalized with a single-stranded oligonucleotide. These DNA-streptavidins (as shown in **Fig. 7.66** – **6**) are synthesized from 5'-thiol-modified oligonucleotides and streptavidin using hetero-bispecific cross-linkers, such as sulfosuccinimidyl-4(*p*-maleimidophenyl)-butyrate (**Fig. 7.66**).[152]

The covalently bound oligonucleotide moiety in (**Fig. 7.66** – **6**) provides a specific recognition domain for the complementary nucleic acid sequence in addition to the four native biotin-binding sites. Thus, it can be used as a versatile molecular adaptor in a variety of applications, ranging from the assembly of nanostructured protein arrays (**Fig. 7.66**) to the generation of laterally-microstructured protein biochips.[146] It is possible to use such DNA-protein biochips as decoding and sensing tools. Such standard biochips decrease the cost of development of devices and also standardize the process. With such biochips it is also possible to fabricate mixed arrays.

7.5.6 DNA Conjugates with Inorganic Materials

DNA can be combined with many chemical functional groups. For example, single-stranded DNA can be attached to electrically-active molecular elements such as metal clusters, fullerenes or certain molecular switches.[153] In 1997, Mirkin's group at Northwestern University in Evanston, Illinois, showed that free-floating target DNA strands with complementary DNA linked to gold nanoparticles are detectable (**Fig. 7.67**).[154] When the target strands bind to the gold-bound complementary strands, they pack the gold particles closer together. That changes the color which the particles reflect, creating a simple color-based detector for specific DNA sequences. They showed that the hybridization was facilitated by freezing and then thawing the solutions, and that the denaturation of these hybrid materials showed transition temperatures over a narrow range that allowed differentiation of a variety of imperfect targets. The transfer of the hybridization mixture to a reverse-phase silica plate resulted in a blue color upon drying that could be detected visually. The system can detect about 10 femtomoles of an oligonucleotide.

Mirkin's DNA detection scheme is already finding industrial applications as a bench-top tool for rapid diagnoses of infectious diseases and the detection of genetic mutations called *single-nucleotide polymorphisms*.[153]

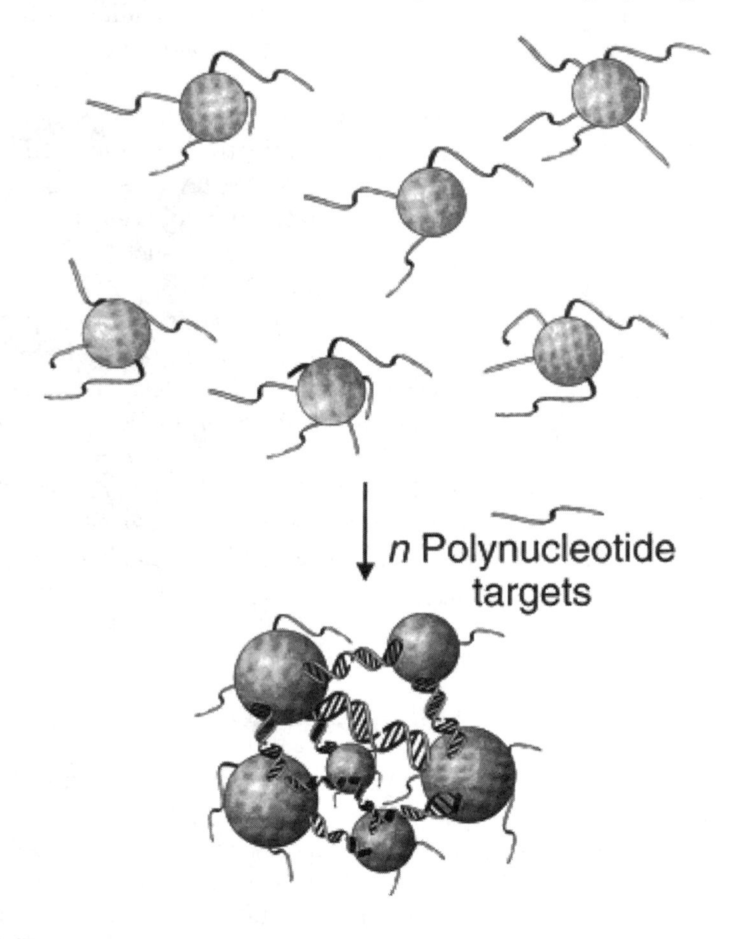

Fig. 7.67 A schematic representation of the concept for generating aggregates signaling hybridization of nanoparticle-oligonucleotide conjugates with oligonucleotide target molecules. Reprinted with permission from the Science Magazine publications.[154]

DNA and its chemical relatives can be combined with inorganic nanoparticles to develop devices which are very suitable as transducing platforms for sensing applications. Williams et al[155] described harnessed the selective binding capabilities of *peptide nucleic acids* (*PNA*) to assemble carbon nanotubes into molecular-scale electronic devices. They reported the development of the first nanotube-based transistor in 1998. For the fabrication of the structure, they scattered nanotubes across a surface which was patterned with small gold electrodes and then used an atomic force microscope to find an individual nanotube connecting two electrodes. They then measured how much current flowed through the nanotube when a voltage was applied to the substrate base which made the nanotube more conductive (the essence of a field effect transistor). Such a device is a perfect transducer for sensing applications. Its dimensions are small so whole transducer can be affected by the target molecules in the ambient.

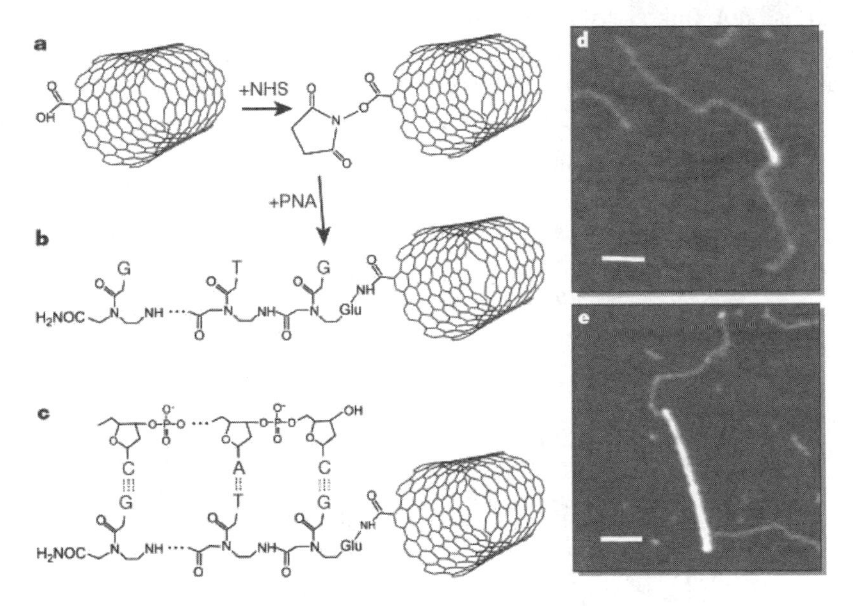

Fig. 7.68 a and b N-hydroxysuccinimide (NHS) esters formed on carboxylated, single-walled carbon nanotubes (SWNTs) are displaced by peptide nucleic acid (PNA), forming an amide linkage. **c** A DNA fragment with a single-stranded, 'sticky' end hybridized by base-pairing to the PNA–SWNT. **d and e** Atomic force microscope (tapping mode) images of PNA–SWNTs. The SWNTs appear as bright lines; the paler strands represent the bound DNA. Scale bars: 100 nm; nanotube diameters: d, 0.9 nm; e, 1.6 nm. Reprinted with permission from the Nature publications.[155]

Hamad-Schifferli et al[156] reported that by guiding the nanoparticles with a radio-frequency (RF) transmitter, they can control whether the DNA fragments exist as two separate single strands or as bound pairs.

In order to demonstrate reversible RF field control, they constructed a DNA hairpin-loop oligonucleotide which was covalently linked to a nanometre-scale antenna. The 38-nucleotide hairpin-loop DNA (**Fig. 7.69-a**) was covalently linked to a 1.4 nm gold nanocrystal by an amine.

A solution of the nanocrystal-linked oligonucleotides was put into an RF field of $f = 1$ GHz. The dehybridization was monitored by the hyperchromicity of DNA which was measured by the optical absorbance at 260 nm in an UV–visible spectrophotometer with an RF pulse of 15 s intervals (**Fig. 7.69-b**). As the RF field was switched on, the absorption intensity increased from 0.22 to 0.25, indicating that dehybridization occurred with the RF field (squares in **Fig. 7.69-b**). When the RF field was switched off, the absorbance returned to its original value. Cycling through the on and off states was repeatable. Control solutions with DNA only (that is, not linked to gold nanocrystals) resulted in no change in the absorbance with RFMF (which are the circles in **Fig. 7.69-b**). This shows that inductive coupling to a covalently-bound metal nanocrystal can reversibly dehybridized DNA with a timescale of at most several seconds.

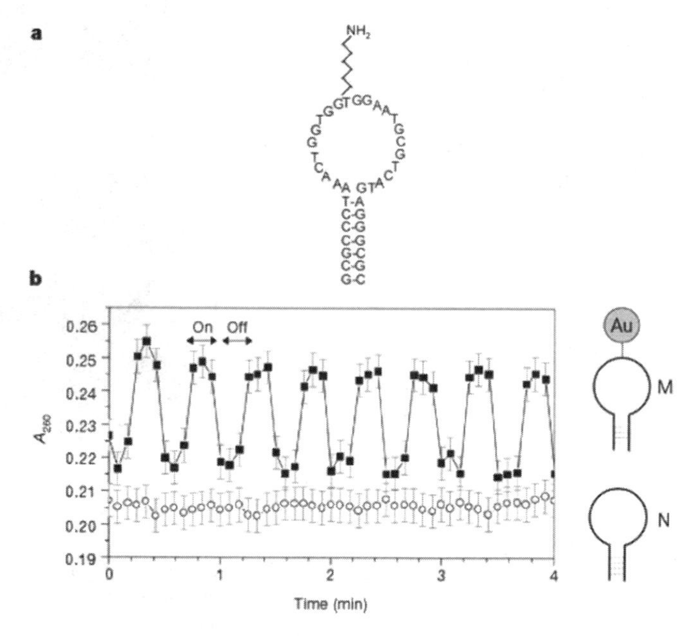

Fig. 7.69 The reversible control of a DNA hairpin-loop oligonucleotide covalently linked to a nanometre scale antenna by switching on and off the RF field. Reprinted with permission from the Nature publications.[156]

DNA strands attached to nanoparticles such as Au nanoparticles are ideal labels for biorecognition and biosensing processes. For instance, the plasmon absorbance features of Au nanoparticles conjugated with DNA can be used for the DNA strands.[157] Similarly tune-able fluorescence properties of conjugates such as these can be also used in biosensing applications.[158]

He et al showed that the sandwich DNA hybridization assay format can be used for the amplification of the Au-amplified SPR measurements. As can be seen in **Fig. 7.70**, after derivatizing the Au surface with a sub-monolayer of 12-mer oligonucleotide (S1) with a sequence which is complementary to half of the target analyte, the target DNA (S2) was introduced, and hybridization led to a very small angle displacement (0.1°) in the SPR reflectivity (**Fig. 7.70**, curve B). The subsequent exposure of the SPR surface to the solution containing Au nanoparticle-tagged S3 probes (S3:Au) led to a pronounced angle shift (**Fig. 7.70**, curve C) - approximately an 18-fold increase in SPR angle shift compared with what was observed in the unamplified assay.

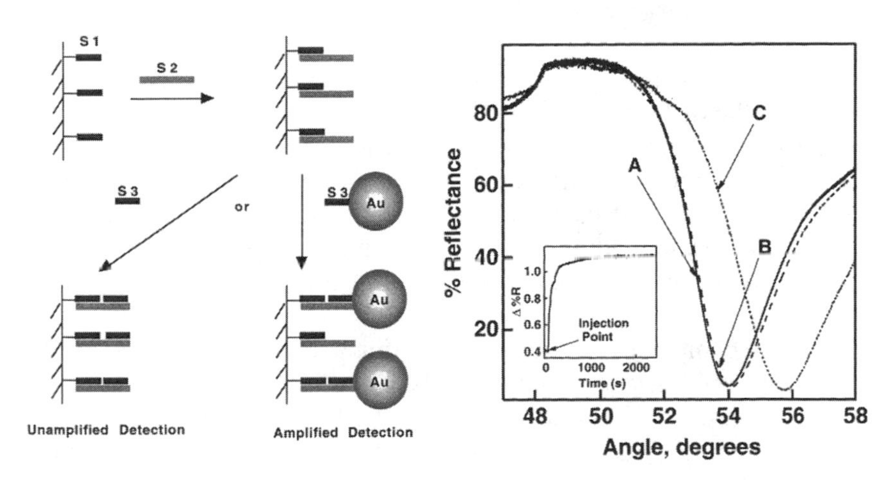

Fig. 7.70 The surface assembly and sensor response (A), after hybridization with its complementary 24-mer target S2 (B), and followed by introduction of S3:Au conjugate (C) to the surface. Inset: the surface plasmon reflectance changes at 53.2° for the oligonucleotide-coated Au film measured during a 60-min exposure to S3:Au conjugates. Reprinted with permission from the American Chemical Society publications.[157]

7.5.7 Bioelectronic Sensors based on DNA

DNA structure can be used to develop electronic and optical sensing devices. Charge carriers can hop along the DNA over distances of at least a few nanometers and in this way DNA can act as a *molecular wire*. As a result, DNA systems represent a novel material for sensing applications based on conductivity changes which can be induced by irradiation and electric field energies.

Early measurements of electron transfer in DNA were performed with a variety of techniques, notably by John Warman et al at Delft University of Technology using microwave conductivity.[159] As a semiconductor DNA shows specific influence on holes and electrons which are generated within its structure. A hole is more stable on a G-C base pair than on an A-T base pair. A hole can be easily established within a G-C base pair; however, as the A-T base pairs have a higher energy, they act as a barrier to a hole transfer. Nevertheless, the hole can *tunnel* from the first G-C site to the second, and can then either hop back to the first G-C pair or move on to the next one. Kelley and Barton from Caltech reported distance-dependencies of DNA conductivity, including the apparent charge tunneling over distances as long as 4 nm within DNA strands.[160] The rate of charge transfer decreases exponentially with the distance travelled. However, when the distance between the G-C base pairs becomes too long for charge carriers to jump electronically (*charge tunneling*), *thermal hopping* becomes the dominant charge-transfer mechanism. Both charge-tunneling and thermal-hopping mechanisms have been verified in experiments, notably by Bernd Giese's group at the University of Basel in Switzerland and by Maibi Michel-Beyerle and co-workers at the Technical University in Munich.[161]

While the combination of charge tunneling and thermal hopping appears to describe the basics, it cannot describe the quantum effects on the conductive charges. For example, the role of the local thermal motions of the bases, the polaron characters of charge, and the distortion by the neighboring DNA structures cannot be explained. In order to overcome some of these problems Zhang et al employed a *tight-binding representation* to show that the electron transport in DNA can also affected directly by the electron-phonon interaction via a twist-dependent hopping.[162] They find that different polaronic properties are exhibited by a two-site model are due to the nonlinearity of the restoring force of the twist excitations, and of the electron-phonon interaction in the model. There are also other models that can be utilized to describe the charge-induced cleaving of DNA as well as the concept of polaron formation and propagation in DNA molecular wires.[163]

The conductive properties of DNA strands can be combined with the other properties of nanoparticles to fabricate transducers for sensing devices. Alebker Kazumov et al placed DNA on carbon-covered electrodes that were fabricated from a thin layer (5 nm) of rhenium which is a superconducting material (**Fig. 7.71**).[164] They found that room temperature dles of DNA have a low resistance, in the order of several tens of kohm, which approaches *the resistance quantum* (described in Chap. 6) of 13 kohm as shown in **Fig. 7.72**. Such a structure can be used as a biosensor by functionalizing the DNA strands or as a gas sensor (as the conductivity of both DNA and carbon nanotubes changes upon exposure to different gas species).

Due to the above mentioned electronic properties, DNA can be efficiently used both in electrical and electrochemical sensing. For instance Xu et al described an electrochemical biosensor based on a DNA modified indium tin oxide (ITO) electrode.[165] They utilized self-assembly and electrochemical techniques to modify the ITO surface using (3-aminopropyl) trimethoxysilane, gold nanoparticles and DNA molecules. The modified electrode was employed to detect mifepristone (a synthetic steroid compound used as a pharmaceutical) with the detection limit of 2×10^{-7} mol/L.

The change of conductance of DNA strands, in contact with other materials, can also be directly employed in biosensing applications. Tsia et al[166] reported on an electrical DNA sensor which was developed using a self-assembled multilayer gold nanoparticle structure between nano-gap electrodes (300 nm apart). Bifunctional organic molecules were utilized to build up the gold nanoparticle monolayer on the wafer substrate. After the hybridization of the target DNA (5'-end thiol-modified probe DNA and 3'-end thiol-modified capture DNA) a second layer of gold nanoparticles was built up through a self-assembly process between the gold nanoparticles and the thiol-modified end of the probe DNA (**Fig. 7.73**). The electrical current through the multilayer gold nanoparticle structure was much greater than that through monolayered gold nanoparticle structure for the same voltage. The concentration of the target DNA in the tested sample solutions ranged from several fM to 100 pM. The linear I-V curves of the multilayer gold nanoparticles structures indicate that the device proposed in this study can detect target DNA concentrations as low as 1 fM.

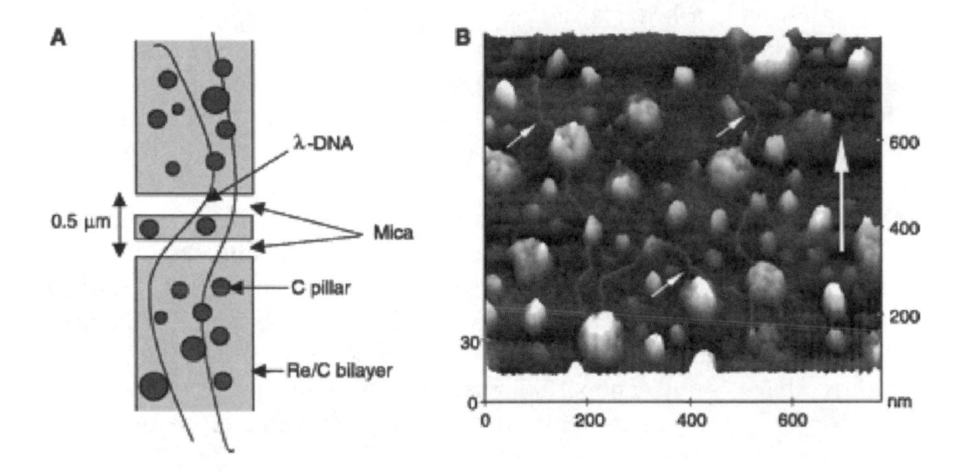

Fig. 7.71 (**A**) A Schematic drawing of the measured sample, with DNA molecules combed between carbon-covered electrodes on a mica substrate. (**B**) An AFM image showing the DNA molecules. Reprinted with permission from the Science Magazine publications.[164]

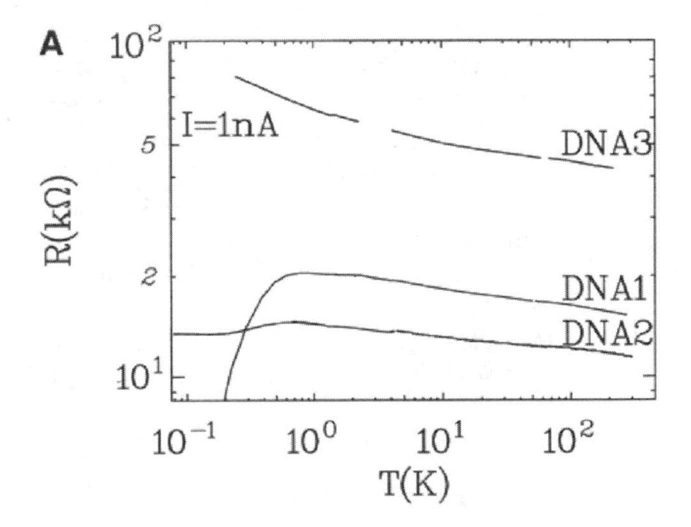

Fig. 7.72 Linear transport measurements on the three samples. DC resistance as a function of temperature (on a large temperature scale), showing the power law behaviour down to 1 K. The sample DNA1, with a 30-µm-wide unetched window, contained approximately 10 combed molecules as estimated from AFM observations; sample DNA2, with a 120-µm-wide window, had about 40 combed DNAs; and sample DNA3 had only a few molecules (probably two or three). Reprinted with permission from the Science Magazine publications.[164]

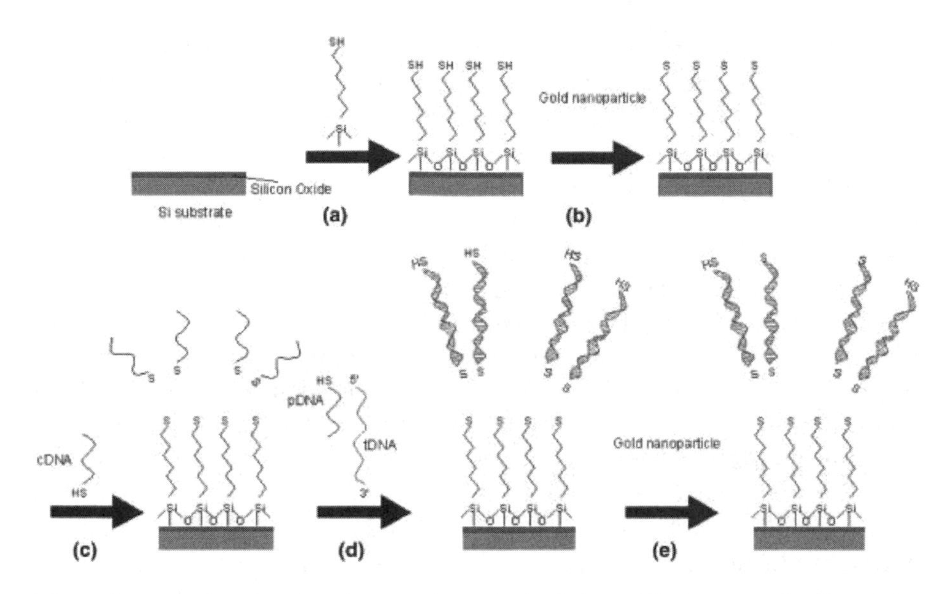

Fig. 7.73 The chemistry of the attachment of thiol-modified DNA oligomers to a self-assembled Au nanoparticles monolayer. (a) Surface modification with THMS, (b) The addition of Au nanoparticles which react with the the thiol molecules, (c) An alkanethiol-cDNA reacts with AuNPs monolayer, (d) tDNA added and hybridized with cDNA and pDNA, thiols react with the multilayered Au nanoparticles, (e) The thiolated-DNA oligomers subsequently react with the Au nanoparticle surface, to yield the self-assembled multilayer of Au nanoparticles. Reprinted with permission from the Elsevier publications.[166]

7.5.8 DNA Sequencing with Nanopores

Nanopores can be used for the detection of specific DNA and RNA strands. One of the first attempts to produce nanopores to detect single DNA strands was made with α-hemolysin, a transmembrananal protein which was inserted into a lipid bilayer.[167] After placing it within a synthetic lipid bilayer it forms a 1.5 nm diameter aqueous channel through the membrane.[168] The α-hemolysin pore remains open at neutral pH and high ionic strength. Furthermore, the α-hemolysin pore passes a steady ionic current in a detectable range, whereas most membrane channels exhibit unstable current levels due to their high sensitivity and spontaneous gating. This steady current flow over a relatively large range ensures a low level of background electrical noise and thus prevents the electrical signals of interest from being masked.

Crystallography has revealed that the structure of the α-hemolysin pore is about 100 Å-long, it is mushroom-shaped and is a homo-oligomeric heptamer, of which the pore has a limiting aperture of 1.5 nm. The transmembranal domain of the channel comprises a 14-strand β-barrel, which is primarily hydrophilic inside and hydrophobic outside.

Such nanopores can be used for decoding and detecting DNA strands. Akeson et al practically showed that single molecules of DNA or RNA can be detected as they are driven through an α hemolysin pore by an applied electric field.[169] During the passage of the DNA and RNA strands, nucleotides within the polynucleotide must pass through the channel pore in a sequential, single-file order because of the limiting diameter of the pore. In order to prove the concept Akeson et al developed an apparatus for their experiment (**Fig. 7.74**). This apparatus is made of an inert materials such as Teflon with conical apertures. Bilayers formed across the aperture of this device exhibited high resistance (>200 GΩ), low noise (<0.6 pA RMS at 5 kHz bandwidth in whole-cell mode, 0.2 pA RMS in patch mode), and were stable for many hours at applied voltages of up to 200 mV. In addition, the bath was in the order of about several tens of microliters in volume, permitting the analysis of small amounts of RNA or DNA.

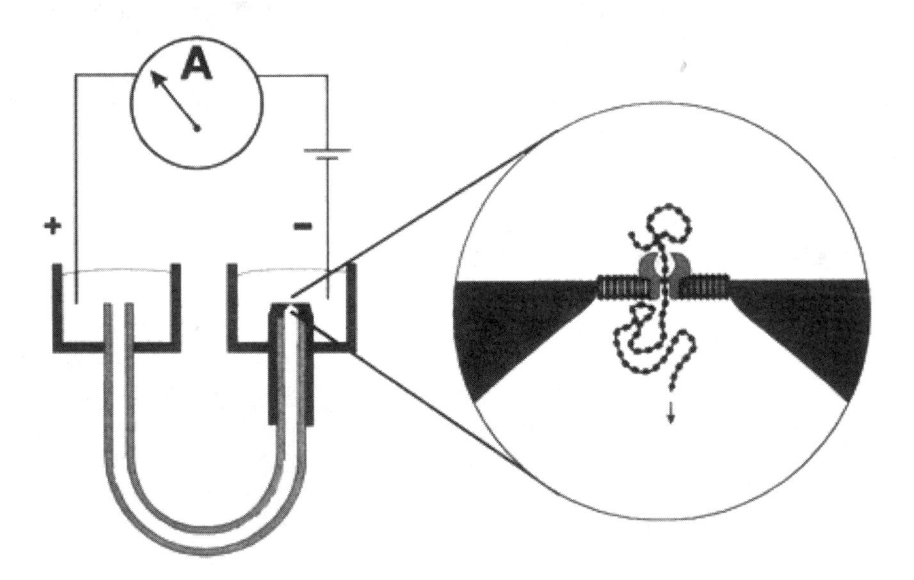

Fig. 7.74 The horizontal bilayer apparatus. A U-shaped Teflon patch tube connects two 70-μl baths milled into a Teflon support (*left*). The baths and the Teflon tube are filled with 1 M KCl buffer. Nucleic acids are driven through the α-hemolysin channel by an applied voltage of 120 mV. Reprinted with permission from the HighWire Press.[169]

Akeson et al demonstrated that this nanopore behaves as a detector that can rapidly discriminate between pyrimidine and purine segments along a strand of RNA. Nanopore-based sensors and the characterization of single molecules represent a method for directly reading the information which is encoded in linear polymers, and can be used for the direct sequencing of individual DNA and RNA molecules.

7.6 Sensors Based on Molecules with Dendritic Arcitectures

Recently, much attention has been given to molecules with dendritic architectures such as dendrimers, dendronized, hyperbranched and brushed-polymers for their applications in sensing.[170-173] They can be used both as sensitive elements, by functionalizing their sites, or as templates for the development of nanostructures.

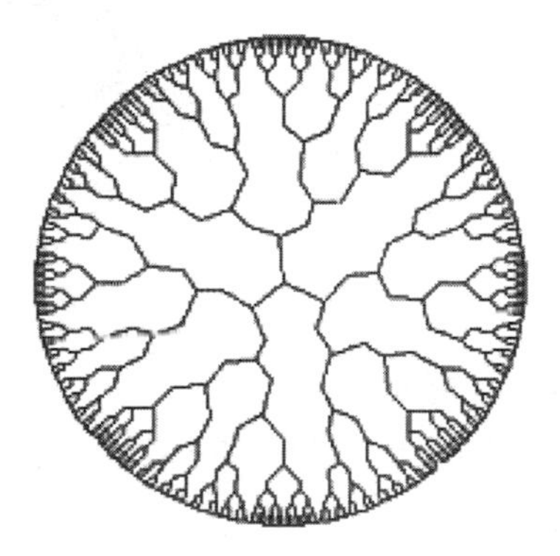

Fig. 7.75 A schematic of a dendrimer.

Dendrimers (**Fig. 7.75**), which are the most famous member of this family, are molecules which are repeatedly branched. A dendrimer is constructed from regular, highly branched monomers leading to a monodisperse, tree-like or generational structure. The molecular architecture of dendritic polymers can be tuned to obtain nanoscale objects with desired properties. The highly compact and branched, as well as the controlled size and plurifunctionality of dendritic polymers make them ideal molecular building blocks for a wide range of interfacial materials involving self-assembled monolayers, Langmuir films, multilayered films, and other surface-confined assemblies.[174] Dendritic macromolecules play an increasingly important role in the materials and surface sciences, where applications involving polymer thin films can benefit from their distinctive chemical and physical properties. Recent investigations have highlighted the use of dendrimers as functional surfaces and as interfacial materials for applications in membranes, adhesives, and microelectronics as well as applications in chemical and biological sensing.

For sensing applications dendrimers can be implemented in the modifications of surfaces, in conjunction with other molecules, to provide multiple functionalities for the surface, and as optically-sensitive particles. As an example, Krasteva et al have developed alcohol and water vapor-sensitive thin-film resistors comprising of gold nanoparticles and different types of organic dendrimers (polyphenylene, poly(propylene imine) and poly(amidoamine)) which were prepared via layer-by-layer self-assembly.[175] The metal nanoparticles were utilized in order to provide the film material with electrical conductivity, and the dendrimers served to cross-link the nanoparticles and to provide sites for the selective adsorption of analyte molecules. The chemical selectivity of this response was controlled by the solubility properties of the dendrimers.

Dendrimers can also be used in functionalizing the nanomaterials. Carbon nanotubes can be covalently functionalized with polyamine dendrimers via carboxylate chemistry.[176] Proteins can then adsorb individually, strongly and noncovalently along nanotube lengths.

Coupling luminescence and dendrimer research topics can lead to valuable new functions.[177] Luminescence measurements are a valuable tool to monitor both basic properties and possible sensing applications. The possibility of conjugating selected chemical units and encapsulating ions in predetermined sites of dendrimers make this happen. The sensing can occur with signal amplification, quenching and sensitization processes, shielding effects, elucidation of dendritic structures and superstructures, and the investigation of dendrimer rotations in solution. For instance, Gong et al[178] synthesized optically-active dendrimers containing a 1,1'-binaphthyl core and cross-conjugated phenylene dendrons. UV and fluorescence spectroscopic studies demonstrated that the energy harvested by the periphery of the dendrimers can be efficiently transferred to the more conjugated core, generating much enhanced fluorescence signal at higher generation (highly branched dendrimers). The fluorescence of these dendrimers can be quenched both efficiently and enantioselectively by chiral amino alcohols. The energy migration and light harvesting effects of the dendrimers make the higher generation dendrimer more sensitive to fluorescent quenchers than the lower ones. Thus, the dendritic structure provides a signal amplification mechanism. As another example, Pugh[179] observed significant enhancement in the fluorescence intensity from 1,1'-bi-2-naphthol (BINOL) to dendritic phenyleneethynylenes containing the BINOL core. This strong fluorescence of the dendrimers allows a very small amount of the chiral materials necessary to carry out the measurements. The light harvesting antennae of the dendrimer funnel energy to the center BINOL unit, whose hydroxyl groups, upon interaction with a quencher molecule, lead to fluorescence quenching. This mechanism makes the dendrimers much more sensitive fluorescence responses than corresponding small molecule sensors. The fluorescence of these dendrimers can be enantioselectively quenched by chiral amino alcohols. It has been observed that the fluorescence lifetime of the generation two dendrimer does not change in the presence of various concentrations of 2-amino-3-phenyl-1-propanol. This demonstrates that the fluorescence quenching is entirely due to static quenching. As such, the formation of non-fluorescent ground-state hydrogen-bonded complexes between the dendrimers and the amino alcohols is proposed to account for the fluorescent quenching.

7.7 Force Spectroscopy and Microscopy of Organic Materials

Mechanical measurements using force microscopy methods can be used for the measurement of the strength of a bond as well as for the analysis of the mechanical properties of biomaterials.

The strength of specific bindings can be measured using force microscopy. The strength of an isolated molecular bond is represented by the maximum force that the bond can withstand before it breaks. These forces also depend on the temperature and the timescale of the measurement.

Such measurements are important as they describe the quality of a sensitive layer, how the components of a sensor are connected to each other, and they also reveal many answers to the questions regarding the connections between cells and their units. In addition, they can describe the protein unfolding process (denaturization) and the hybridization of DNA. Protein unfolding is especially important in the applications of enzymes and antigens in biosensors.

The force microscopy method can be applied for interpreting a range of phenomena such as noncovalent and weak bonds between biological macromolecules. The analysis also gives definitions for the experimentally-defined attractive and adhesive forces in terms of molecular parameters.[180]

Force microscopy can be utilized for the study of biofunctions such as the activities of cells and proteins. For instance, Guthold et al[181] investigated the dynamics of nonspecific and specific *Escherichia coli* RNA polymerase ((RNAP)-DNA) complexes using scanning force microscopy. The force microscopy images that show the dynamics of a transcribing RNAP are presented in **Fig. 7.76**. In the initial image, the arms of the DNA on both sides of the polymerase diffused laterally on the substrate surface but did not move through the polymerase in either direction, indicating a stable stalled elongation complex. Upon the addition of nucleotides, the RNAP began to thread the DNA template in a proccessive and unidirectional manner toward the shorter arm of the DNA (**Fig. 7.76, b-g**). After transcription, the RNAP usually released the DNA.

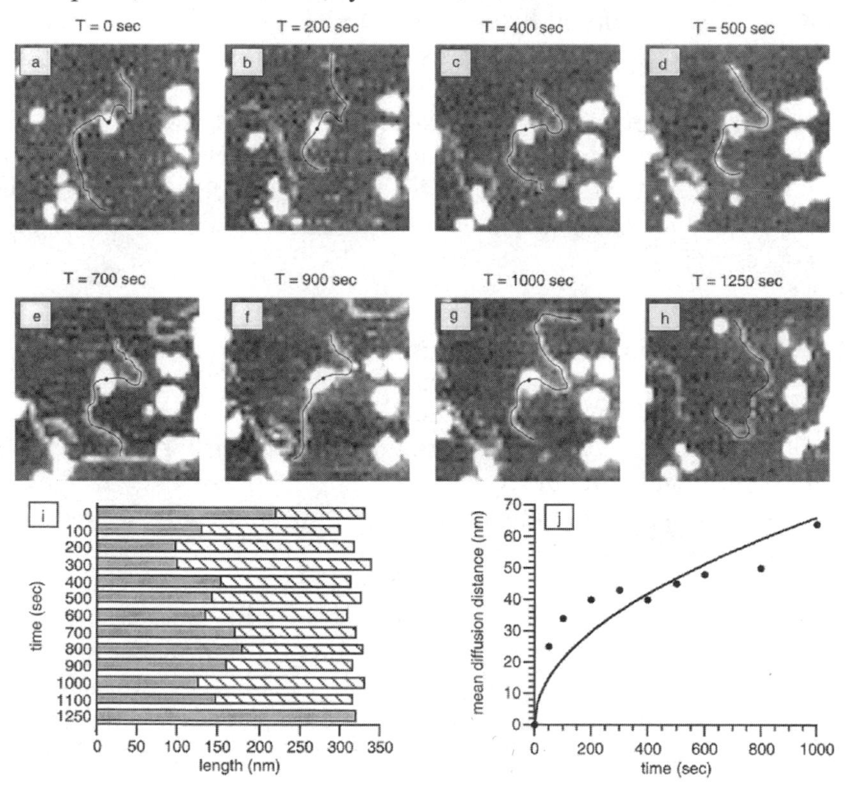

Fig. 7.76 The one-dimensional diffusion of nonspecific binary complexes of RNAP along the DNA. (a-h) An SFM time-lapse sequence of RNAP sliding. Polymerase appears to slide back and forth several times along the DNA molecule before it is released. The elapsed time after the initial observation of the complex is indicated above each image. The DNA contour is traced with a thin line and the center of the RNAP is marked with a dot. Image size: 350 nm. The height color code ranges from 0 nm (black) to 5 nm (white). (i) A bar plot representation of the length of the two DNA arms (hatched and solid bars) in 13 sequential images. (j) A plot of the mean diffusion distance vs. time, where a square root dependence, which is typical for diffusion-controlled motion, is seen. Reprinted with permission from the HighWire Press.[181]

7.8 Biomagnetic Sensors

There are many illustrations of bio-nanotechnology in nature. An extraordinary example is *magnetotactic bacteria*, which are magnetic field sensing (**Fig. 7.77**). These bacteria live in the beds of muddy waters, and have approximately 20 magnetic crystals that are 30 to 120 nanometers in diameters. These crystals make up a miniature bio-compass which is utilized to sense the earth's magnetic field for navigation.[182]

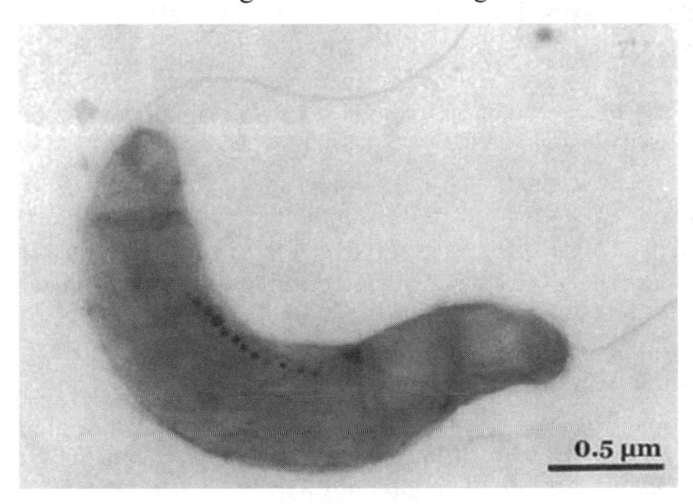

Fig. 7.77 An electron micrograph of a Magnetospirillum gryphiswaldense exhibiting the characteristic morphology of magnetic spirilla. The helical cells are bipolarly flagellated and contain up to 60 intracellular magnetite particles in each magnetosomes, which are arranged in a chain.[182]

Bio functionalization of magnetic nanoparticles such as Fe_3O_4 has been extensively used for the development of biosensors. Such magnetic materials can be used as a pre filtering process for purification of materials to be sensed.[183] Paramagnetic or super-paramagnetic nanoparticles, which can respond to an external magnetic field, provide an efficient media for separating samples linked to the magnetic particles from the liquid suspension. These nanoparticles also exhibit zero hysteresis. The particle-linked molecules can quickly agglomerate or resuspend in the medium in response to a change in the external magnetic force.

Taking advantage of their unique magnetic characteristic and low cost of synthesis, magnetic particles have been widely used as separation tools to purify nucleic acids (i.e., DNA and RNA). For instance, Wang and Kawde showed that DNA probes can functionalize magnetic nanoparticles.

Using an external magnetic field these particles can be separated from the ambient.[184]

Magnetic nanoparticles can also enhance the sensor's performance as mediators. Wang and Kawde also showed that reversible and cyclic magnetic fields stimulate the oxidation of DNA.[184] They realized that positioning an external magnet below the electrode attracts the DNA-functionalized magnetic particles to the surface, and stimulates the oxidation of the guanine nucleobases. Using this phenomenon, they demonstrated that a spatially-controlled DNA oxidation, with an 'ON/OFF' switching of the electron-transfer reaction upon relocating the external magnetic field can be obtained. Interestingly, this process can be reversed and repeated upon switching the position of the magnet, with or without oxidation signals in the presence or the absence of the magnetic field, respectively.

Wang et al, in separate study, showed that magnetically-controlled amplified sensing of DNA can also be conducted using metallic of semiconducting tags.[185] The direct electrical detection of DNA/metal-particle assemblies can be utilized in the detection of DNA hybridization, and could be applied to other bioaffinity assays. As a semiconducting tag, they studied the magnetic triggering of a solid-state electrical transduction of DNA hybridization. They positioned an external magnet below the thick-film electrode which attracts the DNA/particle network and enables the solid-state electrochemical stripping detection of the silver tracer. The hybridization event results in a three-dimensional aggregate structure in which duplex segments link the metal nanoparticles and the magnetic spheres, and that most of this assembly is covered with the silver precipitate. This leads to a direct contact of the metal tag with the surface and enables the solid-state electrochemical transduction using oxidative dissolution of the silver tracer. This method couples the high sensitivity of silver-amplified assays with the effective discrimination against an excess of closely related nucleotide sequences.

7.9 Summary

In this chapter, organic sensors which incorporate nonomaterials in their fabrications and operations were presented. Different types of surface modification techniques were described and, as the major bottom-up strategies for the development of organic sensors such as self-assembly and layer-by-layer techniques were explained. The importance of the surface types in the fabrication and functionalization of surfaces were presented and the applications of typical materials including gold, silicon, metal oxides, carbon as well as conductive and non-conductive polymeric surfaces were described. Along with these descriptions, a large number of examples, discussing the applications of these strategies in corporation with major transcducing platforms (which were introduced in Chap. 3) for biosensing applications, were also presented.

Proteins and their capabilities were described as organic nanodevices. It was seen that proteins, as intelligent assemblies of organic molecules, are able to perform a large number of different tasks ranging from sensing, producing signals, to carrying bioparticles. Some of the typical applications of proteins in the development of sensors were presented and the methods to manipulate their functionalities in order to develop sensing tools were presented. The significance of proteins in sensing applications was further expanded by describing how enzymes and antibodies can be employed in relevant sensing applications.

DNA molecules, as the building blocks of natural entities, were described. It was seen that DNA can be efficiently used as a lock and key element for the development of biochips and selective sensors. It was further presented that DNA strands can also be utilized as elements for the fabrication of sensing structures as they provides ready-made units for the assembly of nanostructures.

The conjugates of metallic and nonmetallic nanoparticles with proteins and sensors were introduced and it was shown that they can be used for the fabrications of sensors and for the modifications of surfaces. It was also observed that DNA's excellent optical and electrical properties can be manipulated in a large number of ways for sensing applications.

In the last part of this section, other concepts in organic sensors such as the usage of molecules with dendritic architectures, the application of the force spectroscopy and microscopy in organic sensing and eventually some examples about magnetic nanomaterials in biosensing were presented.

References

[1] B. R. Eggins, *Chemical Sensors and Biosensors* (John Wiley & Sons, Ltd, Chichester, UK, 2002).

[2] E. Gizeli and C. R. Lowe, *Biomolecular Sensors* (Taylor & Francis, London, UK, 2002).

[3] A. P. F. Turner, I. Karube, and G. S. Wilson, *Biosensors: Fundamentals & Applications* (Oxford University Press, London, UK, 1987).

[4] G. H. Schmid, *Organic Chemistry* (Mosby-Year Book, St. Louis, USA, 1996).

[5] N. Nakajima and Y. Ikada, Bioconjugate Chemistry **6,** 123-130 (1995).

[6] G. W. J. Fleet, J. R. Knowles, and R. R. Porter, Biochemical Journal **128,** 499-& (1972).

[7] P. J. A. Weber and A. G. BeckSickinger, Journal of Peptide Research **49,** 375-383 (1997).

[8] R. E. Galardy, L. C. Craig, J. D. Jamieson, and M. P. Printz, Journal of Biological Chemistry **249,** 3510-3518 (1974).

[9] G. Dorman and G. D. Prestwich, Biochemistry **33,** 5661-5673 (1994).

[10] N. A. Peppas, P. Bures, W. Leobandung, and H. Ichikawa, European Journal of Pharmaceutics and Biopharmaceutics **50,** 27-46 (2000).

[11] B. C. Dave, B. Dunn, J. S. Valentine, and J. I. Zink, Analytical Chemistry **66,** A1120-A1127 (1994).

[12] J. Zhang, Z.-L. Wang, J. Liu, S. Chen, and G.-Y. Liu, *Self-assembled nanostructures* (Kluwer Academic/Plenum Publishing, New York, USA, 2003).

[13] R. J. Jackman, J. L. Wilbur, and G. M. Whitesides, Science **269,** 664-666 (1995).

[14] R. Berger, E. Delamarche, H. P. Lang, C. Gerber, J. K. Gimzewski, E. Meyer, and H. J. Guntherodt, Science **276,** 2021-2024 (1997).

[15] K. Ariga, J. P. Hill, and Q. M. Ji, Physical Chemistry Chemical Physics **9,** 2319-2340 (2007).

[16] M. Shimomura and T. Sawadaishi, Current Opinion in Colloid & Interface Science **6,** 11-16 (2001).

[17] J. M. Lehn, Reports on Progress in Physics **67,** 249-265 (2004).

[18] J. M. Lehn, Angewandte Chemie-International Edition in English **29,** 1304-1319 (1990).

[19] T. Kunitake, Angewandte Chemie-International Edition in English **31,** 709-726 (1992).
[20] A. Ulman, *An Introduction to Ultrathin Organic Films from Langmuir-Blodgett to Self-Assembly* (Academic Press New York, USA, 1991).
[21] G. Decher, Science **277,** 1232-1237 (1997).
[22] P. T. Hammond, Advanced Materials **16,** 1271-1293 (2004).
[23] J. M. Levasalmi and T. J. McCarthy, Macromolecules **30,** 1752-1757 (1997).
[24] X. Y. Liu and M. L. Bruening, Chemistry of Materials **16,** 351-357 (2004).
[25] J. J. Harris, J. L. Stair, and M. L. Bruening, Chemistry of Materials **12,** 1941-1946 (2000).
[26] R. G. Nuzzo and D. L. Allara, Journal of the American Chemical Society **105,** 4481-4483 (1983).
[27] B. Adhikari and S. Majumdar, Progress in Polymer Science **29,** 699-766 (2004).
[28] J. Janata and M. Josowicz, Nature Materials **2,** 19-24 (2003).
[29] A. J. Cunningham, *Introduction to bioanalytical sensors* (Wiley New York, USA 1998).
[30] M. Shamsipur, M. Yousefi, and M. R. Ganjali, Analytical Chemistry **72,** 2391-2394 (2000).
[31] C. R. Lowe and P. D. G. Dean, *Affinity Chromatography* (John Willey and Sons, New York, USA, 1974).
[32] W. H. Scouten, *Affinity Chromatography: Bioselective Adsorption on Inert Matrices* (John Wiley and Sons, New York, USA, 1981).
[33] E. H. Gillis, J. P. Gosling, J. M. Sreenan, and M. Kane, Journal of Immunological Methods **267,** 131-138 (2002).
[34] K. Mosbach, Trends in Biochemical Sciences **19,** 9-14 (1994).
[35] K. Mosbach, Scientific American **295,** 86-91 (2006).
[36] L. I. Andersson, Journal of Chromatography B **739,** 163-173 (2000).
[37] C. D. Liang, H. Peng, A. H. Zhou, L. H. Nie, and S. Z. Yao, Analytica Chimica Acta **415,** 135-141 (2000).
[38] F. L. Dickert, P. Forth, P. A. Lieberzeit, and G. Voigt, Fresenius Journal of Analytical Chemistry **366,** 802-806 (2000).
[39] F. L. Dickert, P. Forth, P. Lieberzeit, and M. Tortschanoff, Fresenius Journal of Analytical Chemistry **360,** 759-762 (1998).
[40] H. S. Ji, S. McNiven, K. H. Lee, T. Saito, K. Ikebukuro, and I. Karube, Biosensors & Bioelectronics **15,** 403-409 (2000).
[41] H. S. Ji, S. McNiven, K. Ikebukuro, and I. Karube, Analytica Chimica Acta **390,** 93-100 (1999).

[42] E. Hedborg, F. Winquist, I. Lundstrom, L. I. Andersson, and K. Mosbach, Sensors and Actuators a-Physical **37-8,** 796-799 (1993).
[43] D. Kriz and K. Mosbach, Analytica Chimica Acta **300,** 71-75 (1995).
[44] S. A. Piletsky, E. V. Piletskaya, A. V. Elskaya, R. Levi, K. Yano, and I. Karube, Analytical Letters **30,** 445-455 (1997).
[45] A. G. MacDiarmid, Synthetic Metals **84,** 27-34 (1997).
[46] H. S. Nalwa, *Handbook of Organic Conductive Materials and Polymers* (Wiley, New York, USA 1997).
[47] T. A. Skotheim, *Handbook of Conducting Polymers, Vols. 1 and 2* (Marcel Dekker, New York, USA, 1986).
[48] J. C. Chiang and A. G. Macdiarmid, Synthetic Metals **13,** 193-205 (1986).
[49] A. G. Macdiarmid, J. C. Chiang, A. F. Richter, and A. J. Epstein, Synthetic Metals **18,** 285-290 (1987).
[50] J. Y. Shimano and A. G. MacDiarmid, Synthetic Metals **123,** 251-262 (2001).
[51] J. Tanaka, N. Mashita, K. Mizoguchi, and K. Kume, Synthetic Metals **29** (1989).
[52] Y. X. Zhou, M. Freitag, J. Hone, C. Staii, A. T. Johnson, N. J. Pinto, and A. G. MacDiarmid, Applied Physics Letters **83,** 3800-3802 (2003).
[53] J. Doshi and D. H. Reneker, Journal of Electrostatics **35,** 151-160 (1995).
[54] D. Xie, Y. D. Jiang, W. Pan, D. Li, Z. M. Wu, and Y. R. Li, Sensors and Actuators B-Chemical **81,** 158-164 (2002).
[55] V. Dixit, S. C. K. Misra, and B. S. Sharma, Sensors and Actuators B-Chemical **104,** 90-93 (2005).
[56] W. Q. Cao and Y. X. Duan, Sensors and Actuators B-Chemical **110,** 252-259 (2005).
[57] E. M. Genies and C. Tsintavis, Journal of Electroanalytical Chemistry **195,** 109-128 (1985).
[58] A. A. Syed and M. K. Dinesan, Talanta **38,** 815-837 (1991).
[59] L. Liang, J. Liu, C. F. Windisch, G. J. Exarhos, and Y. H. Lin, Angewandte Chemie-International Edition **41,** 3665-3668 (2002).
[60] J. Liu, Y. H. Lin, L. Liang, J. A. Voigt, D. L. Huber, Z. R. Tian, E. Coker, B. McKenzie, and M. J. McDermott, Chemistry-a European Journal **9,** 605-611 (2003).
[61] Y. Cao, A. Andreatta, A. J. Heeger, and P. Smith, Polymer **30,** 2305-2311 (1989).
[62] A. Pron, F. Genoud, C. Menardo, and M. Nechtschein, Synthetic Metals **24,** 193-201 (1988).

[63] J. X. Huang, S. Virji, B. H. Weiller, and R. B. Kaner, Journal of the American Chemical Society **125,** 314-315 (2003).
[64] J. X. Huang and R. B. Kaner, Angewandte Chemie-International Edition **43,** 5817-5821 (2004).
[65] C. G. Wu and T. Bein, Science **264,** 1757-1759 (1994).
[66] C. R. Martin, Chemistry of Materials **8,** 1739-1746 (1996).
[67] E. T. Kang, K. G. Neoh, and K. L. Tan, Progress in Polymer Science **23,** 277-324 (1998).
[68] J. S. Tang, X. B. Jing, B. C. Wang, and F. S. Wang, Synthetic Metals **24,** 231-238 (1988).
[69] S. Virji, J. X. Huang, R. B. Kaner, and B. H. Weiller, Nano Letters **4,** 491-496 (2004).
[70] P. D. Dupoet, S. Miyamoto, T. Murakami, J. Kimura, and I. Karube, Analytica Chimica Acta **235,** 255-263 (1990).
[71] Q. L. Yang, P. Atanasov, and E. Wilkins, Biosensors & Bioelectronics **14,** 203-210 (1999).
[72] M. K. Ram, N. S. Sundaresan, and B. D. Malhotra, Journal of Materials Science Letters **13,** 1490-1493 (1994).
[73] N. C. Foulds and C. R. Lowe, Journal of the Chemical Society-Faraday Transactions I **82,** 1259-1264 (1986).
[74] J. C. Cooper and E. A. H. Hall, Biosensors & Bioelectronics **7,** 473-485 (1992).
[75] C. Malitesta, F. Palmisano, L. Torsi, and P. G. Zambonin, Analytical Chemistry **62,** 2735-2740 (1990).
[76] J. H. Kim, J. H. Cho, G. S. Cha, C. W. Lee, H. B. Kim, and S. H. Paek, Biosensors & Bioelectronics **14,** 907-915 (2000).
[77] I. Willner, Science **298,** 2407-2408 (2002).
[78] H. Kawaguchi, Progress in Polymer Science **25,** 1171-1210 (2000).
[79] M. LaBarbera, Science **289,** 1882-1882 (2000).
[80] A. R. Mendelsohn and R. Brent, Science **284,** 1948-1950 (1999).
[81] H. Morgan and D. M. Taylor, Biosensors & Bioelectronics **7,** 405-410 (1992).
[82] C. D. Bain, E. B. Troughton, Y. T. Tao, J. Evall, G. M. Whitesides, and R. G. Nuzzo, Journal of the American Chemical Society **111,** 321-335 (1989).
[83] J. Spinke, M. Liley, H. J. Guder, L. Angermaier, and W. Knoll, Langmuir **9,** 1821-1825 (1993).
[84] S. Lofas, B. Johnsson, A. Edstrom, A. Hansson, G. Lindquist, R. M. M. Hillgren, and L. Stigh, Biosensors & Bioelectronics **10,** 813-822 (1995).

[85] A. N. Naimushin, S. D. Soelberg, D. K. Nguyen, L. Dunlap, D. Bartholomew, J. Elkind, J. Melendez, and C. E. Furlong, Biosensors & Bioelectronics **17,** 573-584 (2002).

[86] F. Caruso, E. Rodda, D. F. Furlong, K. Niikura, and Y. Okahata, Analytical Chemistry **69,** 2043-2049 (1997).

[87] M. Minunni, P. Skladal, and M. Mascini, Analytical Letters **27,** 1475-1487 (1994).

[88] H. Muramatsu, J. M. Dicks, E. Tamiya, and I. Karube, Analytical Chemistry **59,** 2760-2763 (1987).

[89] M. Musameh, J. Wang, A. Merkoci, and Y. H. Lin, Electrochemistry Communications **4,** 743-746 (2002).

[90] Y. H. Lin, F. Lu, Y. Tu, and Z. F. Ren, Nano Letters **4,** 191-195 (2004).

[91] Y. Astier, H. Bayley, and S. Howorka, Current Opinion in Chemical Biology **9,** 576-584 (2005).

[92] B. Alberts, D. Bary, K. Hopkin, A. Johnson, J. Lewis, M. Raff, K. Roberts, and P. Walter, *Essential Cell Biology* (Garland Science, Oxford, UK, 2004).

[93] E. A. Padlan, in *Antigen Binding Molecules: Antibodies and T-Cell Receptors*; *Vol. 49* (1996), 57-133.

[94] J. A. Berzofsky and A. N. Schechter, Molecular Immunology **18,** 751-763 (1981).

[95] S. H. Jenkins, W. R. Heineman, and H. B. Halsall, Analytical Biochemistry **168,** 292-299 (1988).

[96] S. H. Jenkins, H. B. Halsall, and W. J. Heineman, *Advances in Biosensors (Vol. 1)* (JAI Press, London, UK, 1991).

[97] M. Mosbach, in *Biosensors 2000*, San Diego, CA, USA, 2000 (Elsevier, Oxford), 164-167.

[98] A. Shons, J. Najarian, and F. Dorman, Journal of Biomedical Materials Research **6,** 565-& (1972).

[99] C. K. O'Sullivan and G. G. Guilbault, Biosensors & Bioelectronics **14,** 663-670 (1999).

[100] G. Sakai, T. Saiki, T. Uda, N. Miura, and N. Yamazoe, Sensors and Actuators B-Chemical **42,** 89-94 (1997).

[101] K. Yun, E. Kobatake, T. Haruyama, M. L. Laukkanen, K. Keinanen, and M. Aizawa, Analytical Chemistry **70,** 260-264 (1998).

[102] F. Aberl, H. Wolf, C. Kosslinger, S. Drost, P. Woias, and S. Koch, Sensors and Actuators B-Chemical **18,** 271-275 (1994).

[103] C. Kosslinger, S. Drost, F. Aberl, H. Wolf, S. Koch, and P. Woias, Biosensors & Bioelectronics **7,** 397-404 (1992).

[104] G. G. Guilbault, B. Hock, and R. Schmid, Biosensors & Bioelectronics **7,** 411-419 (1992).

[105] M. Plomer, G. G. Guilbault, and B. Hock, Enzyme and Microbial Technology **14,** 230-235 (1992).

[106] B. S. Attili and A. A. Suleiman, Microchemical Journal **54,** 174-179 (1996).

[107] B. S. Attili and A. A. Suleiman, Analytical Letters **28,** 2149-2159 (1995).

[108] A. L. Ghindilis, P. Atanasov, M. Wilkins, and E. Wilkins, Biosensors & Bioelectronics **13,** 113-131 (1998).

[109] E. Katz and I. Willner, Electroanalysis **15,** 913-947 (2003).

[110] A. Brecht, J. Piehler, G. Lang, and G. Gauglitz, Analytica Chimica Acta **311,** 289-299 (1995).

[111] D. Ivnitski, I. Abdel-Hamid, P. Atanasov, and E. Wilkins, Biosensors & Bioelectronics **14,** 599-624 (1999).

[112] P. B. Luppa, L. J. Sokoll, and D. W. Chan, Clinica Chimica Acta **314,** 1-26 (2001).

[113] W. Lukosz, Sensors and Actuators B-Chemical **29,** 37-50 (1995).

[114] J. Homola, Analytical and Bioanalytical Chemistry **377,** 528-539 (2003).

[115] M. Minunni and M. Mascini, Analytical Letters **26,** 1441-1460 (1993).

[116] J. Homola, J. Dostalek, S. F. Chen, A. Rasooly, S. Y. Jiang, and S. S. Yee, International Journal of Food Microbiology **75,** 61-69 (2002).

[117] H. Zhang, K. B. Lee, Z. Li, and C. A. Mirkin, Nanotechnology **14,** 1113-1117 (2003).

[118] R. Wilson and A. P. F. Turner, Biosensors & Bioelectronics **7,** 165-185 (1992).

[119] J. G. Zhao, J. P. Odaly, R. W. Henkens, J. Stonehuerner, and A. L. Crumbliss, Biosensors & Bioelectronics **11,** 493-502 (1996).

[120] C. M. Niemeyer and C. A. Mirkin, *Nanobiotechnology: Concepts, Applications and Perspectives* (Wiley-VCH, Dortmund, Germany, 2004).

[121] M. L. Curri, A. Agostiano, G. Leo, A. Mallardi, P. Cosma, and M. Della Monica, Materials Science & Engineering C-Biomimetic and Supramolecular Systems **22,** 449-452 (2002).

[122] D. Oliver, D. Z. Z. He, N. Klocker, J. Ludwig, U. Schulte, S. Waldegger, J. P. Ruppersberg, P. Dallos, and B. Fakler, Science **292,** 2340-2343 (2001).

[123] L. Muhua, N. R. Adames, M. D. Murphy, C. R. Shields, and J. A. Cooper, Nature **393,** 487-491 (1998).

[124] H. A. Fishman, D. R. Greenwald, and R. N. Zare, Annual Review of Biophysics and Biomolecular Structure **27,** 165-198 (1998).

[125] H. Bayley, O. Braha, and L. Q. Gu, Advanced Materials **12,** 139-142 (2000).

[126] H. Bayley and C. R. Martin, Chemical Reviews **100,** 2575-2594 (2000).

[127] O. P. Hamill, A. Marty, E. Neher, B. Sakmann, and F. J. Sigworth, Pflugers Archiv-European Journal of Physiology **391,** 85-100 (1981).

[128] E. Neher and B. Sakmann, Nature **260,** 799-802 (1976).

[129] B. A. Cornell, V. L. B. BraachMaksvytis, L. G. King, P. D. J. Osman, B. Raguse, L. Wieczorek, and R. J. Pace, Nature **387,** 580-583 (1997).

[130] K. S. Akerfeldt, J. D. Lear, Z. R. Wasserman, L. A. Chung, and W. F. Degrado, Accounts of Chemical Research **26,** 191-197 (1993).

[131] A. Grove, J. M. Tomich, T. Iwamoto, and M. Montal, Protein Science **2,** 1918-1930 (1993).

[132] M. Pawlak, U. Meseth, B. Dhanapal, M. Mutter, and H. Vogel, Protein Science **3,** 1788-1805 (1994).

[133] S. Cheley, G. Braha, X. F. Lu, S. Conlan, and H. Bayley, Protein Science **8,** 1257-1267 (1999).

[134] A. J. Wallace, T. J. Stillman, A. Atkins, S. J. Jamieson, P. A. Bullough, J. Green, and P. J. Artymiuk, Cell **100,** 265-276 (2000).

[135] M. J. Heller, Annual Review of Biomedical Engineering **4,** 129-153 (2002).

[136] G. H. W. Sanders and A. Manz, TRAC-TRENDS IN ANALYTICAL CHEMISTRY **19** 364-378 (2000).

[137] J. Wang, Chemistry-a European Journal **5,** 1681-1685 (1999).

[138] H. KorriYoussoufi, F. Garnier, P. Srivastava, P. Godillot, and A. Yassar, Journal of the American Chemical Society **119,** 7388-7389 (1997).

[139] J. Wang, X. H. Cai, G. Rivas, H. Shiraishi, P. A. M. Farias, and N. Dontha, Analytical Chemistry **68,** 2629-2634 (1996).

[140] K. Hashimoto, K. Ito, and Y. Ishimori, Analytica Chimica Acta **286,** 219-224 (1994).

[141] T. S. Snowden and E. V. Anslyn, Current Opinion in Chemical Biology **3,** 740-746 (1999).

[142] V. S. Y. Lin, K. Motesharei, K. P. S. Dancil, M. J. Sailor, and M. R. Ghadiri, Science **278,** 840-843 (1997).

[143] P. A. E. Piunno, U. J. Krull, R. H. E. Hudson, M. J. Damha, and H. Cohen, Analytical Chemistry **67,** 2635-2643 (1995).

[144] S. Tombelli, R. Mascini, L. Braccini, M. Anichini, and A. P. F. Turner, Biosensors & Bioelectronics **15,** 363-370 (2000).

[145] R. F. Service, Science **298,** 2322-2323 (2002).
[146] C. M. Niemeyer, Chemistry-a European Journal **7,** 3189-3195 (2001).
[147] C. M. Niemeyer, M. Adler, B. Pignataro, S. Lenhert, S. Gao, L. F. Chi, H. Fuchs, and D. Blohm, Nucleic Acids Research **27,** 4553-4561 (1999).
[148] T. Sano, C. L. Smith, and C. R. Cantor, Science **258,** 120-122 (1992).
[149] C. M. Niemeyer, W. Burger, and R. M. J. Hoedemakers, Bioconjugate Chemistry **9,** 168-175 (1998).
[150] C. M. Niemeyer, B. Ceyhan, and D. Blohm, Bioconjugate Chemistry **10,** 708-719 (1999).
[151] C. M. Niemeyer, M. Adler, S. Gao, and L. F. Chi, Angewandte Chemie-International Edition **39,** 3055-3059 (2000).
[152] C. M. Niemeyer, R. Wacker, and M. Adler, Angewandte Chemie-International Edition **40,** 3169-+ (2001).
[153] R. F. Service, Science **277,** 1036-1037 (1997).
[154] R. Elghanian, J. J. Storhoff, R. C. Mucic, R. L. Letsinger, and C. A. Mirkin, Science **277,** 1078-1081 (1997).
[155] K. A. Williams, P. T. M. Veenhuizen, B. G. de la Torre, R. Eritja, and C. Dekker, Nature **420,** 761-761 (2002).
[156] K. Hamad-Schifferli, J. J. Schwartz, A. T. Santos, S. G. Zhang, and J. M. Jacobson, Nature **415,** 152-155 (2002).
[157] L. IIe, M. D. Musick, S. R. Nicewarner, F. G. Salinas, S. J. Benkovic, M. J. Natan, and C. D. Keating, Journal of the American Chemical Society **122,** 9071-9077 (2000).
[158] C. M. Niemeyer, Angewandte Chemie-International Edition **40,** 4128-4158 (2001).
[159] J. M. Warman, M. P. deHaas, and A. Rupprecht, Chemical Physics Letters **249,** 319-322 (1996).
[160] S. O. Kelley and J. K. Barton, Science **283,** 375-381 (1999).
[161] E. Meggers, M. E. Michel-Beyerle, and B. Giese, Journal of the American Chemical Society **120,** 12950-12955 (1998).
[162] W. Zhang, A. O. Govorov, and S. E. Ulloa, Physical Review B **66** (2002).
[163] D. Ramaduraic, Y. Li, T. Yamanakaa, D. Geerpuramc, V. Sankara, M. Vasudevc, D. Alexsona, P. Shic, M. Dutta, M. A. Stroscioa, T. Rajhd, Z. Saponjicd, N. Kotove, Z. Tange, and S. Xuf, in *SPIE conference on Quantum Sensing and Nanophotonic Devices III* (SPIE, 2006).
[164] A. Y. Kasumov, M. Kociak, S. Gueron, B. Reulet, V. T. Volkov, D. V. Klinov, and H. Bouchiat, Science **291,** 280-282 (2001).

[165] J. H. Xu, J. J. Zhu, Y. L. Zhu, K. Gu, and H. Y. Chen, Analytical Letters **34,** 503-512 (2001).
[166] C. Y. Tsai, T. L. Chang, C. C. Chen, F. H. Ko, and P. H. Chen, Microelectronic Engineering **78-79,** 546-555 (2005).
[167] M. Rhee and M. A. Burns, Trends in Biotechnology **25,** 174-181 (2007).
[168] L. Z. Song, M. R. Hobaugh, C. Shustak, S. Cheley, H. Bayley, and J. E. Gouaux, Science **274,** 1859-1866 (1996).
[169] M. Akeson, D. Branton, J. J. Kasianowicz, E. Brandin, and D. W. Deamer, Biophysical Journal **77,** 3227-3233 (1999).
[170] F. Vogtle, S. Gestermann, R. Hesse, H. Schwierz, and B. Windisch, Progress in Polymer Science **25,** 987-1041 (2000).
[171] K. Inoue, Progress in Polymer Science **25,** 453-571 (2000).
[172] O. A. Matthews, A. N. Shipway, and J. F. Stoddart, Progress in Polymer Science **23,** 1-56 (1998).
[173] D. A. Tomalia, Advanced Materials **6,** 529-539 (1994).
[174] D. C. Tully and J. M. J. Frechet, Chemical Communications**,** 1229-1239 (2001).
[175] N. Krasteva, I. Besnard, B. Guse, R. E. Bauer, K. Mullen, A. Yasuda, and T. Vossmeyer, Nano Letters **2,** 551-555 (2002).
[176] J. J. Davis, K. S. Coleman, B. R. Azamian, C. B. Bagshaw, and M. L. H. Green, Chemistry-a European Journal **9,** 3732-3739 (2003).
[177] V. Balzani, P. Ceroni, M. Maestri, C. Saudan, and V. Vicinelli, in *Dendrimers V: Functional and Hyperbranched Building Blocks, Photophysical Properties, Applications in Materials and Life Sciences*; *Vol. 228* (2003), p. 159-191.
[178] L. Z. Gong, Q. S. Hu, and L. Pu, Journal of Organic Chemistry **66,** 2358-2367 (2001).
[179] V. J. Pugh, Q. S. Hu, X. B. Zuo, F. D. Lewis, and L. Pu, Journal of Organic Chemistry **66,** 6136-6140 (2001).
[180] H. Qian and B. E. Shapiro, Proteins-Structure Function and Genetics **37,** 576-581 (1999).
[181] M. Guthold, X. S. Zhu, C. Rivetti, G. L. Yang, N. H. Thomson, S. Kasas, H. G. Hansma, B. Smith, P. K. Hansma, and C. Bustamante, Biophysical Journal **77,** 2284-2294 (1999).
[182] D. Schuler and R. B. Frankel, Applied Microbiology and Biotechnology **52,** 464-473 (1999).
[183] I. M. Hsing, Y. Xu, and W. T. Zhao, Electroanalysis **19,** 755-768 (2007).
[184] J. Wang and A. N. Kawde, Electrochemistry Communications **4,** 349-352 (2002).

[185] J. Wang, D. K. Xu, and R. Polsky, Journal of the American Chemical Society **124,** 4208-4209 (2002).

Index

Accommodation coefficient (deposition) 144
Accuracy (static characteristics) .. 14
Acoustic phonons 331, 335
  Pulse spectroscopy 330, 337
Acoustic plate mode (APM) 126
Acoustic wave transducers 118
Adenosine diphosphate (ADP) ... 423
Adenosine triphosphate (ATP) ... 422
  Phosphatase protein 422
Adsorption 379
  Hydrophobic 379
  Ionic 379
  Van der Waals 379
Alzheimer's 406
Amide 375, 376
Amphiphilic molecules 176, 382
Anharmonic interactions (phonon-phonon interaction) 333
Antibodies 412
  Epitopes 413
  Immunoglobulins 412
  Mutation 412
  Paratopes 413
Antibody nanoparticles conjugates 418
Antimicrobial wound dressings *4*
Aqueous solution techniques (ATS) 173
  Chemical bath deposition (CBD) 173
  Electroless deposition (ED) ... 176
  Liquid phase deposition (LPD) 175
  Successive ion layer adsorption and reaction (SILAR) 175
Atomic force microscopy (AFM) 267
  Constant force mode 268
  Non-contact-mode AFM 299
  Tapping mode or dynamic force mode (DFM) 268
Atom-jellum model 307
Barkhausen effect 36, 362
Bioinformatic 447
Bioluminescence effect 34
Bipolar junction transducer (BJT) 106
Birefringence 54
Bottom-up approach *2*
Bulk acoustic wave (BAW) 119
Bulk-type sensor 372
Cantilever based transducer 123
Capacitive transducer 63
Carbodiimide 376
Carbon surface modification 389
Carboxylic acid 374
Casting 182
  Cold spray 184
  Dip coating 184
  Drop casting 184
  Electrophoresis spray 184
  settling spray 184
  Spary coating 184
  Spin coating 182
  Thermal spary 184
  Thermophoresis spray 184
Chem-7 tests 91
Chemical entrapment 381
Chemical vapor deposition (CVD) 164
  Atmospheric pressure plasma CVD (AP-PCVD) 172
  Atomic layer CVD (ALCVD) 170
  Low Pressure CVD (LPCVD) 168
  Metal-organic CVD (MOCVD) 167
  Plasma-enhanced CVD (PECVD) 168

Remote plasma enhanced CVD (RPECVD) 173
Ultra high vacuum CVD 173
Chemiluminescence effect 34
Chemisorption 306
Volkenshtein theory 318
Chromatin 440
Chromosomes 440
Colloidal suspensions 136
Competitive assay 415
Condensation rate (deposition) 144
Conductormetric transducer 63
Conformal deposition 143
Covalent coupling 372
Activation 374
Critically damped system 20
Cross-linker 373
Dendritic molecules 465
Density functional theory (DFT) 297
Density of states (DOS) 33, 284, 287, 292
In 1D 291
In 2D 289
In 3D 287
Resistance in a 1D structure 295
Spin 358
Surface adsorption 307
Deoxyribonucleic acids (DNA) 436
Base-pairing 440
Complementary DNA (cDNA) 447, 450
DNA-inorganic materials Conjugate 455
DNA-Protein Conjugate 452
Hybridization 443
Molecular wire 459
Mutation 440
Probe 443
Replication 440
Sequencing with nanopores 463
Structure 438
Detection limit (static characteristics) 16
Diatomic lattice 332
Band-pass mechanical filter 332
Diode
1D photodiode 299, 352
Array of photodiodes 268
Laser diode 40
Light-emitting diode 156
Light-emitting diode (LED) 31
One-dimensional light emitting diode 150
Organic light-emitting diode (OLED) 34
Photodiode 31, 77, 107
p-n junction 24, 106
Schottky 108
Semiconductor laser diode 156
Direct assay 415
Dislocation-diffusion theory 146
Dissolution-condensation method 143, 150
DNA microarray 444
DNA technology 442
Doppler effect 34
Drift (static characteristics) 15
Dye-sensitized based devices 26
Dynamic characteristics 13, 17
Dynamic light scattering (DLS) 241
Dynamic range (also dynamic span) 17
Effective refractive index (optical) 67
Effusion cell (Knudsen Cell) 157
Eigen state 286, 335
Electrochemical sensor 417
Affinity biosensor 416
Carbon nanotube 402
DNA 449, 451
Enzyme 426
Glucose monitor (enzyme) 424
pH 93
Zirconia based oxygen sensor 92
Electrochemical transducer 79
Electrochemistry 84
Electro-deposition (electroplating) 179
Electroluminescence effect 31
Electrolyte solution 84
Electromagnetic spectroscopy 211

Absorbance ... 213
Irradiance ... 213
Mlar absorptivity or molar extinction coefficient ... 213
Refractive index ... 212
Tansmittance ... 213
Electron beam lithography (EBL) ... 194
Electronic energy ... 214
Electronic state ... 27
Singlet state ... 27
Triplet state ... 27
Electronic structure ... 27
Electron-phonon interaction ... 334
Endothermic process ... 81
Engines of creation ... 5
Enzyme ... 411, 420
Competitive inhibitor ... 422
Feedback inhibition ... 422
Regulatory molecule ... 422
Substrate ... 420
Enzyme nanopraticle hybrid ... 425
Enzyme-linked immunosorbent assay (ELISA) ... 414, 420, 453
Equilibrium constant ... 80
Error (static characteristics) ... 14
Esters ... 377
Etching (dry) ... 202
Reactive ion etching (RIE) ... 203
Sputter etching ... 203
Vapor phase etching ... 203
Etching (wet) ... 202
Evanescent waves ... 67
Evaporation ... 151
Electron beam evaporation ... 153
Molecular beam epitaxy (MBE) ... 155
Thermal evaporation ... 151
Evaporation-condensation method ... 143, 144
Exothermic process ... 81
Faraday rotation effect ... 54
Faraday-Henry law ... 51
Fermi gas ... 285
Field effect transistor (FET) ... 113, 130, 141, 167, 295, 392
1D structure ... 299
Carbon nanotube - DNA ... 457
Enzyme feld effect transistors (ENFET) ... 118
Ion-sensitive field effect transistor (ISFET) ... 116
Madulation doped ... 167
Metal oxide field effect transistor (MOSFET) ... 113
n-channel enhancement MOSFET ... 113
Piezo FET ... 339
Field emission display (FED) ... 32
Figure of merit ... 40
Film bulk acoustic wave resonator (FBAR) ... 121
First order ordinary differential equation (ODE) ... 18
Fluorescence ... 27
Near-infrared fluorescence ... 28
Focused ion beam (FIB) ... 195
Force spectroscopy and microscopy (organic materials) ... 467
Frank-Van der Merwe growth ... 141
Galvanic cell ... 84
Equilibrium between two half-cells ... 87
Equilibrium within each half-cell ... 87
Half-cell ... 84
Hydrogen electrode ... 85
Salt bridge ... 85
Gas sensor
1D structure ... 302
Additives to sensitive layer ... 323
Conductive polymers (polyaniline) ... 399
Conductometric transducers ... 307
Crystal structure ... 319
Filter ... 328
Magnetic ... 362
Permeable membrane ... 92
SAW response ... 127
SnO_2 ... 64
Surface modification ... 325
Thin film ... 304

Thin film equivalent circuit ... 320
Thin film oxygen sensor 314
Thin film with nanosized grains 317
TiO_2 325
XPS study 234
Gel electrophoresis 442
Gene expression 439, 444
Genetic 436
Genome 439
Giant Magnetostrictive (*GM*) 49
Gibbs free energy 82, 143
Gold nanoparticles (*synthesis*) ... 138
Graetzel cells 26
Grey goo 5
Guanosine diphosphate (*GDP*) .. 424
Guanosine triphosphate (*GTP*) .. 424
Hall effect 36
Hall–Petch (*H–P*) *equation* 346
Hard disk drive (*HDD*) 356
Harmonic approximation (*phonons*) 330
Heaviside function 291
Heterojunction 24
High electron mobility transistors (*HEMT*s) 156
Hooke's law 267, 330
Hysteresis (*static characteristics*) 16
Immobilization 372
Infrared (*IR*) *spectroscopy* 223
Inorganic nanoparticles (synthesis) 136
Interdigital transducers (*IDTs*) ... 65, 125
Interference lithography 200
Interferometric optical transducer 72
International Center of Diffraction Data - ICDD 237
Intersystem crossing 27
Intersystem crossing (electrons) 214
Intrinsically conductive polymers (*ICPs*) 392
Biosensor 400
Conjugation 393
Non-redox doping 394
Photo charge-doping 394
Polyaniline 394
Redox doping 394
Ion implantation 202
Ion plating 163
Ion probe 28
Ion selective electrode (*ISE*) 90
Gas permeable membrane 92
Glass membrane 91
Polymeric membranes 92
Solid-state membrane 91
Junction potential (*electrochemical*) 90
Kerr effect 56
Kronecker delta 292, 335
Lamb wave (or *flexural plate wave-FPW*) 126
Landauer equation 296
Langmuir-Blodgett (*LB*) 176
layer-by-layer (*LbL*) *assembly* ... 384
LIGA 198
Light-dependent resistor (*LDR*) ... 22
Linear representation of a system 17
Linear sweep voltammetry (*LSV*) 99
Standard heterogeneous rate constant 100
Sweep rate 100
Lipid bilayer 430
Liquid crystal display (*LCD*) 33
Mach-Zehnder interferometer 73
Mad Cow disease 406
Magnetic materials
Amplified sensing 470
Filtering 469
Magnetic nanoparticles (*synthesis*) 137
Magneto-mechanical effect (*Magnetostriction*) 48
Magneto-optic Kerr effect (*MOKE*) 55
Magnetotactic bacteria 469
Mangnetoresistive effect 49, 356
Anisotropic magnetoresistance (AMR) 50, 358

Giant magnetoresistance (GMR) 50, 358
Magnetic tunnelling junction (MTJ) 361
Spin valve....360
Mass spectrometry....270
Mass-to-charge ratio 271
Matrix-assisted laser desorption/ionization (MALDI) 273
Radical cations 270
Secondary ion mass spectrometry (SIMS) 270
Time of flight (TOF)....273
Mass Spectrometry 271, 281
Mean free path (*beam*) 156
Measurand....*6*
Metal oxide semiconductor (*MOS*) *capacitor* 111
Metal oxides surface modification 387
Metallic nanoparticles (*synthesis*) 138
Metal-semiconductor junction (*solid state*)....106
Microelectromechanical systems (*MEMS*)....340
Mie theory....241, 352
Minimum detectable signal (*MDS*) 16
Mitosis 427
Molecular beam (*MBE*)....156
Molecular chaperons....406
Molecular imprinting (*MIP*)....391
Moore's law....*2, 3*
Motor protein 427
Multiple-junction cell 25
Nanoelectromechanical systems (*NEMS*)....340
Nanoimprinting 191
Step and flash nanoimprint lithography (SF-NIL)....191
Thermoplastic nanoimprint lithography (T-NIL)....191
Nanolithography....186
AFM 190
dip-pen nanolithography (DPN) 190
STM....190
Nano-patterning....186
Nernst equation....84
Nernst/Ettingshausen effect 38
Nicotinamide adenine dinucleotide (*NAD*) 441
Nicotinamide adenine dinucleotide phosphate (*NADP*) 441
Noise 15
Electronic 15
Generation-recombination....15
Pink....15
Shot....15
Non conducting polymers (*NCPs*) 390
Non-conformal deposition 143
Non-specific binding (*NSB*)....414
Nuclear magnetic resonance (*NMR*) *Spectroscopy* 228
Nucleation site (*thin film deposition*) 144
Number of states (*NOS*)....284
Ohm's law....64
Okazaki fragment....441
Oligonucleotide 444
One-dimensional nano-structured grwoth....143
Optical phonons....332
Optical sensors 348
Cataluminescence....351
Chemiluminescence (CL)....348
Fabry–Perot interference....350
Photoconductor 352
Optical spectroscopy 66
Optical waveguide
Optical fiber 71
Planar....67
Sensitivity....69
Transverse electric (*s*-polarized) 67
Transverse magnetic (*p*-polarized)....67
Optical waveguide based transducer 66

Optical waveguides 67
Organic sensor 371
Organometallic compounds 167
Oscillator
0D....................................... 341
1D....................................... 343
Over damped 20
Oxidizing agent 84
Particle wave duality................ 285
Peltier effect 38
Penetration depth (optical) 69
Percolation threshold................ 141
Periodic bond chain (PBC) theory
.... 144
Phonon 329
Phosphorescence 27
Photoconductive effect 22
Photoconductive Effect............... 22
Photo-crosslinker 379
Photodetector
PIN 167
Photodielectric effect................... 27
Photoelectric effect...................... 21
Photoelectrochemical cell 26
Photoelectrons............................. 21
Photolithography....................... 187
Etching 187
Extreme ultraviolet lithography
(EUV) 188
Lift-off................................. 187
Liquid immersion technique .. 188
Photoluminescence (PL)
spectroscopy.......................... 219
Photoluminescence effect 27
Photon 212
Photoresistor 22
Phototransistor.......................... 167
Photovoltaic effect................. 22, 24
Physical effects 20
Electromagnetic 20
Mechanical............................. 20
Optical................................... 20
Thermal 20
Physical entrapment.................. 380
Physical vapor deposition (PVD)
.... 151
Physisorption 306
Piezoelectric 46
1D structure 338
Acoustic phonon 336
Cantilever 123
FBAR 121
FET 339
Materials............................... 121
QCM..................................... 119
SAW 125
STM stage 265
Piezoresistive effect 43
1D 347
Coefficient.............................. 43
Plasmon resonance.................... 391
Attenuated Total Reflection
(ATR).................................. 76
Dipole 353
Extinction 352
Immunosensor 417
Localized surface plasmon
resonance (LSPR) 356
Nanoparticles 353
Phonon 246
Quadropole 353
Surface plasmon (SP).............. 67
Surface Plasmon Resonance
(SPR)........................... 74, 459
Pockels effect 56
Polymerase chain reaction (PCR)
.... 448
Immuno-PCR 453
Post deposition treatment (thin film)
.... 328
Potential energy surface (PES) .. 305
Precision (static characteristics).. 14
Propagating waves 67
Proteins..................................... 404
α-helix 407
β-sheet 407
Amino acid 404
Amino acid sequence 407
DNA mismatch repair 441
DNA polymerase.... 410, 441, 448
Ferritin.................................. 410
Hemoglobin.......................... 410

Insulin ... 410
Kinesin ... 410, 427
Ligand ... 411
Ligand-protein binding ... 411
Myosin ... 410
Polypeptide ... 405
Proteas ... 409
Quaternary structure ... 409
Recombinant DNS ... 409
Repair polymerase ... 441
Rhodopsin ... 410
RNA polymerase ... 467
Secondary structure ... 409
Serum Albumin ... 410
Tertiary structure ... 409
Topoisomerase ... 410
Trasferrin ... 410
Trypsin ... 409, 410
Pulsed laser deposition (*PLD*) ... 164
Pyroelectric effect ... 47
Quantum dots (*Q-dots*) ... 29, 33, 337
Synthesis ... 136
Quartz crystal microbalance (*QCM*) ... 119
Antibodies ... 417
DNA ... 452
Surface modification ... 402
Quasi-static electrostatic approximation (*photons*) ... 353
Radioimmunoassay (*RIA*) ... 414
Raman spectroscopy ... 245, 330, 337
Dimension measurement ... 333
Hand-held Raman sensor ... 337
Surface enhanced Raman spectroscopy (SERS) ... 246
reciprocal space (*k space* or *Brillouin zone*) ... 286
Redox reactions ... 84
Enzymes ... 424, 425, 426
mediator ... 403
Mediator ... 94, 426
Reducing agent ... 84
Reference electrode ... 87
Saturated calomel ... 87
Silver-silver chloride (Ag/AgCl) ... 87
Reflection high-energy electron diffraction (*RHEED*) ... 158
Reflectometric interference spectrum ... 350
Repeatability (*static characteristics*) ... 16
Reproducibility (*static characteristics*) ... 16
Resistance quantum (*inverse conductance quantum*) ... 295
DNA ... 461
Resistance thermometer ... 42
Resistive transducer (see *Conductormetric transducer*) ... 63
Resolution (*static characteristics*) 14
Response time (*static characteristics*) ... 16
Reverse micelles ... 137
Ribonucleic acid (*RNA*) ... 441
Messenger (mRNA) ... 447
Rotational levels ... 27
Rotational relaxation (*electrons*) ... 214
Rutherford backscattering spectrometry (*RBS*) ... 259
Kinematic factor ... 260
Sauerbrey equation ... 120
Scanning electron microscope (*SEM*) ... 250
Field emission (FE) ... 250
Scanning probe microscopy (*SPM*) ... 263
Scanning tunneling microscopy (*STM*) ... 264
Schottky approximation ... 313
Second order system ... 19
Seebeck effect ... 38
Segmented one dimensional structure ... 150
Selectivity (*static characteristics*). 15
Self assembly ... 381
Self-assembled monolayers (SAMs) ... 382, 386
Thiol SAMs ... 386
Semiconductor nanoparticles (*synthesis*) ... 136

Semiconductor-semiconductor junction (solid state) 106
Sensitivity (static characteristics) 14
Shockley equation 107
Sickle-cell anemia 443
Silanols 388
Silicon dioxide surface modification 387
Silicon surface modification 387
Single-nucleotide polymorphisms 456
Solar cell 167
Sol-gel 184
  Aerogel 185
  Colloidal suspension 185
  Emulsion 185
  Precipitation 185
  Spray pyrolysis 185
solution-liquid-solid (SLS) method 150
Solution-liquid-solid (SLS) method 143
Southern blotting 444
Spin Hall effect (SHE) 38
Spintronics 356
Spontaneous interaction 81
Sputtering 158
  Diode plasma (DC sputterer). 158
  Ion-assisted deposition (IAD)...... 163
  Ion-beam sputtering (IBS) 163
  Magnetic fields (magnetron).. 160
  Reactive sputtering 159
  RF 159
Stability (static characteristics) ... 16
Standard reduction potential 85
Static characteristics 13
Stimulus *6*
Stranski-Krastanov growth 141
Stripping analysis 105
Structural engineering of nanostructures 305
Supersaturation of growth species 144
Surface acoustic wave (SAW).... 119, 125
  Leaky SAW (LSAW) 126
  Molecular imprinting 392
  Rayleigh wave 126
  Shear horizontal waves 126
  Surface modification 402
  Surface-skimming bulk wave (SSBW) 126
Surface-conduction electron-emitter displays (SED) 32
Surface-type sensor (or *affinity sensor*) 372
Tera Hertz (THz) 23
Textured layered (thin film) 322
Thermionic emission 153
Thermistor 42
  Nositive temperature coefficient (NTC) 42
  Positive temperature coefficient (PTC) 42
Thermocouple 40
  Chromel 40
  Constantan 40
  Nicrosil 40
  Nisil 40
Thermodynamics
  First law 81
  Second law 81, 83
Thermoelectric effect 38
Thermoresistive effect 42
Thin film deposition 141
Thin films with nanograins 345
Thiols 378
Thomson effect 38
Time invariant representation of a system 17
Top-down approach *2*
Transducer 13
Transition layer (thin film) 322
Transmembrane protein 429
  Active transport 431
  Extracellular signals 429
  Membrane potential 431
  Nanopores 432
  Passive transport 431
  Patch-clamp sensors 432
  Transmembrane transporter ... 430

Transmission electron microscope (TEM) 255
  High resolution transmission electron microscopy (HRTEM) 256
Ultraviolet-visible (UV-vis) 215
United States' National Nanotechnology Initiative *1*
Vapour-liquid solid (VLS) method 147
Vapour-liquid-solid (VLS) method 143
Verdet constant 54
Vibrational levels 27
Vibrational relaxation (electrons) 214
Villari effect 49
Volmer-Weber growth 141
Voltammetry 94
  Capacitive current 102
  Cottrell's equation 98
  Diffusion current 96
  Double layer capacitance 102
  Electric bi-layer 95
  Fick's first law of diffusion 97
  Fick's second law of diffusion 98
  Voltagram 94
Work function 21
X-ray diffraction (XRD) 237
X-ray lithography (XRL) 197
X-ray photoelectron spectroscopy (XPS) 232
Zeeman effect 54
Zone folding (ZF) model 333

About the Authors

Dr. Kourosh Kalantar-zadeh

Dr. Kalantar-zedeh received his PhD at RMIT University, Melbourne, Australia in 2002 in the field of bioMEMS. He also holds a master of science from Tehran University, Tehran, Iran majoring telecommunications. His fields of intetrst include sensors, nanotechnology, materials sciences, MRI signal processing, ionospheric surveillance systems, electronics, SONAR systems and MEMS. He has published more than 120 scientific papers in refereed journals and in the proceedings of international conferences and holds 4 patents. He is currently a senior lecturer at RMIT University.

Dr. Benjamin Nicholas Fry

Dr. Fry completed his PhD at the University of Utrecht in the Netherlands in the field of Medical Microbiology. He then pursued his research career in the Netherlands and later moved to Australia. Dr. Fry fields of interest include bacterial detection, typing, pathogenesis and vaccine development. He has many publications in the international scientific journals and holds several international patents. Dr. Fry is currently a senior lecturer in the Department of Biotechnology and Environmental Biology at RMIT University, Melbourne, Australia.

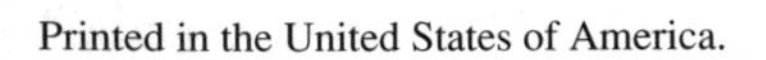
Printed in the United States of America.